# Encapsulation of Active Molecules and Their Delivery System

# Encapsulation of Active Molecules and Their Delivery System

Edited by

Shirish H. Sonawane

Chemical Engineering Department,
National Institute of Technology, Warangal, India

Bharat A. Bhanvase

Chemical Engineering Department, Laxminarayan Institute of Technology, RTM Nagpur University, Nagpur, India

Manickam Sivakumar

Chemical and Environmental Engineering Department,
University of Nottingham, Jalan Broga, Malaysia

Elsevier
Radarweg 29, PO Box 211, 1000 AE Amsterdam, Netherlands
The Boulevard, Langford Lane, Kidlington, Oxford OX5 1GB, United Kingdom
50 Hampshire Street, 5th Floor, Cambridge, MA 02139, United States



**British Library Cataloguing-in-Publication Data**
A catalogue record for this book is available from the British Library

**Library of Congress Cataloging-in-Publication Data**
A catalog record for this book is available from the Library of Congress

ISBN: 978-0-12-819363-1

For Information on all Elsevier publications
visit our website at https://www.elsevier.com/books-and-journals

*Publisher:* Susan Dennis
*Acquisitions Editor:* Anita A. Koch
*Editorial Project Manager:* Andrea Dulberger
*Production Project Manager:* Prasanna Kalyanaraman
*Cover Designer:* Greg Harris

Typeset by MPS Limited, Chennai, India

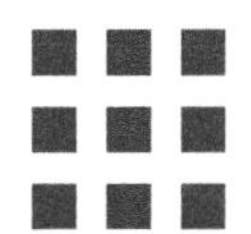

# Contents


List of contributors xiii

1. Current overview of encapsulation 1

*SHIRISH H. SONAWANE, BHARAT A. BHANVASE, MANICKAM SIVAKUMAR AND SHITAL B. POTDAR*

1.1 Introduction to encapsulation 1

1.2 Current trends in the encapsulation processes 4

1.3 Encapsulation of active substrate and their target applications 6

References 7

Further reading 8

2. Physicochemical characterization techniques in the encapsulation of active molecules 9

*SARANG P. GUMFEKAR*

2.1 Introduction 9

2.2 Particle size and its distribution 10

2.3 Surface charges 12

2.4 Imaging of the encapsulated materials 12

2.5 The crystallinity of encapsulation systems 15

2.6 Rheology of the encapsulated materials 17

2.7 Conclusion and outlook 20

References 20

3. Ultrasonic cavitation assisted synthesis of multilayer emulsions as encapsulating and delivery systems for bioactive compounds 23

*JITENDRA CARPENTER, SUJA GEORGE AND VIRENDRA KUMAR SAHARAN*

3.1 Emulsions as encapsulating and delivery system 23

3.2 Stabilization of emulsions 25

3.3 Application of ultrasonication for emulsification 29

3.4 Multilayer emulsion and its application for the encapsulation of bioactive compounds 33

3.5 Case studies 38

References 50

4. Encapsulation of active molecules in pharmaceutical sector: the role of ceramic nanocarriers 53

*JOANA C. MATOS, LAURA C.J. PEREIRA, JOÃO CARLOS WAERENBORGH AND M. CLARA GONÇALVES*

4.1 Nanotechnology in pharmacy and medicine 53

4.2 Ceramic nanoparticles as nanocarriers 55

4.3 Ceramic nanoparticles 56

4.4 Superparamagnetic iron oxide nanoparticles 69

References 80

5. Sonochemical encapsulation of taxifolin into cyclodextrine for improving its bioavailability and bioactivity for food 85

*IRINA KALININA, IRINA POTOROKO AND SHIRISH H. SONAWANE*

5.1 Introduction 85

5.2 Conclusions 100

Acknowledgments 100

References 100

Further reading 102

6. Controlled release of functional bioactive compounds from plants 103

*S.D. TORAWANE, Y.C. SURYAWANSHI AND D.N. MOKAT*

6.1 Introduction 103

6.2 Bioactive compounds 104

6.3 Conclusion 107

References 108

7. Bioactive molecule and/or cell encapsulation for controlled delivery in bone or cartilage tissue engineering 111

*BHASKAR BIRRU, P. SHALINI AND SREENIVASA RAO PARCHA*

7.1 Introduction 111

7.2 Controlled delivery 112

7.3 Cell/biomolecule encapsulation 115

7.4 Bioactive molecule/cell encapsulation for bone and cartilage 122

References 126

8. A review on application of encapsulation in agricultural processes 131

*MAYURI BHATIA*

8.1 Introduction 131

8.2 Encapsulation material 133

8.3 Encapsulation techniques 134

8.4 Encapsulation of active ingredients 135

8.5 Challenges and future prospects 138

8.6 Conclusion 138

References 138

9. Nanofluids-based delivery system, encapsulation of nanoparticles for stability to make stable nanofluids 141

*PARAG THAKUR, SHRIRAM S. SONAWANE, SHIRISH H. SONAWANE AND BHARAT A. BHANVASE*

Nomenclature 142

9.1 Introduction 142

9.2 Encapsulation of nanomaterials 143

9.3 Nanofluid-based delivery system 145

9.4 Targeted drug delivery 148

9.5 Applications of nanofluid-based delivery system 148

9.6 Conclusion 149

References 149

10. Corrosion and nanocontainer-based delivery system 153

*UDAY BAGALE, DIPAK PINJARI, SHRIKANT BARKADE AND IRINA POTOROKO*

10.1 Introduction to corrosion problem 154

10.2 Container approach for corrosion prevention 154

10.3 Different types of container and their method preparation/fabrications 156

10.4 Distribution and performance of container for protective coating 164

10.5 Release of active compounds from container 165

10.6 Case studies 165

10.7 Commercial applications and future prospectus 170

10.8 Conclusion/inference 171

References 171

11. Encapsulation and delivery of active compounds using nanocontainers for industrial applications 177

*SHAILESH A. GHODKE, SHIRISH H. SONAWANE, BHARAT A. BHANVASE AND KALPANA JOSHI*

11.1 Introduction 178

11.2 Nanocontainer synthesis 179

11.3 Control parameters for nanocontainer applications 185

11.4 Active molecules to be delivered 187

11.5 Conclusion and future prospective 190

References 191

12. Virus-like particles: nano-carriers in targeted therapeutics 197

*GUNDAPPA SAHA, PRAKASH SAUDAGAR AND VIKASH KUMAR DUBEY*

12.1 Introduction 197

12.2 Role of virus-like particles as good drug delivery vectors 199

12.3 Virus-like particles: overcoming limitations of other therapeutic approaches 200

12.4 Prerequisite factors in designing of virus-like particles as therapeutics 201

12.5 Immune responses induced by virus-like particles 203

12.6 Current applications of virus-like particles as targeted therapeutics 204

12.7 Conclusion 206

References 206

13. Formulation development and in vitro multimedia drug release study of solid self-microemulsifying drug delivery system of ketoconazole for enhanced solubility and pH-independent dissolution profile 211

*VINOD MOKALE, SHIVRAJ NAIK, TRUPTI KHATAL, SHIRISH H. SONAWANE AND IRINA POTOROKO*

13.1 Introduction 212

13.2 Material and method 212

13.3 Results and discussion 216

13.4 Conclusion 229

Acknowledgment 230

Conflict of interest 230

References 230

14. Molecular recognition, selective targeting, and overcoming gastrointestinal digestion by folic acid–functionalized oral delivery systems in colon cancer 233

*PALLAB KUMAR BORAH AND RAJ KUMAR DUARY*

Nomenclature 234

14.1 Introduction 234

14.2 Structure and function of the folate receptor 235

14.3 Expression of folate receptor in normal and malignant tissues 236

14.4 Folic acid–functionalized uptake of oral delivery systems via folate receptor–mediated endocytosis 237

14.5 Functionalization of folic acid on oral delivery systems 238

14.6 Folic acid–functionalized systems for colon cancer 240

14.7 Oral delivery systems and gastrointestinal digestion 245

14.8 Conclusion 249

Acknowledgments 250

References 250

15. Mathematical modeling and simulation of the release of active agents from nanocontainers/microspheres 257

*ASHISH P. PRADHANE, DIVYA P. BARAI, BHARAT A. BHANVASE AND SHIRISH H. SONAWANE*

15.1 Introduction 258

15.2 Mechanism of release in nanocontainers 258

15.3 Modeling of release of active agents 268

15.4 Simulation of release of active agents 281

15.5 Summary 284

References 284

16. Flavor encapsulation and release studies in food 293

*SHITAL B. POTDAR, VIVIDHA K. LANDGE, SHRIKANT S. BARKADE, IRINA POTOROKO AND SHIRISH H. SONAWANE*

16.1 Introduction 294

16.2 Aroma extraction methods 296

16.3 Encapsulation techniques: conventional and newer approach 300

16.4 Phenomena of encapsulated flavor release 313

16.5 Characterization techniques for encapsulated bioactive compounds 315

16.6 Conclusion and future prospective 317

Acknowledgment 317

References 318

17. Encapsulation and delivery of antiparasitic drugs: a review 323

*SANTANU SASIDHARAN AND PRAKASH SAUDAGAR*

17.1 Introduction: encapsulation and techniques 324

17.2 Need for encapsulated drugs against parasite 327

17.3 Encapsulation of drugs in various parasites 328

17.4 Encapsulated drugs in clinical trials and commercial usage 335

17.5 Summary and future outlook 336

References 336

Index 343

# List of contributors

**Uday Bagale** Department of Food and Biotechnology, South Ural State University, Chelyabinsk, Russian Federation

**Divya P. Barai** Chemical Engineering Department, Laxminarayan Institute of Technology, Rashtrasant Tukadoji Maharaj Nagpur University, Nagpur, India

**Shrikant S. Barkade** Chemical Engineering Department, Sinhgad College of Engineering, Pune, India

**Bharat A. Bhanvase** Chemical Engineering Department, Laxminarayan Institute of Technology, RTM Nagpur University, Nagpur, India

**Mayuri Bhatia** Department of Biotechnology, National Institute of Technology, Warangal, India; Department of Civil Engineering, Indian Institute of Technology Hyderabad, Kandi, India

**Bhaskar Birru** Department of Biosciences and Bioengineering, Indian Institute of Technology Guwahati, Guwahati, Assam, India

**Pallab Kumar Borah** Department of Food Engineering and Technology, School of Engineering, Tezpur University, Tezpur, India

**Jitendra Carpenter** Department of Chemical Engineering, Malaviya National Institute of Technology (MNIT), Jaipur, India

**Raj Kumar Duary** Department of Food Engineering and Technology, School of Engineering, Tezpur University, Tezpur, India

**Vikash Kumar Dubey** School of Biochemical Engineering, Indian Institute of Technology BHU, Varanasi, India

**Suja George** Department of Chemical Engineering, Malaviya National Institute of Technology (MNIT), Jaipur, India

**Shailesh A. Ghodke** Department of Chemical Engineering, Dr. D. Y. Patil Institute of Engineering, Management and Research, Pune, India

**M. Clara Gonçalves** Instituto Superior Técnico, Universidade de Lisboa, Lisbon, Portugal

**Sarang P. Gumfekar** Department of Chemical and Materials Engineering, University of Alberta, Edmonton, AB, Canada

**Kalpana Joshi** Department of Biotechnology, Sinhgad College of Engineering, Pune, India

**Irina Kalinina** Department of Food Technology and Biotechnology, School of Medical Biology, SUSU, Chelyabinsk, Russia

**Trupti Khatal** Department of Pharmaceutical Technology, University Institute of Chemical Technology, North Maharashtra University, Jalgaon, India

**Vividha K. Landge** Chemical Engineering Department, National Institute of Technology, Warangal, India

**Joana C. Matos** Instituto Superior Técnico, Universidade de Lisboa, Lisbon, Portugal; C$^2$TN, Center for Nuclear Sciences and Technologies, Instituto Superior Técnico, Universidade de Lisboa, Lisbon, Portugal

**Vinod Mokale** Department of Pharmaceutical Technology, University Institute of Chemical Technology, North Maharashtra University, Jalgaon, India

**D.N. Mokat** Department of Botany, Savitribai Phule Pune University, Pune, India

**Shivraj Naik** Department of Pharmaceutical Technology, University Institute of Chemical Technology, North Maharashtra University, Jalgaon, India

**Sreenivasa Rao Parcha** Department of Biotechnology, National Institute of Technology Warangal, Warangal, Telangana, India

**Laura C.J. Pereira** C$^2$TN, Center for Nuclear Sciences and Technologies, Instituto Superior Técnico, Universidade de Lisboa, Lisbon, Portugal

**Dipak Pinjari** National Center for Nanoscience and Nanotechnology, University of Mumbai, Mumbai, India

**Shital B. Potdar** Department of Chemical Engineering, National Institute of Technology, Warangal, India

**Irina Potoroko** Department of Food Technology and Biotechnology, School of Medical Biology, SUSU, Chelyabinsk, Russia

**Ashish P. Pradhane** Chemical Engineering Department, Laxminarayan Institute of Technology, Rashtrasant Tukadoji Maharaj Nagpur University, Nagpur, India

**Gundappa Saha** Department of Biosciences and Bioengineering, Indian Institute of Technology Guwahati, Guwahati, India

**Virendra Kumar Saharan** Department of Chemical Engineering, Malaviya National Institute of Technology (MNIT), Jaipur, India

**Santanu Sasidharan** Department of Biotechnology, National Institute of Technology, Warangal, India

**Prakash Saudagar** Department of Biotechnology, National Institute of Technology Warangal, Hanamkonda, India

**P. Shalini** Department of Chemical Engineering, National Institute of Technology Warangal, Warangal, Telangana, India

**Manickam Sivakumar** Chemical and Environmental Engineering Department, University of Nottingham, Jalan Broga, Malaysia

**Shirish H. Sonawane** Chemical Engineering Department, National Institute of Technology, Warangal, India

**Shriram S. Sonawane** Department of Chemical Engineering, Visvesvaraya National Institute of Technology, Nagpur, India

**Y.C. Suryawanshi** Department of Botany, Savitribai Phule Pune University, Pune, India

**Parag Thakur** Department of Chemical Engineering, Visvesvaraya National Institute of Technology, Nagpur, India

**S.D. Torawane** Department of Botany, Savitribai Phule Pune University, Pune, India

**João Carlos Waerenborgh** C$^2$TN, Center for Nuclear Sciences and Technologies, Instituto Superior Técnico, Universidade de Lisboa, Lisbon, Portugal

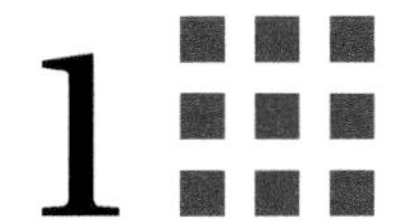

# Current overview of encapsulation

Shirish H. Sonawane[1], Bharat A. Bhanvase[2], Manickam Sivakumar[3], Shital B. Potdar[4]

[1]CHEMICAL ENGINEERING DEPARTMENT, NATIONAL INSTITUTE OF TECHNOLOGY, WARANGAL, INDIA [2]CHEMICAL ENGINEERING DEPARTMENT, LAXMINARAYAN INSTITUTE OF TECHNOLOGY, RTM NAGPUR UNIVERSITY, NAGPUR, INDIA [3]CHEMICAL AND ENVIRONMENTAL ENGINEERING DEPARTMENT, UNIVERSITY OF NOTTINGHAM, JALAN BROGA, MALAYSIA [4]DEPARTMENT OF CHEMICAL ENGINEERING, NATIONAL INSTITUTE OF TECHNOLOGY, WARANGAL, INDIA

**Chapter Outline**

**1.1 Introduction to encapsulation** ..... 1
**1.2 Current trends in the encapsulation processes** ..... 4
**1.3 Encapsulation of active substrate and their target applications** ..... 6
**References** ..... 7
**Further reading** ..... 8

## 1.1 Introduction to encapsulation

Encapsulation dates back to the 19th century. In the year 1963, Harvard Business School prepared a report on encapsulation and underlying phenomena of encapsulation. Fanger published an article in 1974 in the history of encapsulation. Encapsulation is the process of stabilization of active compounds through the structuring of systems capable of preserving their chemical, physical, and biological properties, as well as their release or delivery under established or desired conditions [1]. It is very well known that natural plants, herbs, and food materials have valuable compounds and possess the potential to utilized for various applications, such as in the treatment of diseases like cancer. As an alternative to synthetic preservative and flavouring agents it has been used since the dawn of medicine to treat various illnesses in Ayurveda or food-processing industries as an alternative to synthetic preservatives and flavoring agents. However, these bioactive compounds are subjected to degradation with environmental conditions such as temperature and moisture. So, it is essential to extract and store these valuable compounds. Encapsulation is a promising way to preserve these active compounds.

In the process of encapsulation, either one or the mixture of bioactive material is coated with another single or combination of materials. In encapsulation, two main terminologies

Encapsulation of Active Molecules and their Delivery System. DOI: https://doi.org/10.1016/B978-0-12-819363-1.00001-6

are frequently used the material that is being coated is termed as the active material or core material, and another one is the shell material that is also termed carrier material (shell). The shell material can be in the form of solid, liquid droplets, and gas bubbles to encapsulate liquid or gas inside as a core. The core and shell structure can be in various shapes such as the sphere, microcapsules, microbeads, monocore, multicore, matrix, and multishell. The core and shell material of the encapsulation vary depending on the applications and possess specific characteristics, for example, core materials used in food are essential oils (acting as a preservative and flavoring agent) and shell materials are whey protein, gum arabic, maltodextrin, etc. In the medical field the used, core materials are drugs such as the influenza virus, stem cells, DNA, and insulin and shell materials are polymethylmethacrylate, and the encapsulation methods are copolymerization, solvent displacement, gelation, etc.

Encapsulation is used in various areas for different purposes. In biology, it is useful in the encapsulation of living tissue, individual cells, hormones, enzymes or antibodies, and other biological materials. In the food industry, encapsulation is carried out to add flavor to products that are reduced in various food-processing steps. The extraction of a drug molecule and its encapsulation in the pharmaceutical field give natural drug delivery and also help to reduce the side effects of synthetic drugs. Encapsulation technique is also used in the agricultural field to encapsulate pesticides, fertilizers, and other agrochemicals that allow growers to precisely control the conditions under which the active ingredient is released. Encapsulation can also help to minimize the use of pesticides and reduce their environmental impact. Nanocontainer preparation for corrosion inhibition is also an attractive area gaining the attention of many researchers to use nanoparticles such as titania, zinc molybdate as core material and encapsulate them with different polymers either through layer-by-layer approach or by in situ polymerization [2].

With time the domain of encapsulation received enormous attention, while various methods of extraction and encapsulation have been developed depending upon the application and the characteristics of materials. Various methods of extraction can be listed as hot water bath extraction, Soxhlet extraction [3], microwave-assisted extraction [4], extraction using the maceration [5], supercritical carbon dioxide extraction [6], and ultrasound-assisted extraction. Once the bioactive core material is extracted from a natural source, it is desired to encapsulate using various techniques. The encapsulation techniques can be broadly divided into two main categories, that is, chemical and physical encapsulations. There are three different methods of chemical encapsulation, namely, coacervation, molecular inclusion [7], and cocrystallization [8]. Physical encapsulation is also known as mechanical encapsulation. Spray-drying [9,10], extrusion [11], freeze-drying and vacuum drying, spray-cooling or chilling, and fluidized bed coating are the methods of physical encapsulation. New age techniques of encapsulation include high- and low-energy emulsification techniques [12]. Two types of emulsion are formed, namely, microemulsion and nanoemulsion. The fundamental difference between these two types is the droplet size of emulsion. The size for microemulsion is one to several microns, and it is in the range of few nanometers for nanoemulsion.

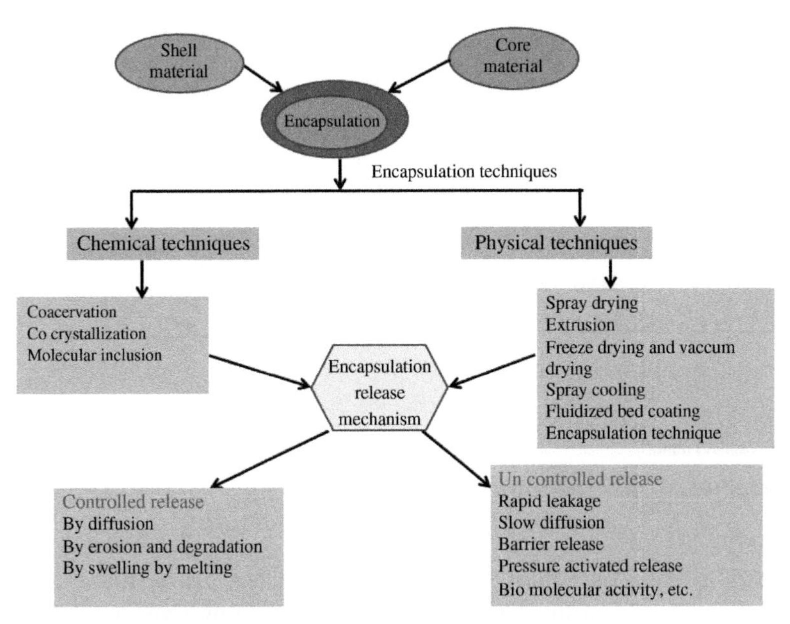


**FIGURE 1–1** Encapsulation techniques and the release mechanisms.

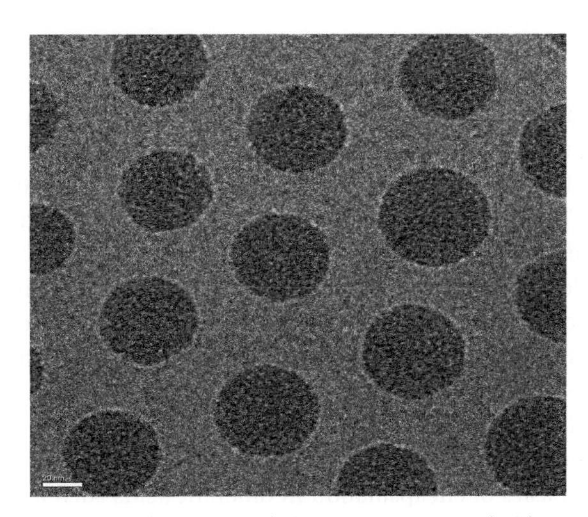

**FIGURE 1–2** Ultrasound-assisted miniemulsion encapsulation process. *Reprinted with permission from Sonawane, et al., J. Phys. Chem. C 114 (2010) 5148–5153.*

The nanoemulsion technology is becoming attractive because of its advantages over microemulsion such as higher encapsulation efficiency and increased product yield (Fig. 1–1).

Ultrasound-assisted miniemulsion is one of the advanced methods of encapsulation. In this method, miniemulsion polymerization is carried out using the ultrasound-assisted technique. The prepared emulsion using ultrasound is shown in Fig. 1–2. Using ultrasound the cavitation technique, it is possible to prepare the inorganic core surrounded by polymeric

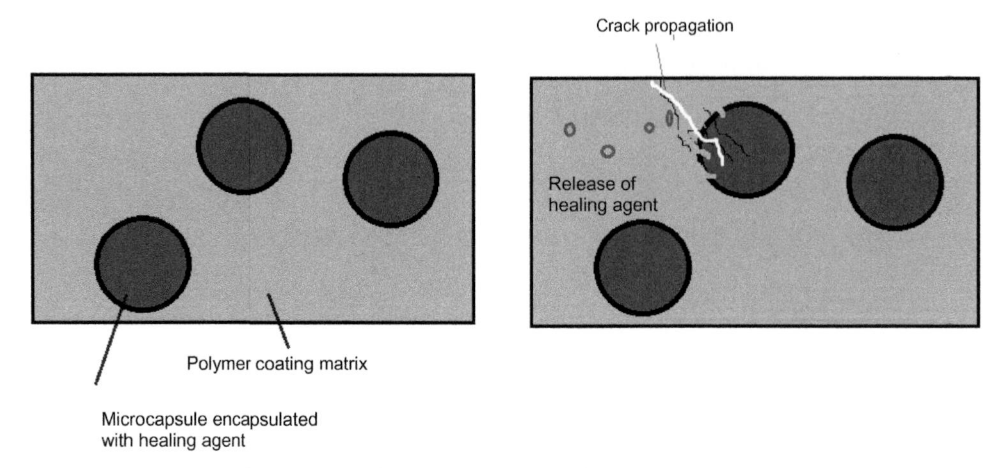


**FIGURE 1–3** Encapsulation of active molecules and the release mechanism.

shell dispersed in the continuous matrix such as water. Using the ultrasound-assisted method, it is possible to encapsulate active material in a tiny size may be in the range of 20–50 nm. The encapsulated material could be used as a photoanode, drug delivery system, etc.

The release of encapsulated core material can occur by two mechanisms: controlled release and uncontrolled (a triggered mechanism) release. The controlled release of the encapsulated core material is the critical step in the process. As shown in Fig. 1–3, the liquid core is encapsulated using a polymeric shell. Based on the crack propagation, the release of active agents occurs from the core. The liberated molecules react with the polymer matrix, and healing occurs. This type of release mechanism is based on mechanical stimulation.

The release of the encapsulated drug in a controlled manner gives the maximum therapeutic efficiency by delivering the drug at the targeted tissue at an optimal amount and in an optimal period. The controlled release of food flavor is desirable as sometimes if an uncontrolled release occurs, the flavor is lost within a short period before reaching the consumer for consumption. The benefits of encapsulation are shown in Fig. 1–4.

The efficiency and benefits of encapsulation attract the researchers and scientists to study encapsulation to overcome limitations such as uncontrolled release in its practical use in many industries. The basics of encapsulation in different fields such as food, pharmaceutical, agriculture, and corrosion have been explained with detailed experimentation and specific examples in the following chapters.

## 1.2 Current trends in the encapsulation processes

The advantages and benefits of encapsulation technology have been applied in various areas of application, from food and beverage industry, waste and environmental technology, plant and animal, medical and pharmaceutical, energy, engineering, etc. Fig. 1–5 shows the functional application areas of encapsulation.

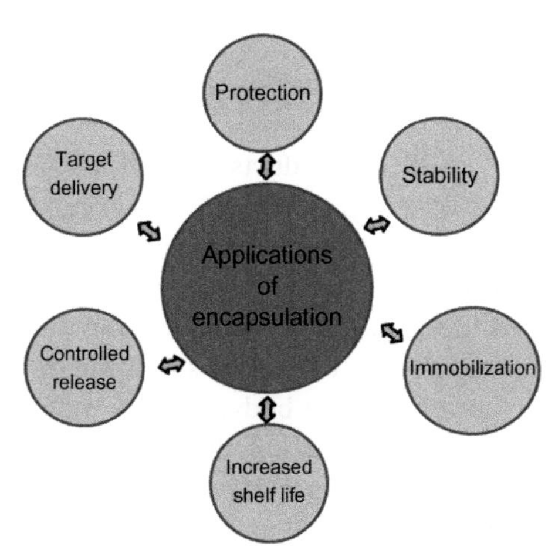


**FIGURE 1–4** Applications of encapsulation.

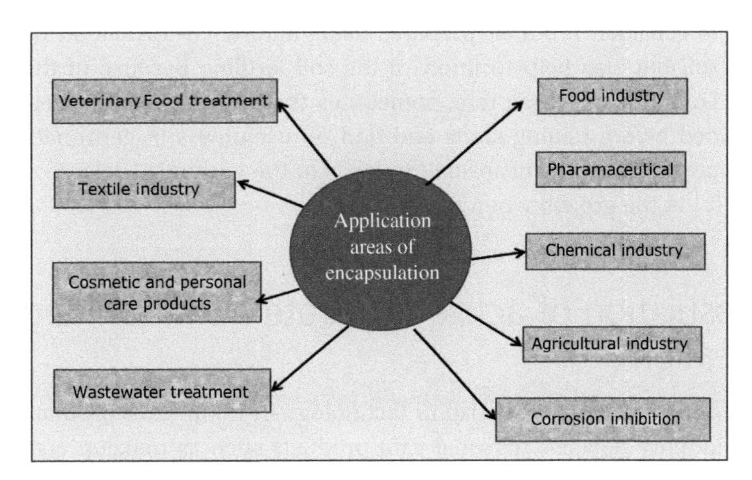


**FIGURE 1–5** Application areas of encapsulation.

As described in the above sections, encapsulation is in use for more than 60 years. In earlier days the focus was on the encapsulation of active material for food and most pharmaceutical drug preparation only. However, the area has become vast, and the encapsulation techniques are used in many sectors starting from dairy products such as personal care product, agricultural product, and surgery tools in medicine.

With the advancement in nanotechnology, scientists are aware of the advantages of nanoscale products over microscale products in terms of using the surface area effectively. So, there is a growing trend in synthesizing nanoscale core–shell emulsion encapsulation. Sometimes, it is found that the single process will not give higher encapsulation efficiency, and thus the combination of encapsulation techniques such as synthesizing core–shell emulsion using ultrasound and making it more stable by storing it in powder form by spray drying facility is more useful. Similarly, the combination of the core material and/or the combination of shell material is now being well practiced because of their unique characteristics.

Minimization of the use of synthetic products in each field is attracting researchers and industries to work in the area of encapsulation. As an example, the chemically prepared drugs have side effects after consumption, and thus, natural drugs are required. The orally administered drugs are dissolved completely and instantly by interacting with the solvent. Thus drug encapsulation provides a layer of protection for the gradual release of its drug content when it is naturally dissolved. Consumers suspect that synthetic preservatives and flavoring agents in food are harmful to the body, so they prefer natural preservatives. The extract of natural products such as essential oils from various food items such as gingerol, soya seed, and safflower have food preservative properties, whereas extracts from fruits act as a flavoring agent and are proved beneficial for health from ancient years. Nearly the same is the scenario for agricultural products, synthetic pesticides, fertilizers, and other agrochemicals, which help in increasing the production at the expense of fertility loss of soil, which is a serious issue to consider. Products prepared using natural encapsulation will not only stop fertility loss of soil but also help to improve the soil fertility. Because of the uncertainty of environmental conditions such as rain, sometimes the active agents in fertilizer and pesticides get degraded before raining starts and find no usefulness in germination and growth. Agrochemicals prepared using encapsulation leads to the controlled release of active materials that are useful in the growth when raining occurs.

## 1.3 Encapsulation of active substrate and their target applications

The research employs microencapsulation technology in home care products such as small microcapsules laundry powder, personal care products such as makeup creams, face wash, lotions, and fragrances, which will neither degrade with environmental temperature nor leave side effects on the skin. Some of the products available in the market now get degraded under specific environmental changes, thereby losing their effects. The encapsulated product will be subjected to target delivery of fragrance such as active materials to release under desired conditions.

Owing to the use of encapsulation in the paint industry self-healing paints which have been designed in a way that paint will have nanoparticles such as zinc, and titania with anticorrosive properties. The encapsulated coatings and cement could make maintenance easier and less expensive. The encapsulated microparticles when mixed into paints and coatings for infrastructure, pipelines, and industrial equipment at risk for corrosion damage.

Microencapsulation is already in use by the pharmaceutical industry, but new encapsulation techniques could expand the possible uses for drug delivery and lead to exciting new therapies. Kreuter in the year 2007, reported that the sustained release of vaccine provides more prolonged immune system stimulation. Besides the vaccine, the sustained release was also applied to other drugs such as five bioactive compounds, protein, DNA [13], and stem cell. Another use could be in the stabilization of vaccines.

Making safer and more effective agrochemicals is another target application of encapsulation technology. It possesses numerous different uses for agrobusiness. Microencapsulation could be used to combine incompatible ingredients into one shelf-stable product to reduce packaging and make storage and application more convenient. It could also be used to reduce the amount of chemical that farmers need to apply. Agricultural products could use microcapsules that open to deliver the product only when it rains.

Encapsulation finds an application in civil engineering for the preparation of self-healing concretes. It is a general observation that the life span of concrete deteriorates, and the underlying reason for this deterioration is due to the cracking mechanism. This phenomenon will not just lead to capital and labor cost, but it is difficult to understand the degree of damage once the construction is complete. Self-healing concretes are synthesized using encapsulation technology. If concrete is formed by a capsule-based self-healing [14], the loss, as mentioned above, of capital and labor can be reduced. In this technique, healing capsules are prepared using materials such as sodium silicate solution; when these encapsulated capsules get ruptured due to cracking, it releases sodium silicate solution into the matrix and thereby reacts with calcium hydroxide (concrete material) to form calcium silicate hydrate (C–S–H) that helps in healing the concrete crack. The crack propagation and healing mechanism are shown in Fig. 1–3. Concrete crack can also be healed using the encapsulation of bacteria such as spores of *Bacillus sphaericus*, with materials such as diatomaceous earth as a carrier (shell) material. The research investigations employ different healing agents such as epoxy, resin, sodium silicate solution, cyanoacrylate with materials such as glass, gelatin, and silica gel as a carrier material [15].

Microencapsulation could also transform remediation methods. The oil and gas industry is already using oil-eating bacteria for oil spill cleanup. Microencapsulation could make this technique more practical and convenient by creating shells that keep bacteria shelf-stable until they are released in the presence of hydrocarbons.

# References

[1] H.D. Silva, M.A. Cerqueira, A.A. Vicente, Nanoemulsions for food applications: development and characterization, Food Bioprocess Technol. 5 (2012) 854–867.

[2] S.E. Karekar, U.D. Bagale, S.H. Sonawane, B.A. Bhanvase, D.V. Pinjari, A smart coating established with encapsulation of zinc molybdate centred nanocontainer for active corrosion protection of mild steel: release kinetics of corrosion inhibitor, Composite Interfaces 25 (9) (2018) 785–808.

[3] Y. Naude, W.H.J. De Beer, S. Jooste, L. Van der Merwe, S.J. Van Ransburg, Comparison of supercritical fluid extraction and Soxhlet extraction for the determination of DDT, DDD and DDE in sediment, Water SA 24 (1998) 205–214.

[4] K. Ganzler, A. Salgo, K. Valkó, Microwave extraction: a novel sample preparation method for chromatography, J. Chromatogr. A. 371 (1986) 299–306.

[5] M.Z. Borhan, R. Ahmad, M.M. Rusop, S. Abdullah, Impact of nanopowders on extraction yield of Centella asiatica, Adv. Mater. Res. 667 (2013) 246–250.

[6] S.S.H. Rizvi, J.A. Daniels, A.L. Benado, J.A. Zollweg, Supercritical Fluid Extraction: Operating Principles and Food Applications, Food Technology, USA, 1986.

[7] Y.H. Cho, J. Park, Encapsulation of flavour by molecular inclusion using β-cyclodextrin: comparison with spray-drying process using carbohydrate-based wall materials, Food Sci. Biotechnol. 18 (1) (2009) 185–189.

[8] C.I. Beristain, A. Vazquez, H.S. Garcia, E.J. Vernon-Carter, Encapsulation of orange peel oil by co-crystallization, LWT–Food Sci. Technol. 29 (7) (1996) 645–647.

[9] Y. Wang, W. Liu, X.D. Chen, C. Selomulya, Micro-encapsulation and stabilization of DHA containing fish oil in protein-based emulsion through mono-disperse droplet spray dryer, J. Food Eng. 175 (2016) 74–84.

[10] B.R. Bhandari, E.D. Dumoulin, H.M.J. Richard, I. Noleau, A.M. Lebert, Flavor encapsulation by spray drying: application to citral and linalyl acetate, J. Food Sci. 57 (1) (1992) 217–222.

[11] G.A. Reineccius, Flavor encapsulation, Food Rev. Int. 5 (2) (1989) 147–176.

[12] E.K. Silva, G.L. Zabot, M.A.A. Meireles, Ultrasound-assisted encapsulation of annatto seed oil: retention and release of a bioactive compound with functional activities, Food Res. Int. 78 (2015) 159–168.

[13] T. Borodina, E. Markvicheva, S. Kunizhev, H. Möhwald, G.B. Sukhorukov, O. Kreft, Controlled release of DNA from self-degrading microcapsules, Macromol. Rapid Commun. 28 (18–19) (2007) 1894–1899.

[14] M. Kessler, N. Sottos, S. White, Self-healing structural composite materials, Compos. Part A: Appl. Sci. Manuf. 34 (8) (2003) 743–753.

[15] G. Souradeep, H.W. Kua, Encapsulation technology and techniques in self-healing concrete, J. Mater. Civil Eng. 28 (12) (2016) 04016165.

# Further reading

T. Jung, Biodegradable nanoparticles for oral delivery of peptides: is there a role for polymers to affect mucosal uptake? Eur. J. Pharm. Biopharma. 50 (1) (2000) 147–160.

T.M. Kauer, J.-L. Figueiredo, S. Hingtgen, K. Shah, Encapsulated therapeutic stem cells implanted in the tumor resection cavity induce cell death in gliomas, Nat. Neurosci. 15 (2) (2011) 197–204.

J. Kreuter, Nanoparticles a historical perspective, Int. J. Pharm. 331 (1) (2007) 1–10.

M. Zeisser-Labouèbe, N. Lange, R. Gurny, F. Delie, Hypericin-loaded nanoparticles for the photodynamic treatment of ovarian cancer, Int. J. Pharm. 326 (1–2) (2006) 174–181.

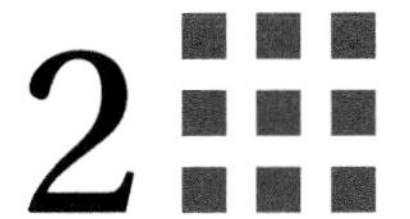

# Physicochemical characterization techniques in the encapsulation of active molecules

Sarang P. Gumfekar

*DEPARTMENT OF CHEMICAL AND MATERIALS ENGINEERING, UNIVERSITY OF ALBERTA, EDMONTON, AB, CANADA*

**Chapter Outline**

**2.1 Introduction** .......... 9
**2.2 Particle size and its distribution** .......... 10
**2.3 Surface charges** .......... 12
**2.4 Imaging of the encapsulated materials** .......... 12
**2.5 The crystallinity of encapsulation systems** .......... 15
**2.6 Rheology of the encapsulated materials** .......... 17
**2.7 Conclusion and outlook** .......... 20
**References** .......... 20

## 2.1 Introduction

Entrapment of an active agent into another substance, which in combination produces particles of size ranging from millimeter to nanometer, is referred to as encapsulation. The internal space that encapsulates active molecules is called core, fill, or payload phase. Similarly, the substance that encapsulates active molecules is called shell, carrier, membrane, coating, or matrix. Encapsulation technology is used in different areas for diverse purposes. For example, encapsulation is used in food technology to protect liquid or solid active agents against environmental parameters such as light, moisture, oxygen, and radicals. In the pharmaceutical domain the delivery of the active molecules is significantly considered while designing the encapsulation system. Various polymeric microstructures are used as the carrier for the encapsulation of different active molecules. Thus it is a multidisciplinary field, where researchers possessing knowledge in the areas of polymers, biomaterials, self-assembly, surface chemistry, biology, and many more collaborate. Researchers have also used the encapsulation technology to prepare intelligent coating which contains encapsulated corrosion

**Encapsulation of Active Molecules and their Delivery System. DOI: https://doi.org/10.1016/B978-0-12-819363-1.00002-8**

inhibitor, which self-releases upon a change in pH [1]. Encapsulated calcium carbonate is used in bio-based self-healing concretes in which, bacterial activity induces the precipitation of calcium carbonate that subsequently heals the crack in the concrete [2]. The investigation of the correlation between the design of an encapsulation system, the method of encapsulation, and the release performance require a series of characterization methods. To determine which characterization is necessary for a specific encapsulation work, it is essential to understand why and how the encapsulation is performed. First, active molecules can be encapsulated by forming a wall around or first making the wall material and then allowing the active molecules to transport inside the wall. Second, it becomes imperative to determine if the undesired leakage takes place or not. Finally, the selectivity of the encapsulation system is important to keep the undesired materials out of the carrier material. Confirmation of these general steps requires extensive characterization of a complete encapsulation system.

While there are a number of independent research articles and exhaustive reviews [3–5] on the encapsulation methods and carrier materials for a variety of applications, yet there is no review that brings together the various characterization techniques and their particular use in the context of encapsulation systems. This chapter provides a guide for using various techniques to characterize different aspects of the encapsulation systems.

## 2.2 Particle size and its distribution

Particle size and its distribution (polydispersity) are the key parameters that strongly correlate with the stability of the encapsulated system. Particle characteristics are also necessary to create an appropriate delivery system. Researchers have used numerous ways to control these parameters, namely, surfactants, ionic macromolecules/polymers, ultrasonic cavitation, and cross-linking of gels [6,7]. The size of particles may vary from 10 nm (e.g., nanoemulsions) to 1 mm (e.g., hydrogel beads). Colloidally stable encapsulated particles often take spherical shape, but other shapes such as the deformed sphere, cylinder, or irregular shapes are also reported, which affect various physical properties and the release mechanisms of active molecules [8]. Polydispersity indirectly indicates the aggregation state of the encapsulated system. Higher polydispersity indicates the presence of aggregates, which may induce instability in the emulsion-based encapsulation systems resulting in a breakdown. The encapsulation system is said to be monodispersed when polydispersity is less than 0.2, while polydispersity less than 0.5 is also considered for pharmaceutical applications [9–11]. In the case of emulsions the droplet size is an important parameter, as a decrease in the droplet size and polydispersity increases the stability of the emulsion. Dynamic light scattering (DLS) is the most common way to measure particle size and polydispersity. Yalcin et al. investigated the stability of gemcitabine hydrochloride-loaded liposomes by measuring the particle size and polydispersity at varying temperatures and time for 6 months [12]. They also reported that the composition of carrier and physicochemical characteristics of bioactive molecules had a significant effect on the particle size of the encapsulated material. A more detailed study on the particle-size distribution of the encapsulated material was performed by Chebil et al. to prepare dispersions exhibiting monomodal dispersity [13]. They used laser sources with two wavelengths, 633 and 466 nm, and detectors that covered a range between 0.015 and 144 degrees to define the broadness of the distribution curve. Eq. (2.1) defines "Span," which can be used to investigate the monomodal or multimodal behavior of the encapsulated particles:

$$\text{Span} = \frac{d(0.9) - d(0.1)}{d(0.5)} \tag{2.1}$$

where $d(0.9)$, $d(0.5)$, and $d(0.1)$ are the particle diameters at 90%, 50%, and 10% cumulative volumes, respectively. Span values lesser than 1.2 were considered to have monomodal distributions in their experiments. Haidar et al. probed the polydispersity of Halofantrine-encapsulated particles using "fingerprinting" particle-size analysis [14]. They demonstrated this analysis by measuring the particle size of certain emulsions as a function of the light-scattering angle. Typically, 90 degrees is used as the scattering angle in the majority of the DLS techniques. The particle size should not change for a monodispersed sample, upon a change in the light-scattering angle. Since the extent of scattering at various angles partly depends on the particle size, the intensity-averaged mean particle size changes for polydispersed samples [15]. Fig. 2–1 shows the particle-size fingerprinting of various lab-prepared

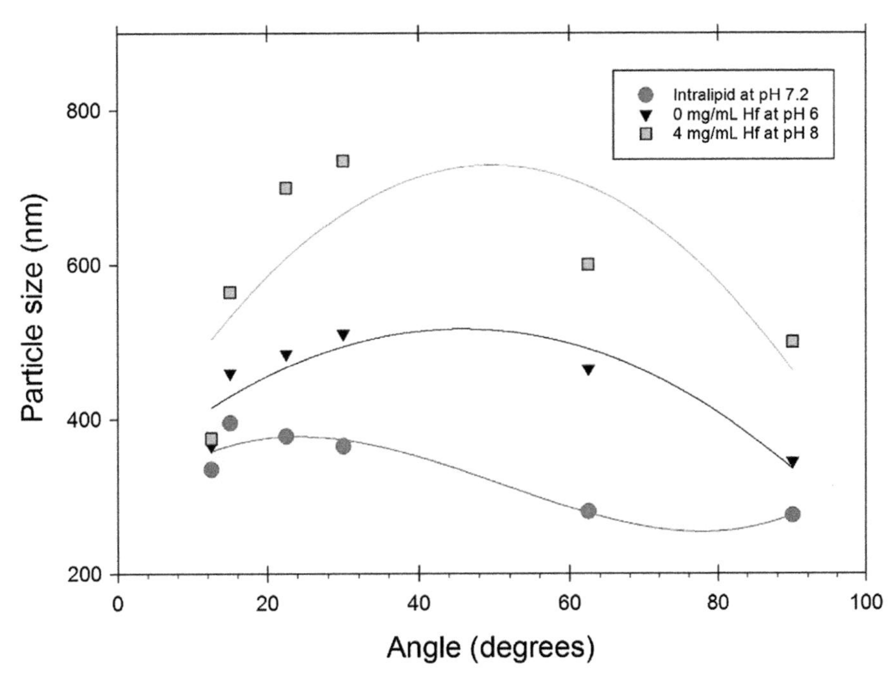


**FIGURE 2–1** Particle-size fingerprinting for various encapsulated materials [14]. Reprinted with permission from I. Haidar, I.H. Harding, I.C. Bowater and A.W. McDowall, Physical characterization of halofantrine-encapsulated fat nanoemulsions, *J. Pharm. Sci.* 108 (6), 2019, 2153–2161.

encapsulations and commercial Intralipid emulsion. It can be seen that the commercial Intralipid (red data points) had relatively more independence on the scattering angle, while lab-made emulsions showed dependence on the scattering angle. This fingerprinting demonstrated that the encapsulated materials had multimodal distributions when they had scattering angle-dependant particle size.

## 2.3 Surface charges

The surface charge of the encapsulated materials significantly correlates with their stability in dispersion. Zeta potential is often used to investigate the surface charges of the encapsulated materials, which indirectly indicates the dominance of electrostatic forces. In the context of encapsulation, researchers have measured zeta potential to investigate the effect of the loading of active molecules on the surface properties of the carrier material. These experiments indicate the stability of the encapsulated materials and reveal the electrostatic interactions between active molecules and carrier material. Generally, it is well-accepted that the particle dispersion is possibly stable if the absolute value of zeta potential is high (whether positive or negative). In biomimetic mineralization of proteins, they are encapsulated in certain materials (cage) to protect and preserve them. Typically, this is achieved by nucleating and further growing a cage material on the protein surface to encapsulate it [16]. In this process the surface charge of the loaded phase plays a vital role in the nucleation and growth of the carrier material. Recently, Maddigan et al. encapsulated proteins in metal–organic frameworks by controlling the surface charges of the proteins [17]. They showed that inducing negative surface charge facilitated encapsulation, while positive surface charge hindered the mineralization process. Fig. 2–2 shows the experimental zeta potential of bovine serum albumin, pepsin, hemoglobin, and myoglobin. It also shows general variation in the zeta potential after chemical functionalization.

Researchers have encapsulated nanoparticles by virus coat proteins to better understand virus assembly and further apply the encapsulated nanoparticles in bioimaging and therapeutic applications [18,19]. Lin et al. investigated the correlation between the encapsulation efficiency of nanoparticles by virus coat proteins and surface charge density of nanoparticles [20]. They revealed that a minimum negative surface charge density is required for the encapsulation to occur. The encapsulation efficiency was found to increase with an increase in the surface charge density of the nanoparticles. Since ionizable groups induce surface charges, the deviation in the number of ionizable groups near the average value is the highest for pH near their p*K*a value. Fig. 2–3 illustrates the encapsulation efficiency concerning changes in the surface charge density in the experiments [21] and models [20]. Gaussian model fits the experimental data well as it considers the charge distribution of the nanoparticles against the fixed-charge model.

## 2.4 Imaging of the encapsulated materials

Electron microscopy (EM) and fluorescence microscopy are arguably one of the most informative and one of the widely used instruments to probe particle morphology, biological

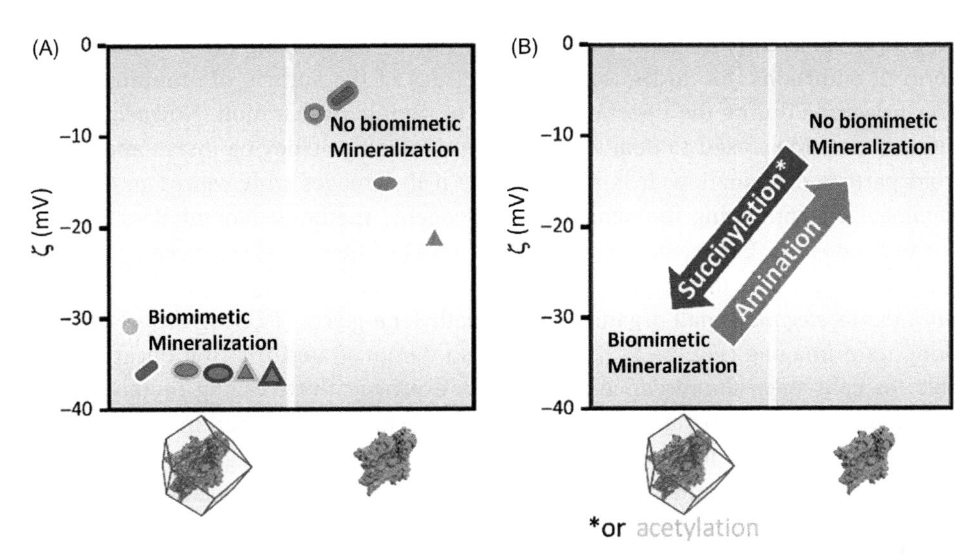


**FIGURE 2–2** (A) The experimental zeta potential of BSA, pepsin, hemoglobin, and myoglobin and (B) general variations in the zeta potential after chemical functionalization [17]. Published by the Royal Society of Chemistry. *BSA*, Bovine serum albumin. Reprinted with permission from N.K. Maddigan, A. Tarzia, D.M. Huang, C.J. Sumby, S.G. Bell, P. Falcaro, et al., Protein surface functionalisation as a general strategy for facilitating biomimetic mineralisation of ZIF-8, *Chem. Sci.* **9** (18), 2018, 4217–4223.

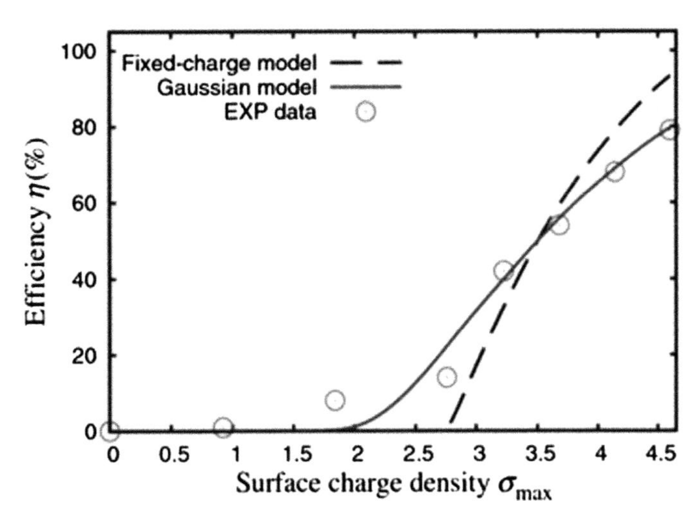


**FIGURE 2–3** The encapsulation efficiency concerning changes in the surface charge density in the case of experiments and models [20,21]. Reprinted with permission from H.-K. Lin, P. van der Schoot and R. Zandi, Impact of charge variation on the encapsulation of nanoparticles by virus coat proteins, *Phys. Biol.* **9** (6), 2012, 066004; M.-C. Daniel, I.B. Tsvetkova, Z.T. Quinkert, A. Murali, M. De, V.M. Rotello, et al., Role of surface charge density in nanoparticle-templated assembly of bromovirus protein cages, *ACS Nano* **4** (7), 2010, 3853–3860.

imaging, and complex emulsions. However, EM can be challenging for the encapsulation in the form of emulsions due to the *soft* characteristics of the surface of emulsion, which can distort or degrade during the rigorous methods of sample preparation. Nowadays, cryogenic techniques are widely used to deal with these problems by achieving instantaneous freezing to avoid particle deformation. It is still arguable if the images truly represent the emulsion morphology after preparing the sample with cryogenic methods. Nonetheless, the state-of-the-art technology in EM has helped to resolve most of these problems, and EM combined with cryogenic cooling techniques is of great help. Recently, Burnett et al. encapsulated *Caenorhabditis elegans* (small organism) in polyethylene glycol (PEG) hydrogels for continuous long-term imaging (Fig. 2–4) [22]. They also demonstrated that the encapsulation was suitable for light-sheet fluorescence microscopy, a provocative imaging method to perform uninterrupted long-term 3D imaging. Organism *C. elegans* was inserted in PEG hydrogel that was rapidly cross-linked to restrict the movement and enable imaging. Their work opens new applications of encapsulation for the imaging of live biological species. Fig. 2–4A shows the movement index of *C. elegans* in PEG hydrogel, and Fig. 2–4B shows high-resolution

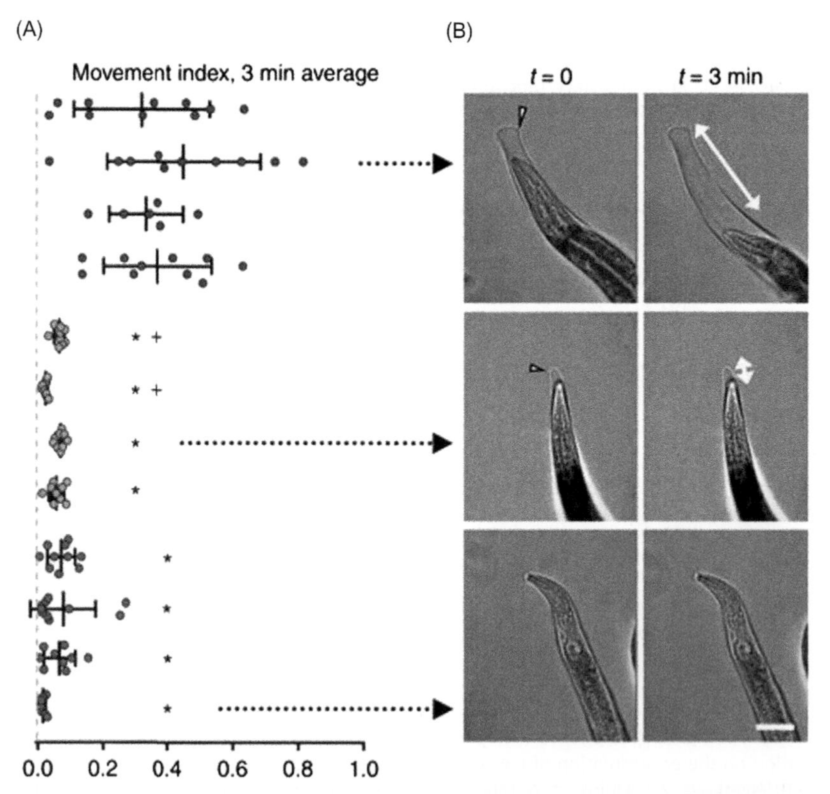


**FIGURE 2–4** (A) The movement index of *Caenorhabditis elegans* in PEG hydrogel and (B) high-resolution microscopy images; the animal can move inside the confined space, as indicated by white arrows [22]. *PEG*, Polyethylene glycol. Reprinted with permission from K. Burnett, E. Edsinger and D.R. Albrecht, Rapid and gentle hydrogel encapsulation of living organisms enables long-term microscopy over multiple hours, *Commun. Biol.* **1** (1), 2018, 73.

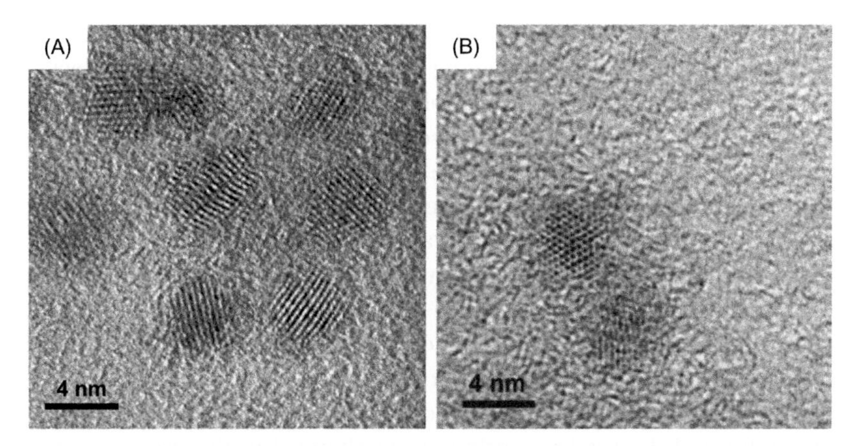


**FIGURE 2–5** Bright-field STEM images showing (A) array of micelle encapsulated quantum dots and (B) individual quantum dots [24]. *STEM*, Scanning tunneling electron microscopy. Reprinted with permission from C.M. Lemon, E. Karnas, X. Han, O.T. Bruns, T.J. Kempa, D. Fukumura, et al., Micelle-encapsulated quantum dot-porphyrin assemblies as in vivo two-photon oxygen sensors, *J. Am. Chem. Soc.* **137** (31), 2015, 9832–9842.

microscopy images; the animal can move inside the confined space, as indicated by white arrows.

Fluorescent markers are often used for in vitro and in vivo biological imaging. Quantum dots can be used as fluorescent markers for biological imaging due to tunable fluorescence and excitation by a single light source. Dubertret et al. encapsulated ZnS–CdSe quantum dots in lipids that were grafted with hydrophilic polymer for in vivo imaging of *Xenopus* embryos [23]. Compared to other materials such as dyes and fluorescent proteins, encapsulated quantum dots provided effective fluorescence, a significant decrease in photobleaching, and colloidal stability in various bioenvironments, and low nonspecific adsorption. Similarly, Lemon et al. employed micelles that were formed by self-assembled quantum dots with Pd(II) porphyrins to quantify dissolved oxygen in the water and in vivo [24]. They performed scanning tunneling EM (STEM) to image quantum dots encapsulated in micelles and studied their state of aggregation. Fig. 2–5 shows the bright-field STEM image of quantum dots encapsulated micelles. The crystalline pattern of the quantum dots, micelle encapsulation, and STEM grid can be distinctly observed.

## 2.5 The crystallinity of encapsulation systems

Researchers have used a variety of methods to investigate the crystal structure of the encapsulated materials, for example, X-ray diffraction (XRD), differential scanning calorimetry (DSC), and Raman spectroscopy. The interaction between the carrier and active molecule, changes in the crystallinity of both materials, and changes in the molecular mass of the carriers can be investigated using crystallinity analysis. Typically, XRD and DSC are used to verify if the active molecules are encapsulated in the carrier material or just aggregated with the carrier material. The successful encapsulation of active molecules in the carrier material is likely not to show the crystalline and melting peaks of individual active molecules, whereas the physical aggregation/mixture of active molecules with carrier material, without encapsulation, shows the characteristics of both active molecules and carrier material [25]. Jenning and Gohla evaluated the encapsulation of retinoids in solid lipid nanoparticles using DSC and wide-angle X-ray scattering [26]. They demonstrated a strong relationship between the degree of crystallinity and retinoids encapsulation. In their study the encapsulation of retinoids was favored in low crystallinity lipids. Further, the metastable polymorphs with several crystal defects improved the encapsulation of retinoids. The encapsulation of active material in the carrier crystal lattice usually exhibits a decrease in the melting point. A linear correlation between the melting temperature and fraction of active molecules indicates the complete diffusion of the drug in the carrier lattice, whereas if the thermal characteristics (e.g., melting temperature) of the carrier matrix are not changed in the presence of active molecules, it may be that the active molecules are not encapsulated in the carrier matrix. Thus DSC measurements can provide useful information on the interaction between the carrier matrix and active molecules to be encapsulated.

Recently, Ahmad et al. used DSC and XRD as complementary tools to support the encapsulation efficiency results of starch-encapsulated catechin [27]. They encapsulated catechin in different types of starch nanoparticles. They observed that the crystallinity of the encapsulated system reduced after performing the nano-encapsulation process, which formed amorphous nanoparticles (Fig. 2–6). The reduction in crystalline character was attributed to the disruption of the long-range ordering of crystalline structure.

De Castro et al. encapsulated the heat exchange material, *n*-docosane, into polyurethane carrier capsules of varying sizes [28]. They reported that the reduction in carrier size led to changes in the crystallinity of encapsulated material and melting/crystallization

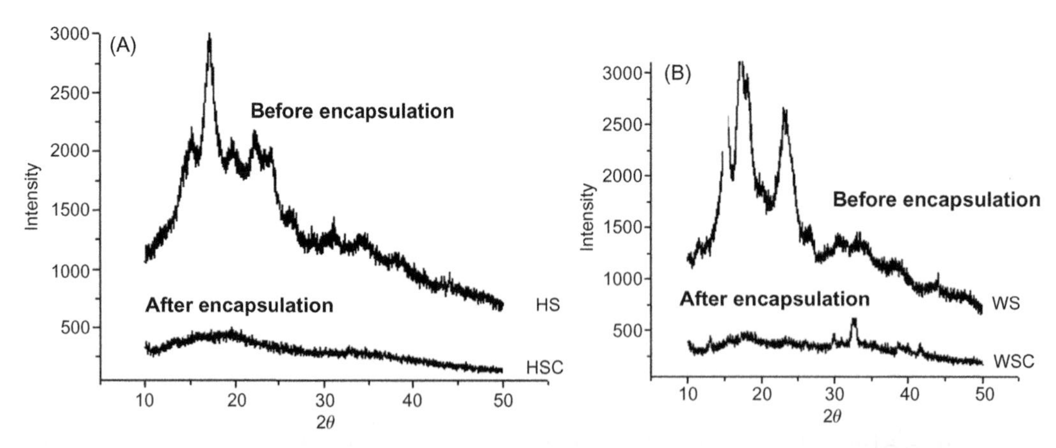


**FIGURE 2–6** Catechin encapsulation by two types of starch. XRD patterns of starch before and after encapsulation: (A) horse chestnut starch and (B) water chestnut starch [27]. *XRD*, X-ray diffraction. Reprinted with permission from M. Ahmad, P. Mudgil, A. Gani, F. Hamed, F.A. Masoodi and S. Maqsood, Nano-encapsulation of catechin in starch nanoparticles: Characterization, release behavior and bioactivity retention during simulated in-vitro digestion, *Food Chem.* **270**, 2019, 95–104.

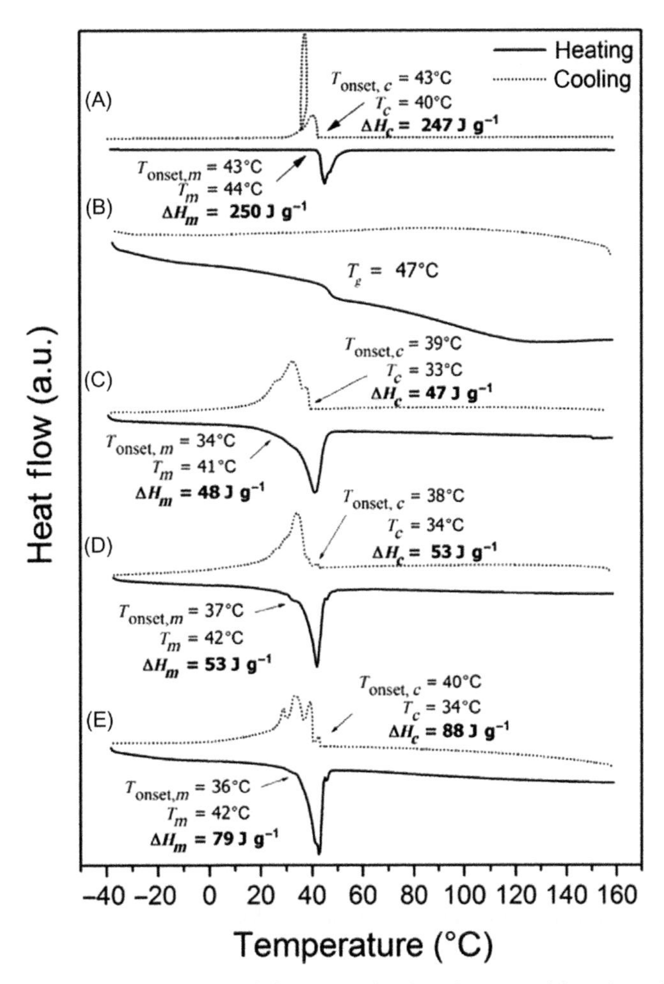


**FIGURE 2–7** Differential scanning calorimetry of (A) active molecule *n*-docosane, (B) carrier polyurethane hollow shell, *n*-docosane loaded into (C) 10 μm capsules, (D) 4 μm capsules, and (E) 2 μm capsules [28]. Reprinted with permission from P.F. De Castro, A. Ahmed and D.G. Shchukin, Confined-volume effect on the thermal properties of encapsulated phase change materials for thermal energy storage, *Chem. – A Eur. J.* **22** (13), 2016, 4389–4394.

temperature. Also, their DSC results showed that the encapsulated active molecule had a highly ordered triclinic phase at low temperature that converted to the face-centered orthorhombic phase before converting to the liquid phase at high temperature. The decrease in carrier microcapsule size increased the enthalpies of melting and crystallization, as seen in Fig. 2–7.

## 2.6 Rheology of the encapsulated materials

Often emulsions are used as a platform for the encapsulation and protection of active molecules. Researchers have used single, multiple, and nanoemulsions to encapsulate various types of molecules [29]. Variations in the composition and preparation technologies for emulsion-based encapsulation lead to different ranges of emulsions with new properties. The final application of the encapsulated system depends on the rheological behavior. Matos et al. encapsulated resveratrol in double emulsion (water-in-oil-in-water) and investigated the effect of emulsion composition on the rheological properties [30]. In general, the double emulsions with higher concentration showed pseudoplastic behavior and a dominant elastic character. The flow behavior of the double emulsions was modeled using the Power Law (Eq. 2.2).

$$\sigma = K(\gamma)^n \quad (2.2)$$

where $\sigma$ is the shear stress (Pa), $\gamma$ is the shear rate ($s^{-1}$), $K$ is the consistency index (Pa $s^n$), and $n$ is the flow behavior index (dimensionless). The majority of the double emulsions showed shear-thinning characteristics at lower $n$ values (Fig. 2–8). This shear-thinning behavior was attributed to the disruption of the network formed at the equilibrium stage.

In the case of the core–shell type of encapsulation the rheological properties of the encapsulated system were found to be dependent on the operational temperature, the concentration of active molecules in the carrier, and the concentration of additives such as surfactants. Yang et al. encapsulated phase change material in various carriers such as polystyrene, polymethyl methacrylate, and polyethyl methacrylate and studied the effect of temperature and concentration of active molecules on the viscosity of the encapsulated systems [31]. They reported that the viscosities of the slurries were not affected by the carrier materials but largely dependent on the temperature; a higher temperature led to a lower viscosity. It was also found that the increase in the concentration of tetradecane (active molecule) increased the viscosity of the encapsulated system.

Recently, the encapsulated materials are drawing attention in "functional foods" domain. Functional foods contain encapsulated "functional" food gradient that may enhance quality, taste, stability, texture, or some other property of the food. Since the encapsulated materials

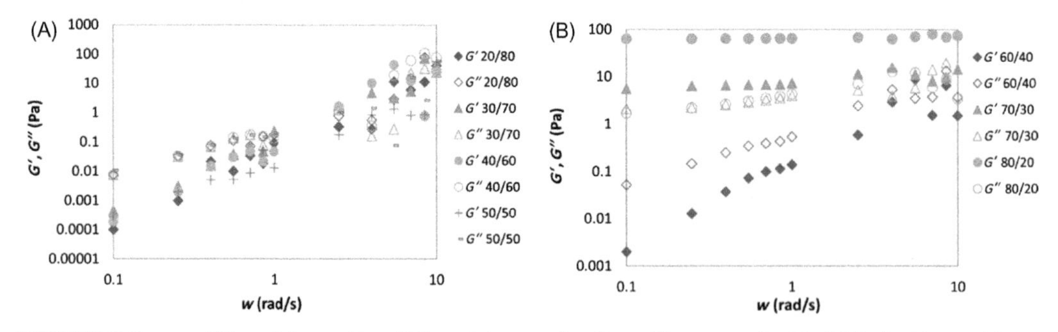


**FIGURE 2–8** Storage ($G'$) modulus and loss ($G''$) modulus as a function of frequency for $W_1/O/W_2$ double emulsions prepared at varying volumetric ratios of $W_1/O$ in $W_2$ [30]. Reprinted with permission from M. Matos, G. Gutiérrez, L. Martínez-Rey, O. Iglesias and C. Pazos, Encapsulation of resveratrol using food-grade concentrated double emulsions: Emulsion characterization and rheological behaviour, *J. Food Eng.* **226**, 2018, 73–81.

have a micrometer or submicron size, they significantly affect the rheology of the final product, which is important for various applications. Comunian et al. encapsulated echium oil and phytosterols by complex coacervation to increase their oxidative stability and added them to yogurt [32]. They obtained storage ($G'$), loss ($G''$) moduli, tan $\delta$ ($G''/G'$), and complex modulus ($((G')^2 + (G'')^2)^{1/2}$) from the rheology experiments and modeled the flow behavior using Herschel–Bulkley equation (Eq. 2.3).

$$\sigma = \sigma_0 + k\gamma^n \tag{2.3}$$

where $\sigma_0$, $k$, and $\gamma$ represent the yield stress (Pa), consistency index (Pa s), and shear rate ($s^{-1}$), respectively. Values of $n$ define the type of fluid: $n = 1$: Newtonian fluid; $n < 1$: pseudoplastic fluid and $n > 1$: dilating fluid. Their results indicated that the apparent viscosity of the yogurts reduced with a higher rate of deformation during shear and remained stable without significant difference throughout the storage modulus. The addition of encapsulated active molecules to the yogurt increased the viscosity of the product without affecting quality. Similarly, Wang et al. encapsulated curcumin in lyotropic liquid crystals and studied the rheological properties of the encapsulated curcumin [33]. Before encapsulation, they dissolved curcumin in ethyl oleate or polyoxyethylene-10-oleyl ether followed by encapsulation in liquid crystals. They studied the temperature evolution of elastic modulus, viscous modulus, and tan $\delta$ upon heating the encapsulated curcumin. Fig. 2–9 shows the effect of temperature dependence of elastic and viscous moduli and tan $\delta$ of curcumin-encapsulated liquid crystals.

Researchers have reported the effect of particle shape and particle-size distribution on the rheological properties of the product [34]. When the encapsulated materials are

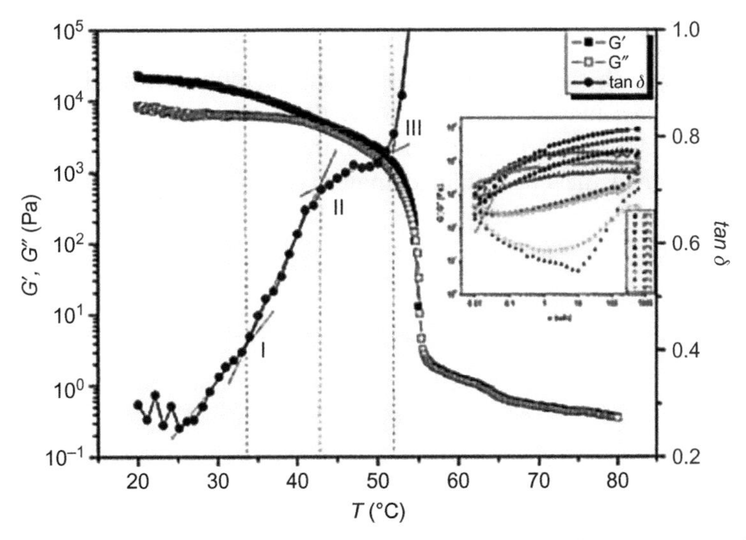


**FIGURE 2–9** Effect of temperature on the elastic and viscous moduli and tan $\delta$ of curcumin-encapsulated liquid crystals [33]. Reprinted with permission from X. Wang, Z. Wang, X. Zhao, L. Zhang and J. Fan, Rheological properties of lyotropic liquid crystals encapsulating curcumin, *J. Dispers. Sci. Technol.* **38** (1), 2017, 132–138.

nonspherical, the viscosity may increase due to extra energy dissipation in the system. Simple empirical equations (Eqs. 2.4 and 2.5) are reported to correlate the shape of the encapsulated particles and intrinsic viscosity.

$$[\eta] = 0.07\ q^{5/3}(\text{for rod-shaped}) \tag{2.4}$$

$$[\eta] = 0.3\ q(\text{for disk-shaped}) \tag{2.5}$$

where $q$ is the axial ratio.

## 2.7 Conclusion and outlook

Physical characterization of the encapsulated systems in a comprehensive manner is essential for the fundamental study as well as industrial applications. Significant literature has reported various types of encapsulation systems and their applications in different fields. However, a particular set of characterization is required to describe the system completely. Many successful formulations for the desired protection and delivery are reported, but the empirical correlations between experimental parameters are still lacking. Nowadays, encapsulation technology is being used beyond conventional encapsulation and delivery of bioactive molecules. A significant number of reports have demonstrated encapsulation-based self-healing technology for various applications. However, an interlinking study between the physical characteristics of encapsulation (viscosity, elastic modulus, strain capacity, etc.) and kinetics of healing is required. One can achieve higher loading capacity, the efficiency of encapsulation, stability, and enhanced released profile using nanosized encapsulation in comparison with microsized encapsulation. Still, improved control on the particle size and prevention of aggregation are required to fetch the benefit of nano-encapsulation. Further investigations can be focused on finding new mechanisms to release the active molecules and improving their efficiency in various systems. Recent advances in the effect of the properties of wall material and storage behavior of the encapsulated materials can be integrated to accelerate their commercialization.

## References

[1] Y. Feng, Y.F. Cheng, An intelligent coating doped with inhibitor-encapsulated nanocontainers for corrosion protection of pipeline steel, Chem. Eng. J. 315 (2017) 537–551. Available from: https://doi.org/10.1016/j.cej.2017.01.064.

[2] M. Alazhari, T. Sharma, A. Heath, R. Cooper, K. Paine, Application of expanded perlite encapsulated bacteria and growth media for self-healing concrete, Constr. Build. Mater. 160 (2018) 610–619. Available from: https://doi.org/10.1016/j.conbuildmat.2017.11.086.

[3] M. Saifullah, M.R.I. Shishir, R. Ferdowsi, M.R. Tanver Rahman, Q. Van Vuong, Micro and nano encapsulation, retention and controlled release of flavor and aroma compounds: a critical review, Trends Food Sci. Technol. 86 (2019) 230–251. Available from: https://doi.org/10.1016/j.tifs.2019.02.030.

[4] T. Alkayyali, T. Cameron, B. Haltli, R.G. Kerr, A. Ahmadi, Microfluidic and cross-linking methods for encapsulation of living cells and bacteria – a review, Anal. Chim. Acta 1053 (2019) 1–21. Available from: https://doi.org/10.1016/j.aca.2018.12.056.

[5] C. Xue, W. Li, J. Li, V.W.Y. Tam, G. Ye, A review study on encapsulation-based self-healing for cementitious materials, Struct. Concr. 20 (1) (2019) 198–212. Available from: https://doi.org/10.1002/suco.201800177.

[6] D.M. Headen, J.R. García, A.J. García, Parallel droplet microfluidics for high throughput cell encapsulation and synthetic microgel generation, Microsyst. Nanoeng. 4 (1) (2018) 17076. Available from: https://doi.org/10.1038/micronano.2017.76.

[7] B.P. Koppolu, S.G. Smith, S. Ravindranathan, S. Jayanthi, T.K. Suresh Kumar, D.A. Zaharoff, Controlling chitosan-based encapsulation for protein and vaccine delivery, Biomaterials 35 (14) (2014) 4382–4389. Available from: https://doi.org/10.1016/j.biomaterials.2014.01.078.

[8] D.J. McClements, Encapsulation, protection, and delivery of bioactive proteins and peptides using nanoparticle and microparticle systems: a review, Adv. Colloid Interface Sci. 253 (2018) 1–22. Available from: https://doi.org/10.1016/j.cis.2018.02.002.

[9] S. Das, W.K. Ng, R.B.H. Tan, Are nanostructured lipid carriers (NLCs) better than solid lipid nanoparticles (SLNs): development, characterizations and comparative evaluations of clotrimazole-loaded SLNs and NLCs? Eur. J. Pharm. Sci. 47 (1) (2012) 139–151. Available from: https://doi.org/10.1016/j.ejps.2012.05.010.

[10] R. Shah, D. Eldridge, E. Palombo, I. Harding, Lipid Nanoparticles: Production, Characterization and Stability, Springer International Publishing, Cham, 2015. Available from: https://doi.org/10.1007/978-3-319-10711-0.

[11] R.M. Shah, D.S. Eldridge, E.A. Palombo, I.H. Harding, Encapsulation of clotrimazole into solid lipid nanoparticles by microwave-assisted microemulsion technique, Appl. Mater. Today 5 (2016) 118–127. Available from: https://doi.org/10.1016/j.apmt.2016.09.010.

[12] T.E. Yalcin, S. Ilbasmis-Tamer, B. Ibisoglu, A. Özdemir, M. Ark, S. Takka, Gemcitabine hydrochloride-loaded liposomes and nanoparticles: comparison of encapsulation efficiency, drug release, particle size, and cytotoxicity, Pharm. Dev. Technol. 23 (1) (2018) 76–86. Available from: https://doi.org/10.1080/10837450.2017.1357733.

[13] A. Chebil, M. Léonard, J.-L. Six, C. Nouvel, A. Durand, Nanoparticulate delivery systems for alkyl gallates: Influence of the elaboration process on particle characteristics, drug encapsulation and in-vitro release, Colloids Surf. B: Biointerfaces 162 (2018) 351–361. Available from: https://doi.org/10.1016/j.colsurfb.2017.11.050.

[14] I. Haidar, I.H. Harding, I.C. Bowater, A.W. McDowall, Physical characterization of halofantrine-encapsulated fat nanoemulsions, J. Pharm. Sci. 108 (6) (2019) 2153–2161. Available from: https://doi.org/10.1016/j.xphs.2019.01.021.

[15] J. Kuntsche, K. Klaus, F. Steiniger, Size determinations of colloidal fat emulsions: a comparative study, J. Biomed. Nanotechnol. 5 (4) (2009) 384–395. Available from: https://doi.org/10.1166/jbn.2009.1047.

[16] A.-W. Xu, Y. Ma, H. Cölfen, Biomimetic mineralization, J. Mater. Chem. 17 (5) (2007) 415–449. Available from: https://doi.org/10.1039/B611918M.

[17] N.K. Maddigan, A. Tarzia, D.M. Huang, C.J. Sumby, S.G. Bell, P. Falcaro, et al., Protein surface functionalisation as a general strategy for facilitating biomimetic mineralisation of ZIF-8, Chem. Sci. 9 (18) (2018) 4217–4223. Available from: https://doi.org/10.1039/C8SC00825F.

[18] M. Manchester, P. Singh, Virus-based nanoparticles (VNPs): platform technologies for diagnostic imaging, Adv. Drug Deliv. Rev. 58 (14) (2006) 1505–1522. Available from: https://doi.org/10.1016/j.addr.2006.09.014.

[19] Y. Ren, In vitro-reassembled plant virus-like particles for loading of polyacids, J. Gen. Virol. 87 (9) (2006) 2749–2754. Available from: https://doi.org/10.1099/vir.0.81944-0.

[20] H.-K. Lin, P. van der Schoot, R. Zandi, Impact of charge variation on the encapsulation of nanoparticles by virus coat proteins, Phys. Biol. 9 (6) (2012) 066004. Available from: https://doi.org/10.1088/1478-3975/9/6/066004.

[21] M.-C. Daniel, I.B. Tsvetkova, Z.T. Quinkert, A. Murali, M. De, V.M. Rotello, et al., Role of surface charge density in nanoparticle-templated assembly of bromovirus protein cages, ACS Nano 4 (7) (2010) 3853–3860. Available from: https://doi.org/10.1021/nn1005073.

[22] K. Burnett, E. Edsinger, D.R. Albrecht, Rapid and gentle hydrogel encapsulation of living organisms enables long-term microscopy over multiple hours, Commun. Biol. 1 (1) (2018) 73. Available from: https://doi.org/10.1038/s42003-018-0079-6.

[23] B. Dubertret, In vivo imaging of quantum dots encapsulated in phospholipid micelles, Science 298 (5599) (2002) 1759–1762. Available from: https://doi.org/10.1126/science.1077194.

[24] C.M. Lemon, E. Karnas, X. Han, O.T. Bruns, T.J. Kempa, D. Fukumura, et al., Micelle-encapsulated quantum dot-porphyrin assemblies as in vivo two-photon oxygen sensors, J. Am. Chem. Soc. 137 (31) (2015) 9832–9842. Available from: https://doi.org/10.1021/jacs.5b04765.

[25] M.J. Cocero, Á. Martín, F. Mattea, S. Varona, Encapsulation and co-precipitation processes with supercritical fluids: fundamentals and applications, J. Supercrit. Fluids 47 (3) (2009) 546–555. Available from: https://doi.org/10.1016/j.supflu.2008.08.015.

[26] V. Jenning, S.H. Gohla, Encapsulation of retinoids in solid lipid nanoparticles (SLN), J. Microencapsul. 18 (2) (2001) 149–158. Available from: https://doi.org/10.1080/02652040010000361.

[27] M. Ahmad, P. Mudgil, A. Gani, F. Hamed, F.A. Masoodi, S. Maqsood, Nano-encapsulation of catechin in starch nanoparticles: Characterization, release behavior and bioactivity retention during simulated in-vitro digestion, Food Chem. 270 (2019) 95–104. Available from: https://doi.org/10.1016/j.foodchem.2018.07.024.

[28] P.F. De Castro, A. Ahmed, D.G. Shchukin, Confined-volume effect on the thermal properties of encapsulated phase change materials for thermal energy storage, Chem. – A Eur. J. 22 (13) (2016) 4389–4394. Available from: https://doi.org/10.1002/chem.201505035.

[29] W. Lu, A.L. Kelly, S. Miao, Emulsion-based encapsulation and delivery systems for polyphenols, Trends Food Sci. Technol. 47 (2016) 1–9. Available from: https://doi.org/10.1016/j.tifs.2015.10.015.

[30] M. Matos, G. Gutiérrez, L. Martínez-Rey, O. Iglesias, C. Pazos, Encapsulation of resveratrol using food-grade concentrated double emulsions: Emulsion characterization and rheological behaviour, J. Food Eng. 226 (2018) 73–81. Available from: https://doi.org/10.1016/j.jfoodeng.2018.01.007.

[31] R. Yang, H. Xu, Y. Zhang, Preparation, physical property and thermal physical property of phase change microcapsule slurry and phase change emulsion, Sol. Energy Mater. Sol. Cell 80 (4) (2003) 405–416. Available from: https://doi.org/10.1016/j.solmat.2003.08.005.

[32] T.A. Comunian, I.E. Chaves, M. Thomazini, I.C.F. Moraes, R. Ferro-Furtado, I.A. de Castro, et al., Development of functional yogurt containing free and encapsulated echium oil, phytosterol and sinapic acid, Food Chem. 237 (2017) 948–956. Available from: https://doi.org/10.1016/j.foodchem.2017.06.071.

[33] X. Wang, Z. Wang, X. Zhao, L. Zhang, J. Fan, Rheological properties of lyotropic liquid crystals encapsulating curcumin, J. Dispers. Sci. Technol. 38 (1) (2017) 132–138. Available from: https://doi.org/10.1080/01932691.2016.1146615.

[34] D.B. Genovese, J.E. Lozano, M.A. Rao, The rheology of colloidal and noncolloidal food dispersions, J. Food Sci. 72 (2) (2007) R11–R20. Available from: https://doi.org/10.1111/j.1750-3841.2006.00253.x.

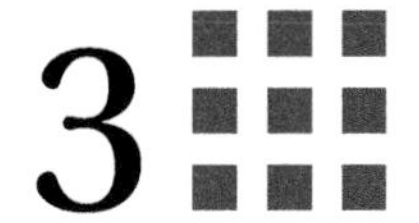

# Ultrasonic cavitation assisted synthesis of multilayer emulsions as encapsulating and delivery systems for bioactive compounds

Jitendra Carpenter, Suja George, Virendra Kumar Saharan

DEPARTMENT OF CHEMICAL ENGINEERING, MALAVIYA NATIONAL INSTITUTE OF TECHNOLOGY (MNIT), JAIPUR, INDIA

**Chapter outline**

**3.1 Emulsions as encapsulating and delivery system** ..... **23**
**3.2 Stabilization of emulsions** ..... **25**
3.2.1 Role of emulsifiers ..... 26
3.2.2 Homogenization ..... 28
**3.3 Application of ultrasonication for emulsification** ..... **29**
3.3.1 Emulsification mechanism ..... 30
3.3.2 Comparison of ultrasonication with high energy homogenizers ..... 31
**3.4 Multilayer emulsion and its application for the encapsulation of bioactive compounds** ..... **33**
3.4.1 Potential approaches for the preparation of multilayered emulsions ..... 34
3.4.2 Applications of multilayer emulsions ..... 36
**3.5 Case studies** ..... **38**
3.5.1 Ultrasonic-assisted synthesis of multilayer emulsions ..... 38
3.5.2 Encapsulation of curcumin in multilayer emulsions ..... 45
**References** ..... **50**

## 3.1 Emulsions as encapsulating and delivery system

Emulsion-based delivery systems have been utilized for the encapsulation of lipophilic bioactive molecules in various food and pharmaceutical products. An emulsion is a homogenous mixture

Encapsulation of Active Molecules and their Delivery System. DOI: https://doi.org/10.1016/B978-0-12-819363-1.00003-X

of two immiscible liquids, which is the dispersion of one liquid (dispersed phase) into another (continuous phase). Emulsions can be classified mainly as (1) oil-in-water (O/W), (2) water-in-oil (W/O) emulsions, and (3) multiple emulsions, which are a combination of both. The type of emulsion, either O/W and/or W/O, is prepared based on their application. In the application of encapsulation and delivery of bioactive molecules, O/W emulsions have been utilized for the development of such products.

O/W emulsions are the colloidal systems that have been utilized as a delivery vehicle for various nutraceuticals such as fatty oils, vitamins, flavors, drugs, and other lipophilic compounds [1]. Emulsion systems have been mainly designed to improve the solubility and bioavailability of lipophilic compounds while incorporating them into the functional foods [1]. Although there are many other approaches such as liposomes, micelles, hydrogels, and other nanodispersions that improve water dispersibility, bioavailability, and chemical stability of the lipophilic compounds, but emulsification has exceptional advantages, particularly in food applications. In food products, O/W emulsions are usually fabricated using food ingredients (including emulsifiers and fatty oils), and also most of the food products are already in the form of emulsions such as beverages, sauces, and desserts which make them relatively easy to incorporate the lipophilic constituents. Moreover, the bioactive compounds can also be incorporated in the core (oil phase) of the emulsion matrix that can improve the stability and functionality and facilitates the controlled release in the human body. The formation of an interfacial layer over the oil droplet carrying lipophilic components protects against their degradation under the environmental conditions. Emulsification technology has been explored in the context of food applications such as milk, creams, dairy, mayonnaise, beverages, and desserts. Thus emulsification as a delivery system is represented as a rapidly emerging area in the food, pharmaceutical industries, etc.

The significant factors that need to be considered before the formation of emulsion for the perfect encapsulation and the controlled release of bioactive are the selection of the oil phase and a suitable emulsifier. The oil phase acts as a carrier for the lipophilic actives such as flavors, nutrients, vitamins, and colors. Generally, highly saturated fats can harm health. Oils with a high polyunsaturated triglyceride content are mainly considered for dietary intake, and oils containing ω-3 and ω-6 fatty acids are often more favorable, as they impart specific health benefits when consumed regularly. Biopolymer-based emulsifier are mainly used for stabilizing the food-grade emulsion, as they contain food ingredients such as amino acids and carbohydrates which could be easily digested in the gastrointestinal (GI) tract of the body [2]. Also, these types of emulsifiers not only prevent the emulsions against physical instability such as coalescence of droplets but also effectively inhibit the oxidation of lipids present in the emulsion. After fulfilling these two the main factor that is to be considered is the emulsion stability at the point of application. The higher the emulsion stability, the better the shelf-life of the encapsulating materials. The stable emulsion helps to maintain the physicochemical properties of the nutraceuticals over a longer storage time and also the behavior of emulsion upon their application to achieve the controlled release of the compounds.

Thus, researchers focus on designing a suitable emulsion system with an enhanced stability to improve the shelf life of the encapsulated nutraceutical compounds. The stability of the emulsion is a crucial parameter that could be affected by small changes in the emulsion composition and also by slight variations in the environmental conditions such as pH, temperature, ionic strength, and lipid autooxidation. Apart from that, it is also necessary to homogenize the emulsion to uniformly distribute the emulsion constituents (emulsifier, oil, and water) for the droplet formation, disruption, and their dispersions in the continuous phase. Thus, it becomes vital to select the best operating condition that helps to achieve the desired stability in an energy-efficient manner. After obtaining the successful encapsulation of bioactive compounds in food emulsions, it is also necessary to evaluate the release properties for their delivery in the physiological conditions of the body. For the mouth, gastric, and intestinal conditions, the delivery environments are quite different, and thus, the release behavior of the compounds will also be different.

## 3.2 Stabilization of emulsions

The significant factor which needs to be considered before the formation of various types of the emulsion is their stability. Higher the stability of emulsion better would be the shelf-life of the materials or products that are encapsulated in the emulsion. The stable emulsion helps to maintain the physicochemical properties of the nutraceuticals or any bioactive compounds over a more extended time. The emulsions (nano and submicron) except microemulsions are usually thermodynamically unstable [2–7]. However, the kinetic stability of the emulsion is affected by various destabilization mechanisms such as creaming, coalescence, flocculation, and Ostwald ripening [6,7]. The thermodynamic stability gives information about the processes or changes taking place before and after the emulsification process, whereas the kinetic stability provides information about the rate at which these changes or processes occur. The emulsions formed for many applications such as food and pharmaceuticals are usually formed nonspontaneously and destabilization occurs at varying kinetic rates. Thus, kinetic stability is of great significance in the emulsification process. The kinetic stability of emulsion depends on various breakdown factors such as creaming, flocculation, coalescence, and Ostwald ripening.

In an emulsion, when two immiscible liquids, that is, oil and water, are mixed, two liquid layers are formed, which are separated by a thin interfacial layer possessing some interfacial forces that minimizes the contact area between the two phases. To form an emulsion, it is required to change the position of this layer by bringing some energy into the system. By supplying adequate energy to the system, the disruption of the interfacial layer occurs, and consequently, the interfacial area between the phases increases. However, when the droplets of oil in the continuous phase are formed under agitation, they continuously move, collide, and may coalesce with each other. After some time, the larger droplets are formed, and two phases get separated as a result of the coalescence of droplets. The use of an emulsifier is then necessary to avoid the droplets from merging.

The interfacial forces are characterized by the Laplace pressure ($\Delta P_L$), which is the pressure difference between the inside and outside of the droplet, across the oil–water interface. It can be expressed by the following equation [8]:

$$\Delta P_L = \frac{2\sigma}{r} \tag{3.1}$$

where $\sigma$ is the interfacial tension between the two phases and $r$ is the droplet radius. The above equation indicates that the droplet size depends on the interfacial tension. It is clear from Eq. (3.1) that if the interfacial tension is higher, the droplets formed will be larger. The Laplace equation indicates that during emulsion formation, the deformation of the droplet is opposed by the pressure gradient between the external and internal side of an interface. The pressure gradient required for emulsion formation is mostly supplied by agitation, and therefore to produce an emulsion of small droplets, very intense agitation/homogenization is required. Therefore, a significant homogenization pressure is required to overcome the interfacial tension and form the emulsion droplets of smaller size. Besides, when a balance between the interfacial tension and the Laplace pressure is achieved, droplet disruption does not occur and attains the maximum reduction in the interfacial tension and droplet size.

As mentioned earlier, emulsions stabilization occurs in two ways, that is, by using a suitable emulsifier and by homogenizing the mixture in a homogenizer that provides shear forces for the droplet disruption and their dispersion in the continuous phase. Relying only on one of them (either emulsifier or homogenizer) would pose a significant challenge to the techno-economic feasibility of food emulsions. The role of emulsifiers is that they adsorb at the oil–water interface and increase the interfacial area, thereby reducing the droplet size of the oil phase in the emulsion. Apart from that, it is also necessary to homogenize the emulsion to uniformly distribute all of the constituents (emulsifier, oil, and water) of an emulsion, required for the disruption and dispersions of the droplets in the continuous phase. Thus, the formation of the emulsion of desired stability in a cost-effective manner depends on the choice of emulsifier and a suitable homogenizer.

### 3.2.1 Role of emulsifiers

Emulsifiers are amphiphilic molecules with both hydrophilic and lipophilic portions adsorbed at the oil–water interface, thus making the emulsion stable. The emulsifier positions itself at the oil/water interface and enhances the interfacial area by reducing the interfacial tension between the phases and facilitates droplet disruption during homogenization. The emulsifier at the interfacial sites creates a stabilizing film to prevent droplet aggregation. An effective emulsifier has two main functions, that is, (1) it rapidly adsorbs at the oil–water interface during homogenization and (2) it forms an interfacial membrane that provides the steric hindrance between the droplets to prevent their coalescence. There are varieties of emulsifiers or stabilizers available that are classified based upon their molecular structure such as small-molecule surfactants (e.g., monoglycerides, polysorbates, lecithins, and fatty acid salts) and macromolecular emulsifiers (biopolymers e.g., proteins and polysaccharides) [9,10].

#### *3.2.1.1 Small-molecule surfactants*

Surfactant molecules possess a hydrophilic head and a hydrophobic tail group, which stabilizes the emulsions by adsorbing at the oil–water interface with the head group in the water and tail group in the oil phase. The hydrophilic portions align themselves with the water phase, while the lipophilic portions align with the lipid phase, as shown in Fig. 3–1A. The surfactants may be ionic, zwitterionic, and nonionic and are widely used in food emulsions. In ionic surfactant the hydrophilic head is either positively (in the case of cationic surfactant) or negatively charged (in the case of anionic surfactant) and stabilizes the emulsions based on the electrostatic interaction with the oil droplets. On the other side, nonionic surfactants possess no charge and usually stabilize the emulsions mainly by steric interactions [10,11]. Usually, nonionic emulsifiers are characterized by the hydrophilic lipophilic balance (HLB) value. A lipophilic emulsifier is classified by a low HLB number ($<7$), which stabilizes the W/O emulsion, and one that is hydrophilic is classified by a high HLB number ($>7$) that stabilizes O/W emulsions [11].

#### *3.2.1.2 Biopolymers*

Biopolymers, mainly the polysaccharides and proteins, are widely used as functional ingredients for various emulsified food products. The majority of food emulsions are formed and stabilized with the combination of polysaccharide and protein. These are the natural and essential food ingredients used in the food emulsion formulations owing to their ability to change the texture of food [12]. The protein–polysaccharide complex can form a thick viscoelastic interfacial film at the oil–water interface (Fig. 3–1B) that improves the stability of emulsions via a combination of electrostatic and steric repulsive forces against various breakdown processes [1,13]. Apart from that, the thick interfacial layer of the protein–polysaccharide complex prevents the attack of various oxidative compounds such as oxygen molecules, free radicals, and transition metal ions, which cause/promotes the lipid oxidation. Therefore, these

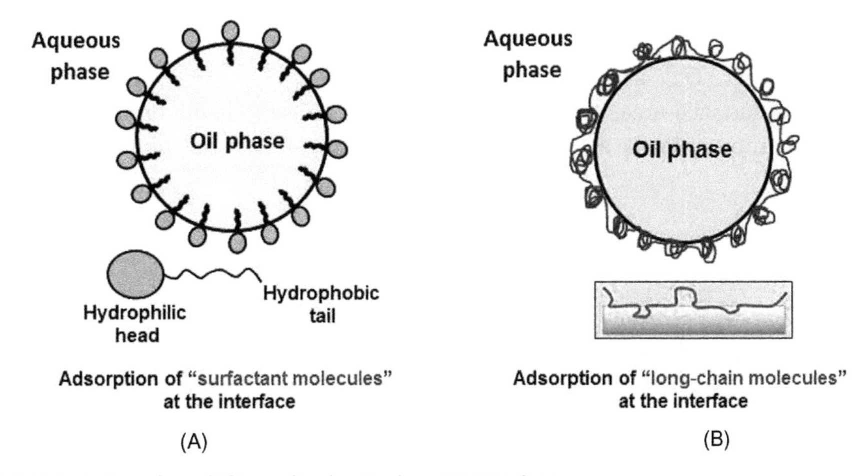


**FIGURE 3–1** Orientation of emulsifier molecules at oil–water interface.

biopolymers not only improve the physical stability of emulsions but also improve the oxidative stability of the lipid phase of an emulsion. Protein and polysaccharide interactions in food emulsions have been studied earlier by many researchers and are being well reviewed in the literature [9,11,13,14]. However, despite extensive research, these biopolymers continue to be one of the challenging areas to explore food hydrocolloids.

### 3.2.2 Homogenization

As discussed, the emulsifier adsorbs at the oil–water interface and subsequently increases the interfacial area by minimizing the interfacial tension between the phases. However, emulsifier molecules are not able to reduce the interfacial forces to a greater extent as required to prepare the uniformly dispersed system. Thus, external energy is required to generate intense disruptive forces for the uniform distribution of all the ingredients and also to break-up bigger oil droplets into smaller droplets [14]. It can be seen from Fig. 3–2 that initially, two immiscible liquids exhibiting an interfacial tension of $\sigma$ are separated because of the differences in the attractive interactions between them. When emulsifiers are added into the solution, they get deposited at the oil–water interface because of their amphiphilic behavior but do not completely stabilize the emulsions as they are not uniformly dispersed into the solution. Upon applying some mechanical shear, the generated disruptive forces emulsify the mixture in two ways: first, it uniformly disperses the emulsifier molecules in the aqueous phase thereby, enhancing the interfacial area between the phases, and second, by the fragmentation of the bigger droplet into smaller-sized droplets [8,14]. Beyond a certain amount of disruptive force, the interfacial tension would not reduce, and droplets of uniform size will be formed, as shown in Fig. 3–2.

The energy required to create an interfacial area, $A$, between the two liquids is $\sigma A$. The relationship between the interfacial area and droplet radius is given in the following equation [8]:

$$A = \frac{6M}{\rho d} \tag{3.2}$$

where $A$ is the interfacial area, $M$ is the mass of the oil phase, $\rho$ is the density of oil, and $d$ is the droplet diameter. With a fixed volume fraction of the oil phase, $A$ is inversely

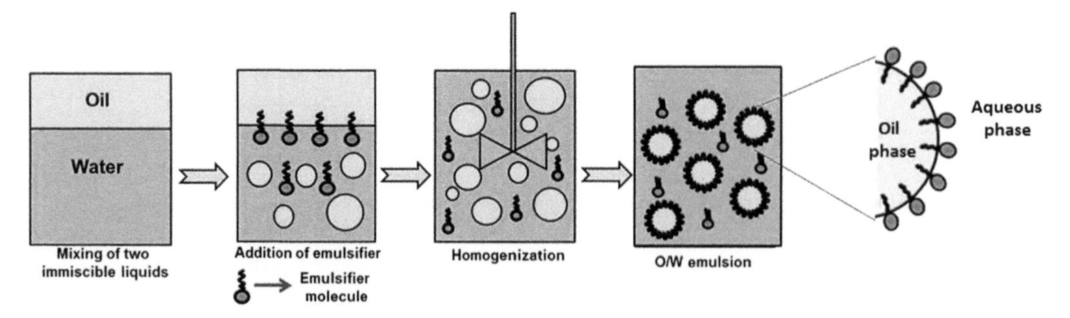


**FIGURE 3–2** Emulsion homogenization process.

proportional to *d*. The emulsions are typically homogenized using mechanical devices known as homogenizers, and numerous research studies have been carried out for the development of an efficient emulsification technique. The most common types of homogenizers used for the production of emulsions on an industrial scale is high-pressure homogenizers (HPHs), microfluidizers, and ultrasonic processors.

## 3.3 Application of ultrasonication for emulsification

In ultrasonication, cavities are generated by passing sound waves through the liquid medium. The range of the sonic spectrum is >16 kHz, which may be subdivided into three main sections as low-frequency and high-power ultrasound (20–100 kHz), high-frequency and medium-power ultrasound (100 kHz to 1 MHz), and high-frequency, and low-power ultrasound (1–10 MHz). Ultrasonic waves consist of rarefaction (expansion) and compression cycles; when these waves are transmitted through a liquid, bubbles/cavities are formed, followed by their rapid growth under the condition of isothermal expansion and finally the adiabatic collapse of these generated cavities, as shown in Fig. 3–3 [6,15,16]. The average distance between the liquid molecules is larger in the rarefaction cycle but smaller in the compression cycle. Cavitation occurs in rarefaction cycles where negative acoustic pressure is sufficiently large to pull apart the liquid molecules from each other, and the distance between the adjacent molecules can exceed the critical molecular distance. At that moment the voidage is created in the liquid, which causes the formation of cavities. Subsequently, in the compression cycle of the sound wave, acoustic pressure is positive, which pushes the molecules together.

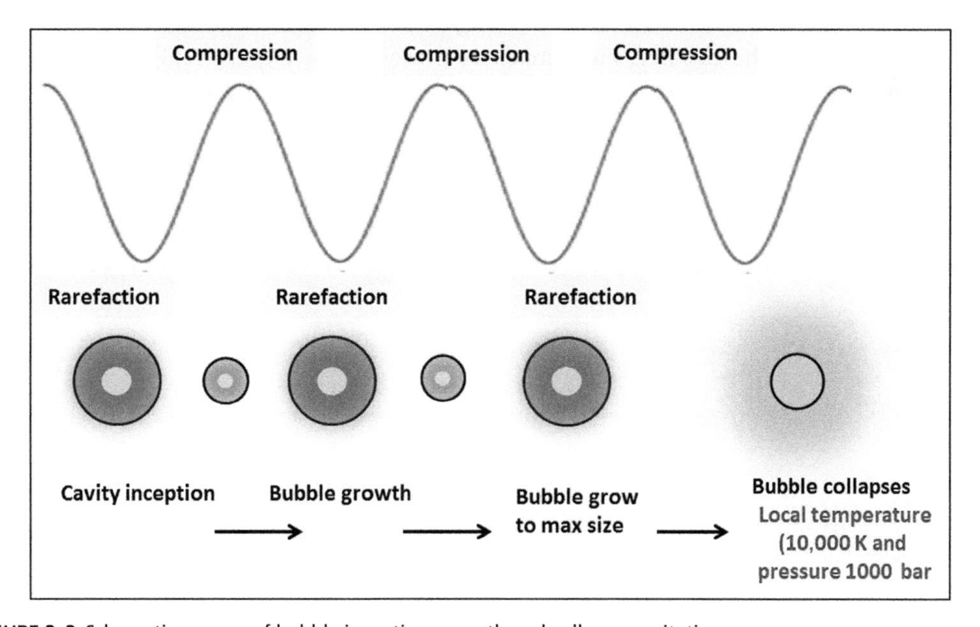


**FIGURE 3–3** Schematic process of bubble inception, growth and collapse: cavitation process.

The cavities will compress (decrease in size) during the compression cycle of the ultrasonic wave, and few of them may collapse in a minimal time interval. This final collapse phase is adiabatic, thus producing high local temperature and pressure conditions.

Ultrasound cavitation is usually produced in two types of reactors, that is, ultrasonic bath and ultrasonic horn. In a bath the ultrasonic transducers are attached to the bath surface that transmits the waves through the liquid medium, whereas an ultrasonic horn consists of a sonotrode or probe which can be immersed in the liquid to be treated. In an ultrasonic reactor, the supplied power converts the line voltage to high-frequency electrical energy. This energy is transmitted to the transducer within the converter, where it is changed into mechanical vibrations. These vibrations generated from the converter are intensified by the ultrasonic probe, that subsequently creates the pressure waves in the liquid. This action creates the numerous vaporous cavities, which expand during the negative pressure excursion, and subsequently collapse violently during the positive excursion. Although this phenomenon, known as cavitation, occurs for a few microseconds, it appears at millions of sites, and the cumulative amount of energy, generated as a result of these cavity collapses is exceptionally high. The larger the probe tip, the larger will be the volume that can be processed but at a lesser intensity. In both the reactors, cavitation is found to occur near the tip of the horn/transducers, and cavitation intensity decreases exponentially when moving away from the horn tip/transducer surface [16].

### 3.3.1 Emulsification mechanism

In comparison to the high-energy homogenizing techniques such as HPHs and microfluidizers, ultrasonication offers an enhanced level of mixing to produce the emulsions. The implosive collapse of cavities dissipates the energy at the micro level for initiating the droplet disruption process. The detailed mechanism is presented in Fig. 3–4 [6]. The first stage shows the generation of primary droplets, where the acoustic field produces the interfacial waves and the instability of which causes the eruption of the oil phase into the water in the form of droplets. The second stage involves the break up of primary droplets, wherein violent and asymmetric implosion of bubbles cause micro-jets, which can effectively break the primary oil droplets. The break up of droplets is primarily controlled by the type and amount of shear applied to droplets. Further, the stability and coalescence of droplets depend on the

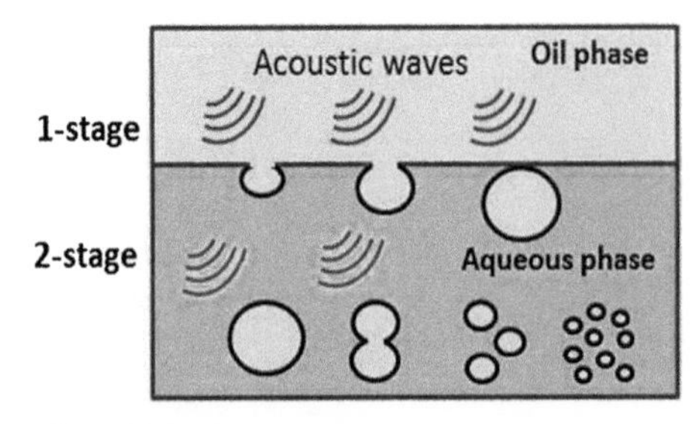


**FIGURE 3–4** Ultrasonication assisted emulsification process.

rate at which the surfactant is adsorbed at the interface of the newly formed droplets and is controlled by the surfactant surface activity and its concentration. Thus, the selection of an appropriate surfactant is much important in producing highly stable nanoemulsions.

### 3.3.2 Comparison of ultrasonication with high energy homogenizers

For many decades, ultrasonication has been utilized for the emulsification because of its potential in producing smaller droplet sizes with superior physical stability. Apart from the ultrasonication process, other high energy devices are used for emulsification such as HPH and microfluidizer, which can generate severe shearing conditions but of different intensities with different modes and geometries.

In HPHs the solution mixture is highly pressurized and forced through a valve that causes a sudden contraction in the flow, as shown in Fig. 3–5. With sufficient throttling the kinetic energy increases that subsequently increases the flow velocity through the valve. The vapor

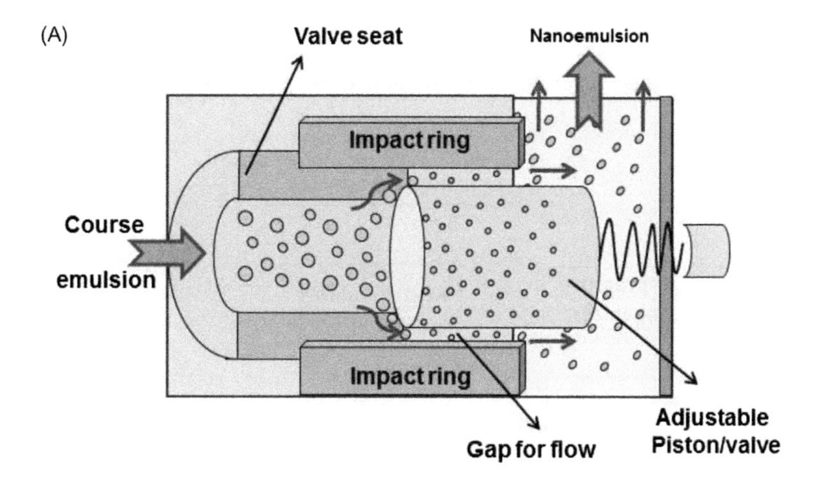


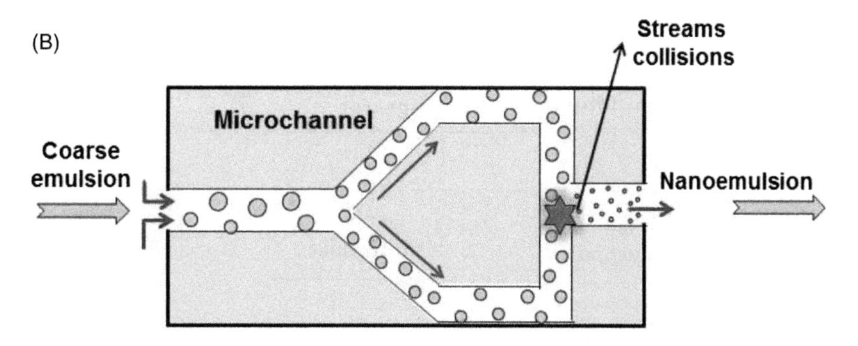


**FIGURE 3–5** Emulsification using (A) high-pressure homogenizer and (B) microfluidizer.

bubbles thus formed expand and further collapse, creating shock waves that can break the interface and thus disperse the droplets into fine sizes. The drawback of this method is that it requires high energy inputs, that is, it usually operates at high pressure ranging from 50 bar to as high as 500 bar or more, which may degrade the emulsion under the extremely high pressure and temperature [17].

In the microfluidizer, two different streams of emulsion mixture are passed at high pressure in the range of 50–1000 bar through a patented fixed geometry interaction chamber (Fig. 3–5) [17–19]. Two jets of liquid emulsions are accelerated at high velocities (400 m/s) from two opposite channels and collide with one another, thus creating tremendous shearing action. In general, inertial forces in the turbulent flow induced by high pressure along with cavitation are the predominant factors that cause the disruptions of droplets into fine sizes [18].

The comparison between ultrasonication and other high energy methods (HPHs and microfluidizers) is shown in Fig. 3–6. Both the high energy techniques, HPH and microfluidizers are capable of producing enormous turbulence and disruptive forces for the desired emulsion droplet size and stability. However, the energy required to generate the disruptive forces is high (operating pressure required to be in the range of 50–500 bar) as compared to ultrasonication [20]. Tang et al. [18] reported a comparison between ultrasonication and microfluidization for

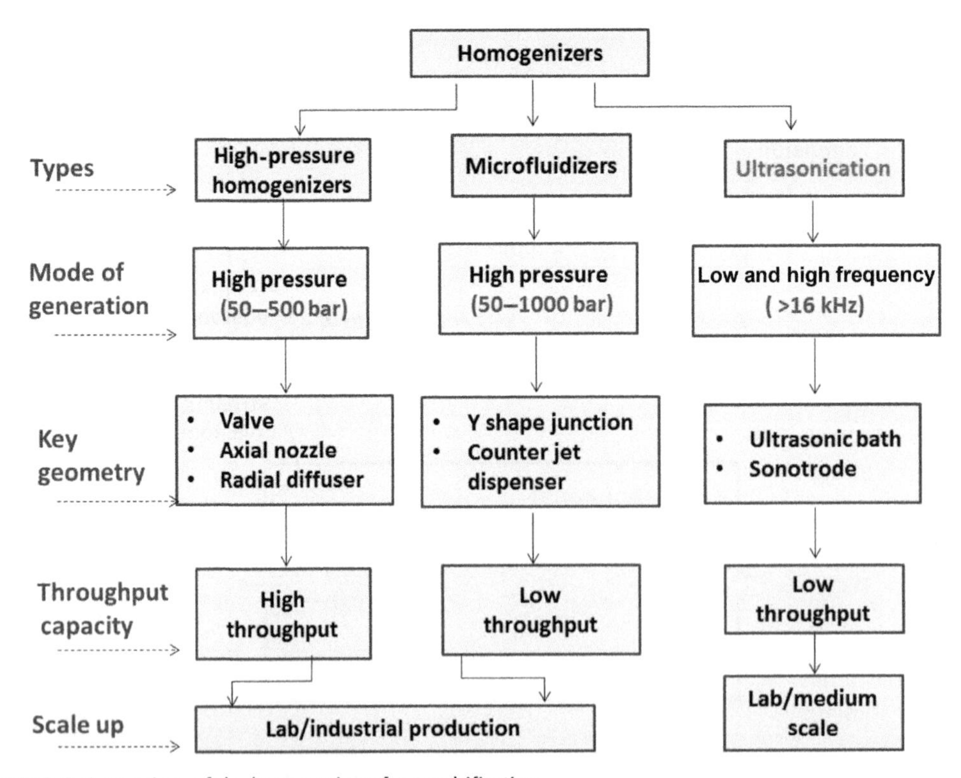


**FIGURE 3–6** Comparison of the homogenizers for emulsification.

the formation of nanoemulsion carrying aspirin. The efficient performance of ultrasonication in terms of emulsion size reduction with minimum energy input was observed. The minimum droplet size of 180 nm was obtained at the delivered energy density of 11 $J/cm^3$ in ultrasonication, which was 12 times lower than microfluidization.

Apart from energy intensiveness, these high energy methods have various other disadvantages, such as it requiring high operating pressure and premixing, complexity in their geometries, line blockages in the case of microchannels, is difficult to clean, and has high maintenance requirements. [6,18,19]. Moreover, the ultrasonication process offers several advantages over other high energy methods. The operational process is easier, has a smaller foot print, simple geometry, and is easy to clean [6]. Some comparative studies revealed that ultrasonication was more effective and energy-efficient as compared to high energy methods for achieving the desired droplet size [6,20,21]. All these facts make ultrasonication the most effective and promising method for the preparation of emulsions.

## 3.4 Multilayer emulsion and its application for the encapsulation of bioactive compounds

In the last few decades, many studies have been reported on the utilization of emulsions for the encapsulation and delivery of bioactive or lipophilic compounds. The microencapsulation in the O/W emulsion matrix has been employed to improve the solubility, stability, and rate of absorption of lipophilic compounds during the physiological transit in the human body. O/W emulsions, mainly the nano and submicron emulsions (<500 nm), possess better kinetic stability against droplet aggregation, coalescence, and phase separations and hence are the suitable systems for the encapsulation and delivery of hydrophobic bioactive compounds [2,22,23].

Previous studies showed that the emulsion-based delivery systems enhance encapsulation and chemical stability before ingestion and subsequently improve its oral bioaccessibility during physiological transit in the human body [24–26]. In the last decade, various studies reported on the fabrication of O/W emulsion prepared using a single emulsifier [27,28] and emulsions consisting of mixed emulsifiers [25,26] for the encapsulation of lipophilic compounds. Although these emulsions formed were highly stable against phase separation, but the stability and physicochemical properties of the encapsulated material was found to be significantly affected by the variations in pH, temperature, and normal storage conditions. Therefore it is essential to design a suitable emulsion matrix system to enhance the stability and shelf-life of the encapsulated compounds. Apart from maintaining the physicochemical properties, it is also necessary to regulate the release of lipophilic compounds within the GI tract of the human body to obtain its nutritional benefits.

Thus, looking at these aspects, multilayer emulsion as a delivery system could be a suitable strategy to overcome the above-stated challenges. Multilayer emulsions typically consist of oil droplets (the core) surrounded by thick layers (the shell) comprising different surfactants and biopolymers (polyelectrolytes). These are formed using a layer-by-layer (LbL) approach

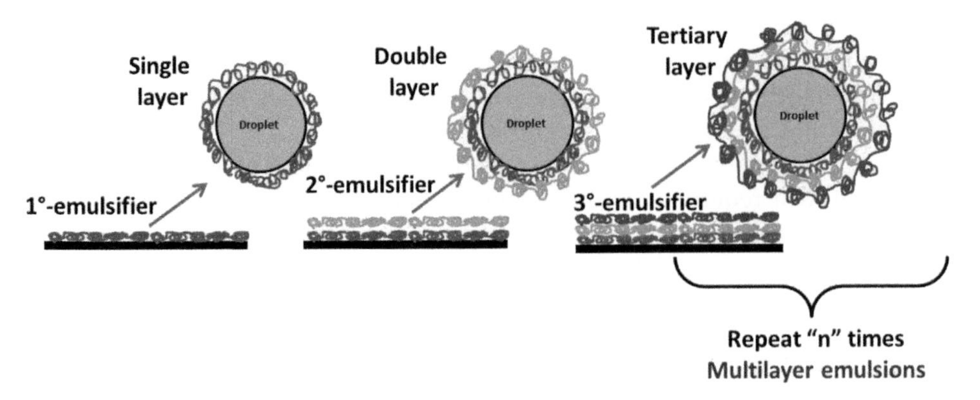


**FIGURE 3–7** Schematic of layer-by-layer deposition of emulsifiers on the droplet surfaces.

that includes the deposition of alternating layers of oppositely charged biopolymers such as protein and polysaccharides over the oil droplets. Depending on the number of interfaces surrounding droplets, the primary, secondary, tertiary, etc., emulsions can be produced [29]. The formation of multilayered membranes on the oil droplets is presented in Fig. 3–7. The electrostatically charged layer of biopolymers forms a thick interfacial membrane at the oil–water interface that helps to inhibit the droplet coalescence and aggregation and thus enhances the physical stability of the emulsion. The multiple coating over the emulsion droplets also helps to hinder the diffusion of various peroxides in the core (oil phase) of the emulsion and thus prevents lipid oxidation. It has been proved that the multilayer emulsions are found to be more stable for a longer storage time against the variations in the physicochemical conditions such as solution pH, temperature, ionic strength, and lipid autooxidation [9,30,31]. Moreover, multilayer emulsions stabilized with food-grade emulsifiers (mainly biopolymers) have advantages such as proper ingestion and consumption of constituents in the GI tract, unlike the emulsions stabilized with synthetic surfactants and organic solvents, which could adversely affect the biological conditions inside the human body.

One of the major advantages of using multilayer emulsions as delivery systems is that the properties of the interfacial layer surrounding oil droplets can be carefully controlled, for example, composition, structure, charge, thickness, permeability, rheology, and environmental responsiveness [32]. This can be achieved by careful control of the system composition and preparation conditions during the formation of multilayer emulsions, for example, emulsifier type and concentration, pH, ionic strength, and the extent of homogenization [9].

### 3.4.1 Potential approaches for the preparation of multilayered emulsions

There are two possible approaches for producing multilayer emulsion: LbL and a mixed approach.

#### *3.4.1.1 Layer-by-layer approach*

In the LbL approach, first, a primary emulsifier (usually a protein) is used to form a primary emulsion (PE) in which the ionic biopolymer rapidly adsorbs at the oil–water interface during homogenization. After that, an oppositely charged secondary biopolymer is added into the PE that adsorbed over the droplet surfaces and produced a double layer emulsion, also called as secondary emulsion (SE). To form some more layers over the droplet surface the above procedure can be repeated with the oppositely charged layer of biopolymers, which stabilizes emulsion based on their electrostatic interaction, as shown in Fig. 3–7. Moreover, the stability of multilayer emulsion depends on the interaction between the biopolymer layers, which, in turn, depends on the pH and ionic strength of the polyelectrolytes or biopolymers [9,29]. The pH of the solution determines the ionization of surface groups, and therefore, the final surface charge density [9]. For effective LBL deposition the pH of the solution should be selected so that the electrical charges on the droplet surface and adsorbing biopolymer are opposite, and the magnitude of the net charge is sufficiently high. For example, anionic biopolymers will not adsorb to the surface of the cationic amino acids group of protein-coated droplets at pH above the isoelectric point of protein (pI ~ 4.5–5) because of the large electrostatic repulsion between them. However, it may adsorb at pH below the isoelectric point of positively charged protein molecules, as the anionic polysaccharide and cationic droplets have opposite charges. It should also be noted that a biopolymer can adsorb to a droplet surface with a similar net charge under certain circumstances. For example, few studies have reported the adsorption of sodium alginate (SA) onto the surface of protein-coated droplets at pH 5 (near to its pI), which was attributed to the presence of some positive patches on the surface of protein-coated droplets [30,31,33].

#### *3.4.1.2 Mixed approach*

The mixed-layer approach consists of the adsorption of protein–polysaccharide complexes formed by two oppositely charged biopolymers during homogenization. This technique first involves the preparation of a bulk aqueous solution of protein–polysaccharide complexes by controlling the pH of the aqueous phase. The resulting mixture of the complex is then used as the emulsifying agent during homogenization. Similarly, as in the LbL technique, here also the pH is the crucial parameter for effective adsorption on the surface of droplets [11].

In both the approaches the thickness of the complexes increases as compared to the conventional nanoemulsions. As a result, the emulsions are more stable against phase separation, droplets coalescence, flocculation, etc., provided that the emulsions are formed with a suitable and sufficient amount of biopolymers along with desired level of homogenization.

Apart from their preparation technique, multilayer emulsions stabilized with biopolymers are prone to destabilization utilizing flocculation, which is mainly of two types: bridging and depletion flocculation. Bridging flocculation occurs when a polyelectrolyte adsorbs to the surface of more than one droplet and links them together as shown in Fig. 3–8 [5,9]. This phenomena generally occurs when the biopolymer concentration is insufficient to thoroughly saturate the droplet surfaces and also when the net charge of the two layers is

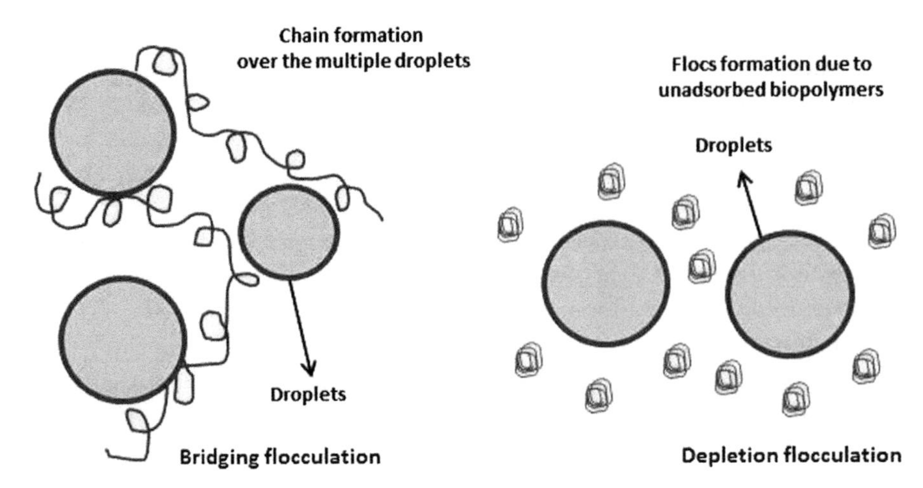


**FIGURE 3–8** Bridging and depletion flocculation.

exceptionally high particularly in the acidic conditions (pH 2–4), which have a tendency to adsorb over multiple droplet surfaces and resulted into the formation of coacervate (gel like structure) of flocs forming network. In the case of depletion flocculation, free biopolymer (unadsorbed) concentration in the continuous phase generates attractive osmotic forces between the droplets. This would promote an increase in the attractive forces between the droplets due to an osmotic process associated with the exclusion of the colloids from a narrow region surrounding the droplets as shown in Fig. 3–8.

Therefore it has been reported that a sufficient amount of biopolymer should be added to prevent bridging flocculation, but not too much to cause depletion flocculation [9]. Hence, it is always necessary to optimize the pH and biopolymer concentration to obtain a highly stable multilayer emulsion.

## 3.4.2 Applications of multilayer emulsions

Multilayer emulsions are the most suitable system for the encapsulation of lipophilic active ingredients, since they can be trapped within the hydrophobic core and protected by the polyelectrolyte shell. Many studies have reported on the preparation of multilayer emulsions for the encapsulation of lipophilic ingredients such as essential oils (e.g., ω-3 oils). Studies have shown that multilayer emulsions can be used to encapsulate oil-soluble flavors, ω-3, and ω-6 fatty acids [30,31,33,34].

Numerous studies have been carried out for the encapsulation of essential fatty oils with different biopolymers using this multilayer technique. For example, Corn oil encapsulated in whey protein isolate (WPI)/lactoferrin layers [35], linseed oil encapsulated in WPI/alginate layers [30], olive oil encapsulated in WPI/alginate layers [33], sunflower oil encapsulated in WPI/tragacanthin layers [36], and menhaden oil encapsulated in WPI/xanthan gum layers

[37]. Most of these studies reported on the preparation of PE using primary emulsifier followed by homogenization and then mixed with the oppositely charged biopolymeric solution. The preparation conditions have been carefully controlled to avoid droplets flocculation which can be done by varying the pH, emulsifier concentration, and level of homogenization. These studies have reported that the chemical stability of the encapsulated lipids was improved on being encapsulated within the multilayer coatings, which was attributed to various reasons: (1) charge repulsion—positively charged coatings may repel cationic peroxides from the droplet surfaces; (2) steric effects—the polyelectrolyte coatings may be impermeable for the diffusion of peroxides because of the extreme packing of the molecules; and (3) antioxidant activity—some polyelectrolytes within the coatings have antioxidant activity (such as proteins). Therefore the improvement in the physical and oxidative stability of emulsions can be attributed to the thick charged layer coatings that increased the repulsive interactions between the droplets and stabilized the emulsions.

In the other case, lipophilic active ingredients such as phenols, antioxidants, vitamins, and drugs can be entrapped within the oil phase surrounded by the nanolaminated coatings of biopolymers. For example, a lipophilic active compound could be dissolved in the oil phase prior before homogenization, and then the charged hydrophilic active biopolymers are incorporated into it to form polyelectrolyte layers surrounding the oil droplets. The encapsulated active molecules would then be retained within the delivery system until they are released at a specifically targeted place in the GI tract of the body. This multilayer technology for the encapsulation of lipophilic compound also helps to maintain their physicochemical properties until it gets released into the human body. Acevedo-Fani et al. [38] reported on the use of multilayer emulsion for the encapsulation of resveratrol, which is an antioxidant phenolic compound mainly found in grapes' skin and berry fruits. The different corn oil emulsions were formed with the biopolymers lactoferrin, alginate, and ε-poly-L-lysine using a LbL technique for the encapsulation of resveratrol. It has been observed that all the emulsions (primary, secondary, and tertiary) were highly stable as the droplet size was not significantly changed during the storage of 4 weeks. This work reported on a comparison in the performance of all the emulsions based on the preservation of antioxidant (resveratrol) and its antioxidant activity during storage. It has been observed that the antioxidant activity of resveratrol in oil was reduced during the storage period, which was attributed to its chemical destabilization under the environmental conditions. However, the encapsulation of resveratrol using multilayered emulsification helps to conserve its antioxidant activity during the storage time of 4 weeks; thus this emulsification process was found to be a superior delivery system for resveratrol.

Moreover multilayer emulsion can also be dried using drying techniques such as freeze and spray drying. This would facilitate the utilization of solid emulsified powders in many food products such as beverages, cereals, and baked foods, etc. to produce the multifunctional, multivitamin nutritional, and healthy food product. An additional advantage of preparing dried multilayer emulsions is that the amount of wall material (hydrophilic biopolymer) required to form stable emulsified powders is considerably lower than the required to stabilize liquid emulsions [39]. Fioramonti et al. [40] carried out the characterization of a freeze-dried microcapsule containing maltodextrin (MD) formed using a multilayer emulsification technique. Multilayer emulsion was prepared by using flaxseed oil stabilized with two oppositely charged biopolymers, that is, WPI and SA. The MD was added at a different concentration to check the leakage of oil during the lyophilization process. It has been observed that the encapsulation efficiency (%EE) of the flaxseed oil (fatty acids) was significantly affected after freeze-drying as the efficiency was only 30%, indicating that a large amount of oil was released during drying. However, on incorporating MD at a concentration of 20 wt.%, the %EE was significantly enhanced to 90%. The advantage of adding MD was that it prevented the microcapsule structure against fissure or breakages during freeze-drying, and thus the oil was perfectly entrapped in the protective double-layer coatings of WPI–SA, and thus did not allow the core material to come onto the surface of the particles.

In summary the effective use of this multilayer technology is to improve the functional properties such as antioxidant, stability to environmental stresses, controlled digestibility, and targeted release profiles of the bioactive compounds. More efforts are still needed for the development of this encapsulating system to improve the bioavailability of bioactive molecules so that they could be easily accessible by the lipase enzymes for their effective digestion.

## 3.5 Case studies

### 3.5.1 Ultrasonic-assisted synthesis of multilayer emulsions

#### *3.5.1.1 Introduction*

In the recent decade the use of the LbL approach for the preparation of emulsions has attracted more interest owing to several advantages, as discussed in the previous section. However, to produce highly stable multilayer emulsion, it becomes necessary to select an appropriate process condition such as the concentration of biopolymers, level of homogenization, charge density by varying pH, and dispersed phase (oil) characteristics to overcome the problem of droplet flocculation such as bridging and depletion. On the other hand, to minimize the use of emulsifiers and achieve the high stability of emulsion against flocculation, emulsions need to be homogenized using a suitable homogenizer. Ultrasonication was proved to be a promising, efficient, and environmentally acceptable technique for the preparation of different emulsions having applications in food, pharmaceutical, and cosmetic products [6,18,21].

In a recent study performed by our group [33], attempts have been made to synthesize stable food-grade multilayer emulsion by utilizing the ultrasonication process. This study includes a detailed investigation on the evaluation of the physicochemical properties of multilayer emulsion such as droplet size, zeta potential, morphology, and physical and oxidative stability of emulsions under variation in the process conditions.

Apart from that the role of ultrasonication on the formation of bilayer emulsion was also explored to understand the effect of sonication on the extent of flocculation (bridging/depletion) during its formation. Moreover, it is also necessary to examine what extent of sonication should be given so that not only the physical stability against flocculation can be enhanced but also an improvement to the oxidative stability of emulsions for a longer storage time could be achieved. Hence, it is necessary to optimize the sonication time while formulating the multilayer emulsions. Some results of the study are discussed below.

#### *3.5.1.2 Experimental*

Before the formation of SE, first the PE was prepared by mixing olive oil (10 wt.%) with an aqueous solution (90 wt.%) of WPI at different pH (4–7) followed by sonication for 15 min (Sonics, 750 W, 20 kHz, 40% amplitude). The SE's were then prepared by mixing PE with an aqueous solution of SA (secondary biopolymer) at a ratio of 1:1 (v/v) and pH was kept the same as that of PE. The prepared SE was further subjected to sonication for 60 s (unless otherwise stated), and the pH was adjusted to the desired value. The final composition of all the constituents in SE formed was 4.93 wt.% oil, 0.443 wt.% WPI, 0.0381 wt.% sodium azide, and 0.1–0.3 wt.% SA. Sodium azide was added to avoid the formation of microbial species in the mixtures. To check the double layer (WPI–SA) interaction and thus the stability of emulsions against flocculation, the effect of pH and SA concentration were thoroughly studied.

All the prepared emulsions were stored at room temperature for further analysis. The emulsions were analyzed using various characterization techniques.

*Particle size and zeta potential measurements*: Analysis was performed using Zetasizer Nano-ZS (Malvern Instruments, United Kingdom) provided with dynamic light scattering and electrophoresis facility.

*Separation and creaming index (CI) measurements*: The separation index (SI) was used to measure the extent of nonemulsified oil or oil layer obtained at the top of PE. The %SI was measured after 7 days of storage. The %SI was determined as follows

$$\%\,\mathrm{SI} = \frac{H_C}{H_T} \times 100 \tag{3.3}$$

where, $H_C$ represents the height of the oil layer obtained at the top and $H_T$ represents the total height of the PE.

The CI was determined by measuring the extent of separation of cream and serum phases after a storage time of 7 days at room temperature. The cloudy and/or transparent phases seen at the bottom of the storage bottles were considered as serum phase. The %CI was measured using the following equation:

$$\%\,\mathrm{CI} = \frac{H_S}{H_T} \times 100 \tag{3.4}$$

where $H_S$ and $H_T$ represent the height of the serum phase and the total height of the emulsion, respectively.

*Peroxide value (PV) measurement*: The analysis was performed by using the ferric thiocyanate method for measuring the concentration of hydroperoxide in emulsions. First, 0.3 mL of the emulsion was added to a mixture of isooctane/2-propanol (1.5 mL, 3:1 v/v) and then vortexed for 30 s, followed by centrifugation for 5 min at 3500 RPM. After that, 0.2 mL of the top solvent layer containing extracted lipids was taken and mixed with 2.8 mL of methanol/1-butanol (2:1 v/v) mixture and 30 μL of 3.94 M ammonium thiocyanate/$Fe^{2+}$ (1:1 v/v) solution. The $Fe^{2+}$ solution was prepared by mixing equal amounts of 0.132 M barium chloride and 0.144 M ferrous sulfate. The barium sulfate gets precipitated, and a clear solution of $Fe^{2+}$ was obtained. The complete mixture of the extracted lipids was kept for 20 min in the dark. The peroxides present in the solution oxidized $Fe^{2+}$ into $Fe^{3+}$, and a faint yellow color appeared in the end. The absorbance was measured at 510 nm using a UV–visible spectrophotometer. The concentration of hydroperoxide in the emulsion mixture was determined using a standard curve of cumene hydroperoxide (CHP) and measured as μM of CHP per gram of oil. All solutions were prepared freshly on a day-to-day basis for analysis.

### *3.5.1.3 Results and discussion*

#### 3.5.1.3.1 Formation of primary emulsion: optimization of sonication time

Initially, the PE was prepared to optimize the required sonication time to obtain a minimum droplet size and SI. The results obtained are shown in Fig. 3–9. It could be observed that with an increase in the sonication time from 3 to 10 min, the droplet size significantly reduced from 473 to 312 nm and further slightly reduced to 308 nm after 15 min of sonication. The reduction in the droplet sizes can be attributed to the physical effects of sonication within the emulsion mixture such as the liquid microjets, and high-intensity turbulence as a result of cavity collapse, which disrupts the droplets. These effects uniformly distribute the protein molecules within the dispersed phase and facilitate their adsorption at the oil–water interfaces. Moreover the %SI was decreased from 20% to almost 0% by increasing the sonication time from 3 to 15 min. At 10 min of sonication time the %SI was 10%, and an oil layer was observed at the top. Though there was no significant change in the droplet size beyond

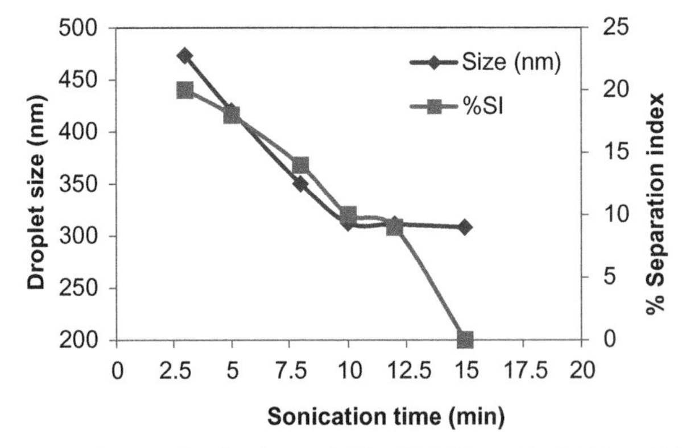


**FIGURE 3–9** Effect of sonication time on droplet size and %SI of PE (10 wt.% oil, 0.9 wt.% WPI, pH 7). *PE*, Primary emulsion; *WPI*, whey protein isolate. *Reprinted (adapted) from J. Carpenter, S. George, V.K. Saharan, A comparative study of batch and recirculating flow ultrasonication system for preparation of multilayer olive oil in water emulsion stabilized with whey protein isolate and sodium alginate, Chem. Eng. Process. – Process Intensification 125 (2018) 139–149, with permission from Elsevier.*

10 min of sonication, higher stability, and perfect dispersion were achieved in 15 min of sonication time, which was considered for further experiments.

#### 3.5.1.3.2 Interaction of alginate–whey protein isolate layers and stability of emulsions

To evaluate the stability of PE and SE, the interaction between SA and WPI coated droplets was monitored using ζ-potential measurements. To establish the interaction between SA molecules and WPI coated droplets, the SA concentration was varied in the range of 0.1–0.3 wt.% at different pH values (4–7). The results obtained are shown in Fig. 3–10. In the PE, ζ-potential of the droplets was varied in the range of +14 to −57 mV by increasing the pH from 4 to 7. Due to the amphiphilic behavior of WPI molecules, the surface charge density changed from positive to negative by varying the pH below its isoelectric point (pI ~ 4.5–4.7) to 7. However, by incorporating SA into PE, the surface charge density over the WPI-coated droplets varied from −39 to −59 mV (at 0.2 wt.% of SA) in the pH range of 4–7. At pH 6 and 7, for all the concentrations of SA, there was no significant difference in

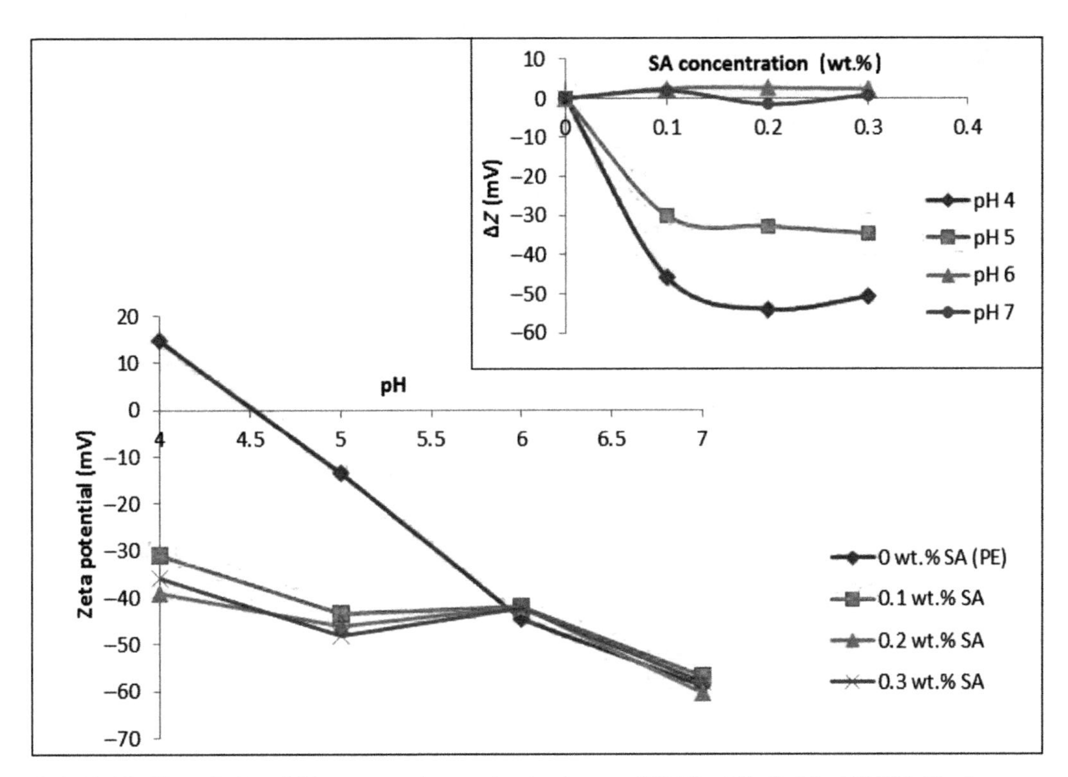


**FIGURE 3–10** Effect of pH and SA concentration on droplet charge of PE (10 wt.% oil, 0.9 wt.% WPI, 15 min sonication) and SE (0.443 wt.% WPI, 4.93 wt.% oil, 60 s sonication). *SA*, Sodium alginate; *SE*, secondary emulsion; *WPI*, whey protein isolate. *Reprinted (adapted) from J. Carpenter, S. George, V.K. Saharan, A comparative study of batch and recirculating flow ultrasonication system for preparation of multilayer olive oil in water emulsion stabilized with whey protein isolate and sodium alginate, Chem. Eng. Process. – Process Intensification 125 (2018) 139–149, with permission from Elsevier.*

the ζ-potential of PE and SE, indicating that limited interaction occurred between WPI-coated droplets and SA molecules because of similar surface charge (net surface charge, $\Delta\zeta \sim 0$), as shown in Fig. 3–6. The droplet size of PE and SE obtained at pH 6 and 7 was similar, and the %CI (phase separation) was also higher, as shown in Fig. 3–11A and B. This indicated that the adsorption of SA molecules over the WPI-coated surfaces was ineffective because of weak interaction, and thus, the SE at pH 6 and 7 destabilizes through depletion flocculation.

At pH 5 the net surface charge of the secondary and primary layers was much higher ($\Delta\zeta = -30$ mV, at 0.1 wt.%, Fig. 3–10), which indicated the adsorption of SA onto the surface of WPI-coated droplets. Under this condition, the positively charged surface ($-NH_3^+$) of WPI-coated droplet could interact with negatively charged groups ($-COO^-$) present in the polysaccharide chains of SA. Moreover the net surface charge ($\Delta\zeta$) was slightly increased up to an optimum SA at a concentration of 0.2 wt.% and then remained constant. Also, on increasing the concentration of SA from 0.1 to 0.2 wt.%, the droplet size of SE was increased from 724 to 836 nm and then remained almost similar at 0.3 wt.% of SA, as shown in Fig. 3–11A. This indicated that the SA molecules at 0.2 wt.% were enough to thoroughly saturate the WPI-coated droplets and thus stabilizes the SE. Also, there was no phase separation, and no droplet flocculation or aggregation was observed in SE at pH 5, which can be seen in Figs. 3–11B and 3–12.

On further reducing the pH to 4 the net charge ($\Delta\zeta$) over the droplets was significantly increased by decreasing the pH from 5 to 4, indicating the presence of a higher density of positively charged protein molecules, which would promote stronger interactions with SA molecules. At pH 4 the extent of interaction was so high that SA starts bridging with more than one droplet at a time. As a result of this, the droplets come closer due to the formation of flocs by the bridging mechanism, and subsequently, the bilayer emulsion destabilizes using bridging flocculation,

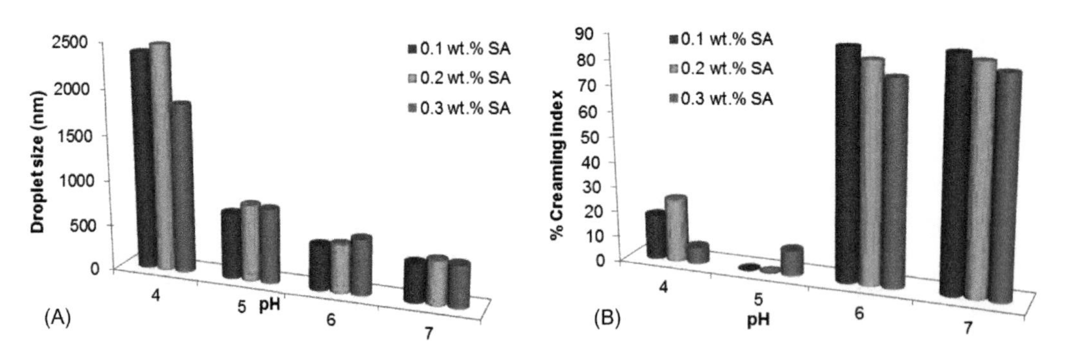


**FIGURE 3–11** Effect of pH and SA concentration on (A) droplet size and (B) creaming index of SE (0.443 wt.% WPI, 4.93 wt.% oil, 60 s sonication). *SA*, Sodium alginate; *SE*, secondary emulsion; *WPI*, whey protein isolate. *Reprinted (adapted) from J. Carpenter, S. George, V.K. Saharan, A comparative study of batch and recirculating flow ultrasonication system for preparation of multilayer olive oil in water emulsion stabilized with whey protein isolate and sodium alginate, Chem. Eng. Process. – Process Intensification 125 (2018) 139–149, with permission from Elsevier.*

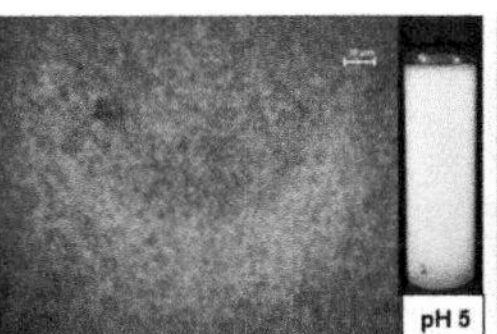

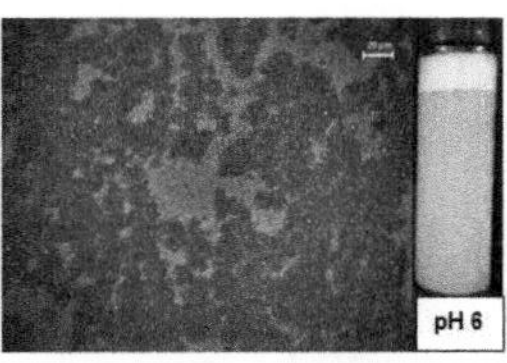

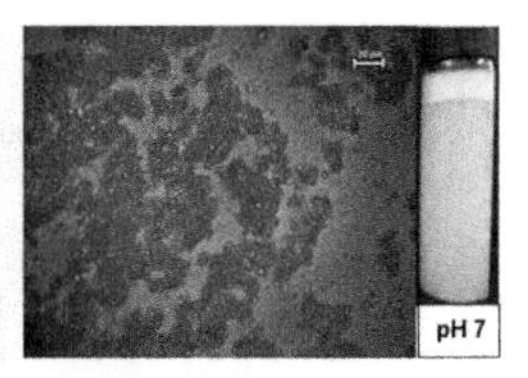


**FIGURE 3–12** Morphology of SE at different pH (0.443 wt.% WPI, 0.2 wt.% SA, 4.93 wt.% oil, 60 s sonication). *SA*, Sodium alginate; *SE*, secondary emulsion; *WPI*, whey protein isolate. *Reprinted (adapted) from J. Carpenter, S. George, V.K. Saharan, A comparative study of batch and recirculating flow ultrasonication system for preparation of multilayer olive oil in water emulsion stabilized with whey protein isolate and sodium alginate, Chem. Eng. Process. – Process Intensification 125 (2018) 139–149, with permission from Elsevier.*

as shown in Fig. 3–12. The droplet size of SE was also found to be higher as compared to that obtained at pH 5 for all the concentrations of SA (Fig. 3–11A), which indicated the occurrence of extensive droplet aggregation at pH 4 as a consequence of bridging flocculation.

Therefore, it has been found that the SE formed at pH 5 and 0.2 wt.% SA was more stable against the depletion and bridging flocculation and showed excellent physical stability against creaming.

#### 3.5.1.3.3 Influence of ultrasonication on the physical and oxidative stabilities of multilayer emulsion (ME)

The effect of sonication time on the physical and oxidative stabilities of SE prepared at optimum conditions (pH 5 and 0.2 wt.% SA) was also studied, and the results are shown in Fig. 3–13. After the addition of SA in PE the droplet size obtained was significantly higher, that is, 6690 nm. Also, a clear serum phase (%CI = 60%) was observed in SE, as shown in Fig. 3–9A, which indicated that the aggregation of droplets in the creaming phase and unadsorbed biopolymers (WPI/SA) remained in the serum phase. To effectively distribute/adsorb SA molecules on the PE droplets, energy is required to overcome the mass transfer resistance caused by the highly viscous aqueous solution of SA.

The stability of SE was improved on sonicating the emulsion. It was observed that the droplet size decreased by sevenfold, that is, from 6690 to 968 nm on sonicating the emulsion for 30 s. The sonication effects disrupt the flocs and disperse SA molecules more uniformly into the system. On further increasing the sonication time, that is, beyond 30 s, the droplet size was very slightly reduced from 968 to 836 nm at 60 s and then remained constant. This indicated that sufficient energy was dissipated into the emulsion within 60 s to obtain the minimum possible droplet size. Moreau et al. [41] also studied the effect of sonication on the formation of SE of corn oil stabilized with β-Lg and pectin complexes. It was reported that the mean particle size of SE was reduced by four times on sonicating the emulsion mixture for 30 s (40% amplitude and 20 kHz frequency). The mean particle size of SE was reduced from 12.9 to 2.6 μm after 30 s of sonication and then remained constant on further sonicating the mixture.

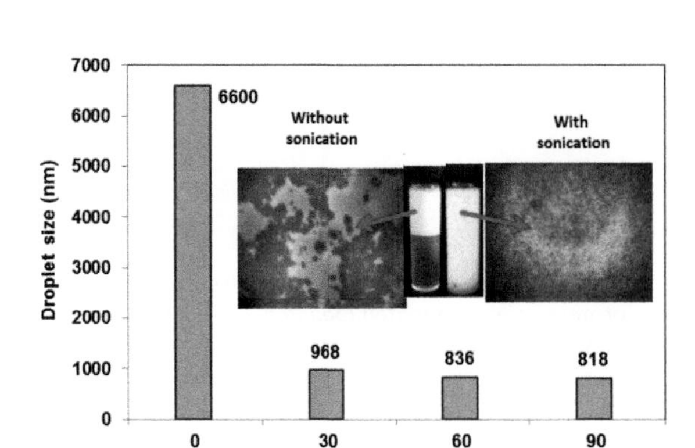


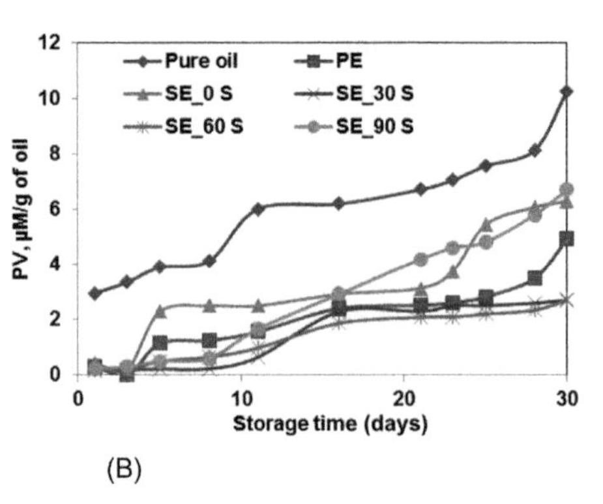


**FIGURE 3–13** Effect of sonication time on (A) physical stability and (B) oxidative stability (PV) of SE (PE: 10 wt.% oil, 0.9 wt.% WPI, pH 7 and SE: 4.93 wt.% oil, 0.443 wt.% WPI, 0.2 wt.% SA, pH 5). *PE*, Primary emulsion; *PV*, peroxide value; *SE*, secondary emulsion; *WPI*, whey protein isolate. *Reprinted (adapted) from J. Carpenter, S. George, V.K. Saharan, A comparative study of batch and recirculating flow ultrasonication system for preparation of multilayer olive oil in water emulsion stabilized with whey protein isolate and sodium alginate, Chem. Eng. Process. – Process Intensification 125 (2018) 139–149, with permission from Elsevier.*

Apart from the physical stability, the oxidative stability of the multilayer emulsion is also a primary concern while formulating them. The lipid oxidation occurs due to the residual or generated peroxides or radicals in the emulsion. Therefore, to observe the lipid oxidation, the PV that is a direct measurement of the concentration of primary oxidation products (lipid hydroperoxide) in the emulsions was measured.

It can be observed from Fig. 3–13B that the PV of pure oil was appreciably higher than that obtained for the oil droplets in the generated emulsions (PE and SE). This indicates that the biopolymers that encapsulate the oil act as a shield and thus prevent the attack of peroxides on the oil molecules. In PE the PV obtained was lower during the first few days of storage, but it increased moderately after 25 days. Moreover, the PV of PE was comparably higher than that of SE prepared in 30 and 60 s of sonication. In the case of SE, the PV of non-sonicated SE (at 0 s) was significantly higher than the sonicated samples. The PV obtained for non-sonicated SE was increased from 0.23 to 6.28 μM CHP/g of oil during the storage time of 30 days. However, the PV of SE prepared at 30 and 60 s of sonication time was found to be 2.4 and 2.7 μM CHP/g of oil, respectively, which was lower than that obtained for pure oil, PE, and non-sonicated emulsion. This demonstrates that the inherent antioxidant activity provided by the stable bilayer system inhibited the attack of peroxides and improved the oxidative stability of the emulsion. However, at a higher sonication time of 90 s, the PV of SE was increased after 15 days. This can be attributed to that, on increasing the sonication time, the rate of generation of free radicals also increased, which lead to faster oxidation of lipid or oil molecules. Thus, it can be concluded that based upon the droplet size and PV measurements, the sonication time of 60 s was found to be optimum for the preparation of SE.

#### *3.5.1.4 Summary*

It has been established that the stability of multilayer emulsion was strongly dependent on the pH, the concentration of SA, and ultrasonication. The SE prepared at pH 5 and 0.2 wt.% of SA was found to be more stable against bridging and depleting flocculation. At pH 5 the electrostatic interaction between biopolymers was strong enough for the formation of a stable bilayer system and thus overcoming bridging and depletion flocculation. The effect of sonication on the stability of SE was also evaluated. It was observed that the droplet size was reduced by seven times, and the extent of flocculation was diminished on sonicating the SE. However, SE emulsion prepared at the highest sonication time (90 s) was not stable against oxidation. Thus, based upon the oxidative and physical stabilities of SE, the optimum sonication time was considered to be 60 s.

### 3.5.2 Encapsulation of curcumin in multilayer emulsions

#### *3.5.2.1 Introduction*

Curcumin is a well-known polyphenol nutraceutical compound obtained from the turmeric plant, which possesses various chemical and biological properties such as anticancer, antioxidant, antiinflammatory, antibacterial, and anticarcinogenic [42,43]. Curcumin comprises of various chemically reactive functional groups that are found to be effective against different cells of diseases such as cancer, arthritis, asthma, atherosclerosis, and heart disease [44]. In the last few decades, researchers have tried to incorporate curcumin into pharmaceutical and food products due to its health benefits. However, there are innumerous challenges for its utilization in the functional foods, medicines, and other products because of its poor water solubility, low bioavailability, and chemical instability against physiological conditions such as pH [28,42,45]. Recently, the microencapsulation of curcumin in the O/W emulsion matrix has been employed to improve its solubility, stability, and rate of the absorption during physiological transit in the human body. The encapsulation technique via multilayer emulsification includes the incorporation of curcumin in the core (oil phase) of the emulsion matrix that improves the stability and functionality and facilitates their release in a controlled manner to the tract in the human body.

In our recent work [46], we have carried out the encapsulation of curcumin in a single layer (WPI) and double layer (WPI–SA) emulsion and studied the performance comparison of both the emulsions based on the EE, antioxidant activity, and release property of curcumin during the storage.

#### *3.5.2.2 Methods for analyzing the encapsulation stability, antioxidant, and release activity*

***Preparation of curcumin loaded emulsions***: The curcumin was first mixed with olive oil [0.2 mg/mL equivalent to 0.022% (w/w)] and thereafter the oil phase consisting of curcumin was mixed with an aqueous solution of surfactants in the same way as described in Section 3.5.1.2, while keeping the process conditions same (pH, the concentration of SA, and sonication time), as obtained in the above case study. The final compositions of PE and SE after the preparation are: PE: 0.0022% (w/w) curcumin, 9.99% (w/w) oil, 0.9% (w/w) WPI and SE: 0.00108% (w/w) curcumin, 4.90% (w/w) oil, 0.443% (w/w) WPI, 0.2% (w/w) SA, 0.0381% (w/w) sodium azide, and the remaining was water. The curcumin loaded PE and SE were stored at room temperature for further analysis of encapsulation stability, antioxidant, and release activity.

*EE*: The %EE was measured as a function of storage time for up to 3 weeks. To measure %EE, a known amount of emulsion was mixed with ethanol, and the mixture was then centrifuged at 4000 RPM for 5 min to allow the release of curcumin. After centrifugation, the upper clear yellow phase containing curcumin was decanted and analyzed using a UV spectrophotometer at 428 nm ($\lambda_{max}$), and its concentration was calculated using the calibration curve. The EE in both the emulsions was determined using the following equation:

$$\%\,\mathrm{EE} = \frac{Cu_e}{Cu_i} \times 100 \tag{3.5}$$

where $Cu_e$ is the concentration of curcumin in the stable phase of emulsions at a specific condition, and $Cu_i$ is the initial concentration of curcumin.

*Antioxidant activity*: Antioxidant activity of the encapsulated curcumin was measured by evaluating its free radical scavenging activity against a well-known DPPH (1,1-diphenyl-2-picrylhydrazyl) free radical. The curcumin containing emulsion was thoroughly mixed with ethanol, and the mixture was then centrifuged at 4000 RPM for 5 min. The upper clear yellow phase of curcumin was mixed with DPPH solution (0.1 mM) in a volume ratio of 2:1. The control solution was also prepared by mixing ethanol with DPPH solution (0.1 mM). All the solutions were kept in the dark at room temperature for 45 min to allow the completion of the reaction between antioxidant (curcumin) and free radicals. The absorbance of all the samples was measured at 518 nm ($\lambda_{max}$) using UV spectrophotometer. The free radical scavenging activity of curcumin was determined as given in the following equation:

$$\%\text{Radical scavenging activity} = \frac{\text{Absorbance}_{\text{control}} - \text{Absorbance}_{\text{sample}}}{\text{Absorbance}_{\text{control}}} \times 100 \tag{3.6}$$

*In vitro release*: To study the release of curcumin under the intestinal condition of a human body, a simulated intestinal fluid (SIF) was prepared by dissolving the pancreatin enzyme (4 mg/mL) and bile salt (25 mg/mL) in a phosphate buffered saline (PBS), while maintaining the pH at 7.5. The emulsions were mixed with SIF in a volume ratio of 1:2 (v/v), and the pH was further adjusted to 7.5 using NaOH solution. The mixtures were incubated for 4 h in an incubator (Remi Lab, India) at 37°C with continuous shaking at 100 RPM. The samples were collected at a regular interval of 1 h, and the concentration of released curcumin was measured through HPLC analysis. The % curcumin released was determined using the following equation:

$$\%\text{Release} = \frac{\text{Concentration of curcumin released at specific time}}{\text{Initial concentration of curcumin}} \times 100 \tag{3.7}$$

### *3.5.2.3 Results and discussion*

#### 3.5.2.3.1 Encapsulation stability, antioxidant, and release activity

The EE of SE was found to be higher than PE during the storage time of 21 days. The %EE of both the emulsions was found to be 100% when measured immediately after the formation of emulsions. However, it gradually decreased to 56% in the case of PE, whereas remained constant at 100% in the case of SE after 21 days, as shown in Fig. 3–14A. In PE, during storage, a thick viscous layer was formed at the top of the emulsion as a result of inter droplet aggregation, as shown in Fig. 3–14B. The extent of aggregation increased during the storage period, which caused a reduction in the concentration of curcumin in the stable dispersed phase of the emulsion. On the other hand, no agglomeration was observed in the case of SE due to its higher stability than PE, as discussed in Section 3.5.1.3.3, which makes it a suitable carrier for the encapsulation of curcumin.

Further, to observe the microstructure of curcumin-loaded SE, the SE was lyophilized using the freeze-drying technique, and the obtained thin layer of the dried emulsion was analyzed using an optical microscope. It can be seen from Fig. 3–14C that each microparticle comprises a tiny glowing hole indicating the presence of oil droplet and black shell

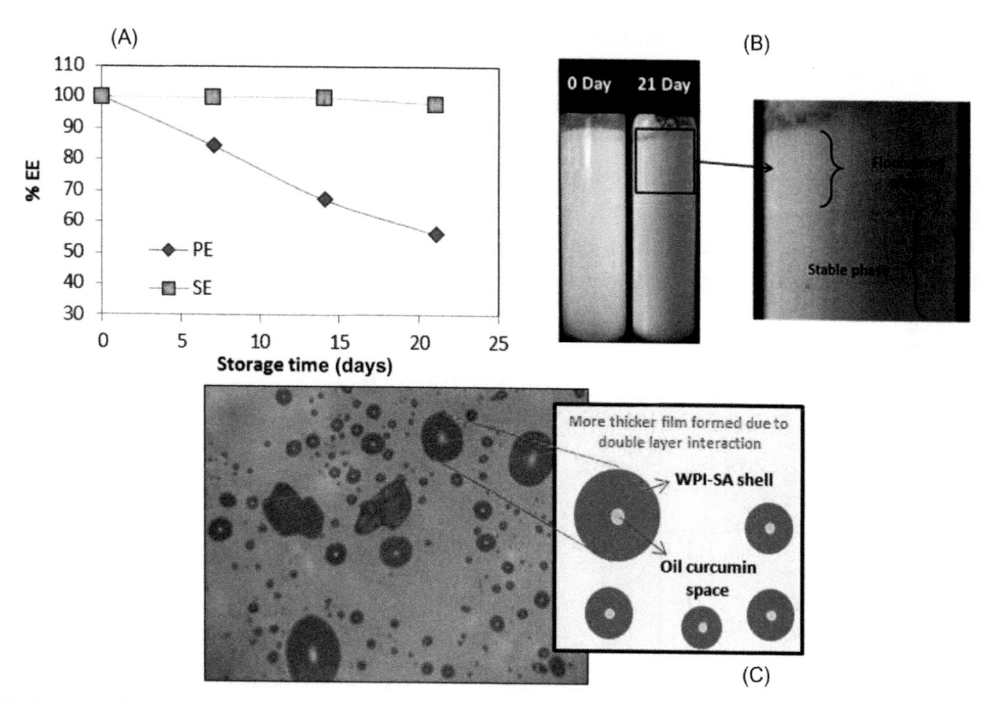


**FIGURE 3–14** (A) Encapsulation efficiency of PE and SE (B) phase aggregated in PE after 21 days, and (C) morphology of SE using optical microscope. *PE*, Primary emulsion; *SE*, secondary emulsion. *Reprinted (adapted) with permission from J. Carpenter, S. George, V.K. Saharan, Curcumin encapsulation in multilayer oil in water emulsion: synthesis using ultrasonication and studies on stability, antioxidant and release activity. Langmuir 35(33) (2019) 10866–10876. ©2019 American Chemical Society.*

formed over each hole indicates the presence of adsorbed biopolymers. Moreover, it can also be seen that each microparticle consists of a single hole and is uniformly dispersed into the system, which indicates that the emulsion droplets were stable against aggregation. Overall, the morphological results confirmed that spherically shaped–SE droplets in a well-dispersed stable form enhanced the encapsulation stability of curcumin than that of PE.

Since curcumin is an antioxidant compound which can deteriorate over time due to changes in the surrounding conditions and therefore it is necessary to monitor its antioxidant activity during its preservation so that the potentiality of curcumin loaded emulsion in the real food/medicinal products can be tested. Therefore, the antioxidant activity of curcumin encapsulated in PE and SE was examined by measuring its free radical scavenging activity against DPPH free radical during the storage.

The DPPH scavenging activity of curcumin encapsulated in PE, SE, and olive oil was measured as a function of storage time, and the results are shown in Fig. 3–15. In the case of curcumin oil (CU-Oil), a significant reduction in the scavenging activity of curcumin was observed during the storage. The scavenging activity of the CU-Oil mixture decreased from 54 to 49% after 21 days of storage. This indicates that the antioxidant activity of curcumin reduced during storage, which can be attributed to the chemical instability of curcumin due to the varying environmental conditions such as pH and temperature. Moreover, the curcumin encapsulated in PE (CU-PE) showed the lowest scavenging activity, that further decreased during the storage period. The maximum scavenging activity of curcumin in CU-PE was observed at 38%, which decreased to 23% after 21 days of storage. The decrease in scavenging activity can be attributed to the decreased EE of curcumin in PE, as shown in Fig. 3–14A. However, the curcumin encapsulated in SE showed almost 80% scavenging activity, that was higher than other encapsulated systems (PE and oil), and no significant reduction in the scavenging activity of curcumin was observed during the storage time of 21

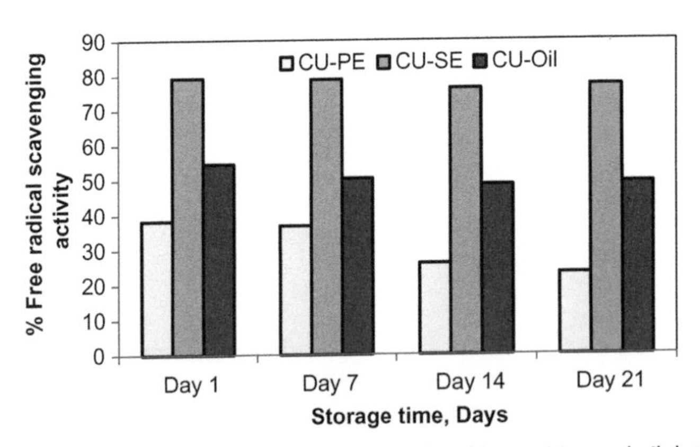


**FIGURE 3–15** Free radical scavenging activity of curcumin encapsulated in emulsion and oil during the storage. *Reprinted (adapted) with permission from J. Carpenter, S. George, V.K. Saharan, Curcumin encapsulation in multilayer oil in water emulsion: synthesis using ultrasonication and studies on stability, antioxidant and release activity. Langmuir 35(33) (2019) 10866–10876. ©2019 American Chemical Society.*

days. Hence, it can be stated that the presence of a double layer shield over the oil droplet containing curcumin facilitated in maintaining the antioxidant property of curcumin during the storage. Moreover, the biopolymers that used to create the encapsulated systems may also act as antioxidant agents and could contribute to the overall scavenging activity against DPPH. Previous studies reported that polypeptides and polysaccharides exhibit specific antioxidant activity [25,26]. Therefore, it can be concluded that the incorporation of secondary emulsifiers not only enhanced the EE but also assisted in maintaining the antioxidant properties of curcumin.

After the successful encapsulation of curcumin in multilayer emulsion, it is necessary to evaluate its bioaccessibility and bioavailability within the GI tract of the human body so that the bioactive compounds can be released and absorbed.

Initially, in vitro digestion was performed under gastric conditions, and it was observed that the emulsions remained stable and thereby not allowing curcumin to release from the emulsified droplets. Hence, no release of curcumin was observed, which might be due to the high resistance of the molecular protein chain against pepsin (gastric enzyme) under the gastric conditions. Therefore, it has been assumed that the emulsion system was stable before the intestinal digestion, and therefore, in vitro release was further studied under the simulated intestinal conditions. A simulated intestinal system was created to perform the in vitro release of curcumin into the intestinal tract of the body. Under the simulated intestinal conditions, the lipid and emulsifier get digested, and curcumin is released. The obtained results are shown in Fig. 3–16.

The maximum release of 63% and 71% were attained after 2 h in PE and SE, respectively, and after that, it remained constant. However, almost 90% of the total curcumin release was observed in the first 1 h only. The higher %release in the case of SE can be attributed to the higher %EE of SE. Under intestinal conditions, the bile salt and pancreatin enzyme degrade

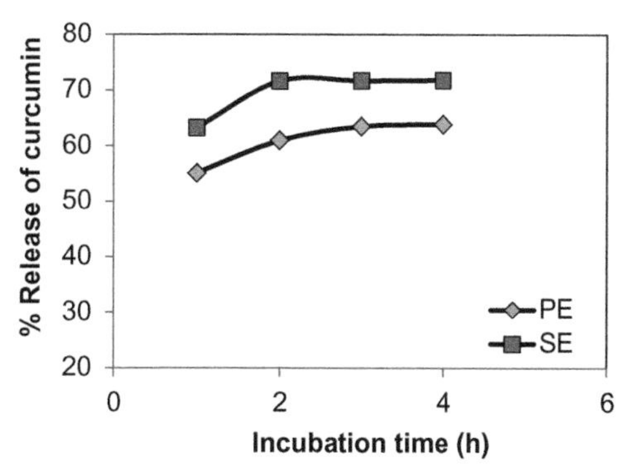


**FIGURE 3–16** In vitro release of curcumin from PE and SE as a function of incubation time under simulated intestinal digestion. *PE*, Primary emulsion; *SE*, secondary emulsion. *Reprinted (adapted) with permission from J. Carpenter, S. George, V.K. Saharan, Curcumin encapsulation in multilayer oil in water emulsion: synthesis using ultrasonication and studies on stability, antioxidant and release activity. Langmuir 35(33) (2019) 10866–10876. ©2019 American Chemical Society.*

the emulsion constituents and release the encapsulated curcumin. The breakdown initiated by the lipolytic activity of pancreatin is activated in the presence of bile salt. The bile salt, which acts as a surface-active agent, preferentially gets adsorbed in the biopolymer coated droplets, or it may also displace the protein and polysaccharides from the interface, thereby allowing lipase present in the pancreatin to enter inside the lipid phase [25]. This promotes the binding of the enzyme with the oil droplets, thereby accelerating the hydrolysis of lipid into monoglycerides and free fatty acids. Apart from lipolysis, the simultaneous proteolysis of the adsorbed protein and polysaccharides resulted in the breakdown of the emulsions and caused the release of curcumin from the core of the lipid phase.

#### *3.5.2.4 Summary*

Multilayer emulsion prepared using a LbL approach was found to be a suitable microencapsulating system with higher EE and improved the chemical stability of the hydrophobic compound such as curcumin. The double-layer emulsion (SE) was found to be resistive against phase separation during storage, that subsequently enhanced the encapsulation stability of curcumin. As a result of improved EE, the curcumin encapsulated in SE possesses higher antioxidant activity than other encapsulating systems. On the other hand, a simulated intestinal digestive model was used for the release of curcumin during the digestion of PE and SE in the presence of pancreatin enzyme, and bile salt and maximum curcumin release of almost 71% and 63% was obtained in SE and PE, respectively. This study confirmed the successful formulation of curcumin encapsulated system for its practical applications, involving the delivery of such a bioactive compound in the intestinal tract of the body.

## References

[1] D.J. McClements, Food Emulsions Principles, Practices, and Techniques Food Emulsions Principles, Practices, and Techniques, second ed., CRC Press, 2005.

[2] I.A.M. Appelqvist, M. Golding, R. Vreeker, N.J. Zuidam, Emulsions as delivery systems in foods, in: J.M. Lakkis (Ed.), Encapsulation and Controlled Release Technologies in Food Systems, second ed., John Wiley & Sons, Ltd, 2016, pp. 129–172. Available from: https://doi.org/10.1002/9781118946893.ch6.

[3] R. Pichot, Stability and characterisation of emulsions in the presence of colloidal particles and surfactants (Ph.D. diss.), University of Birmingham, 2012.

[4] A. Bhushani, C. Anandharamakrishnan, Food-grade nanoemulsions for protection and delivery of nutrients, Nanoscience in Food and Agriculture, 4, Springer, 2017, pp. 99–139.

[5] N. Dasgupta, S. Ranjan, Food nanoemulsions: stability, benefits and applications, An Introduction to Food Grade Nanoemulsions, Springer, Singapore, 2018, pp. 19–48.

[6] M. Sivakumar, S.Y. Tang, K.W. Tan, Cavitation technology – a greener processing technique for the generation of pharmaceutical nanoemulsions, Ultrason. Sonochem. 21 (2014) 2069–2083.

[7] C. Solan, P. Izuierdo, J. Nolla, N. Azemar, M.J. Garcia, Nanoemulsions, Curr. Opin. Colloid. Interface Sci. 10 (2005) 102–110.

[8] J. Zhang, Novel Emulsion-Based Delivery Systems (Ph.D. diss.), 2011.

[9] D. Guzey, D.J. McClements, Formation, stability and properties of multilayer emulsions for application in the food industry, Adv. Colloid Interface Sci. 128–130 (2006) 227–248.

[10] L. Li, In Vitro Gastrointestinal Digestion of Oil-in-Water Emulsions (Ph.D. diss.), Food Technology at Massey University, Auckland, New Zealand, 2012.

[11] M. Ray, R. Gupta, D. Rousseau, Properties and applications of multilayer and nanoscale emulsions, in: C.M. Sabliov, H. Chen, R.Y. Yada (Eds.), Nanotechnology and Functional Foods: Effective Delivery of Bioactive Ingredients, John Wiley & Sons, Ltd, 2015, pp. 175–190. Available from: https://doi.org/10.1002/9781118462157.ch11.

[12] C. Schorsch, M.G. Jones, I.T. Norton, Thermodynamic incompatibility and microstructure of milk protein/locust bean gum/sucrose systems, Food Hydrocolloids 13 (2) (1999) 89–99.

[13] D.J. McClements, Protein-stabilized emulsions, Curr. Opin. Colloid Interface Sci. 9 (2004) 305–313.

[14] O. Sullivan, J. Jonathan, Applications of Ultrasound for the Functional Modification of Proteins and Submicron Emulsion Fabrication (Ph.D. diss.), University of Birmingham, 2015.

[15] P.R. Gogate, Cavitational reactors for process intensification of chemical processing applications: a critical review, Chem. Eng. Process. 47 (2008) 515–527.

[16] P.R. Gogate, Hydrodynamic cavitation for food and water processing, Food Bioprocess Technol. 4 (2011) 996–1011.

[17] S.M. Jafari, Y. He, B. Bhandari, Optimization of nano-emulsions production by microfluidization, Eur. Food Res. Technol. 225 (2007) 733–741.

[18] S.Y. Tang, P. Shridharan, M. Sivakumar, Impact of process parameters in the generation of novel aspirin nanoemulsions—comparative studies between ultrasound cavitation and microfluidizer, Ultrason. Sonochem. 20 (2013) 485–497.

[19] Y.F. Maa, C.C. Hsu, Performance of sonication and microfluidization for liquid–liquid emulsification, Pharm. Dev. Technol. 4 (2) (1999) 233–240.

[20] A. Shanmugam, M. Ashokkumar, Ultrasonic preparation of food emulsions, in: M. Villamiel, A. Montilla, J.V. García-Pérez, J.A. Cárcel, J. Benedito (Eds.), Ultrasound in Food Processing: Recent Advances, John Wiley & Sons, Ltd, 2017, pp. 287–310. Available from: https://doi.org/10.1002/9781118964156.ch10.

[21] B. Abismail, J.P. Canselier, A.M. Wilhelm, H. Delmas, C. Gourdon, Emulsification by ultrasound: droplet size distribution and stability, Ultrason. Sonochem. 6 (1999) 75–83.

[22] S. Parthasarathy, T. Siah Ying, S. Manickam, Generation and optimization of palm oil-based oil-in-water (O/W) submicron-emulsions and encapsulation of curcumin using a liquid whistle hydrodynamic cavitation reactor (LWHCR), Ind. Eng. Chem. Res. 52 (2013) 11829–11837. Available from: https://doi.org/10.1021/ie4008858.

[23] N.J. Zuidam, E. Shimoni, Overview of microencapsulates for use in food products or processes and methods to make them, in: N.J. Zuidam, V. Nedovic (Eds.), Encapsulation Technologies for Active Food Ingredients and Food Processing, Springer, New York, NY, 2010, pp. 3–29. Available from: https://doi.org/10.1007/978-1-4419-1008-0_2.

[24] A.C. Pinheiro, M.A. Coimbra, A.A. Vicente, In vitro behaviour of curcumin nanoemulsions stabilized by biopolymer emulsifiers – effect of interfacial composition, Food Hydrocolloids 52 (2016) 460–467.

[25] T.P. Sari, B. Mann, R. Kumar, R.R.B. Singh, R. Sharma, M. Bhardwaj, et al., Preparation and characterization of nanoemulsion encapsulating curcumin, Food Hydrocolloids 43 (2015) 540–546.

[26] P. Malik, M. Singh, Study of Curcumin Antioxidant Activities in Robust Oil–Water Nanoemulsions, 2017.

[27] H.J. Joung, M.J. Choi, J.T. Kim, S.H. Park, H.J. Park, G.H. Shin, Development of food-grade curcumin nanoemulsion and its potential application to food beverage system: antioxidant property and in vitro digestion, J. Food Sci. 81 (2016) N745–N753.

[28] M. Kharat, Z. Du, G. Zhang, D.J. McClements, Physical and chemical stability of curcumin in aqueous solutions and emulsions: impact of pH, temperature, and molecular environment, J. Agric. Food Chem. 65 (2017) 1525–1532.

[29] G. Bortnowska, Multilayer oil-in-water emulsions: formation, characteristics and application as the carriers for lipophilic bioactive food components—a review, Pol. J. Food Nutr. Sci. 65 (3) (2015) 157–166.

[30] S.A. Fioramonti, M.J. Martinez, A.M.R. Pilosof, A.C. Rubiolo, L.G. Santiago, Multilayer emulsions as a strategy for linseed oil microencapsulation: effect of pH and alginate concentration, Food Hydrocolloids 43 (2015) 8–17.

[31] R. Pongsawatmanit, T. Harnsilawat, D.J. McClements, Influence of alginate, pH and ultrasound treatment on palm oil-in-water emulsions stabilized by β-lactoglobulin, Colloids Surf. A Physicochem. Eng. Asp. 287 (2006) 59–67.

[32] G. Decher, J.B. Schlenoff, Multilayer Thin Films: Sequential Assembly of Nanocomposite Materials, Wiley-VCH, Weinheim, 2003.

[33] J. Carpenter, S. George, V.K. Saharan, A comparative study of batch and recirculating flow ultrasonication system for preparation of multilayer olive oil in water emulsion stabilized with whey protein isolate and sodium alginate, Chem. Eng. Process. – Process Intensification 125 (2018) 139–149.

[34] A.R. Taherian, M. Britten, H. Sabik, P. Fustier, Ability of whey protein isolate and/or fish gelatin to inhibit physical separation and lipid oxidation in fish oil-in-water beverage emulsion, Food Hydrocolloids 25 (5) (2011) 868–878.

[35] A. Teo, S.J. Lee, K.K. Goh, Formation and stability of single and bi-layer nanoemulsions using WPI and lactoferrin as interfacial coatings under different environmental conditions, Food Struct. 14 (2017) 60–67.

[36] F. Azarikia, S. Abbasi, Efficacy of whey protein–tragacanth on stabilization of oil-in-water emulsions: comparison of mixed and layer by layer methods, Food Hydrocolloids 59 (2016) 26–34.

[37] C. Sun, S. Gunasekaran, Effects of protein concentration and oil-phase volume fraction on the stability and rheology of menhaden oil-in-water emulsions stabilized by whey protein isolate with xanthan gum, Food Hydrocolloids 23 (1) (2009) 165–174.

[38] A. Acevedo-Fani, H.D. Silva, R. Soliva-Fortuny, O. Martín-Belloso, A.A. Vicente, Formation, stability and antioxidant activity of food-grade multilayer emulsions containing resveratrol, Food Hydrocolloids 71 (2017) 207–215.

[39] L.A. Shaw, D.J. McClements, E.A. Decker, Spray-dried multilayered emulsions as a delivery method for omega-3 fatty acids into food systems, J. Agric. Food Chem. 55 (8) (2007) 3112–3119.

[40] S.A. Fioramonti, A.C. Rubiolo, L.G. Santiago, Characterisation of freeze-dried flaxseed oil microcapsules obtained by multilayer emulsions, Powder Technol. 319 (2017) 238–244.

[41] L. Moreau, H.-J. Kim, E.A. Decker, D.J. McClements, Production and characterization of oil-in-water emulsions containing droplets stabilized by beta-lactoglobulin-pectin membranes, J. Agric. Food Chem. 51 (2003) 6612–6617.

[42] A. Araiza-Calahorra, M. Akhtar, A. Sarkar, Recent advances in emulsion-based delivery approaches for curcumin: from encapsulation to bioaccessibility, Trends Food Sci. Technol. 71 (2018) 155–169. Available from: https://doi.org/10.1016/j.tifs.2017.11.009.

[43] S. Peng, Z. Li, L. Zou, W. Liu, C. Liu, D.J. McClements, Enhancement of curcumin bioavailability by encapsulation in sophorolipid-coated nanoparticles: an in vitro and in vivo study, J. Agric. Food Chem. 66 (2018) 1488–1497.

[44] G. Bar-Sela, R. Epelbaum, M. Schaffer, Curcumin as an anti-cancer agent: review of the gap between basic and clinical applications, Curr. Med. Chem. 17 (2010) 190–197.

[45] K. Ahmed, Y. Li, D.J. McClements, H. Xiao, Nanoemulsion- and emulsion-based delivery systems for curcumin: encapsulation and release properties, Food Chem. 132 (2012) 799–807.

[46] J. Carpenter, S. George, V.K. Saharan, Curcumin encapsulation in multilayer oil in water emulsion: synthesis using ultrasonication and studies on stability, antioxidant and release activity, Langmuir 35 (33) (2019) 10866–10876.

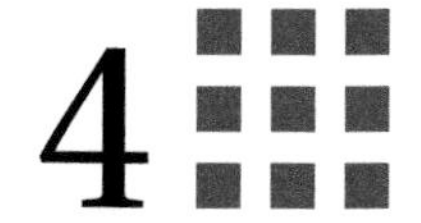

# Encapsulation of active molecules in pharmaceutical sector: the role of ceramic nanocarriers

Joana C. Matos[1,2], Laura C.J. Pereira[2], João Carlos Waerenborgh[2], M. Clara Gonçalves[1]

[1]INSTITUTO SUPERIOR TÉCNICO, UNIVERSIDADE DE LISBOA, LISBON, PORTUGAL
[2]C²TN, CENTER FOR NUCLEAR SCIENCES AND TECHNOLOGIES, INSTITUTO SUPERIOR TÉCNICO, UNIVERSIDADE DE LISBOA, LISBON, PORTUGAL

## Chapter outline

**4.1 Nanotechnology in pharmacy and medicine** ... **53**
**4.2 Ceramic nanoparticles as nanocarriers** ... **55**
**4.3 Ceramic nanoparticles** ... **56**
4.3.1 Silica nanoparticles ... 56
4.3.2 Size ... 67
4.3.3 Mesoporous silica nanoparticles ... 68
4.3.4 Synthesis of mesoporous silica nanoparticles ... 68
**4.4 Superparamagnetic iron oxide nanoparticles** ... **69**
4.4.1 Synthesis, structure and magnetic properties ... 69
4.4.2 Structure and magnetic properties of superparamagnetic iron oxide nanoparticles ... 70
4.4.3 Superparamagnetic iron oxide nanoparticles synthesis ... 72
4.4.4 Superparamagnetic iron oxide nanoparticles coating: protection, stabilization and functionalization ... 74
4.4.5 Core–shell structures and their properties ... 77
4.4.6 Biocompatibility and toxicity of ceramic nanoparticles ... 77
**References** ... **80**

## 4.1 Nanotechnology in pharmacy and medicine

Nanoscience and latter nanotechnology are research fields focusing on the development and manipulation of materials and/or structures in the range scale of 1–100 nm, in at least one

Encapsulation of Active Molecules and their Delivery System. DOI: https://doi.org/10.1016/B978-0-12-819363-1.00004-1

dimension. These nanomaterials/architectures may acquire new and/or improved properties in comparison with the ones exhibited by the same materials at micro- or macroscales [1]. Advances in nanotechnology and molecular biology have allowed the development of nanoparticles (NPs) with properties specifically directed at theranostics. New nanoarchitectures have been produced to enable drug delivery processes in a controlled way, with low immunogenicity, which has led to an improvement in the effectiveness of treatments, with minimal side effects.

NPs date back from 3000 BC. By that time Egyptians ingested Au NPs for mental and body purification. Later, at 10th century BC, Persians used Au and Ag NPs in ceramic glazes [2]. Only the late development in nano-characterization equipment boosts the nanoscience and nanotechnology fields. Today, biomolecular imaging, drug delivery, targeting delivery/imaging, gene delivery, tissue engineering, and stem cell/cell/biomolecules tracking are some of the applications of NPs in pharmaceutical and medicine fields (Fig. 4–1).

NPs may be metallic, ceramic, polymeric, liposomal, dendrimers, micelles at the same time be crystalline or amorphous, plain, or nanostructured and vary in size, size distribution, and surface morphology and chemistry (Fig. 4–2). Finally, NPs properties maybe fine tuned to the design application. Ceramic NPs [where silica ($SiO_2$), titania ($TiO_2$), zinc oxide (ZnO), zirconia ($ZrO_2$), superparamagnetic iron oxide NPs (SPIONs) are some examples] have ground attention in the pharmaceutical and medical fields over the past decades. In these chapter a brief introduction to the topic of NPs in pharmacy and their significance in the relation/application to nanomedicine will be present, followed the exemplification with one of the most representative ceramic NPs—silica-based ones. Silica-based (nonstructured and mesoporous) NPs will be presented and discussed. Case studies illustrating bioconjugation

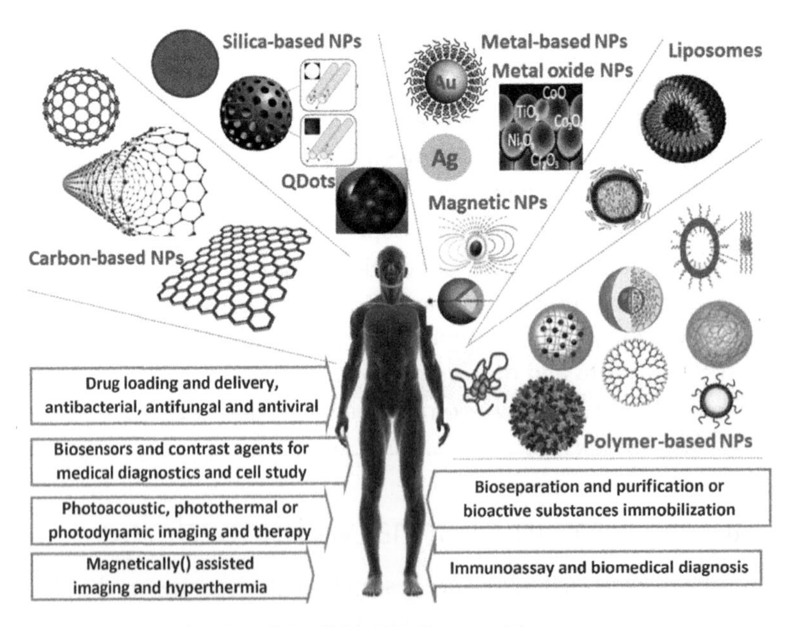


**FIGURE 4–1** NPs in pharmaceutical and medicine fields. *NPs*, Nanoparticles.

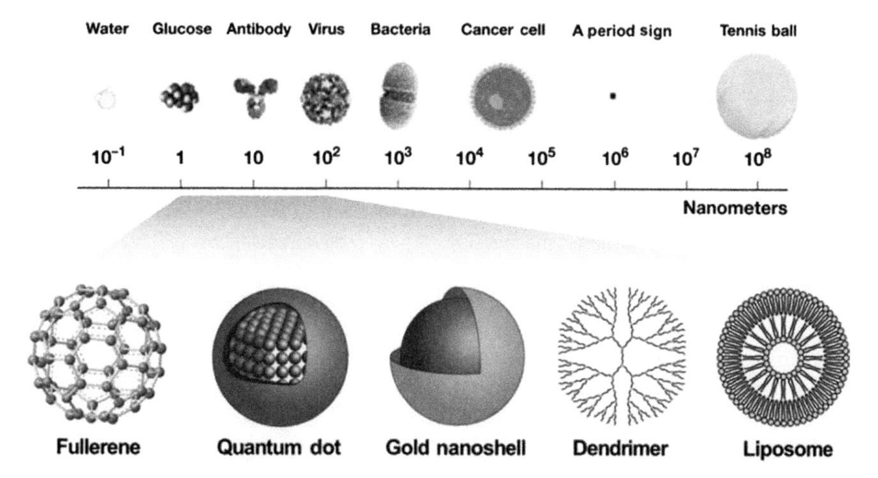


**FIGURE 4–2** Sizes of nanomaterials and biomolecules/organisms comparison [3]. *Reprinted with permission from Chen, G. Hableel, E.R. Zhao and J.V. Jokerst, Multifunctional nanomedicine with silica: role of silica in nanoparticles for theranostic, imaging, and drug monitoring, J. Colloid Interface Sci. 521 (2018) 261–279.*

(with different biomolecules such as DNA, drugs, labels) and decoration (with SPIONs in core–shell structures) are presented for wide spectrum of pharmaceutical applications. Particular attention will be placed on SPIONs for pharmaceutical and biomedical use.

## 4.2 Ceramic nanoparticles as nanocarriers

Ceramic NPs are composed by inorganic compounds such as oxides, carbides, phosphates, and carbonates of metals and semimetals. The well-known ceramics chemical inertia, high vulnerability to pH changes, and high mechanical resistance (constancy in porosity and density, ceramic NPs do not swell) confer them a protection role to several biomacromolecules. By preventing biomacromolecules denaturation, which maybe induced by environmental pH and/or temperature fluctuations, ceramic NPs become a strong candidate as carrier in drug, gene, or protein delivery systems [4–6].

Today, a wide number of engineered NPs provide tools (Table 4–1) to fight a significant number of diseases, namely, bacterial infections, glaucoma, or cancer. Size, shape, porosity, surface chemistry are some of the parameters to be controlled, in order to optimize suitable and efficient drug/gene/protein delivery systems [8]. Comparing with other NP materials, ceramics are much more stable (chemically and mechanically), being their synthesis relatively cheap and well established [9]. Is exactly the synthesis methodology that defines the structural (crystallinity/amorphousness), morphological (shape, size), and reactional performance (based on surface chemistry) of the ceramic NPs.

Silica NPs are largely used in pharmacy as oral delivery carriers (Table 4–2), as therapeutics (in clinical use or under clinical investigation) (Table 4–3) and patented products on the market (Table 4–4).

**Table 4–1** Multifunctional nanoparticles (NPs): materials and functions.

| Component | Material | Function |
|---|---|---|
| Biomedical payload | Imaging agents for optical, MRI, MPI, CT, PET, SPECT, ultrasound imaging (organic dye, QDs, UCNPs, magnetic materials, metal NPs) | Image enhancement |
| | Therapeutic agents (anticancer drugs, DNA, siRNA, hyperthermal/photodynamic materials) | Cancer cell death induction, gene up/downregulation |
| Carrier | Organic (lipid, natural/synthetic polymers) | Multifunctional (protection of payloads, controlled release of drug/gene, biocompatibility, stimuli responsiveness) |
| | Inorganic (hollow metal NPs, hollow metal oxide NPs, C nanostructures, porous, nonporous, core–shell or nanostructured $SiO_2$ NPs) | Multifunctional (imaging ability added to abovementioned functions) |
| Surface modifier | Antibody | Molecular imaging |
| | Aptamer | Target-specific delivery |
| | Peptide/protein | Uptake enhancement |
| | Small molecules | Penetration of barrier |
| | Charge balancing molecules | Signaling transduction |
| | | Stimuli responsiveness |

*MPI*, Magnetic particle imaging; *SPECT*, Single photon emission computed tomography; *UCNPs*, Upconversion nanoparticles
*Source*: Adapted from E.-K. Lim, T. Kim, S. Paik, S. Haam, Y.-M. Huh, K. Lee, Nanomaterials for theranostics: recent advances and future challenges, Chem. Rev. 115 (1) (2015) 327–394 [7].

Another noteworthy ceramic used at nanoscale is iron oxide NPs (SPIONs) and ultrasmall NPs (USPIONs). SPIONs high intrinsic magnetization behavior evidenced great potential in biomedicine being the only clinically approved metal oxide NPs by FDA [12] (Table 4–5). Silica and SPIONs will be discussed in detail in the next sections, where plain and nanostructured architectures are highlighted.

# 4.3 Ceramic nanoparticles

This section presents silica and iron oxide ceramic NPs. Properties, synthesis methods, in situ/ex situ functionalization, final characteristics, and pharmaceutical applications are discussed.

## 4.3.1 Silica nanoparticles

Historically, silica was used in several traditional manufacturing fields, as glass, ceramics, concrete, mortar, sandstone, and silicones. Today, silica is present in an impressive variety of everyday life products, such as glass and tableware, vitroceramics, domestic and industrial water membranes, plastics, paper, paints, silicones, semiconductors, fiber and optical glasses, electronic, optoelectronic, aerospace, and defense products, to mention but a few. In medical field, silica is widely used in dentistry (tooth implants, ceramic pastes), orthopedics (bone implants, scaffolds), dermatology, and specialized medical devices (ophthalmological and bio-glasses,

**Table 4–2** Silica-based oral delivery nanoparticles (NPs).

| Oral delivery system | Silica source | Payload | Coating | Encapsulation method | Release mechanism | In vitro/in vivo/ex vivo |
|---|---|---|---|---|---|---|
| **Nonporous $SiO_2$ NPs** | | | | | | |
| Stober NPs | TEOS | Insulin | PEG 6000<br>PEG 20,000 | Physisorption of insulin to as-synthesized $SiO_2$ NPs—subsequent PEG coating | Passive diffusion | Ex vivo permeation studies with everted rat intestine |
| Stober NPs | TEOS | Insulin | Chitosan | Physisorption of insulin in chitosan suspension to as-synthesized $SiO_2$ NPs | Passive diffusion | In vitro studies of NPs interactions with porcine mucin |
| **Mesoporous $SiO_2$ NPs** | | | | | | |
| MCM-48<br>Ia3d MSM | Luox AS40<br>TEOS/MPTS | Ibuprofen<br>Erythromycin | | Physisorption by immersion | Passive diffusion | In vitro drug release in a simulated body fluid (pH 7.4–7.7) |
| SBA-15 $SiO_2$ | nf | Itraconazole | | Physisorption by immersion | Passive diffusion | In vitro drug release in a simulated gastric fluid (pH 1.2) |
| SBA-15 and MCM-41 functionalized with $-NH_2$ groups | nf | | Bisphosphonates | Electrostatic interaction between drug's phosphate group and silica's amine group at pH 4.8 | Passive diffusion | In vitro drug release in phosphate buffer (pH 7.4) |
| MCM-41 microparticles | TEOS/ triethanolamine | Folic acid | | Impregnation | pH triggered | Yoghurt in vitro drug release in a simulated GIT fluid (pH 2, 4, 7.5) |
| MCM-41 NPs | nf | Rhodamine B | a-CD | Physisorption | Porcine liver esterase triggered | In vitro hydrolysis in HEPES buffer pH 7.5 |
| MCM-48 | TEOS/APTES | Sulfasalazine | Succinylated soy protein isolate | Physisorption and coating | pH/enzyme triggered | In vitro drug release in simulated GIT fluid at pH 1.2, 5, and 7.4 |

(*Continued*)

**Table 4–2 (Continued)**

| Oral delivery system | Silica source | Payload | Coating | Encapsulation method | Release mechanism | In vitro/in vivo/ex vivo |
|---|---|---|---|---|---|---|
| **Hybrid silica microparticles** | | | | | | |
| Core–shell (mesostructured $SiO_2$) | TMOS | Curcumin | | 1. Encapsulation of curcumin in SLN by emulsification/sonication 2. sol–gel | Passive diffusion | In vitro drug release in a simulated GIT fluid (pH 1.2–7.4) |
| Core–shell alginate $SiO_2$ | TMOS/APTMS | LGG | | 1. Preparation of LGG/alginate microgels by electrospraying 2. Mineralization | Erosion of silica shell | In vitro drug release in a simulated GIT fluid (pH 1.2–7.4) |
| **Diatoms silica microparticles** | | | | | | |
| Diatom silica | Fossil | Indomethacin/gentamicin | | Physisorption | Passive diffusion | In vitro drug release in a simulated intestinal fluid (pH 7.2) |
| Diatom silica | Fossil | Mesalamine/prednisone | | Physisorption | Passive diffusion | In vitro drug release in a simulated GIT fluid (pH 1.2–7.4) |

*a-CD*, a-cyclodextrin; *APTES*, (3-Aminopropyl)triethoxysilane; *APTMS*, (3-aminopropyl)trimethoxysilane; *GIT*, Gastrointestinal tract; *HEPES*, 4-(2-hydroxyethyl)-1-piperazineethanesulfonic acid; *LGG*, *Lactobacillus rhamnosus GG; MPTS*, 3-mercaptopropyltrimethoxysilane; *PEG*, polyethylene glycol; *SBA*, Santa Barbara Amorphous; *SLN*, solid lipid nanoparticles; *TEOS*, tetraethyl orthosilicate.
*Source*: Adapted from R. Diab, N. Canilho, I.A. Pavel, F.B. Haffner, M. Girardon, A. Pasc, Silica-based systems for oral delivery of drugs, macromolecules and cells, Adv. Colloid Interface Sci. 249 (2017) 346–362 [10].

**Table 4–3** Silica-based nanoparticles (NPs) therapeutics in clinical use and under clinical investigation.

| Trademark | Formulation | Company | Application | Phase of development |
|---|---|---|---|---|
| C-dots | PEG-coated $SiO_2$ NPs | C-dots Development (United States) | Melanoma (intravenous) | FDA approved 2011 |
| PreveCeutical | $SiO_2$ sol–gel delivery platform | PreveCeutical (Canada) | Platform for nose-to-brain delivery of therapeutic compounds | INC FDA approval |
| Vered | Patented microencapsulation $SiO_2$ NPs | Sol Gel Technologies (Israel) | *Papulopustular rosacea (dermatology)* | Phase II |
| Twin | Patented microencapsulation $SiO_2$ NPs | Sol Gel Technologies (Israel) | *Acne vulgaris* (*dermatology*) | Phase II |
| Sirs-T | Patented microencapsulation $SiO_2$ NPs | Sol Gel Technologies (Israel) | *Acne vulgaris* (*dermatology*) | Phase II |
| Generic | Patented microencapsulation $SiO_2$ NPs | Sol Gel Technologies (Israel) | *Acne vulgaris* (*dermatology*) | Phase III |
| | Ultrasmall silica-based bismuth gadolinium NPs | NH TherAguix (France) | Dual MR–CT guided radiation therapy | Phase I |
| AbsolutMag | Silica NPs, $TiO_2$–$SiO_2$ coated NPs (10, 20, 30 nm, . . ., 20 nm) | Cd Creative Diagnostics (United States) | Theranostic | Phase I |
| DiagNano | Silica magnetic NPs (produced by hydrolysis of orthosilicates in the presence of magnetite) (250 nm–6 nm; 6.0–43 emu/g) | Cd Creative Diagnostics (United States) | DNA/RNA isolation and purification | Phase I |
| AuroLase | PEG-coated silica–gold nanoshells | NanoSpectra Biosciences (United States) | Near-IR light facilitated thermal ablation. Thermal ablation of solid primary and/or metastatic lung tumors | NCT01679470 (Not provided) |
| AuroLase | PEG-coated silica–gold nanoshells | NanoSpectra Biosciences (United States) | MR/ultrasound near-IR light facilitated prostate gland tumors thermal ablation | Phase II |

*PEG*, Polyethylene glycol.

**Table 4–4** Silica-based nanoparticle (NP) patents for medical use.

| Trademark | Formulation | Authors |
|---|---|---|
| **PT20131000062306 PCT/PT2014/000054** | Multifunctional superparamagnetic nanosystem as contrast agent for magnetic resonance imaging and its production method | Gonçalve M.C., Fortes L.M., Martins B.M., Carvalho A.D., Feio G. |
| **WO2011003109, 2011** | Fluorescent silica-based NPs | Bradbury M., Wiesner U., Penate M.O., Ow H., Burns A., Lewis J. |
| **US20100055167 Al, 2010** | Stem cell delivery of antineoplastic medicine | Zhang A., Guan Y., Chen L. |
| **US20107799303, 2010** | Method of preparing silica NPs from siliceous mudstone | Jang H.-D., Chang H.-K., Yoon H.-S. |
| **US20100303716 Al., 2010** | Switchable nano-vehicle delivery systems and methods for making them | Jin S., Oh S., Brammer K, Kong S. |
| **WO2009064964, 2009** | Switchable nano-vehicle delivery systems and methods for making and using them | Jin S., Oh S., Brammer K, Kong S. |
| **US20110092390, 2010** | Methods for making particles having long spin—lattice relaxation times | Marcus C.M. |
| **US20100040693, 2010WO2008018716, 2008** | Silica capsules having nano-holes or nano-pores on their surfaces and method for preparing the same | Chung B.H., Lim Y.T., Kim J.K. |
| **US20100255103, 2010** | Mesoporous silica NPs for biomedical applications | Liong M., Lu J., Tamanoi F., Zink J.I., Nel A. |
| **US20100104650, 2010** | Charged mesoporous silica nanoparticles-based drug delivery system for controlled release and enhanced bioavailability | Lee C.-H., Lo L.-W., Yang C.-S., Mou C.-Y. |
| **US201001361124, 2010 WO2008128292, 2008** | NPs-coated capsule formulation for dermal drug delivery | Prestidge C.A., Simovic S., Eskandar N.G. |
| **US20090263486, 2009** | NPs-stabilized capsule formulation for treatment of inflammation | Prestidge C.A., Simovic S. |
| **US20090181076, 2009** | Drug release from NPs-coated capsules | Prestidge C.A., Simovic S., Eskandar N.G. |
| **WO2009021286, 2009** | Organosilica-encapsulated NPs | Qiao S., Lu G.Q. |
| **WO2009091992, 2009** | Repairing damaged nervous system tissue with NPs | Cho Y.; Shi R., Ivanisevic A., Borgens R. |
| **US20090169482, 2009** | Silica-cored carrier NP | Zhen S., Dai L., Wang R., Qiao T.A., Che W., Harrison W.J. |
| **US20090232899, 2009 WO2005117844, 2005** | Mucoadhesive nanocomposite delivery systems | David A.E., Zhang R., Park Y.J., Yang A.J.-M., Yang V.C. |
| **US20090252811, 2009 WO2005009602, 2005** | Capped mesoporous silicates | Lin V.S.-Y., Lai C.-Y., Jeftinija S., Jeftinija D. M. |
| **EP 20070829819, 2007** | Mesoporous silica particles | Yano T., Sawada T. |
| **WO2005044224, 2005** | Drug delivery system based on polymer nanoshells | Gao J., Al H. |
| **GB2409160(A), 2005 WO2004GB05203, 2005** | A method of engineering particles for use in the delivery of drugs via inhalation | Okpala J. |

*Source*: Based on M.C. Gonçalves, Sol-gel silica nanoparticles in medicine: a natural choice. Design, synthesis and products, Molecules 23 (2018) 2021 [14].

**Table 4–5** Superparamagnetic iron oxide nanoparticles (SPIONs) therapeutics in clinical use and under clinical investigation [13]

| Trademark | Formulation | Company | Application | Phase of development |
|---|---|---|---|---|
| **Endorem** | SPIONs ∅ 150 nm, with magnetite core ∅ 5.6 nm, coated with dextran, and suspended in isotonic glucose solution | **AMAG Pharmaceuticals, Inc. Guerbet S.A.** | MRI negative CA injection for RES (liver, spleen)<br>Liver, spleen (first-generation CA) | FDA approved 2011 |
| **Feridex IV** | SPIONs ∅ 150 nm, with magnetite core ∅ 5.6 nm, coated with dextran, and suspended in isotonic glucose solution | **Advanced Magnetics** | MRI negative CA injection for RES<br>Liver, spleen (first-generation CA) | INC FDA approval |
| **Resovist** | Magnetite NPs (about ∅ 4.2 nm) coated with carbo-dextran, with final ∅ up to 60 nm | **Bayer Schering Pharma AG** (withdrawn from some markets) | Powder for solution for injection MRI intramuscular negative CA<br>Liver, spleen (first-generation CA) | Phase II |
| **Sinerem and Combidex** | SPIONs about ∅ 4–6 nm, coated with dextran | **AMAG Pharmaceuticals, Inc. Guerbet S.A.** (withdrawn from the European market in 2007) | MRI negative CA injection for RES<br>Liver, spleen, lymph mode, bone marrow (second-generation CA) | Phase II |
| **(CTX)-target SPIONs** | CTX-target SPIONs | Clinical trial | MRI negative CA injection for RES (third-generation CA, active targeting SPIONs) | Phase II |
| **AMI-121 Lumirem and Gastromark** | SPIONs coated with silica coating ∅ 300 nm | **AMAG Pharmaceuticals, Inc.** | | Phase II |
| **Abdoscan** | SPIONs coated with poly-styrene ∅ 300 nm | **Amersham** | | |
| **Clariscan** | SPIONs coated with carbo-hydrated PEG ∅ 5–7 nm | **GE Healthcare** (discontinued) | Perfusion, angiography | Phase II |
| **Supravist** | SPIONs coated with carbo-dextran ∅ 3–5 nm | **Bayer Schering Pharma AG** Preclinical | Perfusion<br>Lymph mode<br>Bone marrow | Phase II |
| **MION** | SPIONs coated with dextran ∅ 4–6 nm | Preclinical | Angiography<br>Lymph mode<br>Tumor<br>Infaretion | Phase II |
| **BioPal** | SPIONs coated with various materials N/A | | Animal image | Phase II |

*CA*, Contrast agent; *CTX*, chlorotoxin; *MRI*, Magnetic resonance imaging; *PEG*, polyethylene glycol; *RES*, reticuloendothelial system.
With permission from J. Lodhia, G. Mandarano, N.J. Ferris, P. Eu and S. Cowell, Development and use of iron oxide nanoparticles (Part 1): synthesis of iron oxide nanoparticles for MRI, Biomed. Imaging Intervention J. 6 (2) (2010). Available from: https://doi.org/10.2349/biij.6.2.e12.

scaffolds). New engineered silica materials and devices have emerged with the arising of nanomedicine, where colloidal silica has a pride of place among microparticles and NPs [14].

Silica NPs stand up by their unique properties, such as enormous versatility for surface modification (including surface charge), high payload capacity, and improved blood circulation time, along with lack of toxicity, high biocompatibility, and biodegradability, which give rise to wide range of biomedical and pharmaceutical applications. Silica NPs have proved to be good multifunctional nanoplatforms for drug delivery, gene therapy, and imaging techniques [6,15].

#### *4.3.1.1 Synthesis*

Top-down (when smaller unities are obtained from a single large portion of material) and bottom-up (when atoms/molecules are assembled following an organized order to form particles) are two main approaches for NPs synthesis (Fig. 4–3). Table 4–6 summarizes the most common (bottom-up and top-down) methods to synthesized ceramic NPs.

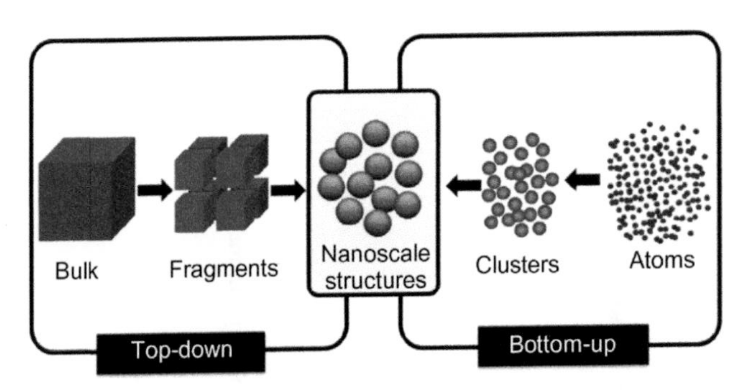


**FIGURE 4–3** Top-down and bottom-up synthesis approach.

**Table 4–6** Top-down and bottom-up main synthesis methods.

| Synthesis approach | Mechanism | Process | Characteristics |
|---|---|---|---|
| Top-down<br>Starting point: Bulk materials<br>Final result: Nanosize particles | Mechanical | High energy milling | Sizes ≈ 200 nm; broad size distribution, time energy consuming (e.g., ceramic raw material) |
| Bottom-up<br>Starting point: Atoms/molecules<br>Final result: Nanosize particles | Chemical: Liquid phase processing | Sol–gel | Sizes ≈ 1–100 nm, good morphology control, medium purity, medium/low yield (e.g., silica, $TiO_2$, $SnO_2$, ZnO, $ZrO_2$, alumina) |
| | | Controlled precipitation | Sizes ≈ 2–100 nm, excellent morphology control, high purity, medium yield (e.g., silica, alumina, ZnO) |
| | Chemical: Gas phase processing | Chemical vapor deposition | Sizes ≈ 1–100 nm, good morphology control (e.g., silica, carbon nanotubes) |
| | | Pyrolysis | Sizes ≈ 10–200 nm, broad size distribution, low morphology control, high yield (e.g., carbon nanotubes, carbon black) |

The synthesis of silica NPs comes from the late 1950s. Kolbe et al. were the firsts to synthesize silica NPs in 1956 [17]. Their synthesis was improved and recognized through Stöber et al. [18], who tune silica NPs' size by playing with the catalyst/precursor ratio.

Silica NPs may be prepared via bottom-up or top-down methodologies. Using a top-down approach, silica NPs is synthesized by electrochemical etching of silicon wafer or by pyrolysis of quartz sand. However, it is commonly accepted that nanosize is not effectively achieved through top-down methodologies [3]. The bottom-up strategy comprises several techniques, such as chemical vapor deposition, wet-chemistry methodologies, flame spray pyrolysis, to mention but a few. Among them, sol–gel stands up as straightforward, easily controllable and easy scale-up process, along with high energy/time saving, and the last but not the least, cost-effective approach (Fig. 4–4).

There are commonly four different via to sol–gel silica NPs synthesis: colloidal route, solution route (acid- or base catalyzed), biomimetic synthesis, and finally templated synthesis [14]. In colloidal route, silica NPs are synthesized in an aqueous medium by supersaturation, polymerization, and silica polymorphs precipitation. The biomimetic synthesis allows silica NPs formation using certain coprecipitation/nucleation agents that can be biologic or biomimetic. In both cases, sol–gel synthesis occurs under neutral or acidic medium.

The most common synthesis route is, nevertheless, the solution route. Metallic salts, metal alkoxides, or other type of organometallic precursors are hydrolyzed and condensed to 0D, 1D, 2D, or 3D sol–gel products (Fig. 4–4). The highly porous amorphous products obtained maybe densified through controlled soft or hard thermal treatments. The most

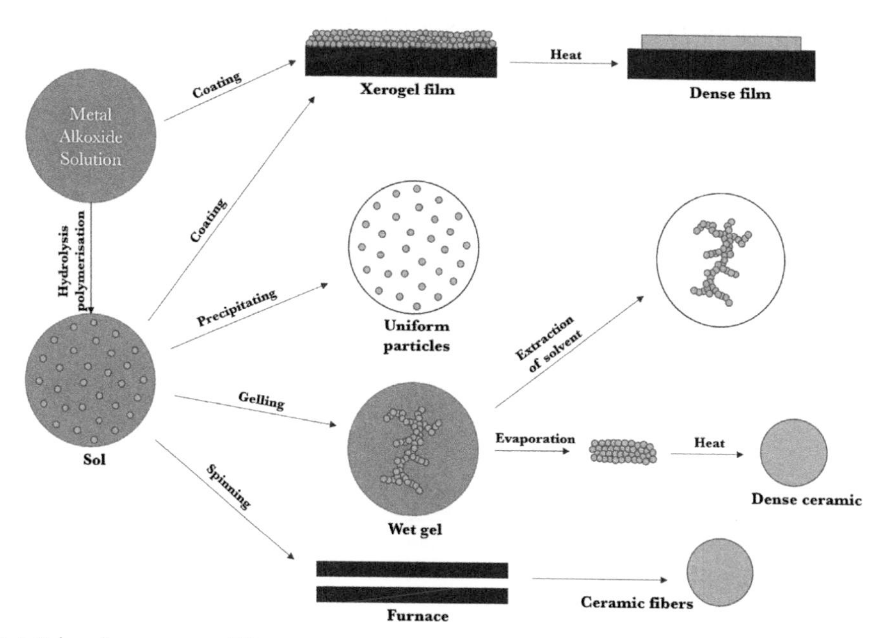


FIGURE 4–4 Sol–gel process versatility.

usual silica source is an alkoxide of silicon, tetraethyl orthosilicate (TEOS). First, the alkoxide hydrolyzes (and produces silicate species and alcohol), and then the silicates condense, to form Si–O–Si network (Fig. 4–5).

Catalyst, pH, water to silica precursor's ratio, water to solvent ratio, solvent and precursor type, chelating agent presence, and temperature are the most important parameters to be controlled in order to optimized NPs structure, size, and morphology [14]. Under basic catalysis, condensation starts only after complete hydrolysis. Silica monomers are negatively charged, and NPs may form. The maximum $SiO_2$ NPs growth occurs at $pH \geq 7$: above that pH value, Ostwald ripening and aggregation are the main growth mechanisms.

Acidic conditions favor the production of 3D gels. Here, the bulks' surfaces show little positive or null surface charge, allowing a bigger interaction between silica species. The hydrolysis is the fastest step, and condensation starts even before the hydrolysis is complete, in terminal silanol groups (resulting chains and network-like gels). Under acidic conditions, silica NPs synthesis is still possible in reverse micelle systems (or water-in-oil microemulsion). Here, hydrolysis and condensation reactions are spatially confined in nanoreactors (dispersed water phase). This process is widely used to synthesize $SiO_2$ NPs due to an easy and effective control of size and morphology. However, the large amounts of surfactants (which may cause cellular membrane rupture) and organic solvents used require extensive washing processes before any biological applications, rendering this method a less *green* one.

Summarizing, Stöber method arises as almost an eco-friendly protocol, since the reactional medium is essentially aqueous and the synthesized NPs are relatively monodisperse, spherical, and electrostatically stable. The classical Stöber method was recently modified and optimized in order to replace ammonia with hydrothermal water ($pH \sim 11$) [19].

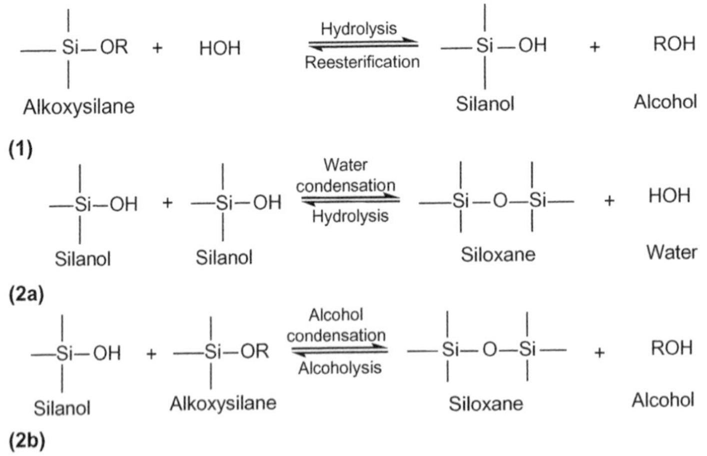

(1) ≡Si–OR (Alkoxysilane) + HOH ⇌ ≡Si–OH (Silanol) + ROH (Alcohol) [forward: Hydrolysis; reverse: Reesterification]

(2a) ≡Si–OH (Silanol) + ≡Si–OH (Silanol) ⇌ ≡Si–O–Si≡ (Siloxane) + HOH (Water) [forward: Water condensation; reverse: Hydrolysis]

(2b) ≡Si–OH (Silanol) + ≡Si–OR (Alkoxysilane) ⇌ ≡Si–O–Si≡ (Siloxane) + ROH (Alcohol) [forward: Alcohol condensation; reverse: Alcoholysis]

**FIGURE 4–5** Hydrolysis and condensation sol–gel reactions.

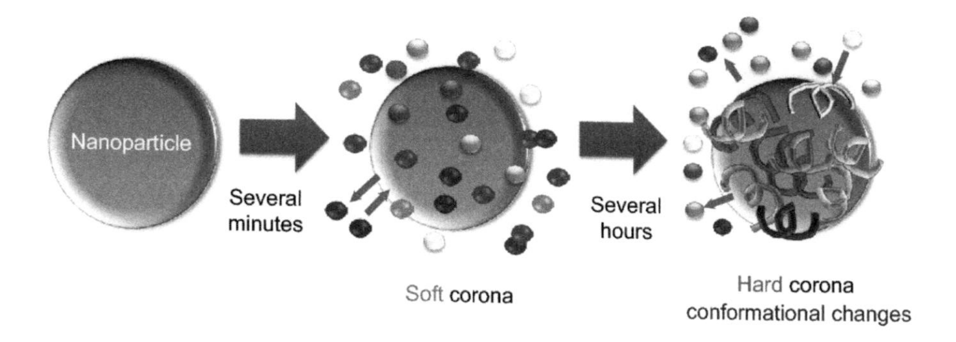


**Protein corona factors**

| NPs | Environment |
|---|---|
| **Material** | Composition |
| **Surface** | Exposure time |
| **Size** | pH |
| **Charge** | Temperature |
| **Shape** | Shear stress |

**FIGURE 4–6** Immune response system to intravenously/muscular injected NPs. *NPs*, Nanoparticles.

### *4.3.1.2 Surface modification*

The biological performance (pharmacokinetics profiles, biodistribution, target recognition, therapeutic efficacy, inflammatory reactions, and toxicity) of intravenously/muscular injected NPs is controlled by a complex array of interrelated physicochemical and biological factors, starting with opsonization, followed by phagocyte ingestion and ending with NPs clearance [14]. Rapid blood clearance limits drugs/gene/therapeutic molecules/markers accumulation at target delivery sites, while NP accumulation in macrophages (within clearance organs) initiates inflammatory responses, inducing toxicity (Fig. 4–6).

NPs surface chemistry is thus pivotal to deceive the immune system and further bioconjugation with biomolecules. The Stöber method (slight modified) allows the silica functionalized through two different approaches: internal or surface functionalization (Fig. 4–7).

In internal functionalization, inorganic or organic molecules are introduced inside the silica NPs by "physical doping" [where second order chemical bounds with silane ($\equiv$Si–OH) or siloxane ($\equiv$Si–o–Si$\equiv$) groups form]. Although this is a simple method, internal functionalization is not so effective; especially when small molecules are concerned (they may escape from the silica host).

To overcome this drawback, silica NPs surface modification is recommended. The introduction of nonhydrolyzable organic groups during the synthesis process (in situ functionalization) gives rise to a hybrid *OR*ganically *MO*dified *SIL*ica matrix (ORMOSIL for short). These hybrid ORMOSIL NPs allow an even easier chemical conjugation/decoration with biomolecules and/or loading with either hydrophilic or hydrophobic drugs/dyes [6].

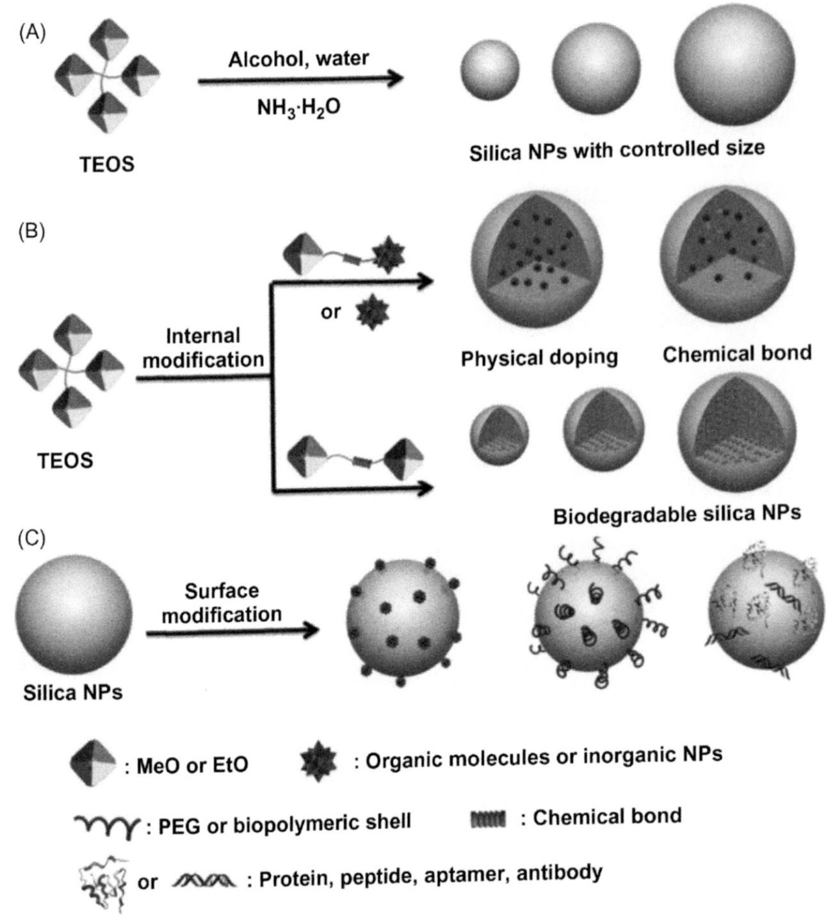


**FIGURE 4–7** Scheme of the design and synthesis of: (A) traditional silica NPs, (B) silica NPs by internal modification, and (C) silica NPs with surface modification [20]. *NPs*, Nanoparticles. *Reprinted with permission from Z. Xu, X. Ma, Y.-E. Gao, M. Hou, P. Xue, C.M. Li, et al., Multifuctional silica nanoparticles as a promising theranostic platform for biomedical applications, Mater. Chem. Front. 1 (2017) 1257*

These ORMOSIL NPs are much more versatile than silica NPs due to the presence of surface organic groups. Tuneable wettability, tailor-made porosity, size, shape, and hardness/complacency are some of the ORMOSIL properties that make them very promising for biomedical applications. Furthermore, the coexistence of both silanol and nonhydrolyzable organic groups allows an easier conjugation or decoration of the NPs surface with molecules of interest such as biomolecules, drugs or dyes. Table 4–7 shows the silanes commonly used for $SiO_2$ NPs surface functionalization.

The amine group ($-NH_2$) is commonly used in gene therapy. Amine confers a positive charge to $SiO_2$ NPs surface, enabling an electrostatically interaction with genes (enhancing their absorption, binding and even protection from enzymatic digestion). This fact increases

**Table 4–7** The most common silanes for surface modification of silica nanoparticles (NPs) and their properties [3].

| Silane for functionalization | Applications |
| --- | --- |
| (3-Aminopropyl)trimethoxysilane<br>(3-Aminopropyl)triethoxysilane<br>**(−NH$_2$)** | • Reduced aggregation<br>• Fluorescent labeling<br>• Surface charge modification<br>• DNA binding and protection from enzymatic cleavage |
| (3-Mercaptopropyl)-trimethoxysilane<br>**(−SH)** | • Conjugate with maleimides<br>• Thiol/disulfide exchange reactions to attach oligonucleotides<br>• Surface charge modification |
| Polyethylene glycol-silane (PEG-silane)<br>**(−PEG)** | • Increased circulation time<br>• Reduced aggregation and increase particle dispersity in aqueous solution |
| Alkylsilane<br>**(Alkyl chain)** | • Hydrophobic coating<br>• Increase ultrasound contrast |
| Carboxyethylsilanetriol<br>**(−COOH)** | • Functionalize silica NPs and provide reactive sites for amine |
| 3-Trihydroxysilylpropyl methylphosphonate<br>**(−PO$_3$)** | • Functionalize silica NPs and provide reactive sites for amine |
| (3-Isocyanatopropyl)-triethoxysilane<br>**(−NCO)** | • Functionalize silica NPs and provide reactive sites for amine |

Reprinted from F. Chen, G. Hableel, E.R. Zhao and J.V. Jokerst, Multifunctional nanomedicine with silica: role of silica in nanoparticles for theranostic, imaging, and drug monitoring, J. Colloid Interface Sci. 521 (2018) 261–279.

significantly the gene (plasmid DNA) chance to reach the cell nucleus and transfect. Besides that, the amine–silica NPs do not cause any damage to the mammal tissues nor present any immunological side effects (which have been frequently reported with viral-mediated gene delivery systems) [14]. $SiO_2$ NPs functionalized with amine and phosphate groups can decrease or relieve inflammation processes and immunomodulatory issues during allergic airway inflammation treatment [21].

$SiO_2$ NPs can easily be grafted, on their surface, with organic molecules (e.g., drugs or dyes) or inorganic NPs (e.g., Au NPs, QDs). To improve the degradation rate and biocompatibility of $SiO_2$ NPs in mammals, polymers/biopolymers, proteins among other types of molecules are used as NPs coating. To improve diagnostic and therapeutic purposes, molecules such as peptides, aptamers, and antibodies can also be conjugated with silica NPs [20].

### 4.3.2 Size

Several efforts have been made in order to achieve an effective multifunctional therapeutic platform for drug delivery, bioimaging, and other medical applications. NPs' size is a critical parameter which that to be optimized during synthesis. Sol–gel Stöber method allows an accurate size control of $SiO_2$ NPs—from tens to hundreds of nanometers—adequate for several pharmaceutical applications. The size effect of silica-based delivery systems in cancer treatment has been studied, and it was concluded that from the three different studied sizes (20, 50, and 200 nm), the best results were achieved with the 50 nm silica NPs. Silica NPs with 50 nm exhibited the highest retention in tumoral sites and the highest efficacy against primary and metastatic tumors in in vivo experiments [22,23].

## 4.3.3 Mesoporous silica nanoparticles

Porous materials have arisen interest in the last decades for wide variety of applications in pharmacy and medicine. Considering the pore size of the material, this can be divided in three different groups: microporous particles (pore sizes less than 2 nm), mesoporous particles (between 2 and 50 nm), and macroporous particles (greater than 50 nm). Mesoporous particles, due to their structural characteristics (such as uniform pore size and a long-range ordered pore structure), have been intensively studied in different fields, from catalysis, adsorption, separation, and sensing to pharmaceutical and biomedical applications.

Mesopores silica NPs (MSNs) may load active molecules/compounds (by physical adsorption or chemical binding) and excels due to their high specific surface area and large pore size [24,25] When compared with plain silica NPs, MSNs have huge surface area to volume ratio [26]. In MSNs the biocompatibility is allied to the tuneable pore size and independent surface functionalization (internal and external pores maybe loaded with different drugs) along with a gating mechanism for pore opening, which make them excellent candidates in drug nanocarrier systems [24,25,27]. Several drug delivery systems have been designed and successful developed for loading cargo such as proteins, DNA, and RNA [27].

Mesoporous materials synthesis dates back to 1970s. However only late in 1992 mesoporous solids were synthesized from aluminosilicate gels using liquid crystal templates. This synthesis was performed by researchers from Mobil Research and Development Corporation who named the new nanostructures by Mobil Crystalline Materials or Mobil Composition of Matter (MCM-41). The pores size can be fine-tuned by surfactant template (usually cationic surfactant). MCM-41 is usually used for drug delivery systems and exhibited hexagonal structure and a pore range from 2.5 up to 6 nm. Other mesoporous materials were also synthesized by varying template or reaction conditions, giving rise to different structural arrangements. For example, MCM-50 has a lamella-like arrangement and MCM-48 a cubic arrangement. The Santa Barbara Amorphous (SBA) type material, first synthesized by the University of California in Santa Barbara, are composed by nonionic triblock copolymers as templates. These materials are different from the MCM materials, since they have thicker silica walls and larger pores (4.6–30 nm). Several types of SBA material appeared with different structures based on the triblock polymers (which determines the symmetry of the mesoporous structure). The SBA-15, with a hexagonal structure, is the most widely used for biomedical applications. Fig. 4–8 shows some different types of mesoporous silica nanoparticles.

## 4.3.4 Synthesis of mesoporous silica nanoparticles

MSNs synthesis can be performed in basic, acidic, or neutral media. The modification of reactional parameters (during the synthesis process) allows the obtention of particles with

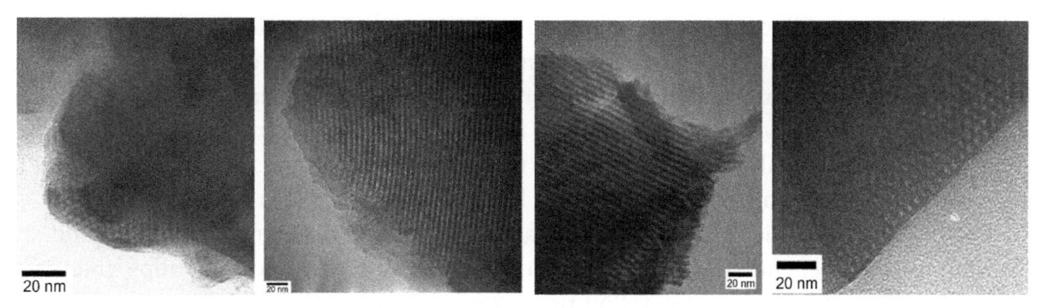


**FIGURE 4–8** Mesoporous silica nanoparticles types.

different sizes and/or shapes. In drug delivery systems the uniform particle size and the pore volume (large enough to enhance the loading capacity) are determinant, so the synthesis control is pivotal in to obtain particles with the desirable characteristics.

The MSNs can be synthesized mainly through four different methods: sol–gel method, template-directed method, microwave-assisted technique, and chemical etching technique.

*Stöber method* is widely used not only to produce spherical monodisperse micron size silica NPs [18] but also the common way to produce MSNs, although laborious and time-consuming process [27].

Other synthesis process widely used is the *template-directed method.* This is the cheapest way to obtain MSNs. Here, the synthesis occurs through a liquid-crystalline template mechanism, where hydrolysis and condensation reactions take place at the surface of nonionic surfactant micelles. The resultant particles are calcinated to remove micellar templates and MSNs form. Parameters such as pH, temperature, concentration of surfactant, and silica source allow the adjustment of MSNs structural characteristics. Templates are usually vesicles, polymeric micelles, and gold or silica NPs.

## 4.4 Superparamagnetic iron oxide nanoparticles

### 4.4.1 Synthesis, structure and magnetic properties

Among others nanoscale drug vehicles, SPIONs emerge in biomedical applications (namely, in cancer therapy and theranostics) due to their magnetic performance. SPIONs present intrinsic high magnetization, good biocompatibility, and reduced toxicity, although SPIONs toxicity in normal cells has been pointed out by scientific communities when they are used in in vivo cancer treatments. The principle behind the development of SPIONs as novel drug delivery carriers is its capacity to be guided by an external magnetic field [28].

SPIONs immobilization in hydrophilic, biocompatible, and functionalized arrays has been a challenge in NP production. Their coating is an important process, since it will reduce aggregation, improve colloidal dispersity/stability, essential for their use as drug delivery carriers [29]. Many studies have been performed to incorporate SPIONs in liposomes, originating magneto-liposomes that are good multifunctional nanoplatforms for both imaging or chemotherapeutic applications combining their properties as contrast agent (CA) for MRI with anticancer drugs [29] Another interesting possibility is encapsulation within silica NPs. Silica coating allows a longtime stability to SPIONs, besides offering great chemical versatility. However, several studies have shown that the magnetic properties of SPIONs may be affected by the coating or the presence of the load (drugs/dyes/others NPs) [28].

The SPIONs magnetic behavior is essentially defined by their morphology, size, and size distribution. This dependency, such as magneto-crystalline and shape anisotropy, interparticle interactions, and magnetic relaxation processes, was analyzed in detail in several studies, by static and dynamic magnetometry and Mossbauer spectroscopy [12,30–32].

## 4.4.2 Structure and magnetic properties of superparamagnetic iron oxide nanoparticles

SPIONs are small particles of magnetite ($Fe_3O_4$) or maghemite ($\gamma$-$Fe_2O_3$) with diameters between 10 and 100 nm. Magnetite is the oldest phase studied, with cubic spinel structure ($AB_2O_4$). Concerning maghemite, this phase shows an intermediate character between *MAG*netite and *HEM*atite. With also a cubic spinel structure, maghemite is chemically more stable than magnetite but presents a lower saturation magnetization. Fig. 4–9 shows both magnetite and maghemite structures, whose properties depend on many physiochemical and synthesis parameters that will influence their size, shape, and saturation magnetization [28,29].

Superparamagnetism is the ability of magnetic NPs to present magnetic behavior with high saturation magnetization and susceptibility, during the application of an external magnetic field. Removing the field, the magnetic behavior is lost, resulting in zero coercivity and magnetic remanence. SPIONs with size below 30 nm, at room temperature, show

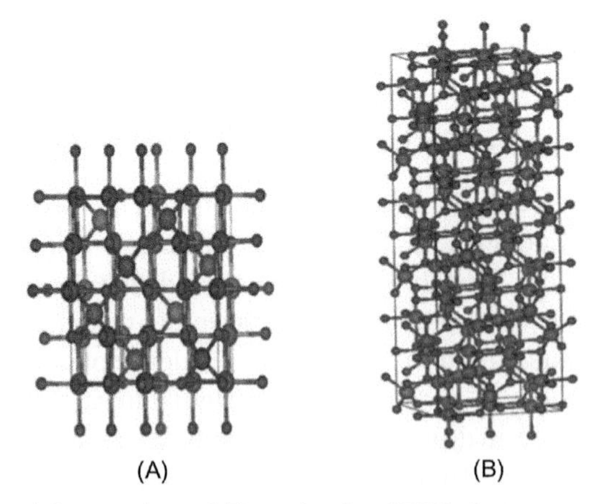


**FIGURE 4–9** SPIONs structures: (A) magnetite and (B) maghemite. *SPIONs*, Superparamagnetic iron oxide nanoparticles.

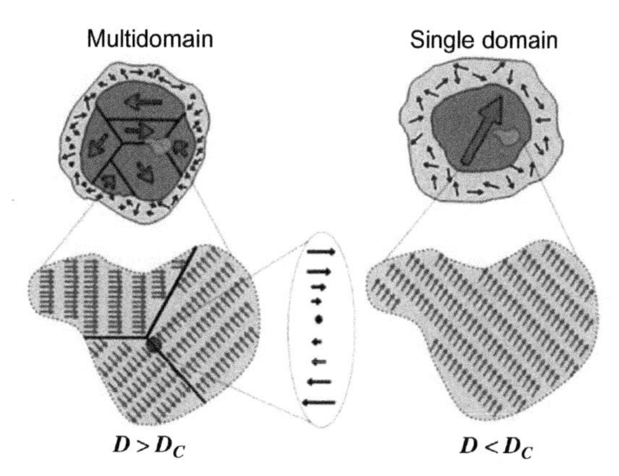


**FIGURE 4–10** Magnetic domains scheme: (A) multidomain ferromagnetic particle for particle sizes larger than the critical diameter ($D > D_c$) and (B) single-domain particle for particle sizes smaller than the critical diameter ($D < D_c$) [33]. *Reprinted with permission from G.F. Goya, V. Grazú and M.R. Ibarra, Magnetic nanoparticles for cancer therapy, Curr. Nanosci. 4 (2008) 1–16*

superparamagnetic behavior. Magnetic iron oxides ($Fe_3O_4$ or $\gamma$-$Fe_2O_3$) below their Curie temperatures (850K and 986K, respectively) lose their permanent magnetic properties, becoming ferromagnetic materials. These ferromagnets in the bulk state show a multidomain magnetic structure (Fig. 4–9). However, reducing the volume of the ferromagnet to the size of the NPs makes a single-domain structure, with a permanent magnetic moment (due to collective spin correlation at the entire NP scale) (Fig. 4–10). The iron oxide NPs present different properties in relation to the bulk ones, which motivate the outstanding interest for their use in pharmacy and nanomedicine [12,34].

In magnetic nanomaterials the single-domain phase dominates at specific dimensions, since the magnetostatic energy equalizes the domain-wall energy. Thus the multidomain phase is transformed in a single-domain phase at nanosize range (Fig. 4–9). $Fe_3O_4$ spherical NPs presents a single domain at critical diameter of 128 nm. This is an important characteristic since SPIONs at a single-domain phase show a high magnetic moment and exhibit superparamagnetism above the blocking temperature, which is a size- and shape-dependent phenomena.

When the magnetic field is applied, the suspended SPIONs magnetic moments tend to align themselves parallel to the applied field, which results in high values of saturation magnetization. However, in the absence of a magnetic field, their magnetic moments align themselves along the easy axis, reverting to their original positions due to longitudinal and transverse relaxivities. Magnetization decreases for SPIONs diameters below 12–10 nm. It was observed that $Fe_3O_4$ NPs, with a diameter of 5 nm, exhibited a low magnetization value of 27 emu/g, due to its reduced size [34] (Fig. 4–11).

Another factor that may influence the magnetic properties of SPIONs is shape. Higher magnetization values and low coercivity values were obtained for spherical magnetite NPs than for ellipsoidal and cubical SPIONs NPs [34]. However, another study reveals that SPIONs with quasicubic morphology present a higher magnetization value (79 emu/g),

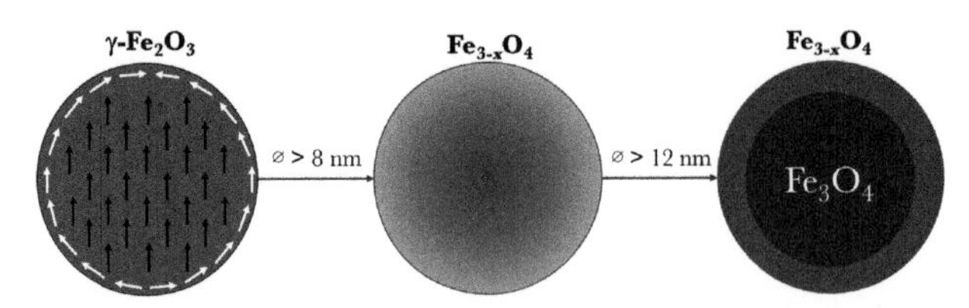


**FIGURE 4–11** Schematic of relation between size and structural composition of SPIONs [34]. *SPIONs*, Superparamagnetic iron oxide nanoparticles. *Reprinted with permission from G. Kandasamy, D. Maity, Recent advances in superparamagnetic iron oxide nanoparticles (SPIONs) for in vitro and in vivo cancer nanotheranostics, International Journal of Pharmaceutics 496 (2015) 191–218*

probably due to an increase in the magnetic core volume or a low surface/volume ratio [35]. A study with octapod-shaped SPIONs showed that these NPs present higher magnetization values ($\approx$71 emu/g) since their core radius is enhanced 2.4 times more than spherical ones. Near bulk magnetization values were obtained for NPs with mixed shape distributions [34,36].

SPIONs magnetic moments appear due to the presence of unpaired 3d electrons in both $Fe^{3+}$ and $Fe^{2+}$ cations (in the cubic FCC lattice). An electron spin coupling between $Fe^{2+}$ and $Fe^{3+}$ ions takes place at octahedral sites, and antiparallel electron spin coupling of $Fe^{3+}$ ions takes place at tetrahedral sites. These superexchange interactions (exchange interaction/coupling between the two sites across the oxygen anions) are responsible for the magnetic behavior of SPIONs. However, at SPIONs surface, the superexchange interactions destruction due to the inclination effect can occur, which is the inclination of the surface atomic spins of the magnetic NPs in a specific angle. This destruction is due to the lack of atoms number required for the single magnetic moment formation and/or less organized spins at the surface when compared to SPIONs core. Other defects that can affect the magnetic properties of SPIONs are oxygen vacancies, edge roughness, cationic positions, Laplace pressure and changes in surface, and/or core chemical ordering and anisotropies [34].

## 4.4.3 Superparamagnetic iron oxide nanoparticles synthesis

There are several methods to produce SPIONs (physical, biological, or chemical). Physical methods encompass gas phase deposition, electron beam lithography, pulsed laser ablation, laser-induced pyrolysis, powder ball milling, and aerosol production. Biological processes allow SPIONs production through fungi, bacteria, or protein activity. Chemical techniques include coprecipitation, microemulsion, hydrothermal synthesis, thermal decomposition, sonochemical synthesis and microwave-assisted synthesis. In this work the focus will be in chemical synthesis.

The *coprecipitation method* is the most used technique in the biomedical field. This method consists in the coprecipitation of ferrous ($Fe^{2+}$) and ferric ($Fe^{3+}$) ions from an aqueous solution (containing salts in 1:2 stoichiometric ratio), in basic medium, under inert atmosphere, and at room or moderate temperatures (70°C–90°C). This method produces magnetite NPs that may be easily oxidized to maghemite. With this approach the adjustment of particle size and size distribution is extremely difficult, being crucial the control of pH, ionic strength, and seed concentration [37]. The coprecipitation method can occur through one of the two topotactic phase transitions: akaganeite phase (crystal nuclei) to goethite phase (arrow-shaped NPs) or ferrous hydroxide phase to lepidocrocite phase (where the SPIONs are formed depending on base rate addition). The base rate addition and the cationic stoichiometric ratio ($Fe^{2+}/Fe^{3+}$) influence the magnetic properties of the synthesized NPs, with the risk of ferromagnetic performance instead of superparamagnetic. Though SPIONs obtained through coprecipitation present low saturation magnetization values (30−50 emu/g) when compared to the bulk material (92 emu/g), due to the particle's polydispersity and lack of a proper crystallinity [34].

The synthesis of SPIONs maybe occur with *microemulsion method* (W/O microemulsion) where the stabilizing agents in continuous oil phase protect the water phase droplets (nanoreactor), giving rise to smaller and more uniform particles. Iron oxide precursor to base ratio, surfactant, and/or solvent concentrations is crucial to control SPIONs' size and shape. The difficulty in removing unreacted chemicals (precursors, base, or surfactants) disallows its application in biomedical applications [34].

*Hydrothermal method* is another methodology, which runs at high temperatures and pressure, inside an autoclave. At the end of the reactional process, temperature is cooldown to room temperature, and the supernatant solution is removed. By adjusting temperature, reactional time, and precursor/coating ratio, SPIONs with controlled size, shape, and magnetic properties are obtained. SPIONs with different morphologies were synthesized using ethylene glycol. The lowest saturation magnetization (66 emu/g) was obtained for flower-like NPs [24]. The disadvantages of the hydrothermal method are the moderate crystallinity of the produced SPIONs and time-consuming process [34].

SPIONs with different sizes and shapes can be produced via *thermal decomposition* (with high boiling points surfactants and organic solvents). Solvent-free thermal decomposition proved to be efficient [34]. The SPIONs physicochemical characteristics and the magnetic properties are modulated by modifying the synthesis parameters such as reaction temperature, surfactants concentration, precursor/surfactants ratio, solvents, and heating rate during reflux and reaction temperature [38,39].

Considering that dopamine shows more affinity to attach with the surface of iron oxide NPs comparing to hydrophobic coatings, Xu et al. [40] developed through *ligand exchange method*, a hydrophilic coating of dopamine attached polyethylene glycol (PEG). To form bilayers on the magnetic NPs surface after concentration optimization, oleic acid was used. However, these conversion methods affect the magnetic properties, colloidal dispersity, and synthesis yield. To solve these problems, hydrophilic SPIONs can be synthesized by one-pot thermolysis method using surfactants and/or solvents [34].

Through the *sonochemical method*, SPIONs are synthesized at room temperature, using the surface of cavitation bubbles produced by ultrasound. Refluxing, irradiation time, and power are used to control SPIONs' size and shape. By increasing the reactional time, at constant frequency (581 kHz) [and comparing with other frequencies (861 and 1141 kHz)], a linear increase of $Fe^{3+}$ was observed with surfactant PEG [34].

Another way to synthesize SPIONs, in short time and cost-effective way, is through *microwave energy*. This methodology consists in providing microwave heating source inside the reaction vessel, promoting the SPIONs synthesis. SPIONs produced with microwave present a low energy surface crystalline facet (which confer a reduced surface reactivity but enhance SPIONs stability) [41]. The surfactants concentration, microwave power, and reaction time influence SPIONs magnetic properties [34].

### 4.4.4 Superparamagnetic iron oxide nanoparticles coating: protection, stabilization and functionalization

Naked SPIONs are stable in high- and low-pH suspensions. However, their rapid aggregation tendency and high surface oxidation in the physiological environment (pH = 7.4) restore its use. At neutral pH, SPIONs are very reactive (due to the huge surface area to volume ratio, high surface energy, and reactivity, with magnetic and long-range attractive van der Waals forces). Aggregation and oxidation decrease SPIONs magnetic performance, biocompatibility but increase toxicity. To overcome these drawbacks, SPIONs surface needs to be coated, which is usually performed with surfactants/capping agents/polymers.

The coupling is performed via end functional groups (either by electrostatic interactions or covalent bonding) increasing the thickness of the new nanostructure. Moreover, it is also important to mention that the surface coating, depending on their nature, amount, composition, and thickness, may affect the inherent magnetic properties of SPIONs. For example, polyoxyethylene (5) nonylphenylether with -OH end group, oleic acid (with either inorganic or organic surfactants), and PEG and dextran polymers (with shorter chains) do not affect the SPIONs magnetization performance [12,42,43], although PEGs and dextrans with longer chains significantly decrease the coated particle magnetization values [12].

Coating is the most commonly used method for SPIONs surface modification, allowing the conjugation of inorganic or organic materials on the surface. Besides preventing SPIONs surface oxidation and/or aggregation, the coating provides the possibility of functionalization, making them promising candidates for biomedicine applications [34,44–46].

There are different vias for SPIONs' coating: in situ coating, postsynthesis adsorption, and postsynthesis grafting. In in situ coating, as in the postsynthesis adsorption methods, occurs uniform encapsulation of SPIONs. In the postsynthesis grafting method, polymer end groups are coupled to SPIONs surface. Coating or grafting are processes used to make SPIONs more stable. However, it is important to highlight that when nonmagnetic materials are used as coating a decrease in saturation magnetization values may occur [47].

The surface modification gives origin to four main different types of nanostructures. Fig. 4–12 shows these different types of iron oxide nanostructures.

The agent most commonly used for the SPIONs surface modification is silica due to their exceptional properties [49]. Table 4–8 summarizes these methodologies, their advantages, and disadvantages.

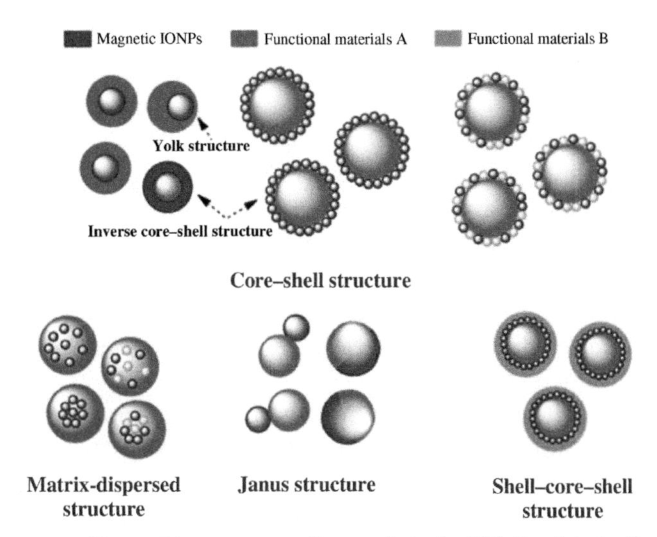


**FIGURE 4–12** Different types of iron oxide nanocomposite morphologies [48]. *Reprinted with permission from W. Wu, Z. Wu, T. Yu, C. Jiang, W.-S. Kim, Recent progress on magnetic iron oxide nanoparticles: Synthesis, surface functional strategies and biomedical applications, Science and technology of advanced materials 16 (2015) 023501.*

**Table 4–8** Advantages and disadvantages of silica-coated superparamagnetic iron oxide nanoparticles synthesis [46].

| Synthesis method | Advantages | Disadvantages |
|---|---|---|
| Stöber method | Controllable silica shell and uniform size, high crystallinity | Lack of understanding of its kinetics and mechanism |
| Microemulsion | Control of the particle size, high homogeneous | Poor yield, large amounts of solvent required, and time-consuming |
| Aerosol pyrolysis | Hermetically coated | Complex experimental conditions |
| Methods based on sodium silicate solution | Crystallinity control and surface area | Depends on preparation method |

The most common approach to synthesize coated silica SPIONs is the classical Stöber method. When SPIONs coating is envisaged, the traditional seed (sodium silicate glass) is replaced by SPIONs. Briefly, SPIONs are uniformly dispersed in ethanol solution and then TEOS is added, followed by the ammonia addition. The ammonia used as catalyst allows size control and inhibits the hydrolysis forming NPs with identical morphology. The particles size increases with the amount of water, TEOS, and ammonia in the reactional medium. Also, increasing the reaction temperature, the silica NPs' sizes increases slightly [50]. Silica-coated SPIONs show columbic repulsion in a colloidal solution, allowing the nanostructures to maintain their chemical stability for longer periods of time. The silanol surfaces can also be modified with other coupling agents that can improve the ability of the magnetic nanosystems to transport small molecules of interest (drugs, dyes, or proteins). It was reported an accumulation of PEGylated silane-coated magnetic iron oxide NPs in murine tumors [47].

This coating also influences the SPIONs magnetic properties due to the difference of thickness of the spin canting surface layer of SPIONs. A study was published where SPIONs were coated with silica and (3-aminopropyl)triethoxysilane, and the obtained magnetization values were, respectively, ~20–30 and 77.7 emu/g [51–53]. Another study, using coated SPIONs with amino–silanes, shows an improvement in cellular uptake efficiency and a decrease in the toxicity comparing with naked SPIONs, silica coated, or dextran coated. This improvement was registered for different cell lines (mouse mesenchymal stem cells, L929, HepG2, PC-3, among others) [54].

The saturation magnetization values of SPIONs coated with nonmagnetic materials suffer a decrease comparing with the values registered in naked SPIONs. Porous silica enhances the ability of drugs transport by the magnetic nanosystems; however, their magnetic properties are degraded. It was reported a reduction on the magnetization values from 85 to 64.7 emu/g or from 79 to 38 emu/g when SPIONs are functionalized with silica. This decrease is even more accentuate when, over silica coating, other subsequent functionalization is made [35].

Ellipsoidal shaped SPIONs coated with silica and mesoporous silica for subsequent functionalization with gold allow a core–shell structure organization for further inclusion of chemotherapeutic drugs for cancer therapy [55].

Another type of coating was performed using polymers. These structures present some important characteristics to make SPIONs suitable as drug delivery vehicles. A natural polymer widely used to SPIONs coating is the dextran. This polymer improves the biocompatibility and biodegradability of the nanosystem, being strongly absorbed on SPIONs surface through hydrogen bonds formed between the hydroxyl groups of the polymer chains and the SPIONs surface cores. Dextran coating or dextran derivatives such as carboxydextran and carboxymethyl dextran have been used in the development of preclinical MRI CAs [47]. Other natural polymers that can be used for SPIONs stabilization are, for example, chitosan, alginate, and pullulan [47].

PEG is a synthetic polymer widely used for nanosystems coating. This is a biocompatible polymer that improves the hydrophilicity, the blood circulation time, and biocompatibility of nanocarriers. It was reported an increase of the $r_2$ relaxivities and image contrast in MRI by using poly(e-caprolactone)-*b*-PEG copolymers surrounding the SPIONs core [49].

Other compounds widely used for incorporation on SPIONs surface are the liposomes. Liposomes are small artificial sphere-shape vesicles that can be developed from cholesterol and natural nontoxic phospholipids. These consist in one or more phospholipid bilayers and were described for the first time in the mid-1960s. Liposomes are very promising structures for drug delivery systems. They are biocompatible, present hydrophilic and hydrophobic character, and they also present a proper size for biomedical applications. The liposome properties are highly dependent of their preparation, lipid composition, surface charge, and size.

The liposomes can be associate with SPIONs by direct encapsulation of the latter within the former or SPIONs can be embed between the lipid bilayer.

The final application determines the spatial location of SPIONs in the liposomes structure. For example, if the SPIONs plus liposomes structure is to be used as drug carrier, it is better to embed the SPIONs between the lipid bilayer so that there is no risk of the drug being affected even before the membrane becomes permeable. However, for MRI tracking, it seems to be beneficial to encapsulate directly the SPIONs in the lumen of liposomes [37].

## 4.4.5 Core–shell structures and their properties

Core–shell NPs can be defined as a structure comprising an inner material, the core, and an outer layer material, the shell. Several combinations of core–shell structures can be made: inorganic–inorganic, inorganic–organic, organic–inorganic, and organic–organic materials. The ideal combination depends on the application. Fig. 4–13 shows different classes of core–shell NPs.

The structure complexity is increased by adding shells, which leads to multishell or onion-like (number of shells superior to two or three) NPs. These structures allow the protection of the deeper layers, and the use of the interface between the core and shell layers that are not compatible with each other. Besides that, both single or multishell NPs can also have solid particle/particles in the central cavity.

## 4.4.6 Biocompatibility and toxicity of ceramic nanoparticles

The nanosystems toxicity is one of the biggest issues concerning the use of NPs in pharmacy and medicine. The NPs high surface energy and high surface reactivity, along with the similarity in size with cellular organelles, may have hazardous effects on cells or tissues [56,57]. Some research works have shown that most of tested NPs have caused oxidative stress and

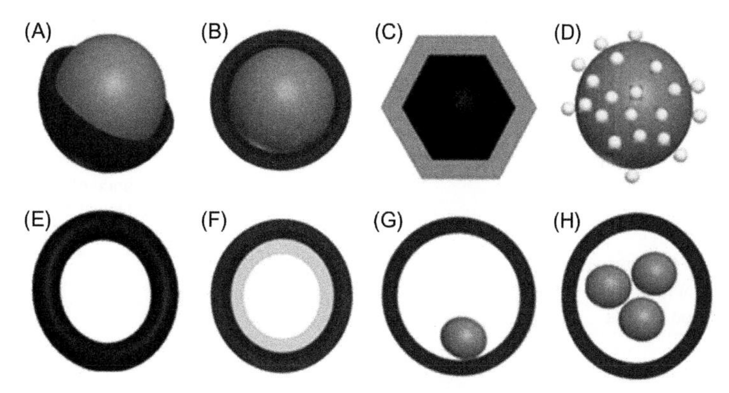


**FIGURE 4–13** Schematic pictures of different structures of core–shell NPs: (A) core–shell NPs, (B) core double-shell particles or core multishell NPs, (C) polyhedral core–shell NPs, (D) core porous–shell NPs, (E) hollow-core shell NPs or single-shell NPs, (F) hollow-core double-shell NPs, (G) moveable-core shell NPs, and (H) multicore shell NPs. *NPs*, Nanoparticles.

inflammation by the reticuloendothelial system. The toxicity promoted by ceramic NPs is different from tissues to cells, and the effects on inflammatory and immunological systems may include oxidative stress or cytotoxic activity in the lungs, liver, heart, and brain among others problems [58,59]. Size, chemical composition/structure, surface chemistry, free radical formation, and dosage are factors that can determine NPs toxicity.

Size of NPs is a very important parameter since it has been observed that the smaller the size of the nanocarriers, the greater their surface area and therefore the greater number of available interactions with the NPs and consequently their toxic effects [60]. Particles smaller than 500 nm can enter in the circulatory system, NPs in range of 100–300 nm are absorbed by intestinal cells, while NPs around 100 nm are essentially absorbed in the lymphatic tissue (Fig. 4–14).

Another important feature to estimate the potential toxicity of NPs is their hydrophobicity versus hydrophilicity character. The degree of molecular absorption interferes in the capacity to penetrate the cell membranes and contribute to NPs' toxicity. NPs-containing

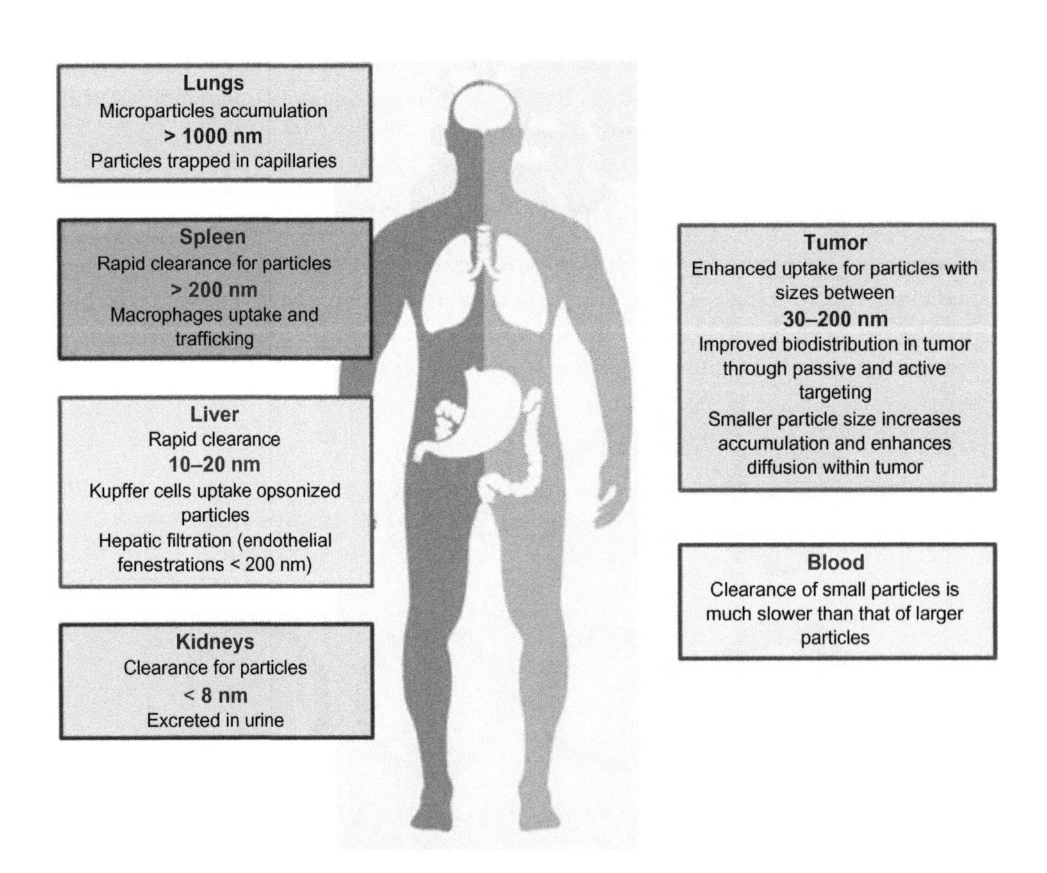


**FIGURE 4–14** NPs in human body. *NPs*, Nanoparticles.

hydrophobic polymers are much more absorbed by cells than the ones containing hydrophilic polymers [61].

Other important characteristic is the existence of reactive species on the NPs surface, which may increase cellular toxicity. One example is the toxicity induced by crystalline silica NPs due to their interactions with reactive oxidative species that can cause lung cancer.

Surface chemical composition is other characteristic that can influence the toxicity caused by the nanocarriers. Modifying their surface chemical composition, the surface structure changes which can also change the nanocarriers properties. One example of this is what happens with SPIONs. It was observed that varying their coating, their toxicity/cytotoxicity also varies. This implies that the surface modification plays a direct role in the nanocarriers toxicity [62]. Also, free radicals of particle surface can cause oxidative stress in cells that gives rise to inflammation, genotoxicity, and cell destruction. The NPs dose in a specific site can also influence their toxicity in human body. It was seen that higher concentrations of NPs (smaller or bigger) could be harmful to health [63]. The interactions between the nanomaterials and the biological systems are very complexes and have a great dependence of the physicochemical properties, such as size, charge, composition, and surface properties, of the former. Slight variations on those characteristics can lead to radically active interactions with living systems, which can affect the biocompatibility, stability, biological performance, and side effects of the nanomaterial [64].

The nanoplatforms size could affect the systemic and lymphatic distribution, tumor penetration, and cellular internalization of the agent of interest and thus affect the nanoplatform performance. Reports have shown that the interendothelial junctions of healthy tissues are smaller than 8 nm, whereas in tumoral tissues, the size range is much higher, from 40 to 80 nm. The size difference between healthy and tumoral tissues should be considered and used since the drug carriers are preferentially uptake via the enhanced permeability and retention effect [20].

Ceramic NPs can easily enter in cells due to their nanosize; however, it is also their size, morphology, and composition that influence the cellular uptake, subcellular localization, and the capacity to catalyze oxidative products. The cellular uptake is essentially made by passive uptake or adhesive interaction, probably due to van der Waals forces, electrostatic charges, steric interactions, or interfacial tension effects that do not origin vesicles [65,66]. This free movement of NPs within the cell gives them direct access to cytoplasm components such as proteins and other organelles, which makes them very dangerous. These can be found in diverse locations inside the cells, such as the outer-cell membrane cytoplasm, mitochondria, lipid vesicles along the nuclear membrane or within the nucleus, and depending on their localization, they can damage DNA or organelles or even cause cell death [67].

Nowadays, the NPs are developed with increasing regard for their biocompatibility due to an improved understanding of the biological impacts. However, considering the mentioned earlier is crucial a deep understanding of the potential risks associated with the exposure to NPs and the effect of several surface coatings used for functionalization. At the present the biocompatibility is measured considering mainly the extent of cytotoxicity observed, thus it is urgent to clearly define criteria to assess the toxicity of NPs more accurately.

# References

[1] M.C. Gonçalves, B. Martins, in: N. Ali, A. Seifalian (Eds.), Multifunctional Core-Shell Nanostructures Nanomedicine, One Central Press Office, Manchester, 2014, pp. 83–110.

[2] M. Mahmoudi, A.M. Sahraian, M.A. Shokrgozar, S. Laurent, Superparamagnetic iron oxide nanoparticles: Promises for diagnosis and treatment of multiple sclerosis, ACS Chem. Neurosci. 2 (2011) 118–140.

[3] F. Chen, G. Hableel, E.R. Zhao, J.V. Jokerst, Multifunctional nanomedicine with silica: role of silica in nanoparticles for theranostic, imaging, and drug monitoring, J. Colloid Interface Sci. 521 (2018) 261–279.

[4] S.S. Mohapatra, S. Ranjan, N. Dasgupta, R.K. Mishra, S. Thomas (Eds.), Applications of Targeted Nano Drugs and Delivery Systems, Nanoscience and Nanotechnology in Drug Delivery, Elsevier, 2019.

[5] D. Singh, P. Dubey, M. Pradhan, M. Rawat, Ceramic nanocarriers: versatile nanosystem for protein and peptide delivery, Expert. Opin. Drug Deliv. 10 (2013) 241–259.

[6] J.C. Matos, A.R. Soares, I. Domingues, G.A. Monteiro, M.C. Gonçalves, ORMOPLEXEs for gene therapy: In vitro and in vivo assays, Mater. Sci. Eng.: C 63 (2016) 546–553.

[7] E.-K. Lim, T. Kim, S. Paik, S. Haam, Y.-M. Huh, K. Lee, Nanomaterials for theranostics: recent advances and future challenges, Chem. Rev. 115 (1) (2015) 327–394.

[8] S.C. Thomas, Harshita, P.K. Mishra, S. Talegaonkar, Ceramic nanoparticles: fabrication methods and applications in drug delivery, Curr. Pharm. Des. 21 (42) (2015) 6165–6188.

[9] C.N.R. Rao, A. Müller, A.K. Cheetham, The Chemistry of Nanomaterials: Synthesis, Properties and Applications, vol. 1, Wiley-VCH Verlag Gmbh & Co., Weinheim, 2004. 1–10, 17–27, 94–110.

[10] R. Diab, N. Canilho, I.A. Pavel, F.B. Haffner, M. Girardon, A. Pasc, Silica-based systems for oral delivery of drugs, macromolecules and cells, Adv. Colloid Interface Sci. 249 (2017) 346–362.

[11] M.L. Foglia, G.S. Alvarez, P.N. Catalano, A.M. Mebert, L.E. Diaz, T. Coradin, et al., Recent patents on the synthesis and application of silica nanoparticles for drug delivery, Recent Pat. Biotechnol. 5 (1) (2011) 54–61.

[12] J.C. Matos, M.C. Gonçalves, L.C.J. Pereira, B.J.C. Vieira, J.C. Waerenborgh, SPIONs prepared in air through improved synthesis methodology: the influence of g-$Fe_2O_3$/$Fe_3O_4$ ratio and coating composition on magnetic properties, Nanomaterials 9 (7) (2019) 943.

[13] J. Lodhia, G. Mandarano, N.J. Ferris, P. Eu, S. Cowell, Development and use of iron oxide nanoparticles (Part 1): synthesis of iron oxide nanoparticles for MRI, Biomed. Imaging Intervention J. 6 (2) (2010). Available from: https://doi.org/10.2349/biij.6.2.e12.

[14] M.C. Gonçalves, Sol-gel silica nanoparticles in medicine: a natural choice. Design, synthesis and products, Molecules 23 (2018) 2021.

[15] E. Bagheri, L. Ansari, K. Abnous, S.M. Taghdisi, F. Charbgoo, M. Ramezani, et al., Silica based hybrid materials for drug delivery and bioimaging, J. Control. Release 277 (2018) 57–76.

[16] M.H. Wakamatsu, R. Salomão, Ceramic nanoparticles: what else do we have to know? Int. Ceram. Rev. 59 (2010) 28.

[17] G. Kolbe, Das komplexchemische verhalten der kieselsäure (Ph.D. Thesis), 1956.

[18] W. Stöber, A. Fink, E. Bohn, Controlled growth of monodisperse silica spheres in micron size range, J. Colloid Interface Sci. 26 (1) (1968) 62–69.

[19] J.C. Matos, I. Avelar, M.B.F. Martins, M.C. Gonçalves, Greensilica® vectors for smart textiles, Carbohydr. Polym. 156 (2017) 268–275.

[20] Z. Xu, X. Ma, Y.-E. Gao, M. Hou, P. Xue, C.M. Li, et al., Multifuctional silica nanoparticles as a promising theranostic platform for biomedical applications, Mater. Chem. Front. 1 (2017) 1257.

[21] V. Marzaioli, C.J. Groß, I. Weichenmeier, C.B. Schmidt-Weber, J. Gutermuth, O. Groß, et al., Specific surface modifications of silica nanoparticles diminish inflammasome activation and in vivo expression of selected inflammatory genes, Nanomaterials 7 (11) (2017) 355.

[22] X. Wu, M. Wu, J.X. Zhao, Recent development of silica nanoparticles as delivery vectors for cancer imaging and therapy, Nanomedicine 10 (2) (2014) 297–312.

[23] J.W. Kim, L.U. Kim, C.K. Kim, Size control of silica nanoparticles and their surface treatment for fabrication of dental nanocomposites, Biomacromolecules 81 (2007) 215–222.

[24] E. Poorakbar, A. Shafiee, A.A. Saboury, B.L. Rad, K. Khoshnevisan, L. Ma'mani, et al., Synthesis of magnetic gold mesoporous silica nanoparticles core shell for cellulase enzyme immobilization: improvement of enzymatic activity and thermal stability, Process. Biochem. 71 (2018) 92–100.

[25] M. Liong, J. Lu, F. Tamanoi, J.I. Zink, A. Nel, Mesoporous Silica Nanoparticles for Biomedical Applications, Google Patents, 2018.

[26] A.E. Nel, L. Mädler, D. Velegol, T. Xia, E.M. Hoek, P. Somasundaran, et al., Understanding biophysicochemical interactions at the nano–bio interface, Nat. Mater. 8 (7) (2009) 543–557.

[27] R. Narayan, U.Y. Nayak, A.M. Raichur, S. Garg, Mesoporous silica nanoparticles: a Comprehensive review on synthesis and recent advances, Pharmaceutics 10 (2018) 118.

[28] S.A. Wahajuddin, Superparamagnetic iron oxide nanoparticles: magnetic nanoplatforms as drug carriers, Int. J. Nanomed. 2012 (7) (2012) 3445–3471.

[29] M.R. Faria, M.M. Cruz, M.C. Gonçalves, A. Carvalho, G. Feio, M.B.F. Martins, Synthesis and characterization of magnetoliposomes for MRI contrast enhancement, Int. J. Pharm. 446 (2013) 183–190.

[30] K.M. Krishnan, Biomedical nanomagnetics: a spin through possibilities in imaging, diagnostics, and therapy, IEEE Trans. Magn. 46 (2010) 2523–2558.

[31] V. Kuncser, P. Palade, A. Kuncser, S. Greculeasa, G. Schinteie, Engineering magnetic properties of nanostructures via size effects and interphase interactions, in: V. Kuncser, L. Miu (Eds.), Size Effects in Nanostructures, Springer, Berlin, 2014, pp. 169–237.

[32] E. Tombácz, R. Turm, V. Socoliuc, L. Vékás, Magnetic iron oxide nanoparticles: recent trends in design and synthesis of magnetoresponsive nanosystems, Biochem. Biophys. Res. Commun. 468 (2015) 442–453.

[33] G.F. Goya, V. Grazú, M.R. Ibarra, Magnetic nanoparticles for cancer therapy, Curr. Nanosci. 4 (2008) 1–16.

[34] G. Kandasamy, D. Maity, Recent advances in superparamagnetic iron oxide nanoparticles (SPIONs) for in vitro and in vivo cancer nanotheranostics, Int. J. Pharm. 496 (2015) 191–218.

[35] L. Wortmann, S. Ilyas, D. Niznansky, M. Valldor, K. Arroub, N. Berger, et al., Bioconjugated iron oxide nanocubes: synthesis functionalization, and vectorization, ACS Appl. Mater. Interfaces 6 (2014) 16631–16642.

[36] W. Chen, P. Yi, Y. Zhang, L. Zhang, Z. Deng, Z. Zhang, Composites of aminodextran-coated $Fe_3O_4$ nanoparticles and graphene oxide for cellular magnetic resonance imaging, ACS Appl. Mater. Interfaces 3 (2011) 4085–4091.

[37] C.A. Monnier, D. Burnand, B. Rothen-Rutishauser, M. Lattuada, A. Petri-Fink, Magnetoliposomes: opportunities and challenges, Eur. J. Nanomed. 6 (4) (2014) 201–215.

[38] D. Maity, S.G. Choo, J. Yi, J. Ding, J.M. Xue, Synthesis of magnetite nanoparticles via a solvent-free thermal decomposition route, J. Magn. Magn. Mater. 321 (2009) 1256–1259.

[39] D. Maity, J. Ding, J.-M. Xue, Synthesis of magnetite nanoparticles by thermal decomposition: time, temperature surfactant and solvent effects, Funct. Mater. Lett. 1 (3) (2008) 189–193.

[40] C. Xu, J. Xie, N. Kohler, E.G. Walsh, Y.E. Chin, S. Sun, Monodisperse magnetite nanoparticles coupled with nuclear localization signal peptide for cell-nucleus targeting, Chemistry—Asian J. 3 (2008) 548–552.

[41] E. Carenza, V. Barcelo, A. Morancho, J. Montaner, A. Rosell, A. Roig, Rapid synthesis of water-dispersible superparamagnetic iron oxide nanoparticles by a microwave-assisted route for safe labeling of endothelial progenitor cells, Acta Biomater. 10 (2014) 3775–3785.

[42] M. Darbandi, F. Stromberg, J. Landers, N. Reckers, B. Sanyal, W. Keune, et al., Nanoscale size effect on surface spin canting in iron oxide nanoparticles synthesized by the microemulsion method, J. Phys. D: Appl. Phys. 45 (2012) 195001.

[43] C. De Montferrand, L. Hu, Y. Lalatonne, N. Lièvre, D. Bonnin, A. Brioude, et al., $SiO_2$ versus chelating agent@ iron oxide nanoparticles: interactions effect in nanoparticles assemblies at low magnetic field, J. Sol-Gel Sci. Technol. 73 (2014) 572–579.

[44] A. Mohammadi, M. Barikani, M. Barmar, Effect of surface modification of $Fe_3O_4$ nanoparticles on thermal and mechanical properties of magnetic polyurethane elastomer nanocomposites, J. Mater. Sci. 48 (2013) 7493–7502.

[45] S.N. Sun, C. Wei, Z.Z. Zhu, Y.L. Hou, S.S. Venkatraman, Z.C. Xu, Magnetic iron oxide nanoparticles: synthesis and surface coating techniques for biomedical applications, Chin. Phys. B 23 (3) (2014).

[46] N. Zhu, H. Ji, P. Yu, J. Niu, M.U. Farooq, M.W. Akram, et al., Surface modification of magnetic iron oxide nanoparticles, Nanomaterials 8 (2018) 810.

[47] S. Laurent, A.A. Saei, S. Behzadi, A. Panahifar, M. Mahmoudi, Superparamagnetic iron oxide nanoparticles for delivery of therapeutic agents: opportunities and challenges, Expert. Opin. Drug Deliv. 11 (9) (2014) 1449–1470.

[48] W. Wu, Z. Wu, T. Yu, C. Jiang, W.-S. Kim, Recent progress on magnetic iron oxide nanoparticles: Synthesis, surface functional strategies and biomedical applications, Sci. Technol. Adv. Mater. 16 (2015) 023501.

[49] M. Abbas, B.P. Rao, M.N. Islam, S.M. Naga, M. Takahashi, C. Kim, Highly stable-silica encapsulating magnetite nanoparticles ($Fe_3O_4/SiO_2$) synthesized using single surfactantless-polyol process, Ceram. Int. 40 (2014) 1379–1385.

[50] L. Zhao, J.G. Yu, B. Chang, X.J. Zhao, Preparation and formation mechanism of monodispersed silicon dioxide spherical particles, Acta Chim. Sin. 61 (2003) 562–566.

[51] B.K. Sodipo, A.A. Aziz, Non-seeded synthesis and characterization of superparamagnetic iron oxide nanoparticles incorporated into silica nanoparticles via ultrasound, Ultrason. Sonochem. 23 (2015) 354–359.

[52] B.K. Sodipo, A.A. Aziz, A sonochemical approach to the direct surface functionalization of superparamagnetic iron oxide nanoparticles with (3-aminopropyl)triethoxysilane, Beilstein J. Nanotechnol. 5 (2014) 1472–1476.

[53] M.N. Islam, M. Abbas, B. Sinha, J.-R. Joeng, C. Kim, Silica encapsulation of sonochemically synthesized iron oxide nanoparticles, Electron. Mater. Lett. 9 (2013) 817–820.

[54] X.-M. Zhu, Y.-X.J. Wang, K.C.-F. Leung, S.-F. Lee, F. Zhao, D.-W. Wang, et al., Enhanced cellular uptake of aminosilane-coated superparamagnetic iron oxide nanoparticles in mammalian cell lines, Int. J. Nanomed. 7 (2012) 953–964.

[55] Y. Chen, H. Chen, D. Zeng, Y. Tian, F. Chen, J. Feng, et al., Core/shell structured hollow mesoporous nanocapsules: a potential platform for simultaneous cell imaging and anticancer drug delivery, ACS Nano 4 (2010) 6001–6013.

[56] G.E. Marchant, Small is beautiful: what can nanotechnology do for personalized medicine? Curr. Pharmacogenomics Personalized Med. 7 (2009) 231–237.

[57] A.A. Shvedova, V.E. Kagan, B. Fadeel, Close encounters of the small kind: adverse effects of man-made materials interfacing with the nano-cosmos of biological systems, Annu. Rev. Pharmacol. Toxicol. 50 (2010) 63–88.

[58] C. Muhlfeld, P. Gehr, B. Rothen-Rutishauser, Translocation and cellular entering mechanisms of nano-particles in the respiratory tract, Swiss Med. Wkly. 138 (2008) 387–391.

[59] N.R. Yacobi, H.C. Phuleria, L. Demaio, C.H. Liang, C.A. Peng, C. Sioutas, Nanoparticle e effects on rat alveolar epithelial cell mono-layer barrier properties, Toxicol. Vitro 21 (2007) 1373–1381.

[60] I. Linkov, F.K. Satterstrom, L.M. Corey, Nanotoxicology and nanomedicine: making hard decisions, Nanomed. Nanotechnol. Biol. Med. 4 (2008) 167–171.

[61] R.P. Schins, R. Du, D. Hohr, A.M. Knaapen, T. Shi, C. Weishaupt, Surface modification of quartz inhibits toxicity, particle uptake, and oxidative DNA damage in human lung epithelial cells, Chem. Res. Toxicol. 15 (2002) 1166–1173.

[62] Q. Feng, Y. Liu, J. Huang, K. Chen, J. Huang, K. Xiao, Uptake, distribution, clearance, and toxicity of iron oxide nanoparticles with different sizes and coatings, Sci. Rep. 8 (2018). Available from: https://doi.org/10.1038/s41598-018-19628-z.

[63] S. Singh, T. Shi, R. Du, C. Albrecht, D. van Berlo, D. Hohr, Endocytosis, oxidative stress and IL-8 expression in human lung epithelial cells upon treatment with ne and ultra ne $TiO_2$: role of the specific surface area and of surface methylation of the particles, Toxicol. Appl. Pharmacol. 222 (2007) 141–151.

[64] S. Harper, C. Usenko, J.E. Hutchison, B.L.S. Maddux, R.L. Tanguay, In vivo biodistribution and toxicity depends on nanomaterial composition, size, surface functionalization and route of exposure, J. Exp. Nanosci. 3 (2008) 195–206.

[65] T. Xia, M. Kovochich, J. Brant, M. Hotze, J. Sempf, T. Oberley, Comparison of the abilities of ambient and manufactured nanoparticles to induce cellular toxicity according to an oxidative stress paradigm, Nano Lett. 6 (2006) 1794–1807.

[66] M. Geiser, B. Rothen-Rutishauser, N. Kapp, S. Schurch, W. Kreyling, H. Schultz, Ultrafine particles cross cellular membranes by nonphagocytotic mechanisms in lungs and in cultured cells, Environ. Health Perspect. 113 (2005) 1555–1560.

[67] D. Singh, S. Singh, J. Sahu, S. Srivastava, M.R. Singh, Ceramic nanoparticles: Recompense, cellular uptake, and toxicity concerns, Artif. Cells Nanomed. Biotechnol. 44 (2016) 401–409.

5

# Sonochemical encapsulation of taxifolin into cyclodextrine for improving its bioavailability and bioactivity for food

Irina Kalinina[1], Irina Potoroko[2], Shirish H. Sonawane[3]

[1]SOUTH URAL STATE UNIVERSITY, CHELYABINSK, RUSSIA [2]DEPARTMENT OF FOOD TECHNOLOGY AND BIOTECHNOLOGY, SCHOOL OF MEDICAL BIOLOGY, SUSU, CHELYABINSK, RUSSIA [3]CHEMICAL ENGINEERING DEPARTMENT, NATIONAL INSTITUTE OF TECHNOLOGY, WARANGAL, INDIA

**Chapter Outline**

**5.1 Introduction** ..... 85
5.1.1 Properties of taxifolin and chemical interaction with the components of food matrix ..... 86
5.1.2 Effective delivery strategies for nutriceuticals ..... 89
5.1.3 Studying the properties of taxifolin–β-cyclodextrine conjugates, obtained by sonochemical approach placing in the food matrix ..... 92
**5.2 Conclusions** ..... 100
**Acknowledgments** ..... 100
**References** ..... 100
**Further reading** ..... 102

## 5.1 Introduction

Ultrasonic treatment in the food industry has a wide range of applications. For the most part, it is connected with the demonstrated effects resulting from the collapse of a cavitation bubbles [1–4]. Some of the most useful ultrasound applications are extraction, emulsification, viscosity change, modification of protein, crystallization, sterilization, drying, etc. This chapter reviews the possibilities of ultrasonic treatment during the encapsulation of biologically active substances (BASs), antioxidants, for increasing their bioavailability of the cell systems of the human body.

Encapsulation of Active Molecules and their Delivery System. DOI: https://doi.org/10.1016/B978-0-12-819363-1.00005-3

Natural antioxidants are widely used for the pharmaceutical correction of the oxidative stress [5,6].

Taxifolin is a popular antioxidant this is commonly used for various purposes. Possessing some useful properties taxifolin has certain restrictions on the use in the food industry and biomedicine because of its low solubility and bioavailability. This results in the following problem of the biological effects of taxifolin detected by in vitro analysis cannot be transferred to in vivo conditions [7–11].

Low solubility (around 20 mg/L in the water at ambient temperature) leads to minimal and slow absorption of taxifolin, which is an essential factor restricting its bioavailability [12]. Besides, similar to the majority of flavonoids, taxifolin is chemically labile and quickly degrades when exposed to alkaline conditions [13,14]. This calls for searching the ways to modify the properties of taxifolin to ensure the highest manifestation of its bioactive and pharmacophoric effects.

### 5.1.1 Properties of taxifolin and chemical interaction with the components of food matrix

Using taxifolin for enriching the food products defines the necessity of its adaptation to the multicomponent food matrix of the product. According to its chemical nature, taxifolin is a flavononol, a leader among the most popular antioxidant which including vitamins C, E and β-carotone in large quantity. As part of phenolic compounds (polyphenols), taxifolin is found in numerous plants, berries, fruits, vegetables, edible oils, nuts, medicinal herbs, and plants, and in different types of red wine [9,10].

The material for obtaining taxifolin of high purity is 70–80% raw taxifolin, which is extracted from the roots of Siberian larch-tree (*Larix sibirica* Ledeb.) and Gmelin larch-tree [*Larix gmelinii* (Rupr.)], growing in Russia on the ecological territory of Siberia and the Far East.

Taxifolin belongs to class 6 of the Safety data sheet, indicating that it is nontoxic. Animal tests proved that when administered orally up to 10 g/kg of wet weight, there was no malfunctioning of the animal body. The European Food Safety Authority published a scientific opinion on the safety of applying dihydroquercetin (taxifolin) as a new food ingredient. In Russia, the requirements for dihydroquercetin are regulated by GOST 33504-2015 (Food additives. Dihydroquercetin. Technical conditions).

The biological activity of taxifolin is determined by its native, that is, natural form, due to which the molecule performs its biological destiny. Such a form of taxifolin (Fig. 5–1) presupposes the spatial arrangement of its functional groups (stereochemistry) with the optically active functional components, enantiomers, denoted by the light where the plane-polarized light is rotating.

The molecular structure of taxifolin is determined by the chirality and spatial arrangement of its functional groups with optically active functional components (Fig. 5–2), which are really sensitive to the process of their mutual transformation (racemization), which leads

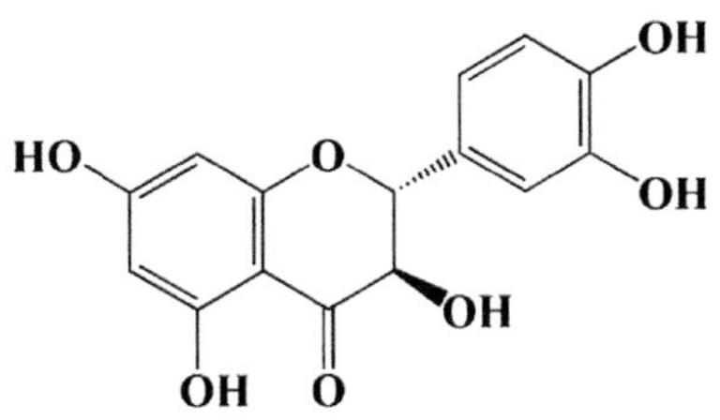

**FIGURE 5–1** Chemical structure of taxifolin.

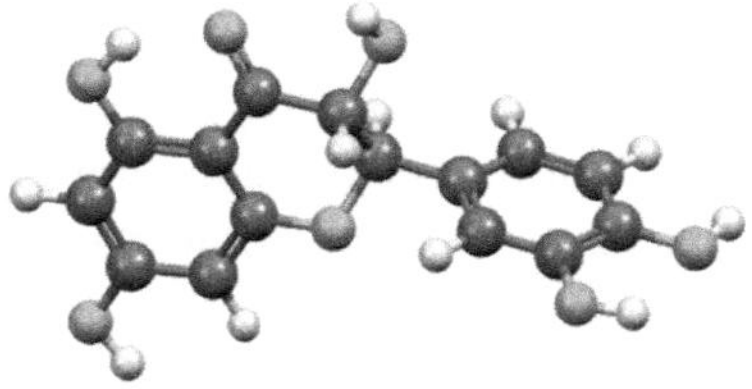

**FIGURE 5–2** Spatial structure of a taxifolin molecule.

to the disappearance of optical activity, the loss of natural, native, and biological activity of taxifolin.

In plants, taxifolin presents a nonenzymatic antioxidant, where the monomeric form that actively participates in the fermentative processes of plant bodies, in the biosynthesis, that is, in the formation of related flavonoid molecules (phenol compounds or polyphenols), hormonal processes of plant growth, which are characteristic of taxifolin metabolic processes directed at the growth and survival of the plant body. In recent decades, taxifolin is at the center of attention due to its unique pleiotropic biological properties.

Research teams from the United States, Croatia, the Netherlands, and Germany have contributed significantly to the development of the theory of interaction between plant antioxidants of polyphenol nature and macronutrients of the food matrix and have proved the importance of this interaction [15–17].

Interactions between polyphenols and food system components are predominantly based on noncovalent hydrophobic interactions [18]. Less frequently, the interaction between proteins and plant phenols can lead to the creation of covalent bonds [19]. Hydrogen bonds are formed predominantly in the case of interaction between polyphenols and proteins [20,21], as well as polyphenols and carbohydrates [22].

Many macromolecules of the food matrix having a complex porous structure can detect polyphenols and as a result, change their availability for absorption. Besides, these interactions with the food matrix can redefine the role of polyphenols. The components of the food system can protect polyphenols from oxidation during their movement through the gastrointestinal tract and deliver them into the intestine more intact [23,24]. Moreover, it is expected that delivering and release of polyphenols can create a positive antioxidant environment in the gastrointestinal tract. Polyphenols as antioxidant compounds, can prevent the oxidation of lipids and vitamins.

The systematization of the available information about the interaction of BASs of polyphenol nature with food matrix components points at the dual interdependent character of this interaction (Fig. 5–3). The consequences of these interactions can be multiple and play both a positive and negative part in the bioavailability of BAS and macromolecules of the food product.

Studies [22–25] have indicated the positive effects of creating polyphenol complexes with macromolecules of the food matrix. Polyphenols can get transported to the lower parts of the gastrointestinal tract, where they free from complex structures by exposing them to various ferments and microorganisms that are naturally present in the colon, which display their bioactive properties.

However, some scholars argue that the bioavailability of polyphenol can get reduced due to the formation of stable complexes with macromolecules, primarily with proteins and carbohydrates.

In any case, it is beyond doubt that the interaction of polyphenols with food ingredients is of great importance and should be taken into account when developing enriched functional and specialized products based on polyphenol BAS.

It is for this reason that the development and using in vitro simulation systems for studying the effect of the food matrix and the process of digestion on BAS properties seems to be of great importance. Using these simulations will lead to better understanding and predicting BAS activity in the human body and consequently will help define their potential biopharmacophoral effects.

Besides, there is a need to develop a proven multilevel test strategy, in vitro, in vivo, and in silico, for assessing the potential BAS benefit for human health.

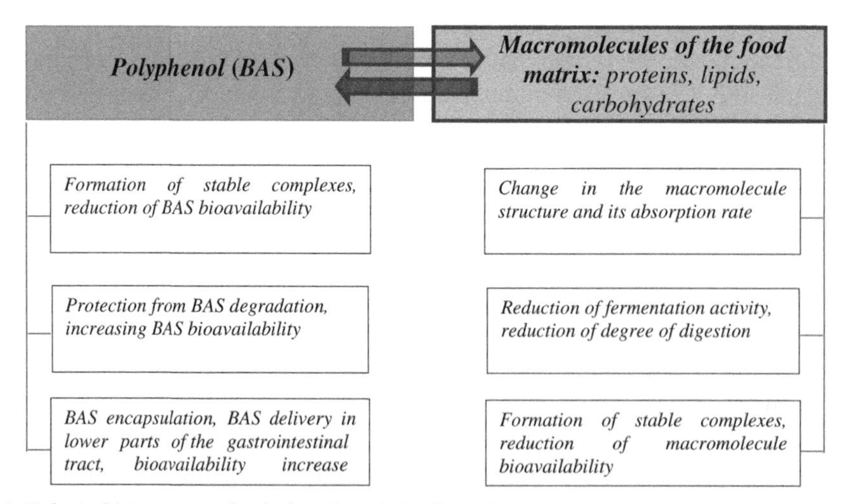


**FIGURE 5–3** Biological interaction of polyphenols with food matrix.

## 5.1.2 Effective delivery strategies for nutriceuticals

Nowadays, several strategies have been developed for improving the bioavailability of nutriceuticals through designing and constructing a food matrix. For this purpose different systems are used (microemulsions, microgels, liposomes, nanoemulsions, multi-emulsions, and microclusters), where each of them has its advantages and downsides.

There exist several requirements for the targeted delivery of nutriceuticals:

- The system is to be made of ingredients with the use of processing methods that are accessible and economically viable.
- The system must be compatible with the food matrix, that is, it should not produce a negative effect on the expiry date, exterior, rheological properties, or taste.
- The delivery system is to be sustainable and functional within the food matrix limits exposed to pH changes, temperature, dehydration, and mechanical aggregation.
- The delivery system must have functional characteristics suitable for specific products; as well as dispersibility, possibility, protection from chemical degradation, taste marking, bioavailability increase, or release control.

The manufacturers of food enriched with nutriceuticals should take into consideration all the factors mentioned above factors at each stage of the technological process (from designing the food matrix to obtaining the finished product).

#### *5.1.2.1 Possibility of encapsulating taxifolin into cyclodextrine for its effective delivery through the food matrix*

This section focuses on using encapsulation processes as an effective instrument for the targeted delivery of an active nutraceutical-taxifolin.

Cyclodextrins (CDs) were applied as a carrier of taxifolin. CDs are complex natural cyclical glucose oligomers, having an approximate toroidal structure or a form of a truncated cone, with a hollowness inside (Fig. 5–4). CD molecules consist of D-(+)-glucopyranose

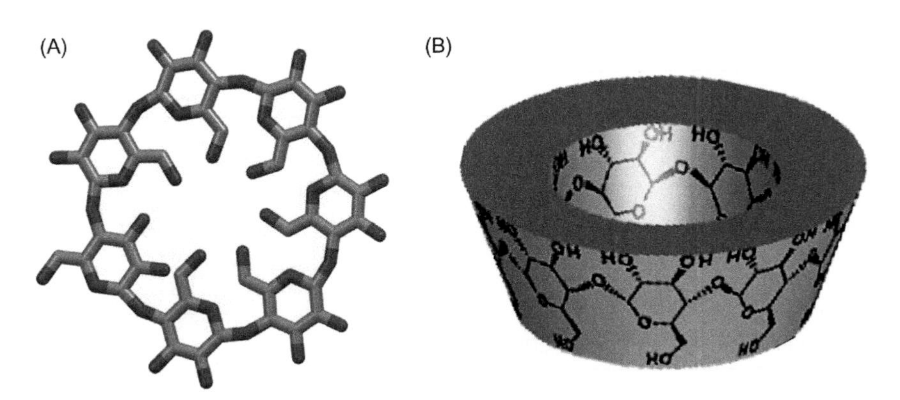


**FIGURE 5–4** Structure of a *βCD* molecule: (A) side view and (B) front view.

chains linked by means of an α-1,4-glucoside bond. The structure of βCD molecule is represented in Fig. 5.4 [26,27].

The structure of CD for using it as a container is its main advantage. The internal cavity of CD has hydrophobic properties that determine its ability to form water-soluble complexes with hydrophobic compounds.

CD consisting of six glucopyranose residues are called αCD; seven glucopyranose residues are referred to as βCD; eight glucopyranose residues as γCD; etc. Although the most popular are CDs having 13 and more chains, and the most available and obtained under industrial conditions are αCD, βCD, and γCD [8]. The main parameters of CD molecules and their physical and chemical properties for practical implementation are given in Table 5–1.

All CDs present colorless crystal substances that melt with decomposition at 260°C–300°C [31,32].

βCD has a higher decomposition temperature in comparison with other CDs. The poor solubility of βCD greatly determines the stability of its conjugates in the air because of the slower speed of liquid phase CD formation as a result of the absorption of water vapor. βCD has the longest storage time of its conjugates compared to α- and γCDs [28].

At present, studies are focused on using CD to increase the solubility, bioavailability, and bioactivity of medicines or BASs, as reflected in more than 50,000 articles on this theme. Various approaches have also been proposed for obtaining CD with "guests" of various natural and molecular size: *n*-propanol, *para*-iodophenol, *trans*-cinnamic acid, methylparaben, 4,4′-diaminobiphenyl, curcumin, brazilin, resveratrol, L-arginine, etc. [28–32]. Most nutriceuticals in conjugates with complex ethereal bonds are released in the colon, whereas in the upper parts of the gastrointestinal tract, they are slightly hydrolyzed [33].

In food technology, CDs are authorized food additives and are used as stabilizers. βCD is obtained from starch processed by using CD transferase. Based on the analysis of the available data, we chose βCD as a nutriceutical carrier for the delivery of taxofolin.

**Table 5–1** Main characteristics of cyclodextrins [26–30].

| Property | αCD | βCD | γCD |
|---|---|---|---|
| Number of glucopyranose chains | 6 | 7 | 8 |
| Molecular weight (Da) | 972 | 1135 | 1297 |
| External diameter (Å) | 14.6 | 15.4 | 17.5 |
| Internal diameter (Å) | 4.7–5.3 | 6.0–6.5 | 7.9–8,3 |
| Height (Å) | 7,9 | 7.9 | 7.9 |
| Water solubility at 25°C (g/100 mL) | 14,5 | 1,85 | 23.2 |
| Decomposition temperature (°C) | 278 | 299 | 267 |
| Maximum water content in a hydrate (mol/mol CD) | 6.1 | 12.3 | 15.5 |

*CD*, Cyclodextrine.

### *5.1.2.2 Simulation of taxifolin complexes with β-cyclodextrine based on quantum chemical calculations*

Understanding chemism and the mechanism of nutriceutical incorporation into the food matrix based on quantum chemical calculations will ensure its sustainability and effectiveness of its properties in the transportation process in the human body.

We conducted molecular simulation via open access source www.chemosophia.com and international databases. QSAR-analysis and MOPS algorithm were the main tools during the simulation. MOPS algorithm searches for the most thermodynamically stable structures of the complexes were carried out by the iterative analytical sorting of newly generated set complexes, and the sorting is already described [34–38]. Due to this method, the most thermodynamically stable complex structure corresponds to global energy minimum, while energy measuring of the structures and evaluation of Hessian vibrational modes are carried out based on the continual properties of the solvent in MERA force field.

We simulated taxifolin complexes with βCD using the MOPS algorithm. The structures of their main components were taken from the London South Bank University database. After we examined the structure of βCD, we revealed the inside cavity, which makes it possible in the encapsulation of low-molecular weight components of the mixtures in an aquatic environment. In other words, the βCD structure makes it possible for the penetration of substances with dimensions corresponding to cavity size. The inner part of βCD consists of oxygen atoms of glycosidic bonds and hydrogen atoms of glucose CH-group; therefore, it is better to perform encapsulation by making hydrophobic interactions with a molecule integrated inside the cavity.

Taking into account the van der Waals radius of inner part atoms, the dimensions of the βCD cavity correspond well to the dimensions of dihydroxyphenyl substituent taxifolin, which make it possible for a complete or partial (with dihydroxyphenyl substituent) penetration of taxifolin.

The simulation of taxifolin complexes with βCD shows that taxifolin fits into the βCD cavity with dihydroxyphenyl substituent that tightly shapes to the inner part of the cavity atoms, forming a series of hydrophobic interactions with CH-fragments of monomeric glucose links. Hydrophilic OH-groups of dihydroxyphenyl substituents are located outside, which makes possible their hydration, increasing the solubility of taxifolin in the polar solvents. Benzodihydropyrone cycle taxifolin, oriented almost perpendicularly to the dihydroxyphenyl substituent, closes the entrance to the tube, forming a series of hydrogen bonds between oxygen and hydrogen atoms of OH- and CO- groups of taxifolin and oxygen and hydrogen atoms of OH groups of monomeric glucose chains.

The dotted lines in Fig. 5–5 show a total of five bonds discovered in the complex.

The distances between hydrogen atoms and oxygen of hydrogen bonds equal to 2.047, 2.144, 2.343, 2.242, 1.959 Å, proving the effectiveness of these interactions. As a result, the most hydrophobic part of T molecule, phenyl substituent is inside the tube, and more hydrophilic one, benzodihydropyrone cycle, and OH groups of the dihydroxyphenyl substituent are

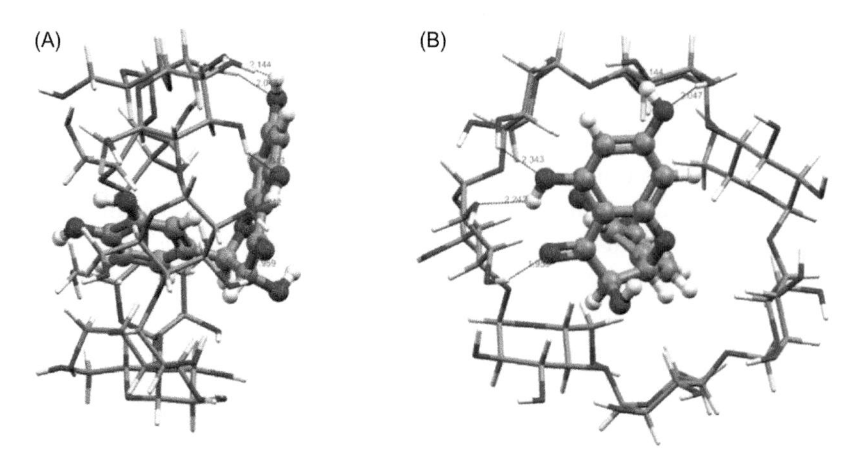


**FIGURE 5–5** Molecular model of taxifolin-βCD conjugates: (A) side view (dihydroxyphenyl substituent inside the tube), (B) front side view (series of hydrogen bonds of the benzodihydropyrone cycle and OH groups of monomeric glucose chains).

outside; thus, providing the best solubility in water and solutions of alcohol compared to pure taxifolin.

### 5.1.3 Studying the properties of taxifolin–β-cyclodextrine conjugates, obtained by sonochemical approach placing in the food matrix

#### *5.1.3.1 The methodology of sonochemical taxifolin–β-cyclodextrine conjugates*

The general approach to the design of experiments was based on the complex analysis of available literature using open-access databases.

In the aggregated form, the approach and description of experimental conditions are given next (Fig. 5–6).

The novelty of this approach is determined by the ultrasonic treatment as an intensifying factor for the synthesis of βCD conjugates (T–βCD).

The reference samples were the solutions of native taxifolin and T–βCD conjugate, that were obtained from the mechanical mixing of corresponding amounts of T and βCD in a water–alcohol solution (10% volume of ethanol) at 40°C.

At this stage, the following approaches were applied to obtain T–βCD conjugates:

- *Solvent:* water–alcohol solution (10% volume of ethanol).
- *Components ratio:* T:βCD was set as 3:1 by the molar mass (*this ratio is based on patent information, literature, and from our research*).

*Methodology for obtaining* the simulation solutions of T–βCD conjugate:

- *Reference samples*: specific amounts of taxifolin and βCD were dissolved in 200 mL of a solvent and mixed at the speed of 200 r/min during 30 min at 40°C.

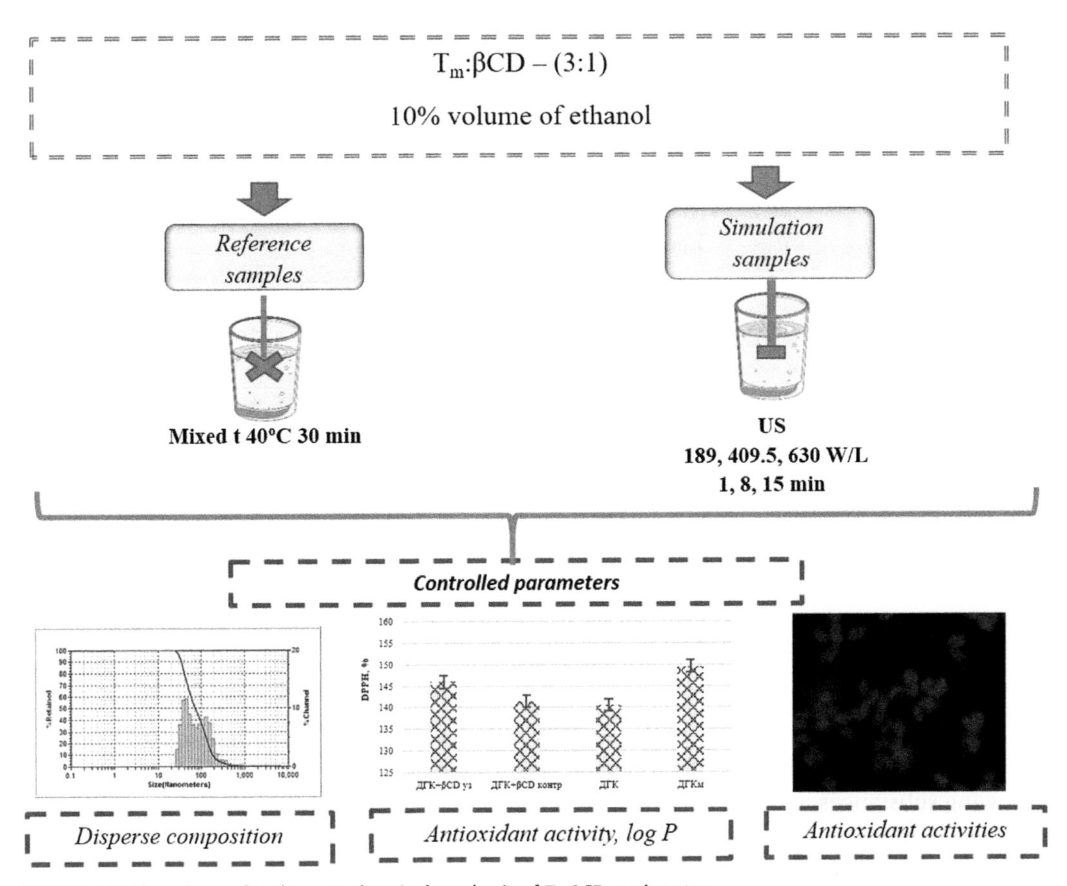


FIGURE 5–6 The scheme for the sonochemical synthesis of T–βCD conjugates.

- *Simulation samples*: certain amounts of taxifolin and βCD are dissolved in 1000 mL of solvent and were exposed to ultrasonic treatment at fixed modes in a coolant jacket with the temperature control not exceeding 40°C.
- *Ultrasonic exposure modes*: frequency, 22 ± 1.65 kHz; intensity, no less than 10 $W/cm^2$; power 189, 409.5, and 530 W/L; exposure time, 1, 8, and 15 min.

To optimize the ultrasound settings while obtaining T–βCD complexes, we evaluated samples from different modes by general AOA (DPPH approach) as the most critical bioactivity parameter of a food ingredient. We formed an array of experimental data and carried out two-factor planning taking power and time of USE as variable factors and general AOA (DPPH) value as a dependent variable (Fig. 5–7).

We studied the physical and chemical properties of the obtained conjugates. The key indicators were: solubility in different solvents, dispersion composition, antioxidant activity, and log *P*.

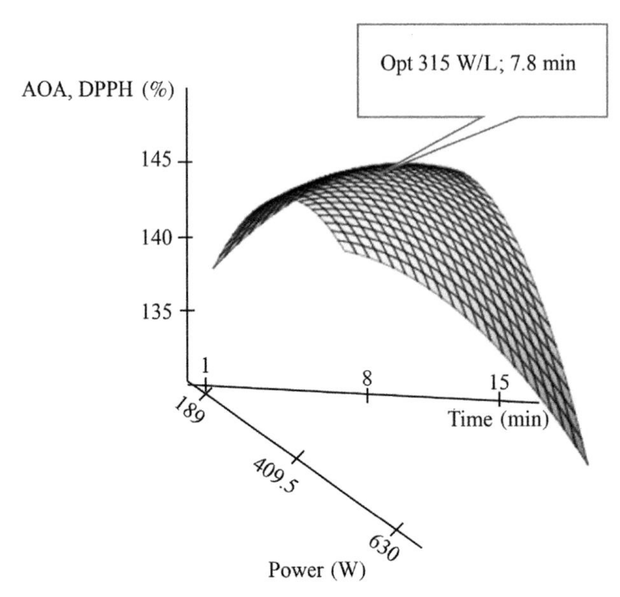


**FIGURE 5–7** Simulation of the process of sonochemical conjugation of T–βCD with increased bioactivity. $Y = -6.033 \cdot 10^{-5}X_1^2 - 0{,}039X_1 - 2.43\ 10^{-3}X_1X_2 + 0.057X_2^2 + 1.383X_2^2 + 133.692$.

For studying the solubility of conjugates for a possible comparative analysis, we took the following substances as elution solvents: distilled water at 20°C; water–alcohol solution (10% of ethanol volume); 96% ethanol; corn oil.

The solubility results of the taxifolin conjugates and native taxifolin are given in Table 5–2.

The obtained results prove that the solubility of T–βCD samples conjugates increases significantly.

Thus, the solubility of reference conjugate T–$\beta CD_{ref}$ in comparison to the reference taxifolin increased by 2.25 times in water–alcohol solution, and in corn oil, it roughly doubled. In comparison to the reference, the T solubility of T–$\beta CD_{us}$ conjugates obtained by USE increased six times for water; 3.4 times for water–alcohol solution four times for corn oil. The least increase in solubility was observed for alcohol (ethanol, 96%), for T–$\beta CD_{ref}$—1.26 times, and T–$\beta CD_{us}$—1.75 times. This is likely due to the critical level of solubilization of taxifolin in ethanol.

Then, we analyzed the disperse composition of conjugate solutions (Fig. 5–8). The results show that the reference sample has a particle size fraction of 379 nm, which may be due to the processes of agglomeration of compounds or the addition of free taxifolin particles to T–βCD complexes.

The results of antioxidant activity (DPPH) evaluation and log $P$ samples of taxifolin and βCD conjugate solutions as the main characteristic of these complexes with their bioavailability and bioactivity are presented in Fig. 5–9.

**Table 5–2** Results of defining the solubility of T–βCD conjugates.

| Samples | Solubility after 30 min (%) | | | |
|---|---|---|---|---|
| | **Water 20°C** | **Water–alcohol solution (10% of ethanol volume)** | **Ethanol 96%** | **Corn oil** |
| Reference T (98.9% of pure T) | 0.08 | 0.16 | 0.47 | 0.16 |
| T–$\beta CD_{ref}$ | 0.18 | 0.31 | 0.59 | 0.32 |
| T–$\beta CD_{us}$ | 0.48 | 0.54 | 0.82 | 0.64 |

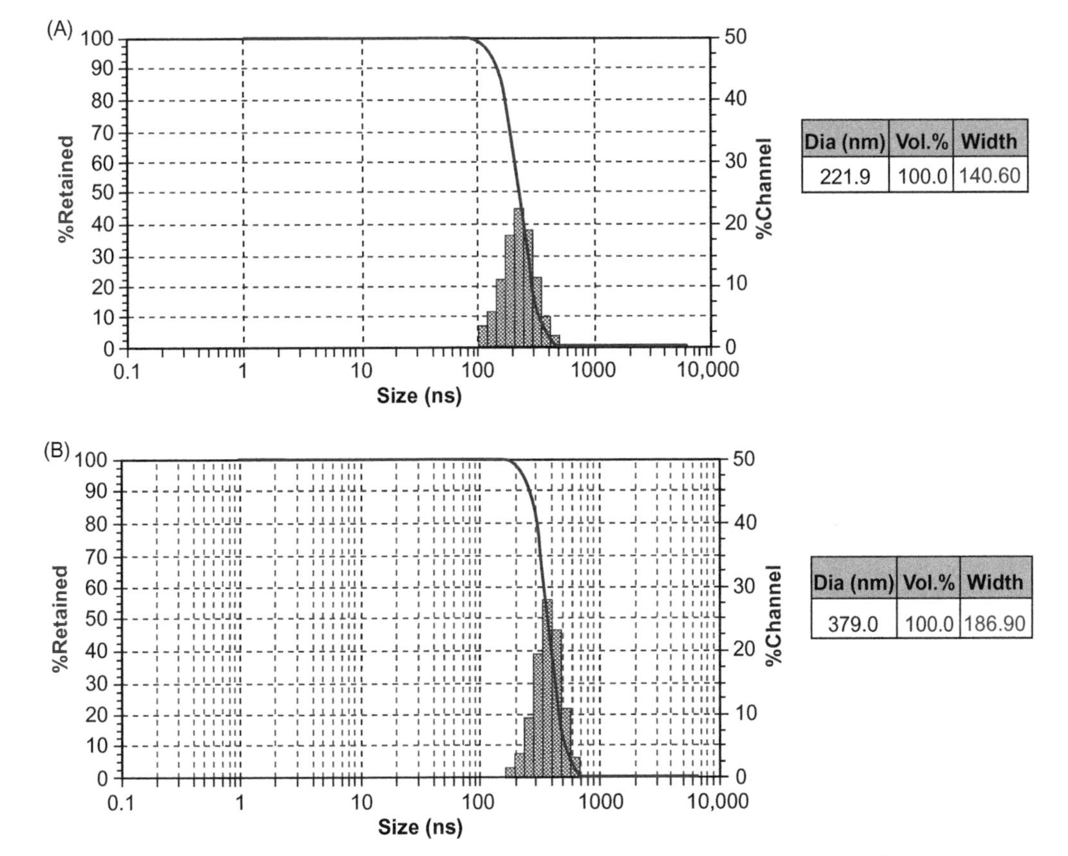


**FIGURE 5–8** The dispersed composition of T–βCD conjugates solvents (0.1%) obtained by using the method of laser dynamic light scattering: (A) T–$\beta CD_{us}$, (B) T–$\beta CD_{ref}$.

There was a slight decrease in the antioxidant activity of the reference solution of T–βCD conjugate in comparison with the initial taxifolin solution. This can be explained by the screening of the OH groups of taxifolin in the formation of the conjugate, which did not participate in the implementation of the antioxidant effect. However, the antioxidant status of

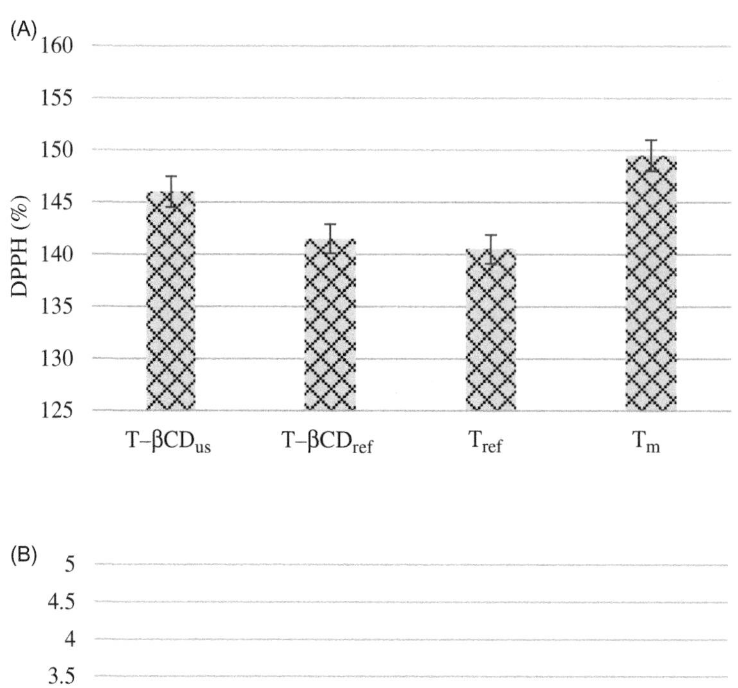


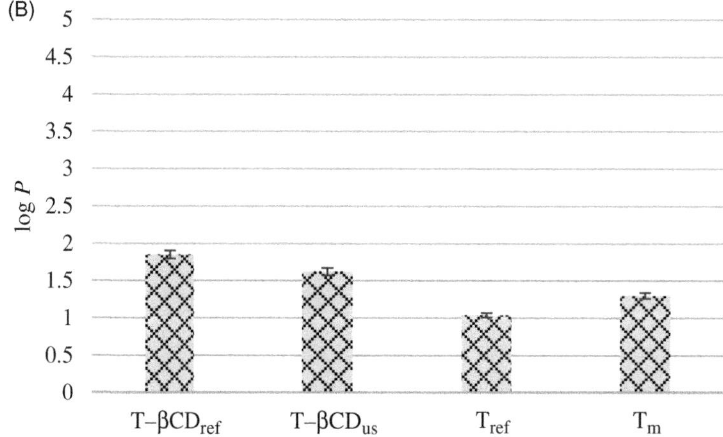


**FIGURE 5–9** Total antioxidant status (A) [DPPH (%)] and permeability, (B) (log *P*) 0.1% of taxifolin solutions and T–βCD conjugates.

T–βCD conjugates was slightly higher than the initial taxifolin, which may be due to the higher antioxidant status of taxifolin the reference sample.

The antioxidant status of the T–$\beta CD_{us}$ conjugate sample obtained with the use of ultrasound exceeded the value of T–$\beta CD_{ref}$. that during the ultrasonic synthesis of conjugates in hydroalcoholic solution T–βCD complexes are formed, in which a smaller number of hydroxyl OH- groups are screened inside the CD cavity and remain active. Besides, it may be explained by an increase in the monomeric fraction of taxifolin exposed to ultrasound, as described above. Besides, one should not deny the overall effect of increasing the solubility of substances exposed to ultrasound, which could also lead to an increase in antioxidant status of the T–$\beta CD_{us}$ image.

The data presented in Fig. 3–33B indicates a distinct impact of the process of the conjugation of taxifolin on its potential permeability through cell membranes, expressed by the log $P$ index. The values of this index for the T–$\beta CD_{us}$ conjugate are more than 1.8 times higher than taxifolin and 1.4 times higher than taxifolin (ref). The comparison of log $P$ values for the two types of conjugates demonstrates the applicability of ultrasound in the synthesis of this type of complexes. The value of log $P$ for T–$\beta CD_{us}$ is 14.2% higher than for T–$\beta CD_{ref}$. The obtained results correlate with the results of similar studies presented in the literature.

### *5.1.3.2 Studying the bioavailability and bioactivity of taxifolin and β-cyclodextrine conjugates in vitro using cell cultures models*

Studying BAS, irrespective of the subsequent purpose of their usage, usually requires evaluation of their biological activity. The evaluation methods alternative to standard tests on the experimental animals, namely, cell cultures models are widely used in the biochemical and toxicological studies [39,40].

Such methods allow reducing the cost significantly and time of preliminary study of new biologically active additives, food ingredients, especially at the stage of their preclinical trials in addition to solving ethical problems associated with the mass use, and death of experimental animals.

Another advantage of in vitro models is the ability to work directly with human cell cultures, which makes the data more adequate when they are projected onto the human body. Besides, using cell cultures makes it possible to define the nature of the biological activity of the studied compounds directly at the cellular level and taking into account the complex synergistic and/or multidirectional effects of the mixture of chemical entities.

In vitro studies were carried out on the cell cultures of unclassified human neuroblastoma SH-SY5Y to evaluate the efficiency of the antioxidants activity of the obtained food ingredients on T basis.

Neuroblastoma cells are a classical experimental model for the study of various mechanisms of proliferation, differentiation, and apoptosis. According to PubMed, at least two neuroblastoma reviews are published weekly, and the total number of publications has approached 37,000, and increasing annually by almost 1500 studies.

The presented studies have shown that neuroblastoma cells are sensitive to a wide range of BASs and can also act as an adequate model for assessing their bioactivity [41].

The intensity evaluation of the generation of reactive oxygen species (ROS) in neuroblastoma cells in the reference sample and the presence of the studied substances was based on the use of fluorescent dye, 2,7-dichlorodihydrofluorescein-diacetate (DCFH2-DA). When it penetrates the cells, the dye is hydrolyzed with the participation of intracellular esterases. The hydrolyzed form of the dye (DCFH2) does not penetrate through the cytoplasmic membrane of the cells and remains inside, and localized mainly in the cytosol.

The dye is oxidized by various free radicals and becomes an indicator of overall oxidative stress. The dye was added into the cells for 30 min at 37°C and the dye concentration was 10 μm. After 2 h of incubation, the cells were washed (the dye was added 30 min before washing). Then Herbology ester (PMA, phorbol-12-myristate-13-acetate) was added to the cell culture for inducing the generation of reactive oxygen species generation.

All the experiments were performed in Hank's balanced salt solution medium with HEPES added to maintain the buffer capacity. The solution contained (in mm): 156 NaCl, 3 KCl, 2 $MgSO_4$, 1 $Na_2HPO_4$, 1.25 $KH_2PO_4$, 2 $CaCl_2$, 10 glucose and 10 HEPES, at pH 7.4. The fluorescence registration was performed using a fluorescent tablet reader Tecan Spark 10 M with a thermostatic chamber.

A system of an inverted motorized fluorescent microscope Leica DMI6000B, with a monochrome CCD camera HAMMAMATSU C9100, an illumination source (Leica EL6000) with a high-pressure mercury lamp HBO 103W/2, was used to visualize the cell cultures using fluorescent microscopy. To visualize the nuclei, a fluorescent dye Hoechst 3334, which penetrates the cell membranes and binds to DNA, was used. A fluorescent cube A (Leica, Germany) with an excitation filter BP 340–380, a dichroic mirror 400 and an emission filter LP 425 were used to excite the fluorescence (Hoechst 33342). To excite the fluorescence of 2,7-dichlorofluorescein (oxidized form of dye, DCM), we used fluorescent cube L5 (Leica, Germany) with an excitation filter BP 480/40, dichroic mirror 505, and an emission filter BP 527/30. The shooting was performed using the lens Leica HCX PL APO lambda blue 63.0 × 1.40 oil. To compare the fluorescence intensities of the dye in cultures incubated with samples, the survey was carried out at the same settings that were used to shoot the reference sample. Since 2,7-dichlorofluorescein is subjected to photo-oxidation under excessive irradiation, the cells were being shot in the transmitted light, and the fluorescence was registered with the same parameters of amplification, exposure, and intensity of light flux.

The results of the dye fluorescence measurement indicate that the intensity is proportional to the number of ROS formed, as shown in Fig. 5–10. The fluorescence intensity with PMA was taken as a reference unit.

The data presented in the figures show that the sample 3, T–$\beta CD_{us}$ conjugate (0.68 with reference to PMA), is the most active in the blocking of ROS, which is generally apparent and correlates with the results of current and the previous studies.

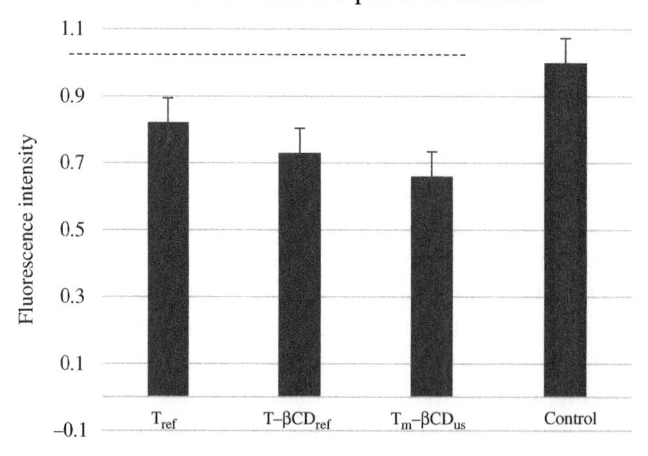


**FIGURE 5–10** The intensity of ROS generation in 3 h period (with PMA is taken as a reference unit, and all other values are relative to the reference). *PMA*, Phorbol-12-myristate-13-acetate; *ROS*, reactive oxygen species.

**Table 5–3** Microphotographs of neuroblastoma cells.

| A the start of testing | After 3 h |
| --- | --- |
| Taxifolin reference | |
| 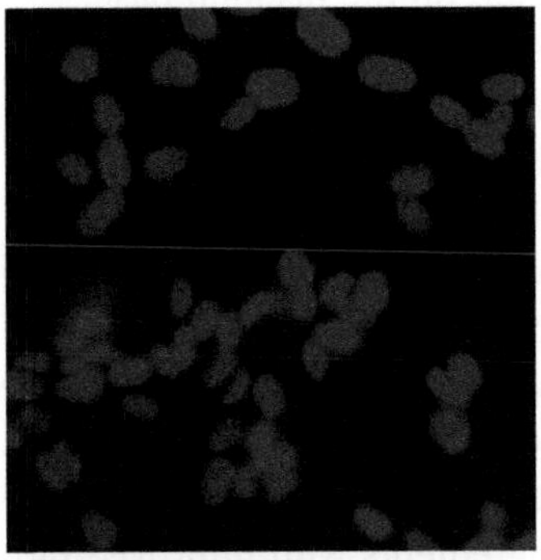 | 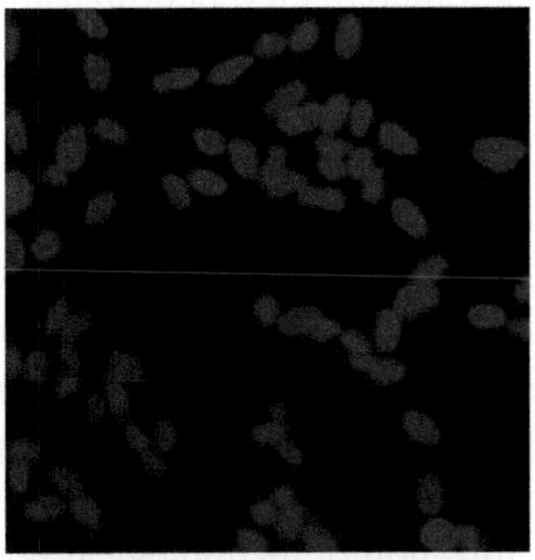 |
| T−$\beta CD_{ref}$ | |
| 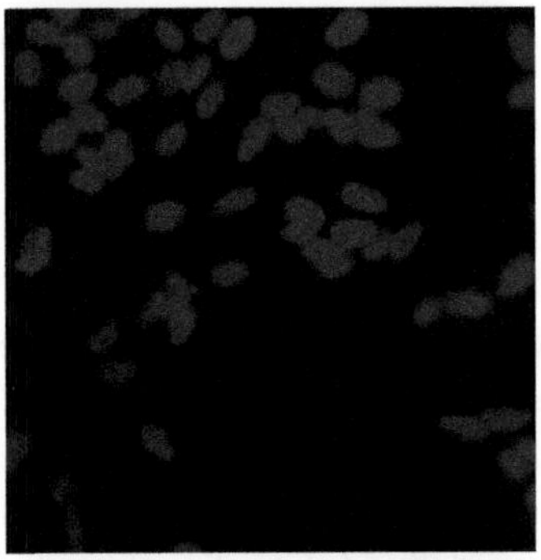 | 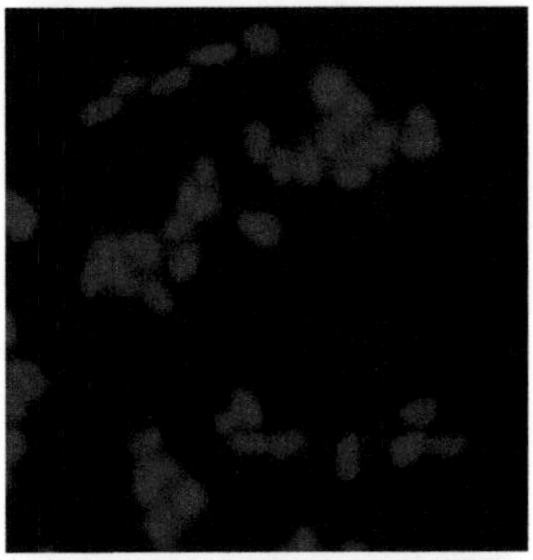 |
| T−$\beta CD_{us}$ | |
| 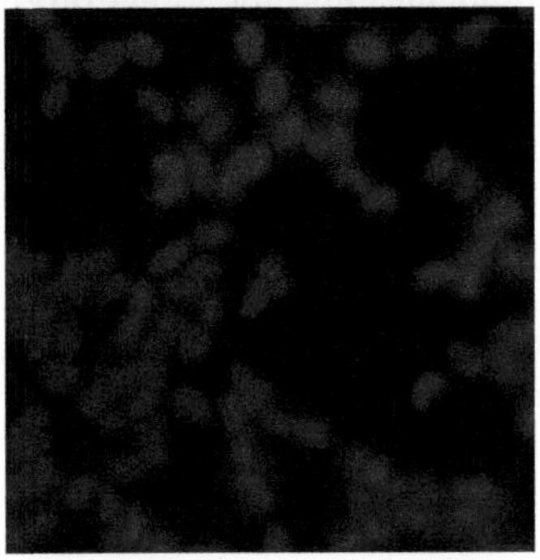 | 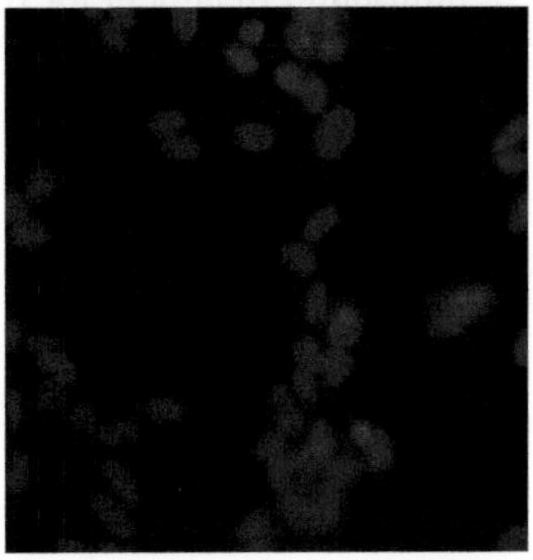 |

*CD*, Cyclodextrine.

A high total antioxidant in combination with permeability through the cell membrane is of high priority when using this model for the biologically active substance, and taxifolin conjugate possessed these important characteristics (antioxidant status—143%, log *P*—1.85). The T−$\beta CD_{ref}$ sample (fluorescence intensity is 0.73 to the reference PMA) was slightly inferior based on the results of the studies.

The studies of antiradical activity with the neuroblastoma cell culture model prove the efficacy and are generally associated with the results of predictive studies. Table 5–3 presents microphotographs, showing the changes in the fluorescence of cell cultures of undifferentiated neuroblastoma in the process of ROS PI inhibition on the taxifolin base.

The following data show that the fluorescence intensity at the end of the study was minimal for the $T{-}\beta CD_{us}$ 3-conjugate sample. The effectiveness of ROS inhibiting for conjugate was higher by about 20% to the reference taxifolin sample and by 7% to $T{-}\beta CD_{ref}$. It generally agrees with the results of physical and chemical studies.

Thus the studies of antiradical activity with the neuroblastoma cell culture model prove the efficacy and are generally associated with the results of reported studies.

The results of physical and chemical studies prove the viability and feasibility of the developed technique to improve the effectiveness of the delivery of plant antioxidants. Physical and chemical studies as well as mathematical simulations, we carried out the optimization of technological regimes for the production of food ingredients with improved bioavailability and bioactivity characteristics.

The research perspectives are connected with transferring the obtained results into real technological processes of food production.

## 5.2 Conclusions

To assess the effectiveness of the obtained AOA food ingredients based on taxifolin, we conducted a series of *in vitro* experiments on the cell cultures of undifferentiated SH-SY5Y human neuroblastoma. The presented studies proved that neuroblastoma cells demonstrate sensitivity to a wide range of biologically active substances and can act as an adequate model of their bioactivity assessment. Based on the obtained results, we have defined the rational modes of ultrasound treatment for regulating the synthesis of taxifolin conjugates and the target properties of the complexes. The proposed approach enables the effective delivery of biologically active compounds into the systems of the human body.

## Acknowledgments

This chapter was written with the support from the Government of RF (Resolution no. 211 of 16.03.2013), Agreement no. 02.A03.21.0011, and subsidies for the fulfillment of a fundamental part of a state order under Project no. 40.8095.2017/BCh and grant RFBR 18-53-45015.

## References

[1] M. Ashokkumar, Applications of ultrasound in food and bioprocessing, Ultrason. Sonochem. 25 (2015) 17–23.

[2] M. Ashokkumar, D. Sunartio, et al., Modification of food ingredients by ultrasound to improve functionality: a preliminary study on a model system, Innovative Food Sci. Emerg. Technol. 9 (2) (2008) 155–160.

[3] T.J. Mason, F. Chemat, M. Ashokkumar, 27: power ultrasonics for food processing, Power Ultrason. (2015) 15–843.

[4] O. Krasulya, S. Shestakov, V. Bogush, et al., Applications of sonochemistry in Russian food processing industry, Ultrason. Sonochem. 21 (2014) 2112–2116.

[5] H.M. Rawel, D. Czajka, S. Rohn, J. Kroll, Interactions of different phenolic acids and flavonoids with soy proteins, Int. J. Biol. Macromol. 30 (2002) 137–150.

[6] D. Venkat Ratnam, D.D. Ankola, V. Bhardwaj, D.K. Sahana, M.N.V. Ravi Kumar, Role of antioxidants in prophylaxis and therapy: a pharmaceutical perspective, J. Control. Release 113 (3) (2006) 189–207.

[7] I.V. Kalinina, I.Y. Potoroko, N.V. Popova, D.G. Ivanova, A.V. Nenasheva, Regulation of homeostasis with products enriched by antioxidants in athletes from low-intensity sports, Hum. Sport Med. 18 (4) (2018) 110–116.

[8] L.-J. Yang, W. Chen, S.-X. Ma, Y.-T. Gao, R. Huanga, S.-J. Yana, et al., Host–guest system of taxifolin and native cyclodextrin or its derivative: preparation, characterization, inclusion mode, and solubilization, Carbohydr. Polym. 85 (3) (2011) 629–637.

[9] V.S. Rogovskii, A.I. Matiushin, N.L. Shimanovskii, et al., Antiproliferative and antioxidant activity of new dihydroquercetin derivatives, Eksp. Klin. Farmakol. 73 (2010) 39–42.

[10] Y. Zu, W. Wu, X. Zhao, et al., Enhancement of solubility, antioxidant ability and bioavailability of taxifolin nanoparticles by liquid antisolvent precipitation technique, Int. J. Pharm. 471 (2014) 366–376.

[11] I.Y. Potoroko, I.V. Kalinina, N.V. Naumenko, et al., Sonochemical micronization of taxifolin aimed at improving its bioavailability in drinks for athletes, Hum. Sport Med. 18 (3) (2018) 90–100.

[12] V.A. Bhattaram, U. Graefe, C. Kohlert, M. Veit, H. Derendorf, Pharmacokinetics and bioavailability of herbal medicinal products, Phytomedicine 9 (3) (2002) 1–33.

[13] S. Scalia, M. Mezzen, Incorporation of quercetin in lipid microparticles: effect on photo- and chemical-stability, J. Pharm. Biomed. Anal. 49 (1) (2009) 90–94.

[14] S. Tommasini, D. Raneri, R. Ficarra, M.L. Calabrò, P. Ficarra, Improvement in solubility and dissolution rate of flavonoids by complexation with β-cyclodextrin, J. Pharm. Biomed. Anal. 35 (2) (2004) 379–387.

[15] L. Jakobek, Interactions of polyphenols with carbohydrates, lipids and proteins, Food Chem. 175 (2015) 556–567.

[16] D.D. Schramm, M. Karim, H.R. Schrader, R.R. Holt, N.J. Kirkpatrick, J.A. Polagruto, et al., Food effects on the absorption and pharmacokinetics of cocoa flavanols, Life Sci. 73 (2003) 857–869.

[17] B. Chanteranne, F. Branca, A. Kardinal, K. Wahala, V. Braesco, P. Ladroite, et al., Food matrix and isoflavones bioavailability in early post menopausal women: a European clinic study, Clin. Interventions Aging 4 (2008) 711–718.

[18] Z. Yuksel, E. Avci, Y.K. Erdem, Characterization of binding interactions between green tea flavonoids and milk proteins, Food Chem. 121 (2010) 450–456.

[19] J. Kroll, H.M. Rawel, S. Rohn, Reactions of plant phenolics with food proteins and enzymes under special consideration of covalent bonds, Food Sci. Technol. Res. 9 (2003) 205–218.

[20] A. Shpigelman, G. Israeli, Y.D. Livney, Thermally-induced protein-polyphenol co-assemblies: beta lactoglobulin-based nanocomplexes as protective nanovehicles for EGCG, Food Hydrocolloids 24 (2010) 735–743.

[21] R.A. Frazier, E.R. Deaville, R.J. Green, E. Stringano, I. Willoughby, J. Plant, et al., Interactions of tea tannins and condensed tannins with proteins, J. Pharm. Biomed. Anal. 51 (2010) 490–495.

[22] F. Saura-Calixto, Dietary fiber as a carrier of dietary antioxidants: an essential physiological function, J. Agric. Food Chem. 59 (2011) 43–49.

[23] R.S. MacDonald, K. Wagner, Influence of dietary phytochemicals and microbiota on colon cancer risk, J. Agric. Food Chem. 60 (2012) 6728–6735.

[24] K.M. Tuohy, L. Conterno, M. Gesperotti, R. Viola, Up-regulating the human intestinal microbiome using whole plant foods, polyphenols, and/or fiber, J. Agric. Food Chem. 60 (2012) 8776–8782.

[25] H. Palafox-Carlos, J.F. Ayala-Zavala, G.A. Gonzalez-Aguilar, The role of dietary fiber in the bioaccessibility and bioavailability of fruit and vegetable antioxidants, J. Food Sci. 76 (2011) R6–R15.

[26] V. Menocci, J.G. Goulart, P.R. Adalberto, O.L. Tavano, D.P. Marques, J. Contiero, et al., Cyclodextrin glycosyltransferase production by new *Bacillus* sp. strains isolated from Brazilian soil, Braz. J. Microbiol. 39 (4) (2008) 682–688.

[27] K. Uekama, Design and evaluation of cyclodextrin-based drug formulation, Chem. Pharm. Bull. 52 (8) (2004) 900–915.

[28] Z. Li, M. Wang, F. Wang, Z. Gu, G. Du, J. Wu, et al., γ-Cyclodextrin: a review on enzymatic production and applications, Appl. Microbiol. Biotechnol. 77 (2) (2007) 245.

[29] Z. Li, J. Zhang, M. Wang, Z. Gu, G. Du, J. Li, et al., Mutations at subsite3 in cyclodextrin glycosyltransferase from *Paenibacillus macerans* enhancing α-cyclodextrin specificity, Appl. Microbiol. Biotechnol. 83 (3) (2009) 483–490.

[30] T. Endo, H. Ueda, Large ring cyclodextrins: recent progress, ChemInform 37 (2006) 38.

[31] J. Kaulpiboon, P. Pongsawasdi, Purification and characterization of cyclodextrinase from *Paenibacillus* sp. A11, Enzyme Microb. Technol. 36 (2–3) (2005) 168–175.

[32] T. Xie, Y. Yue, B. Song, S. Qian, Increasing of product specificity of γ-cyclodextrin by mutating the active domain of α-cyclodextrin glucanotransferase from *Paenibacillus macerans* sp. 602-1, Chin. J. Biotechnol. 29 (9) (2013) 1234–1244.

[33] R.S. Singhal, J.F. Kennedy, S.M. Gopalakrishnan, A. Kaczmarek, C.J. Knill, P.F. Akmar, Industrial production, processing, and utilization of sago palm-derived products, Carbohydr. Polym. 72 (1) (2008) 1–20.

[34] A.O. Griewank, Generalized descent for global optimization, J. Opt. Theor. 34 (1981) 11–39.

[35] R.V. Pappu, R.K. Hart, J.W. Ponder, Analysis and application of potential energy smoothing and search methods for global optimization, J. Phys. Chem. 102 (48) (1998) 9725–9742.

[36] L. Piela, J. Kostrowicki, H.A. Scheraga, The multiple-minima problem in the conformational analysis of molecules. Deformation of the potential energy hypersurface by the diffusion equation method, J. Phys. Chem. 93 (8) (1989) 3339–3346.

[37] R.A.R. Butler, E.E. Slaminka, An evaluation of the sniffer global optimization algorithm using standard test functions, J. Comput. Phys. 99 (1993) 28–32.

[38] V. Potemkin, A. Pogrebnoy, M.A. Grishina, Technique for energy decomposition in the study of "receptor-ligand" complexes, J. Chem. Inf. Model. 49 (6) (2009) 1389–1406.

[39] K.L. Maier, E. Mateikova, H. Hinze, et al., Different selectivities of oxidants during oxidation of methionine residues in the α-1-proteinase inhibitor, FEBS Lett. 250 (3) (1989) 221–226.

[40] M. Balls, J.H. Fentem, The use of basal cytotoxicity and target organ toxicity tests in hazard identification and risk assessment, ATLA 20 (1992) 368–389.

[41] J.L. Bell, A. Malyukova, M. Kavallaris, G.M. Marshall, B.B. Cheung, TRIM16 inhibits neuroblastoma cell proliferation through cell cycle regulation and dynamic nuclear localization, Cell Cycle 12 (6) (2013) 889–898.

# Further reading

Y. Li, D.J. Yang, S.L. Chen, S.B. Chen, A.S. Chan, Process parameters and morphology in puerarin, phospholipids and their complex microparticles generation by supercritical antisolvent precipitation, Int. J. Pharm. 359 (2008) 35–45.

T.J. Mason, P. Cintas, Sonochemistry, in: J.H. Clark, D.J. Macquarrie (Eds.), Handbook of Green Chemistry and Technology, Blackwell Science, Oxford, 2008, pp. 372–393.

6

# Controlled release of functional bioactive compounds from plants

S.D. Torawane, Y.C. Suryawanshi, D.N. Mokat

*DEPARTMENT OF BOTANY, SAVITRIBAI PHULE PUNE UNIVERSITY, PUNE, INDIA*

**Chapter outline**

**6.1 Introduction** ..... **103**
**6.2 Bioactive compounds** ..... **104**
6.2.1 Phenolic compounds ..... 104
6.2.2 Polyamines ..... 105
6.2.3 Flavonoids ..... 105
6.2.4 Tannins ..... 107
**6.3 Conclusion** ..... **107**
**References** ..... **108**

## 6.1 Introduction

Plants and herbs used to treat various diseases. Plants are capable of producing a massive amount of different bioactive compounds. Plants enclosing useful phytochemicals might enhance the necessities of the human body by acting as natural antioxidants. Numerous higher plant components that are itemized can be grouped as essential, substances, plant volatiles, natural sweeteners, plant toxins, carcinogens, coloring materials, estrogenic agents, natural hallucinogens, phytotoxins, or allelochemicals [1]. In agriculture, farmers are interested in finding ways of increasing the levels of bioactive compounds in the crops, such as aroma, color, and flavor. Despite this fact, bioactive compounds are beneficial for human health, and there is more scientific research needs to be carried out before we can begin to make science-based dietary recommendations [2]. There are enough recommendations for consuming food sources rich in bioactive compounds. From a practical perspective this translates to recommend a diet, rich in a variety of fruits, vegetables, whole grains, legumes, oils, and nuts. Currently, the pharmaceutical industry is looking into organic plants with higher contents of bioactive compounds. It is known that the interaction between different food metabolites and their biosynthetic pathways can contribute to the healthy or harmful characteristics of their food [3].

Encapsulation of Active Molecules and their Delivery System. DOI: https://doi.org/10.1016/B978-0-12-819363-1.00006-5

## 6.2 Bioactive compounds

In recent years, intense attention is focused on the use of bioactive compounds for preventing various diseases. The bioactive compounds are types of phytochemicals that typically occur in small amounts in plants and foods (such as fruits, vegetables, nuts, oils, and whole grains) and capable of modulating metabolic processes and resulting in the promotion of better health [4]. Bioactive compounds have actions in the body that may promote good health; and hence they are intensively studied to evaluate their effects and benefits on human health. The bioactive compounds different from nutrients as they are not essential, and there are no recommended daily intake values [5]. Currently, there are significant and growing amounts of scientific evidence demonstrating the utility of various active compounds in the prevention of cancer, heart, and other diseases [6,7]. Some of them are discussed below.

### 6.2.1 Phenolic compounds

Phenolic compounds are the class of secondary metabolites, and are recognized due to their antioxidant activity. This antioxidant activity can confer food quality and beneficial potential to human health. Free radicals are harmful molecules that can damage human cells. Antioxidants are the reducing agents that neutralizes these free radicals [8,9]. The stability and functionality of phytochemicals in the human body depends on the presence of these bioactive compounds. The phenolic compounds formed from the Shikimate pathway is responsible for the primary and secondary metabolite production [10].

Resveratrol is a type of natural phenol, produced by several plants. Piceatannol is a metabolite of resveratrol found in red grapes, wine, white tea, passion fruit, and Japanese knotweed. The high amount of ornithine decarboxylase present in the cell is the critical indicator of tumor cells. Both the resveratrol and piceatannol can reduce the ornithine decarboxylase levels in Caco-2 colorectal cancer cells [11,12]. Serotonin and dopamine play an important role in neuropsychiatric disorders, for example, depression, schizophrenia, and Parkinson's disease (PD). Some of the phenolic compounds affect the production of biogenic amine by inhibiting lactic acid bacterial growth [13]. Usually, the polyphenols are mostly absorbed in the large intestine (90%) by gut microbiota or excreted. Comparatively, polyphenols are less absorbed by the small intestine (10%). The absorption of polyphenol in the human intestine is complicated, and it depends on the bioavailability of human gut microbiota [14].

Tea and coffee are the most consumed drinks in the world, which contain various polyphenolic compounds. Tea is a rich source of epicatechin, catechins, epicatechin gallate, epigallocatechin, compounds that can enhance various effects in humans [15]. Caffeine is the principal bioactive compound in coffee [16,17]. It is necessary to understand that vegetables and dietary fruits provide other bioactive compounds, besides polyphenols, which help the human body to prevent chronic diseases, such as cancer, diabetes, hypertension, obesity, and cardiovascular disease [18].

## 6.2.2 Polyamines

Polyamines are the small and abundant organic compounds having more than two amino groups. Mainly three types of polyamines: putrescine, spermidine, and hermospermine. Polyamines are the essential compounds in plant physiology, that play an important role in growth, cell proliferation, synthesis of nucleic acids and proteins, resistance, membrane stability, apoptosis, senescence, and differentiation [19,20]. The mutation of polyamines can cause critical damage to the plants. Mammals obtain polyamines from food or synthesize at their own. Ornithine decarboxylase in mammals and arginine decarboxylase in plants are responsible for polyamines synthesis [21]. In plants and mammals, spermidine is formed from spermidine synthase. The oxidation of polyamines occurs via ammine oxidase. Abiotic and biotic stresses form oxidative stress, that is mainly responsible for the accumulation of reactive oxygen species (ROS) [22,23]. These ROS is harmful if present in higher amounts. ROS damage the nucleic acids and oxidizing proteins of cells. The polyamines is important constituent in the ROS accumulation by acting as a ROS scavengers and osmoprotectors. The polyamines avoid the damage of cells cased by stress in plants by spermidine, putrescine, and spermine that are positively charged at physiological pH, and can bond to negatively charged membrane cell groups [24,25]. These positively charged polyamines can also interact with the negatively charged molecules, such as RNA and DNA through electrostatic bonding [26]. DNA structure stabilizes by the bonding mainly with spermine and spermidine. A high amounts of polyamine are also undesirable as they can cause headaches, hypertension, hypotension, and other allergic diseases. The polyamines intake as well as the control of some diseases have been studied by many research groups [27,28].

The regular consumptions of polyamines and phenolic compounds rich food have been reported to improve the health of humans. The gut microbiota produces enzymes that different from human gut enzymes, which may enhance or reduce the metabolite activity of human guts. The bioactive compounds and the gut microbiota are involved in a complex metabolism. Interactions between these components are essential for human responses to food-based interventions, and to achieve desired health improvement [29–31].

## 6.2.3 Flavonoids

Flavonoids are the secondary metabolite and the class of polyphenols, that are widely distributed in plant-based food. Firstly, in 1930, flavonoids were isolated from lemon juice. From 1930 untill now, about 6000 different types of flavonoids have been discovered. In plants, flavonoids play various important functions such as plant pigmentation, protection from UV light, modulating enzymatic action, and gene regulation. They are also responsible for the aroma and colors of flowers [32]. Flavonoids are formed primarily for plants' protection against oxidative stress. They also play a significant role in biotic and abiotic stress. The flavonoids are found in a large numbers in plants, that include citrus fruits, cherries, berries, grapes, arugula, onions, soybeans [33].

**Table 6.1** Flavonoids' subclasses and their compounds with their natural sources [34].

| Flavonoids subclasses | Examples | Natural source |
|---|---|---|
| Anthocyanins | Pelargonidin, cyanidin, delphinidin, peonidin, petunidin, and malvidin | Fruits, vegetables, nuts and dried food, medicinal plants, and others |
| Flavonols | Epicatechin, catechin, gallocatechin, ECG, EGC, and EGCG | Medicinal plants and others |
| Flavonones | Naringenin, naringin, hesperidin, hesperetin, and eriodicytol | Fruits, medicinal plants, and others |
| Flavones | Sinensetin, isosinensetin, nobiletin, tangeretin, luteolin, apigenin, chrysin, baicalein, galangin, and repoifolin | Fruits, medicinal plants, and others |
| Flavonols | Kaempferol, quercetin, fisetin, isorhamnetin, myricetin, rutin, and morin | Fruits, vegetables, medicinal plants, and others |
| Isoflavones | Daidzein, genistein, daidzin, glycitein, and genistein | Legumes and medicinal plants |
| Chalcones | Phloretin, phlioridzin, arbutin, and chalco naringenin | Fruits, vegetables, medicinal plants, and others |

*ECG*, Epicatechin gallate; *EGC*, epigallocatechin; *EGCG*, EGC gallate.

According to structures, flavonoids are categorized into the following subgroups: anthocyanins, flavonols, flavones, flavanols or catechins, and flavanones. Flavonoids' subclasses and their compounds are shown in Table 6.1.

A recent study shows that no total daily intake of flavonoids can be estimated because it varies from cultures to nations. Recent evidences suggests that a flavonoids-rich diet may be beneficial for human health. Various reports and controlled trials suggest that flavonoids can be used to cure chronic diseases, including coronary heart disease, type 2 diabetes mellitus, cancer, PD. Almost all flavonoid subclasses act as antioxidants, among which catechins and flavones are the most powerful. Flavonoids can produce various proinflammatory substances and enzymes, such as cyclooxygenase, tumor necrosis factor $\alpha$, and lipoxygenase [35,36].

Anthocyanins are water-soluble flavonoids that are mostly present in fruits and berries of various plants. Anthocyanins reduce blood cholesterol levels in patients with a high risk of cardiovascular diseases. It was reported that low-density lipoprotein cholesterol level was reduced in 122 patients after supplementing with anthocyanins (320 mg/day). Anthocyanins also seem to protect against cataracts [37,38].

Flavanols or catechins are 3-hydroxy derivatives of flavanones also play a significant role in the amelioration of cardiovascular disease risk factors. Catechins could manage the protein and RNA expression of fatty acid metabolism enzymes in the liver. Quercetin also plays a critical role in blood pressure regulation and vascular tone [39].

Prevention of cancer causing activation is one of the most critical mechanisms by which flavonoids can exert their chemoprevention effects. The consumption of high flavonoids can reduce ovarian cancer risk (18% risk reduction), aerodigestive tract cancer (33% risk reduction), and lung cancer (16% risk reduction) [40,41].

Kaempferol was recently reported, and this flavonoid possesses antiproliferative and apoptosis inducing activities in several cancer cell lines [42,43].

Quercetin glycosides, rutin, and isoquercitrin are the types of flavonoids having distinct features in upregulating the production of intracellular antioxidant enzymes such as glutathione peroxidase (GPx), superoxide dismutase (SOD), catalase (CAT), and glutathione in a 6-hydroxydopamine (6-OHDA−) induced PC-12, and rat pheochromocytoma cells [44,45]. When oxidative stress occurs, then flavonoids can protect neuronal cells through the elevation of intracellular antioxidant enzymes. It seems that a regular consumption of foods, rich in flavonoids, reduces the risk of neurodegenerative diseases, and counteracts or delays the onset of age-related cognitive disorders. Future studies are required for flavonoid research that includes trials to understand the bidirectional relation between flavonoid metabolism and the microbiome.

### 6.2.4 Tannins

Tannins are categorized into two types: hydrolysable and condensed. Hydrolysable tannins are composed of a monosaccharide core (most often glucose) with several catechin derivatives attached, while the condensed tannins are large polymers of flavonoids [46]. Condensed tannins are more stable and have less potential to cause toxicity. Tannins indiscriminately bind to larger tannins and proteins are used as astringents in cases of diarrhea, skin bleedings, and transudates. Tannins are very widely distributed in plants. Examples of the plant families associated with the presence of tannins are *Polygonaceae* (knotweed family) and *Fagaceae* (beech family) [47].

## 6.3 Conclusion

Plants are the rich sourced of bioactive compounds. Plants produce a long range of bioactive compounds via secondary metabolites pathway. The bioactive compounds may elicit an extended range of diverse effects in animals and men, eating plants, depending upon their species and intake amount. Plants with potent bioactive compounds are often characterized as both poisonous and medicinal, and a beneficial or an adverse result may depend on the amount eaten, and the context of intake. For typical food and feed plants with bioactive compounds with less pronounced effects, the intakes are usually regarded as beneficial. At present, it should be suggested to eat abundant dietary sources of bioactive compounds from fruits and vegetables, legumes, spices, nuts, and herbs every day.

# References

[1] A.G. Atanasov, B. Waltenberger, E.-M. Pferschy-Wenzig, T. Linder, C. Wawrosch, P. Uhrin, et al., Discovery and resupply of pharmacologically active plant-derived natural products: a review, Biotechnol. Adv. 33 (2015) 1582–1614.

[2] A. Hohtola, Bioactive compounds from northern plants, in: M.T. Giardi, G. Rea, B. Berra (Eds.), Bio-Farms for Nutraceuticals: Functional Food and Safety Control by Biosensors, Advances in Experimental Medicine and Biology, Springer US, Boston, MA, 2010, pp. 99–109.

[3] G.J. Gil-Chávez, J.A. Villa, J.F. Ayala-Zavala, J.B. Heredia, D. Sepulveda, E.M. Yahia, et al., Technologies for extraction and production of bioactive compounds to be used as nutraceuticals and food ingredients: an overview, Compr. Rev. Food Sci. Food Saf. 12 (2013) 5–23.

[4] P.M. Kris-Etherton, K.D. Hecker, A. Bonanome, S.M. Coval, A.E. Binkoski, K.F. Hilpert, et al., Bioactive compounds in foods: their role in the prevention of cardiovascular disease and cancer, Am. J. Med. 113 (2002) 71–88.

[5] M.J. Gibney, H. Vorster, F. Kok, Introduction to Human Nutrition (The Nutrition Society Textbook), Wiley-Blackwell, Chichester, 2002.

[6] F.B. Hu, Plant-based foods and prevention of cardiovascular disease: an overview, Am. J. Clin. Nutr. 78 (2003) 544S–551S.

[7] R. Béliveau, D. Gingras, Role of nutrition in preventing cancer, Can. Fam. Phys. 53 (2007) 1905–1911.

[8] N. Balasundram, K. Sundram, S. Samman, Phenolic compounds in plants and agri-industrial by-products: antioxidant activity, occurrence, and potential uses, Food Chem. 99 (2006) 191–203.

[9] F. Cardona, C. Andrés-Lacueva, S. Tulipani, F.J. Tinahones, M.I. Queipo-Ortuño, Benefits of polyphenols on gut microbiota and implications in human health, J. Nutr. Biochem. 24 (2013) 1415–1422.

[10] M.S. Hussain, S. Fareed, M. Saba Ansari, A. Rahman, I.Z. Ahmad, M. Saeed, Current approaches toward production of secondary plant metabolites, J. Pharm. Bioallied Sci. 4 (2012) 10.

[11] R. Nowak, M. Olech, N. Nowacka, Chapter 97 – Plant polyphenols as chemopreventive agents, in: R.R. Watson, V.R. Preedy, S. Zibadi (Eds.), Polyphenols in Human Health and Disease, Academic Press, San Diego, CA, 2014, pp. 1289–1307.

[12] H. Piotrowska, M. Kucinska, M. Murias, Biological activity of piceatannol: leaving the shadow of resveratrol, Mutat. Res. Mutat. Res. 750 (2012) 60–82.

[13] H.A. Gomez-Gomez, C.V. Borges, I.O. Minatel, A.C. Luvizon, G.P.P. Lima, Health benefits of dietary phenolic compounds and biogenic amines, Bioact. Mol. Food (2019) 3–27.

[14] G. Gross, D.M. Jacobs, S. Peters, S. Possemiers, J. van Duynhoven, E.E. Vaughan, et al., In vitro bioconversion of polyphenols from black tea and red wine/grape juice by human intestinal microbiota displays strong interindividual variability, J. Agric. Food Chem. 58 (2010) 10236–10246.

[15] H. Asakura, T. Kitahora, Antioxidants and polyphenols in inflammatory bowel disease: ulcerative colitis and Crohn disease, Polyphenols: Prevention and Treatment of Human Disease, Academic Press, 2018, pp. 279–292.

[16] M.G. Ferruzzi, The influence of beverage composition on delivery of phenolic compounds from coffee and tea, Physiol. Behav. 100 (2010) 33–41.

[17] A.B. Sharangi, Medicinal and therapeutic potentialities of tea (*Camellia sinensis* L.) – a review, Food Res. Int. 42 (2009) 529–535. Available from: https://doi.org/10.1016/j.foodres.2009.01.007.

[18] K.B. Pandey, S.I. Rizvi, Plant polyphenols as dietary antioxidants in human health and disease, Oxid. Med. Cell. Longev. 2 (2009) 270–278.

[19] A.W. Galston, R.K. Sawhney, Polyamines in plant physiology, Plant Physiol. 94 (1990) 406–410.

[20] T. Kusano, T. Berberich, C. Tateda, Y. Takahashi, Polyamines: essential factors for growth and survival, Planta 228 (2008) 367–381.

[21] M.C. Chibucos. Biogenic Polyamines in the Plant-Pathogenic Oomycete *Phytophthora sojae*: First Functional Analysis of Polyamine Transport (Ph.D.), Bowling Green State University, Bowling Green, OH, 2004.

[22] A. Cona, G. Rea, R. Angelini, R. Federico, P. Tavladoraki, Functions of amine oxidases in plant development and defence, Trends Plant. Sci. 11 (2006) 80–88.

[23] R. Wimalasekera, F. Tebartz, G.F.E. Scherer, Polyamines, polyamine oxidases and nitric oxide in development, abiotic and biotic stresses, Plant Sci. 181 (2011) 593–603.

[24] S.S. Hussain, M. Ali, M. Ahmad, K.H.M. Siddique, Polyamines: natural and engineered abiotic and biotic stress tolerance in plants, Biotechnol. Adv. 29 (2011) 300–311.

[25] P. Sharma, A.B. Jha, R.S. Dubey, M. Pessarakli, Reactive Oxygen Species, Oxidative Damage, and Antioxidative Defense Mechanism in Plants Under Stressful Conditions [WWW Document], 2012.

[26] T.N.J.I. Edison, R. Atchudan, M.G. Sethuraman, Y.R. Lee, Reductive-degradation of carcinogenic azo dyes using *Anacardium occidentale* testa derived silver nanoparticles, J. Photochem. Photobiol. B: Biol. 162 (2016) 604–610.

[27] V. Vijayanathan, T. Thomas, A. Shirahata, T.J. Thomas, DNA condensation by polyamines: a laser light scattering study of structural effects, Biochemistry 40 (2001) 13644–13651.

[28] A.C. Childs, D.J. Mehta, E.W. Gerner, Polyamine-dependent gene expression, Cell. Mol. Life Sci. 60 (2003) 1394–1406.

[29] A.J. Parr, G.P. Bolwell, Phenols in the plant and in man. The potential for possible nutritional enhancement of the diet by modifying the phenols content or profile, J. Sci. Food Agric. 80 (2000) 985–1012.

[30] A.G.P. Samaranayaka, E.C.Y. Li-Chan, Food-derived peptidic antioxidants: a review of their production, assessment, and potential applications, J. Funct. Foods 3 (2011) 229–254.

[31] A.-M. Boudet, Evolution and current status of research in phenolic compounds, Phytochemistry 68 (2007) 2722–2735.

[32] A. Kozlowska, D. Szostak-Węgierek, Flavonoids—food sources, health benefits, and mechanisms involved, Bioact. Mol. Food (2019) 53–78.

[33] K. Das, A. Roychoudhury, Reactive oxygen species (ROS) and response of antioxidants as ROS-scavengers during environmental stress in plants, Front. Environ. Sci. (2014) 2.

[34] A.N. Panche, A.D. Diwan, S.R. Chandra, Flavonoids: an overview, J. Nutr. Sci. 5 (2016) 1–15.

[35] J.A. Ross, C.M. Kasum, Dietary flavonoids: bioavailability, metabolic effects, and safety, Annu. Rev. Nutr. 22 (2002) 19–34.

[36] B.L. Tan, M.E. Norhaizan, W.-P.-P. Liew, H. SulaimanRahman, Antioxidant and oxidative stress: a mutual interplay in age-related diseases, Front. Pharmacol. 9 (2018) 1–28.

[37] W.J. Craig, Phytochemicals: guardians of our health, J. Am. Diet. Assoc. 97 (1997) S199–S204.

[38] D. Pascual-Teresa, D.A. Moreno, C. García-Viguera, Flavanols and anthocyanins in cardiovascular health: a review of current evidence, Int. J. Mol. Sci. 11 (2010) 1679–1703.

[39] J.G. Gormaz, N. Valls, C. Sotomayor, T. Turner, R. Rodrigo, Potential role of polyphenols in the prevention of cardiovascular diseases: molecular bases, Curr. Med. Chem. 23 (2) (2016) 115–128.

[40] X. Hua, L. Yu, R. You, Y. Yang, J. Liao, D. Chen, et al., Association among dietary flavonoids, flavonoid subclasses and ovarian cancer risk: a meta-analysis, PLoS One 11 (2016) e0151134.

[41] H.D. Woo, J. Kim, Dietary flavonoid intake and smoking-related cancer risk: a metaanalysis, PLoS One 8 (2013) e75604.

[42] G. Galati, P.J. O'Brien, Potential toxicity of flavonoids and other dietary phenolics: significance for their chemopreventive and anticancer properties, Free. Radic. Biol. Med. 37 (2004) 287–303.

[43] G. Rusak, H.O. Gutzeit, J.L. Müller, Structurally related flavonoids with antioxidative properties differentially affect cell cycle progression and apoptosis of human acute leukemia cells, Nutr. Res. 25 (2005) 143–155.

[44] K.B. Magalingam, A. Radhakrishnan, N. Haleagrahara, Rutin, a bioflavonoid antioxidant protects rat pheochromocytoma (PC-12) cells against 6-hydroxydopamine (6-OHDA)-induced neurotoxicity, Int. J. Mol. Med. 32 (1) (2013) 235–240.

[45] A.S. Chesser, V. Ganeshan, J. Yang, G.V. Johnson, Epigallocatechin-3-gallate enhances clearance of phosphorylated tau in primary neurons, Nutr. Neurosci. 19 (1) (2016) 21–31.

[46] A.A. Shahat, M.S. Marzouk, 13 – Tannins and related compounds from medicinal plants of Africa, in: V. Kuete (Ed.), Medicinal Plant Research in Africa., Elsevier, Oxford, 2013, pp. 479–555.

[47] A. Bernhoft, Bioactive compounds in plants – benefits and risks for man and animals, Norwegian Acad. Sci. Letters (2010) 11–18.

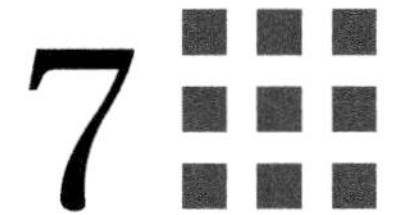

# Bioactive molecule and/or cell encapsulation for controlled delivery in bone or cartilage tissue engineering

Bhaskar Birru[1], P. Shalini[2], Sreenivasa Rao Parcha[3]

[1]DEPARTMENT OF BIOSCIENCES AND BIOENGINEERING, INDIAN INSTITUTE OF TECHNOLOGY GUWAHATI, GUWAHATI, ASSAM, INDIA [2]DEPARTMENT OF CHEMICAL ENGINEERING, NATIONAL INSTITUTE OF TECHNOLOGY WARANGAL, WARANGAL, TELANGANA, INDIA [3]DEPARTMENT OF BIOTECHNOLOGY, NATIONAL INSTITUTE OF TECHNOLOGY WARANGAL, WARANGAL, TELANGANA, INDIA

**Chapter outline**

**7.1 Introduction** .... **111**
**7.2 Controlled delivery** .... **112**
**7.3 Cell/biomolecule encapsulation** .... **115**
7.3.1 Essential requirements for encapsulation .... 115
7.3.2 The design aspect of hydrogels for encapsulation .... 118
7.3.3 Multifunctional cell/bioactive molecule encapsulation system .... 120
**7.4 Bioactive molecule/cell encapsulation for bone and cartilage** .... **122**
7.4.1 Bioactive molecules .... 122
7.4.2 Encapsulation of cells .... 124
**References** .... **126**

## 7.1 Introduction

Tissue engineering (TE) is a promising alternative for the development of engineered grafts for repair or replacement of tissue. A biomaterial is essential to regenerate tissue in the laboratory, and various natural or synthetic biomaterials were tested for their suitability in the application of TE. Ideally, biomaterials are biodegradable, bioresorbable, and biocompatible and also should possess mechanical strength of in vivo tissue. An essential aspect in

**Encapsulation of Active Molecules and their Delivery System. DOI: https://doi.org/10.1016/B978-0-12-819363-1.00007-7**

the selection of biomaterial in bone or cartilage TE is to induce cell differentiation into bone or cartilage and its mechanical strength must mimic the native tissue. The limitations of autografts and allografts for the repair of the bone defect are proposed as an alternative substitute using TE. Thus bone TE (BTE) has emerged for the development of engineered bone graft by combining cells with biomaterial scaffolds. Cell, gene, and bioactive molecule delivery should be precisely controlled to facilitate a favorable environment for the formation of neo-tissue. Stem cells are being widely used for bone and cartilage TE, owing to their ability to differentiation into the desired lineage provided necessary supplements and conditions. A porous, biodegradable scaffold having excellent mechanical strength is essential for the scaffold to be used in BTE. The mechanical strength of the scaffold varies from bone to cartilage. The interconnected porous structure facilitates good nutrient transport and cell invasion and colonization. The porous architecture, mechanical strength, and tissue formation of the scaffold in proportion to the degradation of the scaffold are vital factors to be considered in the selection of biomaterial for BTE. Various porous scaffolds such as metals, bioactive glass, and synthetic and natural polymers in the range of micro and macro were developed for BTE.

The goal of scaffold fabrication is to deliver cell, gene, and bioactive molecules is to repair or replace defected tissue. A biomaterial interaction with biomolecules and cells may induce cell adhesion, migration, and even stimulate cell proliferation and differentiation [1]. Bioactive molecules and cells encapsulated within the scaffold and controlled delivery to the target site are crucial for reconstructing or repairing tissue. The loading of a sufficient number of cells without lasting their innate characteristic features at the damaged site is crucial in scaffold-based TE. Scaffold-based biomolecule delivery has emerged in the field of TE, in particular, and research on the delivery of growth factor using a variety of scaffolds is intense for bone and cartilage TE. Biomolecules or cells can be delivered to the damaged site in various ways, and encapsulation is widely used for delivering the growth factors and cells to repair or reconstruct the damaged tissue effectively.

This chapter outlines the importance of biomolecules/cell encapsulation in TE and regenerative medicine. Briefly, the improvements in encapsulation in terms of advanced techniques and their efficiency for biomolecule/cell delivery have been discussed. The concept of controlled delivery to the target site is a detrimental factor in bone and cartilage repair. Cell and gene therapy solely depends on the controlled delivery of a cell or gene at the damaged site. The localization and systematic delivery of molecule/cell augment for the treatment of disease or repair of damaged tissue are essential. We intended to provide the overall significance of encapsulation of molecules for the effective delivery systems in one place and also summarize the achievements in the field of TE and regenerative medicine.

## 7.2 Controlled delivery

It is obvious that controlled drug delivery is widespread in everyday life. The use of medication, including nasal spray, coated pills, oral strips, and injectable drugs, etc. that are tailored for controlled delivery [2]. However, drug delivery to the target site by obviating the toxic effect on other parts of the body, in the long run, is quite challenging. The research in drug delivery aimed to develop an efficient, controllable, localized, long-term durable delivery is a potential approach in treating diseases. The toxic effect and unknown cellular and molecular interaction with other cells/organs in the body are major limitations of drugs. Cell or gene delivery to the target site is a promising alternative strategy to treat disease or reconstruct tissue compared to drug-assisted treatments. The controlled delivery of biofactors to the targeted site can be achieved through different approaches, and especially dose and systematic delivery in terms of its regulation in spatiotemporal fashion is an important aspect, otherwise abnormal tissue growth takes place. The understanding of molecular mechanisms on tissue development strategies through elaborated research in the field of bone or cartilage TE related to the delivery of growth factors attained a greater interest, owing to their significant role in stem cell differentiation into bone or cartilage regeneration. Several signaling cascades were involved in the tissue regeneration, which is also dependent on many factors such as scaffold properties, growth factors, cell–material and cell–cell interactions, physical cues, and bio mimicking microenvironment such as in vivo were played a vital role in shaping the future of tissue constructs. Thus significant research is taking place to elucidate the importance of scaffold features for bone or cartilage regeneration and biochemical, physical cues regulation. In the few decades, the role of biomolecules in bone or cartilage regeneration and molecular events pertaining to differentiation potential of stem cells in the controlled environment has attracted more attention. Herein, we discuss in detail the approaches followed for the delivery of bioactive molecules for bone and cartilage TE.

Given the significance of local delivery of bioactive molecules, various technologies have been evolved for spatial regulation of bioactive factor delivery for bone regeneration. The developed technologies are presented in the schematic diagram of Fig. 7–1, and the use of technology relies on the degree of control and complexity [3]. Physical entrapment of bioactive molecules in the biomaterial scaffold is the most common method used for controlled delivery, and the diffusion phenomenon or scaffold degradation controls the delivery of molecule. Importantly, the affinity of molecules with the scaffold material also plays a significant role in delivery, and higher affinity resulted in the slow delivery of molecules. If the molecule is bound covalently to the material, the delivery could occur when the material degrades, or bonds are broken. The concept of bioactive molecule delivery is not meant for the delivery of the molecule, primarily intended to protect the activity of molecule along with the controlled delivery. Various stimuli techniques have been used to deliver the biofactors in response to applied stimuli. Nanotechnology tools have widely been demonstrated to introduce the stimuli response given delivery, and most widely studied the concept of nanomaterial-based drug or gene delivery for cancer treatment [4].

Diffusion-based delivery is the easiest method; however, targeted and controlled delivery could not be achieved. Bioactive factors usually dissolved into solutions and freeze dried to make a scaffold resulted in safe guarding of the biomolecule till it reaches the cell, and enzymes cannot act upon the molecule directly till the degradation of the material. Several factors could influenced the delivery mechanism: the pore size of the scaffold, degradation rate, swelling over time, scaffold-active molecular interactions, and physiological conditions given in the system

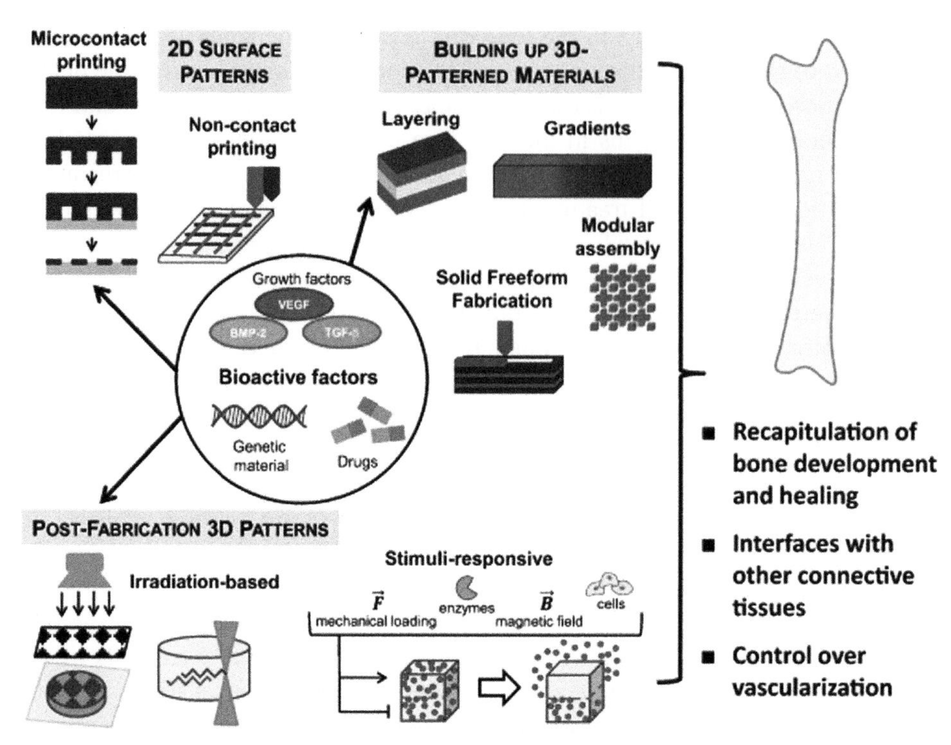


**FIGURE 7–1** Technologies used for bioactive molecule delivery for bone healing and development. *Adapted with permission from J.E. Samorezov, E. Alsberg, Spatial regulation of controlled bioactive factor delivery for bone tissue engineering, Adv. Drug Deliv. Rev. 84 (2015) 45–67. Elsevier ©2015.*

are essential to be considered to govern the delivery of bioactive molecule [5]. Biomaterial–molecular interactions resulted in a sustained release of the molecule, which can be modulated according to the desired period of delivery. Intermolecular interactions based on the affinity or charge is a very effective delivery mechanism compared to diffusion alone. Growth factors BMP-2, VEGF, FGF-2, and TGF-β1 widely studied for bone regeneration, wherein the controlled delivery has governed using negatively charged biomaterials due to their net positive charge [6–9]. These growth factors have an affinity for heparin; and hence, heparin and its derivatives were widely explored for the controlled delivery based on their affinity interactions in BTE applications [10].

BMP-2 controlled delivery for BTE is highly exploited, mainly various biomaterial scaffolds have been studied for the delivery such as collagen I, chitosan, poly(lactic-glycolic acid) (PLGA), polycaprolactone (PCL), and silk hydrogel for bone regeneration [11–15].

Various technologies have been developed to create the micropatterns on scaffolds and hydrogels for the cell or growth factor encapsulation, which were discussed in detail by Samorezov and Alsberg [3]. 2D surface patterns were well studied for bone or cartilage TE. The lacuna existed in 2D models in this strategy could not mimic the exact niche to develop tissue constructs. Generally, the three-dimensional (3D) environment existing in in vivo, facilitates cell–cell cross talk and also nutrient uptake, and waste disposal can be provided to favor cell growth for tissue development. The concept of 3D cell culture has arisen to develop engineered tissue constructs and disease models for drug screening to mimick the in vivo environment. 3D printing is the best choice in producing the scaffold matrices according to the shape and size required for grafting, where cell culture can be done by providing appropriate growth factors for tissue regeneration. However, there is a great challenge persisted in developing high potential approaches for cell adhesion and provided mechanical cues that are essential to be considered for developing functional tissue. Given this interest, the surface patterning has attracted attention to improve the surface roughness for magnifying cell adhesion over the surface. Thus 3D printing had been mostly used as the technique for generating a micropattern-based scaffold matrices or hydrogels for a variety of TE applications.

## 7.3 Cell/biomolecule encapsulation

The immobilization of the molecule/cell in biocompatible material refers to encapsulation. Therapeutic molecules also can be encapsulated in materials to deliver the target site systematically for more extended periods. The concept of encapsulation has attracted attention to develop a protective layer from the immunosuppressive mechanism, which causes graft rejection. Conventionally, a variety of immunosuppressive drugs are used to avoid graft rejection, and the best way to minimize the usage of these drugs in the long run after graft implantation is encapsulation. The concept of bioencapsualtion was proposed by Thomas Chong, wherein cells were encapsulated in ultrathin polymer membrane to aid the immune protection and were coined as "artificial cells" [16]. The encapsulation of islet cells, introduced into a diabetic rat model, was the first successful approach, wherein the cells were viable with controlled glucose level. In 1998 the encapsulated cells were used in human trials. Though the encapsulated therapeutic molecules were administered to the target site for treatment, there is a need to inject encapsulated cells for sustained controlled delivery. On top of this, cell encapsulation is essential when genetically modified or nonhuman cells needed in the case of limited availability of donor cell [17].

### 7.3.1 Essential requirements for encapsulation

It is essential to understand the process inferred in the encapsulation mechanism, and the stability of the encapsulated molecule needs to be considered. Some critical factors that are highly essential to be considered before encapsulation are the choice of material, mechanical strength and durability, permeability, cell source, chemical stability of the material, biosafety, and long-term survival [18]. The permeability of the membrane should be in the limits of cutoff value of immune cells for immune protection. In the past two decades, enormous research has been taken place in the selection of biomaterial to the development of smart materials for target delivery systems via the encapsulation approach. Various natural and synthetic polymers have been studied for their suitability to encapsulate the cells/ biomolecules and also the cell behavior and viability have been demonstrated concerning their desired functional activity at the in vivo level. Polymers such as alginate, agarose, polyethylene glycol (PEG), chitosan, polyvinyl alcohol, silk, and methacrylate gel have been assessed for encapsulation [2]. In particular, hydrogels widely studied for cell encapsulation. Stem cell–based therapy has attracted the attention to treat various diseases. However, stem cell alone could not be viable in the in vivo environments, as the signaling cascades can not respond to the newly given cells in the human system. Consequently, cell death would occur, which is a setback existing in stem cell–based therapy. There is an alternative essential for supporting the survival and growth of stem cells, which could be possible by providing essential growth factors. A variety of hydrogels have been developed to encapsulate the cells together with appropriate growth factors [19]. All the polymer materials and hydrogels should possess the following characteristic features:

- The material should be nonimmunogenic protect from the host immune system.
- To provide immune protection, they should have finely tuned porosity for blocking immune-related cells and allowing essential nutrients for cell growth.
- Controlled degradation is essential to an increase in the life time of the cell.
- Low adsorption of proteins.

In encapsulated cells the porous structure of material would act as the primary gateway for an incoming and outgoing molecule (Fig. 7–2). The nutrients required for cell metabolism generally will be obtained from the surrounding environment and also metabolic waste needs a release from the encapsulated matrix. Mostly, hydrogel materials are highly recommended for the cell encapsulation owing to desirable mechanical strength, finely tuned porosity, tunable degradation, possible to design for stimuli responsive delivery, and traceable by introducing the fluorescent molecule [20]. Numerous studies have suggested that a high surface area–to-volume ratio plays a vital role in developing material for encapsulation. Microbeads and hydrogel matrix should be designed in such a way to aid good nutrient transport, waste removal, blocking the immune cells, and importantly high surface area–to-volume ratio for prolonged cell survival in the encapsulated matrix [21]. For the encapsulation of molecules and cells, the chosen material has to possess excellent mechanical strength in such a way that once implanted; it should not be broken under the physiological conditions and also not to be dissolved. The broad spectrum of studies on microencapsulation over other conventional scaffold–based encapsulation demonstrates the superiority of microencapsulation for cell encapsulation [17]. The advantages of microencapsulation have attracted attention to the development of various technologies, and even microfluidic platforms have been developed for microencapsulation. The essential parameters required for cell/bioactive molecule encapsulation are depicted in Fig. 7–3.

Biocompatibility of material is another vital consideration for cell encapsulation. Very few clinical grade polymers have been approved, which are essential to be used for the cell

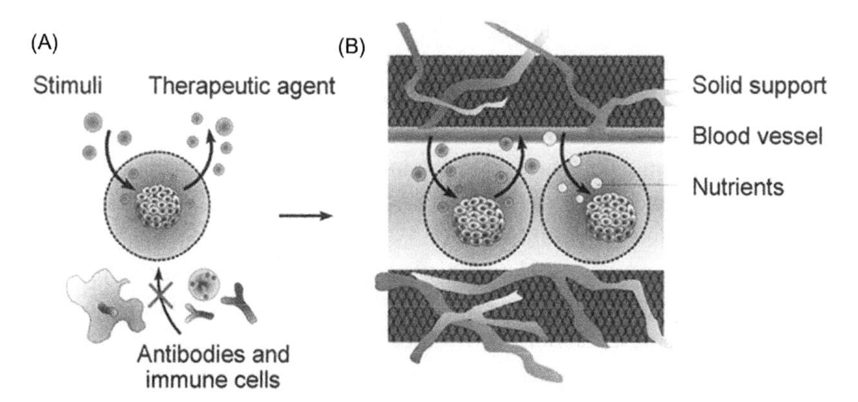


**FIGURE 7–2** Cell encapsulation (A) Diffusion of nutrients through a membrane and blocking the immune cells and antibodies. (B) Optimal nutrient availability to the encapsulated cell through the prevascularized system. *Reprinted with permission from G. Orive, R.M. Hernández, A.R. Gascón, R. Calafiore, T.M. Chang, P. De Vos, et al., Cell encapsulation: promise and progress, Nat. Med. 9 (2003) 104–107. Springer Nature ©2003.*

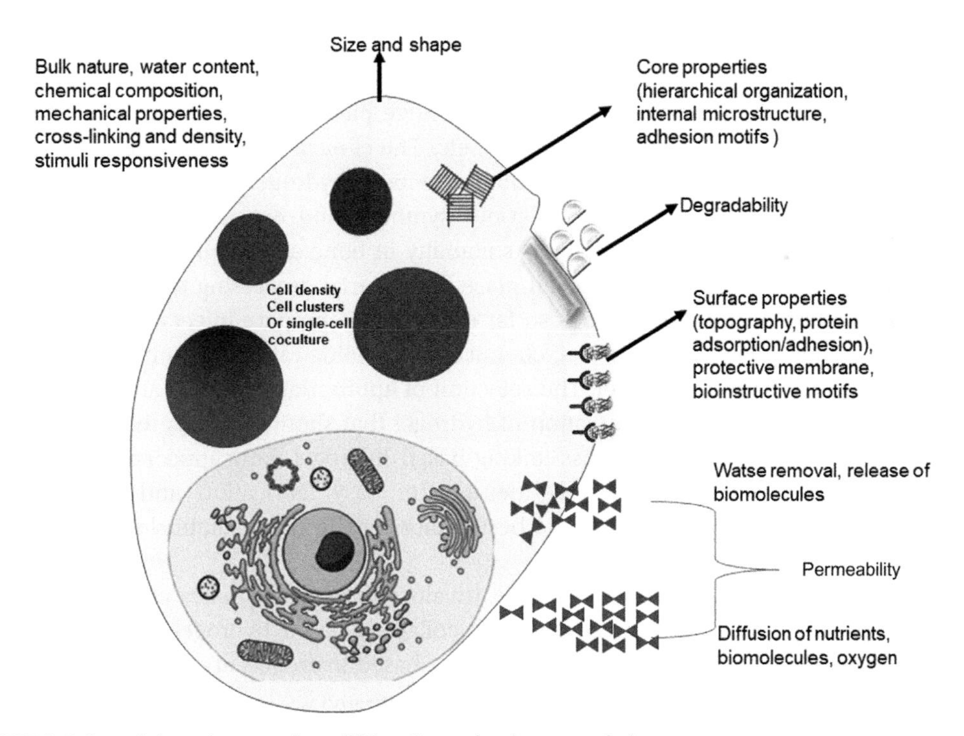


**FIGURE 7–3** Essential requirements for cell/bioactive molecule encapsulation.

encapsulation and any other graft implantation. Alginate has been explored well for its suitability to encapsulate the cells/biomolecules. However, the scientific findings were concluded that there is a great need to minimize the endotoxins and protein content, which play a significant role in the biocompatibility of the material. Reproducibility of material and fabrication technology for acquiring the same size and shape for encapsulation is highly recommended, which could pave the way for commercializing the technology for immediate use, and easily can be translated into clinical use.

Cell adhesion motifs are essential to be considered for cell/biomolecule encapsulation. For this, several cationic polymers are being explored for encapsulation. In particular, the biomolecule charge usually considered to select the cationic/anionic material. Electrostatic interaction is one of the best approaches for the adhesion of cell or bioactive molecules. This could pave the way for cell encapsulation. The presence of cell-binding motifs such as RGD sequence (arginine, glycine, and aspartate) over material also considered owing to the cell adhesive properties of the material.

### 7.3.2 The design aspect of hydrogels for encapsulation

Recently injectable hydrogel concept is the research trend owing to its benefits such as minimally invasive procedure and can avoid the complication arises in defect management. The clinicians ideally prefer to have a minimally invasive strategy for the treatment of various diseases since the general surgery procedure causes huge pain, chances of infection, the need for anesthesia before surgery, heavy blood loss, etc. The clinicians over the world are looking for an alternative strategy, and in the meantime, injectable hydrogels are the potential substitutes in TE and regenerative medicine. Various synthetic and natural materials have been used for gelation and well studied for their suitability in bone and cartilage TE. Importantly, a significant amount of research has taken place for material processing to develop a variety of hydrogel platforms, and the methods so far employed to prepare injectable hydrogels are described in detail in Fig. 7–4. Physical, chemical, and biological methods have been developed for the preparation of hydrogel. The selection of appropriate material and type of fabrication plays a vital role in the formulation of hydrogel that should possess the characteristic features of native tissue. Physical cross-linking-based hydrogel fabrication mostly preferred over chemically cross-linked hydrogels, because of easy fabrication and no chemicals involved in the fabrication made as to the best choice for the encapsulation of cell/bioactive molecule.

Natural polymers such as polysaccharide (hyaluronic acid, alginate, chitosan, agarose, and gellan gum), protein (fibroin, gelatin, and collagen) based hydrogels have extensively investigated for TE application. Synthetic polymers–based hydrogels are also widely reported for TE applications. Depending on the desirability, tissue-specific hydrogels using a combination of natural and synthetic polymers can be developed in various TE and regenerative medicine applications. Hydrogels can absorb water due to the presence of hydrophilic polymers. Swelling and retaining its original structure are the common features of hydrogels and will not dissolve in water. Hydrogels are biocompatible, and its porous structure facilitates

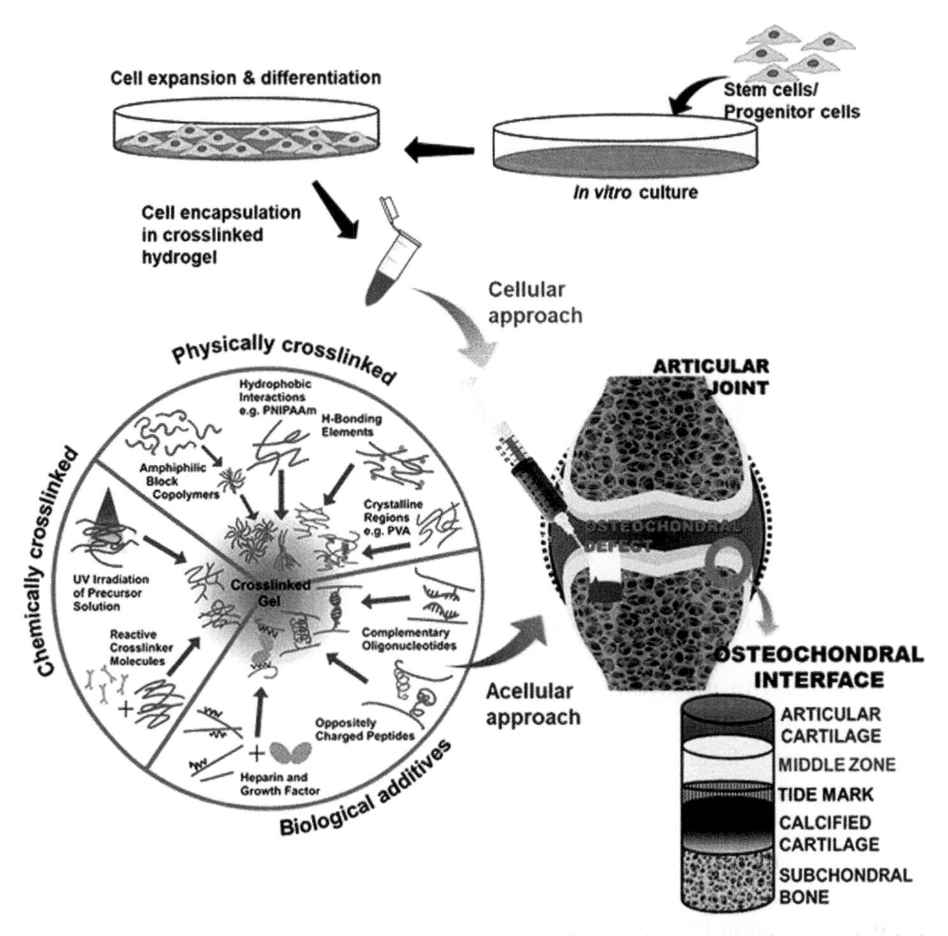


**FIGURE 7–4** Schematic illustration of various methods for injectable hydrogel preparation for osteochondral tissue engineering. *Reprinted with permission from Y.P. Singh, J.C. Moses, N. Bhardwaj, B.B. Mandal, Injectable hydrogels: a new paradigm for osteochondral tissue engineering, J. Mater. Chem. B 6 (2018) 5499–5529 [22]. The Royal Society of Chemistry ©2018; E.S. Place, J.H. George, C.K. Williams, M.M. Stevens, Synthetic polymer scaffolds for tissue engineering, Chem. Soc. Rev. 38 (2009) 1139–1151 [23]. The Royal Society of Chemistry ©2009.*

good nutrient, and oxygen transfer within the hydrogel, and also uniform cell distribution in the hydrogel could be achieved through encapsulation. In particular, hydrogels have widely used for cartilage repair owing to their mechanical properties in line with cartilage mechanical strength, and especially it is a minimally invasive approach to fill the defect of any size. Natural polymers—based hydrogels have been widely accepted due to their higher biocompatibility, and non-immunogenesity. Alginate, hyaluronic acid, fibrin-based hydrogels have been used for articular cartilage repair and regeneration. The reproducibility of chemical and mechanical properties is an important consideration in choosing the polymer for hydrogel fabrication. Mostly, synthetic polymers have a greater advantage in reproducing their chemical and physical properties, which is quite challenging in the case of natural polymers. However, all the synthetic polymers do not have biocompatibility feature, which is quite an important aspect of choosing a material for any TE and regenerative medicine applications [24].

Controlled delivery of bioactive molecules to mimic natural microenvironment in bone and cartilage setting is a significant factor to be considered in the fabrication of hydrogel. Pore size and degradation kinetics can be tailored in the hydrogels to attain the controlled delivery of biofactor. The combination of scaffold and hydrogel can also be used for the controlled delivery of bioactive molecule, where degradation kinetics plays a role in a controlled delivery. Besides this controlled delivery, tissue formation concerning the degradation of the scaffold is essential to be considered in the fabrication of hydrogel/scaffold for any TE application. The success rate of the hydrogel is dependent on the systematic delivery of the bioactive molecule to the target site. The overall dose of the bioactive molecule, neutralization of the immune system, controlled delivery, and possible delivery of bioactive molecules to the nontargeted site are the influencing parameters to decide the efficacy of the approach.

The mechanism of the loading of the bioactive molecules into the hydrogel also plays a crucial role in the release of a biomolecule at the target site, which in turn affects the efficiency of therapeutic action. Physical and chemical methods are used for the loading of the bioactive molecules in the hydrogel network for encapsulation. The controlled delivery of bioactive molecules in the physical encapsulation can be achieved through the diffusion mechanism, which is facilitated by the porous network of the hydrogel. However, this pore-mediated diffusion cannot sustain long-term delivery of bioactive molecule at the target site. Thus chemical immobilization of bioactive molecule can be used to overcome the limitation of short-term release of the bioactive molecule, and also its stimuli-responsive approach could be adopted for the long-term sustained release to improve the efficacy of therapeutic action. The degradation kinetics of hydrogel polymer also controls the delivery of the bioactive molecule in both physical and chemical immobilization of bioactive molecules, which ideally prolongs the delivery period of the molecule. However, the strong chemical cross-linking of a hydrogel polymer network with a bioactive molecule resists the delivery at the target site. Thus chemical composition and cross-linking mechanism thoroughly need to be understood in developing hydrogel for the controlled delivery of a bioactive molecule.

### 7.3.3 Multifunctional cell/bioactive molecule encapsulation system

The dynamic microenvironment existing in the human body influences the cell's behavioral response. The materials used for biological application and studies about the clinical investigations should consider the physical and chemical conformational changes of materials under the applied mechanical and chemical stimuli. Thus numerous materials have been studied for their conformational and chemical modifications under applied stimuli. Stimuli-responsive materials have been developed to introduce chemical or physical modifications under applied stimuli. These changes are also associated with changes in the polymeric chains of scaffold or hydrogel. Generally, the changes in polymeric chains occur due to changes in pH, temperature, mechanical forces, electrical, and magnetic field. In general, this type of stimuli has been studied for the controlled delivery of bioactive molecule or drugs. Numerous in situ hydrogels have been proposed for the localized delivery of molecules/cells to physiological changes in pH. In particular, encapsulated systems are being well studied with the response to temperature, magnetic, and light stimuli.

Temperature-responsive hydrogels have been fabricated using thermoresponsive polymers. The gelation mechanism relies on temperature, especially polymers, exhibit lower critical solution temperature, and hydrogel formation takes place with increasing temperatures. These hydrogels are of special interest in biomedical application owing to minimally invasive implantation using in situ hydrogel formation in response to physiological temperature. The liquid state allows for easy injection, and gelation permits for conformational modifications at the defect site are the reasons for wider biomedical applications. The release of water from hydrogel could occur after gelation would allow a physical change in terms of the shape and size of the hydrogel matrix. The size and shape of the lesion site may not be filled with hydrogel due to the structural changes of the hydrogel. This also may impair the delivery of the bioactive molecule through a diffusion mechanism and also affects the cell viability in the encapsulated system. To overcome this problem, dual-gelling macromeres (methacrylate thermogelling) have been introduced in the fabrication of hydrogel for the encapsulated system. The advantages of these hydrogels are biointegration and biodegradation. The functionalized mono-acryloxyethyl phosphate hydrogels were used to encapsulate mesenchymal stem cells for bone regeneration, and the results have demonstrated an enhanced alkaline phosphatase activity, calcium mineralization, and collagen formation affirming the improved bone growth [25]. The functionalized hydrogels using temperature-responsive materials for cell encapsulation provided improved cell proliferation and differentiation. The tri block polymer poly(ethylene oxide)-*b*-poly(propylene oxide)-*b*-poly(ethylene oxide) possessing thermal responsive properties was used to encapsulate TGF-β1 for cartilage regeneration.

Photoactivated encapsulation systems also have a particular interest in biomedical applications. No chemical cross-linkers are required for gelation, and similar to temperature-activated hydrogels, hydrogel formation occurs under exposed light. The most commonly UV light cross-linking approach has been widely used for the encapsulation, and in particular, spatiotemporal control on cells/bioactive molecules is the advantage of encapsulation. The clinically acceptable curing time is considered for in situ gelling and requires low energy. The polymerization under applied light can be achieved in the physiological pH and temperature, which promotes the use of photoactivated encapsulation systems for clinical settings. The essential requirement is to select an appropriate photo initiator, which should be biocompatible and irradiate with visible light. The photo initiators may produce free radicals, which are generally toxic to cells. The cell viability, cross-linking, mechanical, and degradation properties are ideally considered for photoactivated encapsulation system. The prolonged exposure of UV light effects on the cell encapsulation system also needs to be evaluated.

The magnetic stimuli-responsive encapsulation system is another approach to design system for the encapsulation of cells/bioactive molecules. Smart materials have been developed by introducing magnetic nanoparticles in microscale hydrogels, which respond to applied magnetic stimuli. 3D constructs can be developed using magnetic nanoparticle–based hydrogels under applied magnetic stimuli, which have been investigated for TE and regenerative medicine. Multilayered structures have been employed for 3D construct development, wherein magnetic nanoparticles are added in layered assembly. The encapsulated cells in this system could be used to study their differentiation ability to magnetic stimuli. The molecular mechanism involved in cell differentiation can also be investigated using a cell encapsulated system. The encapsulated cells using magnetic nanoparticles in hydrogel delivery depend on the degradation of hydrogel. The fibers manipulated with magnetic coating have a potential application due to the easy fabrication of a variety of scaffolds systems in different sizes and shapes. This enables the usage of these scaffolds for various TE applications.

## 7.4 Bioactive molecule/cell encapsulation for bone and cartilage

### 7.4.1 Bioactive molecules

The incorporation of the bioactive molecule into the scaffold is essential, which aids in cell adhesion, proliferation, differentiation, and subsequently, it is possible to regenerate a functional tissue. For example, mitogens induce cell division, growth factors stimulate cell proliferation, and morphogens are involved in tissue generation. A few bioactive molecules encapsulated systems for bone and cartilage TE are indicated in Table 7–1. TGFF-1 was

**Table 7–1** Encapsulation of bioactive molecules for bone and cartilage tissue engineering.

| Bioactive molecules | Scaffold/hydrogel | Functional activity | Reference |
|---|---|---|---|
| TGF-β3 | Fibrin hydrogel | Higher production of glycosaminoglycan and collagen formation | [26] |
| TGF-β3 | Poly(lactide-*co*-caprolacton) scaffold | Formation of a hyaline cartilage-specific lacunae structure which inhibits the hypertrophy of chondrocyte differentiation | [27] |
| Mesenchymal stem cells and anti-BMP-2 monoclonal antibody | Alginate microspheres | Higher expression of Runx2 and alkaline phosphatase activity | [28] |
| BMPs | α-Tricalcium phosphate/ poly(D,L-lactide-*co*-glycolide) composite | Enhanced osteogenesis confirmed with higher alkaline phosphatase activity | [29] |
| Dexamethasone and BMP-2 | PLGA | Improved bone regeneration | [30] |

*PLGA*, Poly(lactic-glycolic acid).

encapsulated in peptide-based hydrogels for the controlled delivery to enhance cartilage regeneration [31]. Bone morphogenic proteins have been widely studied for BTE owing to their participation in osteogenesis of stem cells. The hydrogels were injected into a scaffold, wherein the controlled delivery of bioactive molecule can be achieved by the hydrogel, which also improved the mechanical strength. These scaffolds are suitable for hard TE. The ceramic, metal, and polymer-based scaffolds combined with hydrogel, have been attracted the attention for developing such scaffolds for BTE applications [32]. Calcium phosphate (CaP) scaffolds encapsulated BMP-2 have shown enhanced cell proliferation and ALP activity, and the same scaffolds subcutaneously implanted in rats exhibited remarkable bone regeneration. Along with BMP-2, other biomolecules also were investigated for different TE applications using encapsulation techniques in the desired scaffolds. In another study, mesoporous silica nanoparticles (MSN) were used for encapsulating peptide derived from BMP-7. MG63 cells adhesion and proliferation were enhanced in MSN encapsulated system, and a notable improvement in the controlled release of the peptide was observed. Human mesenchymal stem cells (hMSCs) were also checked for osteogenic differentiation ability in the MSN encapsulated systems; alkaline phosphatase activity, calcium mineralization, and collagen formation were shown higher compared to bare MSN. This demonstrated that the encapsulated system provided a controlled release of the peptide, which improved the osteogenic differentiation of hMSCs [33]. Self-assembling peptide hydrogels have been widely investigated for the kinetic release of bioactive molecules, are nonimmunogenic and nonpathogenic, and also can form scaffold architecture at physiologic pH and temperature [34,35]. For cartilage regeneration, bone marrow stromal cells were encapsulated in hydrogel wherein TGF-β1 release kinetics was controlled using self-assembly peptide-based hydrogel [31]. Silk fibroin (SF)/CaP/PLGA nanocomposite scaffolds have been investigated for a vascular endothelia growth factor (VEGF) sustained delivery for improved angiogenesis along with the regeneration of bone tissue [36]. This could have a potential application in BTE owing to the ability of encapsulated VEGF scaffolds that could address the major limitations of vascularity, which is the key attention of tissue engineers who looked into developing engineered graft with good vascularization. Gene-activated scaffolds have emerged to substitute the growth factor embedded scaffolds, especially in regards to encapsulate growth factor in scaffolds for a variety of TE applications. In particular, osteochondral TE that represents both bone and cartilage regeneration, have been well investigated for nonviral gene delivery using nanoparticle encapsulated scaffolds. TGF-β3 and BMP-2 plasmid DNAs were encapsulated in collagen-II and collagen-I–Hap scaffolds, respectively, using CaP/PEI nanoparticles to induce mesenchymal stem cells differentiation into subchondral bone and articular cartilage regeneration [37]. Gene transfection efficiency was shown higher in the nonviral gene-activated matrix in which cells were encapsulated using nanoparticles compared to 2D transfection. The chondrogenic growth factor TGF-β1 and bone-related growth factor BMP-2 were encapsulated using mineral coated HAp nanoparticles for spatiotemporal controlled delivery of these growth factor–related pDNAs. The porcine MSCs were shown higher transfection in the encapsulated matrices. They also showed upregulation of chondrogenic and osteogenic gene expression, which indicated that bone and cartilage regeneration successfully could be achieved via encapsulation based on nonviral gene-activated based scaffolds [38]. The pDNAs of TGF-β3 and BMP-2 were encapsulated individually and in a combination of both in alginate hydrogels, wherein cells were also encapsulated in the scaffold. Both pDNAs encapsulated scaffolds are proved to be potential candidates for endochondral bone repair [39]. Poly(lactic-*co*-glycolic acid) microparticles were used to encapsulate polypeptide TP580 for inducing osteogenesis and were imbedded in poly(propylene fumarate) network. These scaffolds were implanted in rabbit femoral defect models, where TP580 enhanced the regeneration of endochondral bone [40]. The significant research outcomes in bioactive molecule/cell encapsulation were employed in either degradable [PLGA, poly(L-lactide) (PLA)] or nondegradable carriers (HAp, silica, gold, and dendrimer). Bone homeostasis is maintained by osteoblast and osteoclast cells, incorporating nanocarrier encapsulated growth factor and inhibitor could resulted in osteoblast proliferation and modulated osteoclast proliferation [41]. Many scientific studies have tried to elucidate the combinatorial growth, and inhibitor factors controlled delivery for clinical BTE applications [42]. BMP-2 growth factor coated with PLGA nanoparticles encapsulated in fibrin-based hydrogel showed the ectopic bone regeneration, which is modulated by the controlled release of growth factor for enhanced bone formation in rat calvarial defect model [43].

## 7.4.2 Encapsulation of cells

In TE and regenerative medicine, differentiated or undifferentiated cells, can be used for implantation at the defect site either for repair or regeneration. The scaffold architecture is crucial for cell invasion since the cell colonization throughout the scaffold is desirable for any application and also scaffold provide uniform distribution of nutrients to reach all the cells over and the core of the scaffold and removal of metabolic waste is also an important consideration, which leads to toxic to healthy cells. It is essential to look into the properties of scaffold, such as porosity, interconnectivity, pore size, and crystalline nature of the material, and these parameters could be decided based on the TE application [44]. Nanocomposite scaffolds have greater advantages over microporous scaffolds such as enhanced cell adhesion and protein or ligand absorption owing to the availability of high surface area–to-volume ratio. Cellular cross talk and cell adhesion were improved on the nanostructured surface scaffolds compared to flat surface scaffolds. Pore size and porosity are also significantly contributing to cell adhesion, and interconnected porous structure allows the cell colonization and even nutrient distribution throughout the scaffold. Along with the characteristic features of the scaffold loaded cells must be viable, can synthesize the extracellular matrix (ECM), and must maintain the tissue homeostasis. Numerous cell types of porcine adipose–derived MSCs, human umbilical cord blood derived MSCs, mouse osteoblast cells, human bone marrow stem cells, induced pluripotent stem cells, and human dental pulp stem cells have been studied for various TE applications [45–49]. Various cell encapsulated systems used for bone and cartilage TE are given in Table 7–2. Cell-laden scaffolds have been extensively studied for bone and cartilage TE along with a variety of TE applications. Though scaffold possesses all the essential properties such as good porous

**Table 7–2** Cell encapsulated systems for bone or cartilage tissue engineering.

| Cell type | Material type | Bone/ cartilage TE | Reference |
|---|---|---|---|
| Osteoblast | Injectable RGD-modified PEG hydrogels | Bone | [50] |
| Murine embryonic stem cells | Alginate | Bone | [51] |
| Mouse BMSCs | Methacrylated glycolchitosan-collagen | Bone | [52] |
| iPSCs | Alginate microbeads | Bone | [53] |
| hMSCs | Gelatin methacryloyl-nanosilicate | Bone | [54] |
| Osteoblasts | Alginate, gelatin, and nanocrystalline cellulose composite hydrogel | Bone | [55] |
| iPSMSCs | Calcium phosphate cement | Bone | [53] |
| Chondrocytes | Collagen-II/hyaluronic acid | Cartilage | [56] |
| Porcine auricular chondrocytes | CMP-TA and CS-TA | Cartilage | [57] |
| Bone marrow stem cells and chondrocytes | PCL–PEG–PCL | Cartilage | [58] |
| Human umbilical cord blood stem cells | Calcium phosphate scaffolds | Bone | [59] |
| Mesenchymal stem cells | Nano-calcium phosphate scaffolds | Bone | [60] |

*BMSCs*, Bone marrow stromal cells; *CMP-TA*, carboxymethyl pullulan-tyramine; *CS-TA*, chondroitin sulfate-tyramine; *hMSCs*, human mesenchymal stem cells; *iPSCs*, induced pluripotent stem cells; *iPSMSCs*, induced pluripotent stem cell–derived mesenchymal stem cells; *PCL*, polycaprolactone.

structure, interconnectivity, mechanical strength, pore size, and surface roughness, it is difficult to use them without any additional nanocomposites in TE, which is quite challenging. Thus nanocomposite scaffolds have emerged to mitigate the limitations associated with microporous scaffolds. The scientific reports on the use of nanocomposite scaffolds have shown better cell colonization over scaffolds and would have great potential applications in TE. In particular, nanoparticles such as HAp, and CaP generally improve the mechanical strength, and higher surface area–to-volume ratio provides improved cell attachment. These intrinsic features of nanoparticle addition are quite necessary for developing the scaffolds for BTE applications. The incorporation of nanoparticle and constructing layer-by-layer structures in scaffolds ideally are suitable for osteochondral TE application, wherein the scaffold can mimic the mechanical and physical environment of bone and cartilage in one scaffold. Various methodologies have undertaken to produce functionalized biomimetic scaffolds for bone and cartilage TE applications. However, cell encapsulation is a prominent technique that protects them the immune system and makes them biocompatible. Thus cell encapsulation has attracted attention to bone and cartilage TE. Laponite nanoparticles modified with polyacrylate were used to encapsulate primary osteoblast cells, and gelation was improved with the addition of laponite nanoparticles to the SF solution. This hydrogel has shown enhanced osteogenic differentiation and can be used for irregular bone defects [61]. The prolonged life of mesenchymal stem cells can be achieved through cell encapsulation is a potential therapy for osteoarthritis, wherein the immunomodulatory, noninflammatory, and trophic factors would prevent the mechanism against osteoarthritis [62]. An injectable alginate–CaP load-bearing hydrogel was used to encapsulate induced pluripotent stem cell–derived mesenchymal stem cells (iPSMSCs) for bone regeneration, and the results of this study confirmed that CaP-based hydrogels are the potential candidatures for bone, dental application, and in particular highly recommended for cranial defects [53]. The collagen-II/hyaluronic acid in situ hydrogels were used to encapsulate chondrocytes and growth factor TGF-β1; further cell proliferation and viability, ECM synthesis, glycosaminoglycan production, and chondrogenic gene expression studies were investigated. It was found that these hydrogels were potential candidates for cartilage TE applications [56]. Hyaluronic acid/PEG hydrogel with adequate mechanical strength was developed for cartilage TE, where hyaluronic acid plays an important role in chondrocytes adhesion, differentiation, and its ideal biomaterial for cartilage regeneration [26]. In situ hydrogels have extensively investigated for cell encapsulation, and significant research has undertaken on cartilage regeneration. The encapsulated hydrogels have shown higher cell viability and proliferation, extracellular matrix synthesis and confirmed cartilage regeneration. Thus hydrogels have potential applications for cartilage TE. The mechanical and degradation properties of hyaluronic acid hydrogels can be tailored using composite hydrogels. An injectable hydrogel modified with fibrin has attracted attention since the protein that improves cell adhesion, proliferation, and differentiation. Fibrin-based scaffolds have extensively reported for cartilage TE application, and fiber-reinforced scaffolds have greater advantages to employ in BTE applications due to the increased mechanical strength of scaffold. Encapsulated cells were introduced into these scaffolds and found out enhanced cell viability, proliferation, and differentiation. The alginate microbeads were used for the encapsulation of human embryonic stem cells, and further encapsulated microbeads were fused in the CaP cement for BTE application. The encapsulated cells have shown higher viability, proliferation, and differentiation; especially, the osteogenic potential of these stem cells that have checked for gene expression and ECM synthesis [63]. Human umbilical cord blood stem cells encapsulated in alginate–fibrin microbeads injectable scaffolds have investigated for bone regeneration; and gene expression of ALP, osteocalcin, and Runx2 showed higher by the encapsulated cells. The release of stem cells at the targeted site is fast compared to the normal scaffold, which triggers the osteogenic differentiation. The fibrin addition to the alginate tailored mechanical strength of the injectable scaffold and cell attachment enhanced due to the presence of protein [63].

# References


[1] R.A. Perez, H.W. Kim, Core-shell designed scaffolds for drug delivery and tissue engineering, Acta Biomater. 21 (2015) 2–19.

[2] J.L. Pedraz, D. Emerich, P. De Vos, G. Orive, L.U. Wahlberg, D. Poncelet, et al., Cell encapsulation: technical and clinical advances, Trends Pharmacol. Sci. 36 (2015) 537–546.

[3] J.E. Samorezov, E. Alsberg, Spatial regulation of controlled bioactive factor delivery for bone tissue engineering, Adv. Drug Deliv. Rev. 84 (2015) 45–67.

[4] A. Gangrade, B.B. Mandal, Injectable carbon nanotube impregnated silk based multifunctional hydrogel for localized targeted and on demand anticancer drug delivery, ACS Biomater. Sci. Eng. 5 (2019) 2365–2381.

[5] M.K. Nguyen, E. Alsberg, Bioactive factor delivery strategies from engineered polymer hydrogels for therapeutic medicine, Prog. Polym. Sci. 39 (2019) 1235–1265.

[6] P. Taddei, V. Chiono, A. Anghileri, G. Vozzi, G. Freddi, G. Ciardelli, Silk fibroin/gelatin blend films cross-linked with enzymes for biomedical applications, Macromol. Biosci. 13 (2013) 1492–1510.

[7] Q. Li, J. Wang, H. Liu, B. Xie, L. Wei, Tissue-engineered mesh for pelvic floor reconstruction fabricated from silk fibroin scaffold with adipose-derived mesenchymal stem cells, Cell Tissue Res. 354 (2013) 471–480.

[8] T. Chlapanidas, M.C. Tosca, S. Faragò, S. Perteghella, M. Galuzzi, G. Lucconi, et al., Formulation and characterization of silk fibroin films as a scaffold for adipose-derived stem cells in skin tissue engineering, Int. J. Immunopathol. Pharmacol. 26 (2013) 43–49.

[9] R.S. Hayden, M. Vollrath, D.L. Kaplan, Effects of clodronate and alendronate on osteoclast and osteoblast co-cultures on silk-hydroxyapatite films, Acta Biomater. 10 (2014) 486–493.

[10] J.M. Oliveira, S.S. Silva, P.B. Malafaya, M.T. Rodrigues, N. Kotobuki, M. Hirose, et al., Macroporous hydroxyapatite scaffolds for bone tissue engineering applications: Physicochemical characterization and assessment of rat bone marrow stromal cell viability, J. Biomed. Mater. Res., A 91 (2009) 175–186.

[11] F. Dehghani, N. Annabi, Engineering porous scaffolds using gas-based techniques, Curr. Opin. Biotechnol. 22 (2011) 661–666.

[12] N. Siddiqui, K. Pramanik, E. Jabbari, Osteogenic differentiation of human mesenchymal stem cells in freeze-gelled chitosan/nano β-tricalcium phosphate porous scaffolds crosslinked with genipin, Mater. Sci. Eng. C 54 (2015) 76–83.

[13] B.B. Mandal, A. Grinberg, E. Seok Gil, B. Panilaitis, D.L. Kaplan, High-strength silk protein scaffolds for bone repair, Proc. Natl. Acad. Sci. U.S.A. 109 (2012) 7699–7704.

[14] N.K. Mekala, R.R. Baadhe, S.R. Parcha, Study on osteoblast like behavior of umbilical cord blood cells on various combinations of PLGA scaffolds prepared by salt fusion, Curr. Stem Cell Res. Ther. 8 (2013) 253–259.

[15] R.E. Billo, C.R. Oliver, R. Charoenwat, B.H. Dennis, P.A. Wilson, J.W. Priest, et al., A cellular manufacturing process for a full-scale biodiesel microreactor, J. Manuf. Syst. (2014). Available from: https://doi.org/10.1016/j.jmsy.2014.07.004.

[16] T.M.S. Chang, Semipermeable microcapsules, Science (80–) 146 (1964) 524–525.

[17] A. Murua, A. Portero, G. Orive, R.M. Hernández, M. de Castro, J.L. Pedraz, Cell microencapsulation technology: towards clinical application, J. Control. Release 132 (2008) 76–83.

[18] G. Orive, R.M. Hernández, A.R. Gascón, R. Calafiore, T.M. Chang, P. De Vos, et al., Cell encapsulation: promise and progress, Nat. Med. 9 (2003) 104–107.

[19] G. Choe, J. Park, H. Park, J.Y. Lee, Hydrogel biomaterials for stem cell microencapsulation, Polymers (Basel), 10, 2018, pp. 1–17.

[20] L. Gasperini, J.F. Mano, R.L. Reis, Natural polymers for the microencapsulation of cells, J. R. Soc. Interface 11 (2014) 20140817.

[21] J. Schrezenmeir, J. Kirchgessner, L. Gerö, L.A. Kunz, J. Beyer, W. Mueller-Klieser, Effect of microencapsulation on oxygen distribution in islets organs, Transplantation 57 (1994) 1308–1314.

[22] Y.P. Singh, J.C. Moses, N. Bhardwaj, B.B. Mandal, Injectable hydrogels: a new paradigm for osteochondral tissue engineering, J. Mater. Chem. B 6 (2018) 5499–5529.

[23] E.S. Place, J.H. George, C.K. Williams, M.M. Stevens, Synthetic polymer scaffolds for tissue engineering, Chem. Soc. Rev. 38 (2009) 1139–1151.

[24] A. Rey-Rico, H. Madry, M. Cucchiarini, Hydrogel-based controlled delivery systems for articular cartilage repair, Biomed. Res. Int. 2016 (2016) 1–12.

[25] M.P. Lutolf, J.L. Lauer-Fields, H.G. Schmoekel, A.T. Metters, F.E. Weber, G.B. Fields, et al., Synthetic matrix metalloproteinase-sensitive hydrogels for the conduction of tissue regeneration: engineering cell-invasion characteristics, Proc. Natl. Acad. Sci. U.S.A. 100 (2003) 5413–5418.

[26] F. Yu, X. Cao, Y. Li, L. Zeng, B. Yuan, X. Chen, An injectable hyaluronic acid/PEG hydrogel for cartilage tissue engineering formed by integrating enzymatic crosslinking and Diels–Alder "click chemistry", Polym. Chem. 5 (2013) 1082–1090.

[27] S. Hee, S. Hyun, Y. Jung, TGF-β3 encapsulated PLCL scaffold by a supercritical $CO_2$ – HFIP co-solvent system for cartilage tissue engineering, J. Control. Release 206 (2015) 101–107.

[28] A. Moshaverinia, S. Ansari, C. Chen, X. Xu, K. Akiyama, M.L. Snead, et al., Biomaterials Co-encapsulation of anti-BMP2 monoclonal antibody and mesenchymal stem cells in alginate microspheres for bone tissue engineering, Biomaterials 34 (2013) 6572–6579.

[29] A. Sharma, F. Meyer, M. Hyvonen, S.M. Best, R.E. Cameron, N. Rushton, Osteoinduction by combining bone morphogenetic protein (BMP)-2 with a bioactive novel nanocomposite, Bone Jt. Res. 1 (2012) 145–151.

[30] D. Hoon, C. Ho, I. Hwan, H. Jae, K. Park, D. Keun, Fabrication of core–shell microcapsules using PLGA and alginate for dual growth factor delivery system, J. Control. Release 147 (2010) 193–201.

[31] P.W. Kopesky, S. Byun, E.J. Vanderploeg, J.D. Kisiday, D.D. Frisbie, A.J. Grodzinsky, Sustained delivery of bioactive TGF-β1 from self-assembling peptide hydrogels induces chondrogenesis of encapsulated bone marrow stromal cells, J. Biomed. Mater. Res., A 102 (2014) 1275–1285.

[32] C.Y. Xu, R. Inai, M. Kotaki, S. Ramakrishna, Aligned biodegradable nanofibrous structure: a potential scaffold for blood vessel engineering, Biomaterials 25 (2004) 877–886.

[33] S. Wei, M. Wang, Y. Deng, R. Zhang, Z. Luo, Q. Zhao, et al., Peptide-laden mesoporous silica nanoparticles with promoted bioactivity and osteo-differentiation ability for bone tissue engineering, Colloids Surf., B: Biointerfaces 131 (2015) 73–82.

[34] M.C. Branco, D.J. Pochan, N.J. Wagner, J.P. Schneider, Macromolecular diffusion and release from self-assembled β-hairpin peptide hydrogels, Biomaterials 30 (2009) 1339–1347.

[35] K. Rajangam, M.S. Arnold, M.A. Rocco, S.I. Stupp, Peptide amphiphile nanostructure-heparin interactions and their relationship to bioactivity, Biomaterials 29 (2008) 3298–3305.

[36] M. Farokhi, F. Mottaghitalab, M.A. Shokrgozar, J. Ai, J. Hadjati, M. Azami, Bio-hybrid silk fibroin/calcium phosphate/PLGA nanocomposite scaffold to control the delivery of vascular endothelial growth factor, Mater. Sci. Eng. C 35 (2014) 401–410.

[37] Y. Lee, H. Wu, C. Yeh, C. Kuan, H. Liao, H. Hsu, et al., Enzyme-crosslinked gene-activated matrix for the induction of mesenchymal stem cells in osteochondral tissue regeneration, Acta Biomater. 63 (2017) 210–226.

[38] A. Mcmillan, M. Khanh, T. Gonzalez-Fernandez, E. Alsberg, Dual non-viral gene delivery from microparticles within 3D high-density stem cell constructs for enhanced bone tissue engineering, Biomaterials 161 (2018) 240–255.

[39] T. Gonzalez-Fernandez, E.G. Tierney, G.M. Cunniffe, F.J. O'Brien, D.J. Kelly, Gene delivery of TGF-β3 and BMP2 in an MSC-laden alginate hydrogel for articular cartilage and endochondral bone tissue engineering, Tissue Eng., A 22 (2016) 776–787.

[40] E.L. Hedberg, H.C. Kroese-Deutman, C.K. Shih, R.S. Crowther, D.H. Carney, A.G. Mikos, et al., Effect of varied release kinetics of the osteogenic thrombin peptide TP508 from biodegradable, polymeric scaffolds on bone formation in vivo, J. Biomed. Mater. Res., A 72 (2005) 343–353.

[41] R. Nanomedicine, E.A. Seifalian, G.G. Walmsley, A. Mcardle, R. Tevlin, A. Momeni, et al., Nanotechnology in bone tissue engineering, nanomedicine nanotechnology, Biol. Med. 11 (2015) 1253–1263.

[42] A. Tautzenberger, A. Kovtun, A. Ignatius, Nanoparticles and their potential for application in bone, Int. J. Nanomed. 7 (2012) 4545–4557.

[43] C. Qiao, K. Zhang, H. Jin, L. Miao, C. Shi, X. Liu, et al., Using poly(lactic-*co*-glycolic acid) microspheres to encapsulate plasmid of bone morphogenetic protein 2/polyethylenimine nanoparticles to promote bone formation in vitro and in vivo, Int. J. Nanomed. 8 (2013) 2985–2995.

[44] B. Bhaskar, R. Owen, H. Bahmaee, Z. Wally, P.S. Rao, G.C. Reilly, Composite porous scaffold of polyethylene glycol (PEG)/polylactic acid (PLA) support improved bone matrix deposition in vitro compared to PLA-only scaffolds, J. Biomed. Mater. Res., A. 106 (2018) 1334–1340.

[45] J.C. Moses, S.K. Nandi, B.B. Mandal, Multifunctional cell instructive silk-bioactive glass composite reinforced scaffolds toward osteoinductive, proangiogenic, and resorbable bone grafts, Adv. Healthc. Mater. 7 (2018) 1701418.

[46] B. Bhaskar, R. Owen, H. Bahmaee, P.S. Rao, G.C. Reilly, Design and assessment of a dynamic perfusion bioreactor for large bone tissue engineering scaffolds, Appl. Biochem. Biotechnol. 185 (2018) 555–563.

[47] B. Birru, N.K. Mekala, S.R. Parcha, Improved osteogenic differentiation of umbilical cord blood MSCs using custom made perfusion bioreactor, Biomed. J. 41 (2018) 290–297.

[48] B. Bhaskar, N.K. Mekala, R.R. Baadhe, P.S. Rao, Role of signaling pathways in mesenchymal stem cell differentiation, Curr. Stem Cell Res. Ther. 9 (6) (2014) 508–512.

[49] B. Birru, N.K. Mekala, S.R. Parcha, Mechanistic role of perfusion culture on bone regeneration, J. Biosci. 44 (2019) 23.

[50] J.A. Burdick, K.S. Anseth, Photoencapsulation of osteoblasts in injectable RGD-modified PEG hydrogels for bone tissue engineering, Biomaterials. 23 (2002) 4315.

[51] Y.S. Hwang, J. Cho, F. Tay, J.Y.Y. Heng, R. Ho, S.G. Kazarian, et al., The use of murine embryonic stem cells, alginate encapsulation, and rotary microgravity bioreactor in bone tissue engineering, Biomaterials 30 (2009) 499–507.

[52] C. Arakawa, R. Ng, S. Tan, S. Kim, B. Wu, M. Lee, Photopolymerizable chitosan – collagen hydrogels for bone tissue engineering, J. Tissue Eng. Regen. Med. 11 (2017) 164–174.

[53] P. Wang, Y. Song, M.D. Weir, J. Sun, L. Zhao, C.G. Simon, et al., A self-setting iPSMSC-alginate-calcium phosphate paste for bone tissue engineering, Dent. Mater. 32 (2016) 252–263.

[54] A. Paul, V. Manoharan, D. Krafft, A. Assmann, J.A. Uquillas, R. Shin, et al., Nanoengineered biomimetic hydrogels for guiding human stem cell osteogenesis in three dimensional microenvironments, J. Mater. Chem. B 4 (2016) 3544–3554.

[55] K. Wang, K.C. Nune, R.D.K. Misra, The functional response of alginate-gelatin-nanocrystalline cellulose injectable hydrogels toward delivery of cells and bioactive molecules, Acta Biomater. 36 (2016) 143–151.

[56] L.-S. Kontturi, E. Järvinen, V. Muhonen, E.C. Collin, A.S. Pandit, I. Kiviranta, et al., An injectable, in situ forming type II collagen/hyaluronic acid hydrogel vehicle for chondrocyte delivery in cartilage tissue engineering, Drug Deliv. Transl. Res. 4 (2014) 149–158.

[57] F. Chen, S. Yu, B. Liu, Y. Ni, C. Yu, Y. Su, et al., An injectable enzymatically crosslinked carboxymethylated pullulan/chondroitin sulfate hydrogel for cartilage tissue engineering, Sci. Rep. 6 (2016) 20014.

[58] C. Ko, K. Ku, S. Yang, T. Lin, S. Peng, Y. Peng, In vitro and in vivo co-culture of chondrocytes and bone marrow stem cells in photocrosslinked PCL–PEG–PCL hydrogels enhances cartilage formation, J. Tissue Eng. Regen. Med. 10 (2016) E485–E496.

[59] L. Zhao, M.D. Weir, H.H.K. Xu, Biomaterials Human umbilical cord stem cell encapsulation in calcium phosphate scaffolds for bone engineering, Biomaterials 31 (2010) 3848–3857.

[60] P. Wang, L. Zhao, J. Liu, M.D. Weir, X. Zhou, H.H.K. Xu, Bone tissue engineering via nanostructured calcium phosphate biomaterials and stem cells, Bone Res. 2 (2014) 14017.

[61] D. Su, L. Jiang, X. Chen, J. Dong, Z. Shao, Enhancing the gelation and bioactivity of injectable silk fibroin hydrogel with laponite nanoplatelets, ACS Appl. Mater. Interfaces 8 (2016) 9619–9628.

[62] C. Vinatier, J. Guicheux, Cartilage tissue engineering: from biomaterials and stem cells to osteoarthritis treatments, Ann. Phys. Rehabil. Med. 59 (2016) 139–144.

[63] M. Tang, W. Chen, M.D. Weir, W. Thein-Han, H.H.K. Xu, Human embryonic stem cell encapsulation in alginate microbeads in macroporous calcium phosphate cement for bone tissue engineering, Acta Biomater. 8 (2012) 3436–3445.

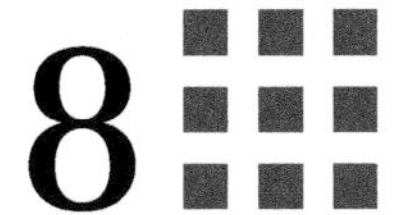

# A review on application of encapsulation in agricultural processes

Mayuri Bhatia[1,2]

[1]DEPARTMENT OF BIOTECHNOLOGY, NATIONAL INSTITUTE OF TECHNOLOGY, WARANGAL, INDIA [2]DEPARTMENT OF CIVIL ENGINEERING, INDIAN INSTITUTE OF TECHNOLOGY HYDERABAD, KANDI, INDIA

## Chapter outline

**8.1 Introduction** ..... 131
**8.2 Encapsulation material** ..... 133
**8.3 Encapsulation techniques** ..... 134
**8.4 Encapsulation of active ingredients** ..... 135
**8.5 Challenges and future prospects** ..... 138
**8.6 Conclusion** ..... 138
**References** ..... 138

## 8.1 Introduction

Agriculture sector contributes a major part in India's GDP as livelihood of 58% of the pollution depends on it. There has been a constant effort for increasing the yield to meet the increasing demand. The food production has to be increased 1.3 times to feed the 70% increased demand [1]. To meet the challenge, farmers are indiscriminately using hazardous chemical based agro-products. These include pesticides, insecticides, and fertilizers, which degrade the soil quality and favorable microbiota. Soil contamination affects the groundwater quality by penetrating in the soil layers. The chemicals also have the tendency to accumulate that might affect the yield and productivity of the seeds in the long run. Crop produce in such polluted soil causes contaminants to even transfer into the feed paving the path to enter in the food chain. The micro- and macro-pathogens tend to develop resistance toward these agrochemicals, and the latter becomes ineffective compelling the use of a higher dose leading to increased risk [1,2].

Encapsulation of Active Molecules and their Delivery System. DOI: https://doi.org/10.1016/B978-0-12-819363-1.00008-9

The studies conducted in the recent times are focused on a thoughtful implication of the pesticides and fertilizers, précised targeting of pathogens, prompt delivery and replenishment of nutrients, improvement of soil fertility, and degradation of the implied chemicals into nontoxic species.

Encapsulation has found its vivid applicability in combating the possible environmental and associated human risk. The technique involves enclosure of target molecules into another walled material for targeted application. The technique was initially developed as a part of biotechnology where it was used for the encapsulation of cell to separate it from the metabolites. It has been also applied as matrix for enzymes to efficiently reuse the bioactive agent. The technique then found its application in pharma and food industries for drug delivery, flavoring, texture, preserving, packaging, etc. The technology advancement has recently empowered the agricultural processes with encapsulation, to reduce the use of hazardous agrochemicals, for improved seed germination and soil quality, regular nutrient dose, water retention, etc. The chemicals, nutrients, growth promoters, seeds, etc., are encapsulated in a biocompatible material that can be natural or synthetic in origin. The encapsulators are attached to a stimulus for triggering the release of the encapsulate [3–7]. Fig. 8–1 depicts the biotic and abiotic factors influencing the active molecules and the permeable protection covering provided through encapsulation.

Over the years conventional methods were implicated for sustainable farming. Fig. 8–2 depicts the technology advancement from conventional spraying to controlled release system using membranes and matrices to encapsulation as an epitome for molecule delivery and safety. The method of encapsulation has also been divided into macro-, micro-,

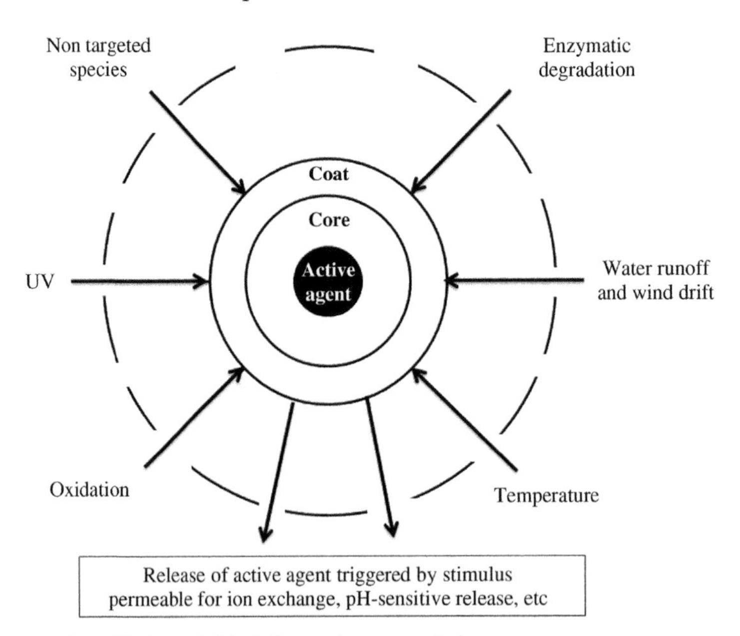


**FIGURE 8–1** Protection from biotic and abiotic factors by encapsulation.

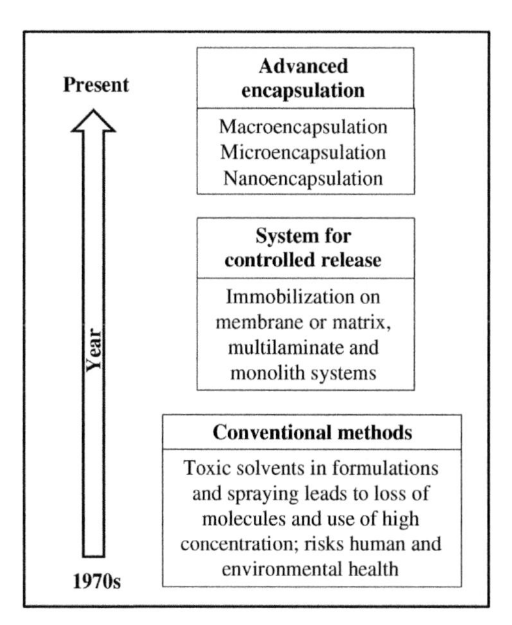


**FIGURE 8–2** Encapsulation technology advancement from conventional spraying to controlled release system to an epitome for molecule delivery and safety.

and nanoencapsulation. The size of particles varies from a few micrometers to nanometers. Based on the application, the size of the encapsulate is decided, where nanometeric size is used for targeted drug-delivery system within the living organisms, and micro- and macro-sized particles cover the broad spectrum of the application [6,8]. The chapter discusses the role of encapsulation in various agricultural processes for intensified sustainable production.

## 8.2 Encapsulation material

Several studies have been conducted on diversified material for multifarious characteristics. The material used for encapsulation forms a coat and is also regarded as shell, membrane, matrix, capsule, or carrier based on the respective application. The active agent is present inside the core [6,9]. The core protects the ingredients against degradation by environmental factors, restricts toxicity caused by mammals and plants, reduces evaporation and leaching, improves targeted action against microbes, insects and pests, lowers the environmental risk, and makes the handling easy for toxic chemicals.

The active agent can be solid, gaseous, or liquid in nature. Encapsulations abet the improved delivery at controlled rate under specific conditions, simultaneously protecting the active agent against external factors [10]. Thus, the encapsulator properties play a significant role. The availability, degradability, permeability, etc., are considered before selecting the material for encapsulation. The encapsulator must have barrier properties to restrict the influence of external factors on the inner molecules [7].

Polymers have contributed for ages as encapsulators due to ease of modification, compatibility nontoxic nature, and permeability. The natural molecules used as encapsulators are chitosan, alginate, lignin, starch, modified polysaccharides, carboxymethyl cellulose, gum arabic, gelatin, etc., have been implicated for targeted and controlled release of micromolecules. Synthetic polymers such as polyvinyl alcohol, polystyrene, and polyacrylamide, however, have also been implicated for similar applications. The pesticides, insecticides, herbicides, and growth regulators including amino acids, metals, and other macro- and micro-nutrients are encapsulated in these polymers. The polymers are tuned for controlled release based on seasonal conditions, desired release rate, effective implication of formulation, etc. This helps in reducing the chemical usage and their interaction with non-target species, limits leaching, and the degradation and volatilization of the encapsulate [11,12].

The other category of polymers is superabsorbents with high retention capacity and is specifically used in arid regions. The examples of superabsorbents include polyacrylic acid, polyvinylchloride, and polyacrylamide. The stimuli used to trigger the release are pressure, temperature, pH, UV, visible light, enzymes, microbiota-based chemicals. These hydrogels superadsorbents are synthesized using four different methods [13]. In solution polymerization, solvents are used to dissolve the monomers, and in the presence of initiator at a certain temperature, polymerization is carried out. The monomers can also be polymerized in the presence of detergents and free radicals initiators to impart certain shape and size to the final product by the selection of a surfactant, called suspension. The monomers can be grafted on a suitable substrate that acts as a matrix and is polymerized to yield a copolymer with different properties of hydrogel and is called graft polymerization. Emulsion polymerization and photopolymerization are also used for synthesizing application-based encapsulator. The encapsulator preserves the stability and shelf-life of bioactive molecules. It even lowers the pace of degradation until desired use [14].

## 8.3 Encapsulation techniques

The active molecules can be encapsulated using various measures based on the desired end product. For example, if the required end product is to be in the specific shape and size, suspension polymerization can be used, whereas the hydrophilic polymers can be converted to hydrophobic polymers by cross-linking them to other polymers with desired properties. Thus, based on the objective, the material and method of synthesis is selected [15].

The major objective for encapsulation remains immobilization, protection, stability, functionalization, and controlled release. It has marked its applications in food, cosmetics, agriculture, and pharma industries. The general principle of encapsulation comprises three steps. The first step is the selection of nature of core, that is, solid or liquid. The solid cores comprise a matrix containing powder or crystals of the encapsulate, and the encapsulate in solvents, surfactants, suspension, emulsion forms the liquid cores. The next step is to disperse the liquid cores by spray method, forming emulsion, or droplet extrusion [7,16].

The spray drying method comprises of nebulized slurry in a vertical chamber through which the hot air is circulated, which dries the product and collects it at the bottom of the chamber. Similar to this method, spray cooling is performed by circulating cold air that solidifies the product. Spray drying method had been implicated for encapsulation of fertilizer by ethyl cellulose (EC) and polyhydroxybutyrate with EC [17,18].

Droplet extrusion is another method that aids the encapsulation process. The material to be encapsulated is mixed with encapsulator and passed through a die of desired shape and size into a solution that forms the hardened shell. This method is generally used to encapsulate seeds for increasing their shelf-life. The encapsulator and the hardening solution used for the purpose can be sodium alginate and calcium chloride. It has also been implicated for synthesizing urea-based fertilizers [19].

Coacervation method has been implicated for herbicide encapsulation. This method involves polymer deposition around a macromolecule through changes in the physiochemical characteristics of the solution. Ediphenphos [20] was applied as microencapsulate pesticide synthesized by the forementioned process.

Other techniques for microencapsulation include emulsification, fluidized bed, droplet-gelation, thermal gelation, solvent evaporation, and polycondensation. These are used based on the desired end product [21].

## 8.4 Encapsulation of active ingredients

This section will discuss the various studies related to the encapsulation of agrochemicals and other agricultural nutrients.

Herbicides are used to inhibit the growth of weeds partially or completely. Weeds grow on the in the fertile soil that are specifically used for cop production. These acquire the nutrients and other physical requirements including light and water, obstructing their availability. They have easy seed dispersal ability to invade the field and release toxins from roots and leave hindering seed germination for the desired crop and also affect the proliferation of nearby vegetation. Weeds also create a favorable habitation for other plague-forming species [3,4].

Herbicides are, therefore, used in a huge amount to combat the plague. These chemicals once sprayed on the cultivation zones are difficult to remove. These either accumulate in the soil or get diluted are washed off through rains or even the wind drifts some amount leading to environmental pollution and risks human health. The associated drawbacks can be circumvented by thoughtful application. The chemicals used as herbicides are mixed with an adjuvant or coadjuvant before its application to activate or modify their activity. This restricts the use of toxic solvents that were implicated as conventional methods [4].

Nanoencapsulation of pesticides have also been introduced as a smart technology. Pesticides are used for restricting the pest growth. The conventional formulations required high volume of solvents similar to herbicides [12].

There is high loss of chemical due to photo- and biodegradation, runoff, volatilization, and absorption by soil and grown plants that inflict on human such as eye irritation, respiratory, and dermal illness as well as cause the imbalance in eco-system.

Thus, various systems were designed for controlled release before encapsulation. These systems were capable of overcoming the lacunas such as volatilization and runoff. These were able to provide sustainable effectiveness of the chemical and reduced direct exposure of the applicant. These systems were classified into membrane, matrix, laminate, and monolith. In membrane and matrix systems, the active agent is bound to a membrane or a matrix and is released on variation in the osmotic pressure, pH variation, ion exchange, etc. The active agents can be loaded in multilaminate system, where the agent is present inside the central lamina protected by external lamina. The monolith system involves inert polymeric dispersion containing herbicide that disperses into the soil by concentration gradient [4].

An advancement in the technology led to microencapsulation of herbicides. There have been studies on use of urea-formaldehyde microcapsules for the encapsulation of a known plaguicide, acetochlor. In the study, the microcapsules were modified using calcium-based lignosulfonate microcapsules [22] for higher encapsulation efficiency and decreased particle size [3].

Starch has been a preferred coating material for the encapsulation of various plaguicides, including plant oils [23], monoterpenes [24], and 2, 4, 5-trichlorophenoxyacetic acid herbicide [25]. Neem seed oil and chlorpyrifos are another active molecules used as plaguicides. Their varying efficiency due to change in encapsulators were studied. Starch, guar gum, and both were implemented with cross-linking to urea formaldehyde [5]. Cyclodextrin, a biopolymer synthesized from starch by enzymatic conversion, is vividly used in encapsulation process. It was found to improve physiochemical properties of bentazon, a herbicide, for a better herbicidal activity [26].

Chitosan has also been used as coating material for encapsulation. Fertilizers made of nitrogen potassium and phosphorus is coated in chitosan to provide stability, and the outer coat consists of superabsorbent polymeric gel that retains water [7]. The coats provide enhanced stability, slow-release rate, and ambient conditions to protect the fertilizer from degradation and further helps in better interaction with the soil. Similarly, cellulose and its derivates have significantly performed well as encapsulators [18]. And a combinational polymer of EC and PHB was capable to reduce urea dissolution [17].

There have been studies on plant materials that act as insecticides. One such study was conducted by Amoabeng et al. in the year 2014, where extracts of various plants including *Ageratum conyzoides*, Siam weed, tobacco, and *Ricinus communis* (castor oil plant) were grinded to make a mixture. This mixture was tested against the efficiency of emamectin benzoate, a synthetic insecticide [27]. Boursier et al. used neem extract as the active agent in fighting the plaguicides [28]. Fouad et al. tested the plant extract synthesized from various species of Brazilian Cerrado. The effect of extract was then studied on eggs of insects [29].

An improvement in the application of these bioactive agents calls for the implication of nanobased formulation and encapsulation to enhance the surface properties and for a superior plaguicidal action without compromising human and environment safety. This section will follow with some recent examples of bioactive molecule nanoencapsulation.

Azadirachtin was emulsified with ricinoleic acid and coated by carboxymethyl chitosan for increased shelf-life and to maintain the efficacy of botanical insecticide in the external environment [30]. Rotenone; carvacrol, thymol, and eugenol; and curcumin are bioactive insecticide, pesticide, and bactericide extracted from plants, essential oils, and turmeric, respectively. Rotenone was encapsulated in chitosan and polymeric nanoparticles by Lao et al. (2010) and Martin et al. (2013). Carvacrol was also encapsulated in coat made of chitosan nanoparticles and chitosan/cyclodextrin [31,32]. Nanoclay is another potential encapsulator with water retention ability [33], and its nanofilm was used for the encapsulation of thymol [34]. Curcumin encapsulated in hydroxypropylcellulose nanoparticles possess bactericidal activity [35].

Nanocarriers can also be categorized as inorganic and organic. Micelles, vesicles, and liposomes are synthesized from organic polymers for controlled delivery of antimicrobial agents that are responsive to change in temperature and polarity. Inorganic group consists of silica, zinc, and iron oxide nanoparticles, which are sensitive to light, electricity, and other stress conditions for controlled release of antimicrobial agents. The examples of liposome bound molecules are adriamycin, cytarabines, and fluorouracil for bactericidal activity [16,36].

Microorganisms such as diazotrophic bacteria, mycorrhizal fungi, actinomycetes, *Pseudomonas*, and *Trichoderma* are widely used for improving soil quality and to protect crops from pests. These are encapsulated in a suitable agent to provide protection from the attack of other microbial species and environmental factors that degrade or restrict their functioning. Conventionally powdered and liquid inoculants were applied to the crops, whereas granulated formulations were used for soil fertility. These were immobilized to silica, clay, and other plant growth media. Peat has been successfully used as a carrier but the peat-coated inoculants release carbon dioxide as the peat decomposes. Hence, microencapsulation was developed as advancement in the technology. Microbeads are one of the types that protect the cells from external stress and increase their viability and efficacy [21].

Alternatively, the plant growth promoting bacteria can be trapped in gel-based beads, for example, alginate beads, for prolonged survival. Also, these beads provide nutrition to the cells and protects them bacteriophage. Other plant growth promoting bacteria are broadly applied in arid regions for improving plantation and reforestation. These also help in promoting mangrove restoration, phytodegradation, and remediation of soils. *Azospirillum brasilense* and *Bacillus pumilus* are the examples of plant growth promoting bacteria that work in association with mycorrhiza fungi and rhizobia for nitrogen fixation and growth of plants and crops [21,37].

## 8.5 Challenges and future prospects

Though the encapsulation techniques have immense potential, there are a few challenges to be met. The cost-effectiveness of successfully encapsulated molecules need to be explained. The coating material cost is generally higher than the encapsulate. The techniques used for encapsulation are also energy-intensive process. Thus, to scale up the production, the process has to be optimized without compromising the achieved efficiency at smaller scale. The effectiveness of active ingredient from bio-origin might vary from source to source. The extraction of molecules from the source again incurs labor and technology. Degradation of coating material in both micro- and nano-forms remains debatable, where the chronic impacts will be known over the time. The nanocarriers, if not degraded, might form a new class of pollutants, and due to their size, it may become a serious concern associated to ecological fitness. Sole natural polymers have lesser stabilizing efficiency, and copolymers or polymeric blends use chemicals with intricate degradation process. The dose, nutrient demand, and stability vary in different seasons; hence, proper examination of the conditions before implications is needed. Therefore, these challenges can be circumvented by further studies and defines the potential areas for future research.

## 8.6 Conclusion

Biotic and abiotic stress degrades the agrochemicals and reduces their efficacy by altering the functioning. Conventional methods of agrochemical application have been improved by implementing the encapsulation process for preserving the agrochemicals from overuse and degradation. The technological advancement has transformed the agriculture sectors by the establishment of micro- and nano-tools that preserves the active molecule and aids in systematic release. The controlled release is triggered by an external stimulus, including pH and temperature changes, ion exchange, and osmotic and hydraulic pressure and enhance nutrient absorption. Encapsulation of bio- and chemi-active molecules helps in better microbicidal activity. Nanoencapsulators have also been implicated as degradation enhancers, but their fate in the environment remains unknown [38]. Further research is required to study their life cycle, exposure in the food chain, degradation, and overall impact on the biota.

## References

[1] P.L. Kashyap, X. Xiang, P. Heiden, Chitosan nanoparticle based delivery systems for sustainable agriculture, Int. J. Biol. Macromol. 77 (2015) 36–51.

[2] Y. Bashan, L. de-Bashan, Encapsulated formulations for microorganisms in agriculture and the environment, PGPR Formulation (2016) 4–6.

[3] E.V.R. Campos, J.L. de Oliveira, L.F. Fraceto, Applications of controlled release systems for fungicides, herbicides, acaricides, nutrients, and plant growth hormones: a review, Adv. Sci. Eng. Med. 6 (4) (2014) 373–387.

[4] F. Sopeña, C. Maqueda, E. Morillo, Controlled release formulations of herbicides based on micro-encapsulation, Ciencia e investigación agraria 36 (1) (2009) 27–42.

[5] S.G. Kumbar, A.R. Kulkarni, A.M. Dave, T.M. Aminabhavi, Encapsulation efficiency and release kinetics of solid and liquid pesticides through urea formaldehyde crosslinked starch, guar gum, and starch + guar gum matrices, J. Appl. Polym. Sci. 82 (11) (2001) 2863–2866.

[6] N. Zuidam, E. Heinrich, N. Zuidam, V. Nedovic, Encapsulation Technologies for Food Active Ingredients and Food Processing, 1, Springer, The Netherlands, 2010, pp. 3–31.

[7] J. Chen, S. Lü, Z. Zhang, X. Zhao, X. Li, P. Ning, et al., Environmentally friendly fertilizers: a review of materials used and their effects on the environment, Sci. Total Environ. 613 (2018) 829–839.

[8] K.G.H. Desai, H. Jin Park, Recent developments in microencapsulation of food ingredients, Drying Technol. 23 (7) (2005) 1361–1394.

[9] V. Nedovic, A. Kalusevic, V. Manojlovic, S. Levic, B. Bugarski, An overview of encapsulation technologies for food applications, Proc. Food Sci. 1 (2011) 1806–1815.

[10] P. de Vos, M.M. Faas, M. Spasojevic, J. Sikkema, Encapsulation for preservation of functionality and targeted delivery of bioactive food components, Int. Dairy J. 20 (4) (2010) 292–302.

[11] M. Curcio, N. Picci, Polymer in agriculture: a review, Am. J. Agric. Biol. Sci. 3 (2008) 299–314.

[12] J.L. de Oliveira, E.V.R. Campos, M. Bakshi, P. Abhilash, L.F. Fraceto, Application of nanotechnology for the encapsulation of botanical insecticides for sustainable agriculture: prospects and promises, Biotechnol. Adv. 32 (8) (2014) 1550–1561.

[13] W.E. Rudzinski, A.M. Dave, U.H. Vaishnav, S.G. Kumbar, A.R. Kulkarni, T. Aminabhavi, Hydrogels as controlled release devices in agriculture, Des. Monomers Polym. 5 (1) (2002) 39–65.

[14] U. Lesmes, D.J. McClements, Structure–function relationships to guide rational design and fabrication of particulate food delivery systems, Trends Food Sci. Technol. 20 (10) (2009) 448–457.

[15] P. T. d Silva, L.L.M. Fries, C. R. d Menezes, A.T. Holkem, C.L. Schwan, É.F. Wigmann, et al., Microencapsulation: concepts, mechanisms, methods and some applications in food technology, Ciência Rural 44 (7) (2014) 1304–1311.

[16] B. Huang, F. Chen, Y. Shen, K. Qian, Y. Wang, C. Sun, et al., Advances in targeted pesticides with environmentally responsive controlled release by nanotechnology, Nanomaterials 8 (2) (2018) 102.

[17] M.M. Costa, E.C. Cabral-Albuquerque, T.L. Alves, J.C. Pinto, R.L. Fialho, Use of polyhydroxybutyrate and ethyl cellulose for coating of urea granules, J. Agric. Food Chem. 61 (42) (2013) 9984–9991.

[18] M. Fernández-Pérez, F. Garrido-Herrera, E. González-Pradas, M. Villafranca-Sánchez, F. Flores-Céspedes, Lignin and ethylcellulose as polymers in controlled release formulations of urea, J. Appl. Polym. Sci. 108 (6) (2008) 3796–3803.

[19] E.I. Pereira, C.C. da Cruz, A. Solomon, A. Le, M.A. Cavigelli, C. Ribeiro, Novel slow-release nanocomposite nitrogen fertilizers: the impact of polymers on nanocomposite properties and function, Ind. Eng. Chem. Res. 54 (14) (2015) 3717–3725.

[20] K. Tsuji, Microencapsulation of pesticides and their improved handling safety, J. Microencapsulation 18 (2) (2001) 137–147.

[21] R.P. John, R. Tyagi, S. Brar, R. Surampalli, D. Prévost, Bio-encapsulation of microbial cells for targeted agricultural delivery, Crit. Rev. Biotechnol. 31 (3) (2011) 211–226.

[22] W. Bai, Y. Wang, X. Song, X. Jin, X. Guo, Modification of urea-formaldehyde microcapsules with lignosulfonate-Ca as Co-polymer for encapsulation of acetochlor, J. Macromol. Sci., A 51 (9) (2014) 737–742.

[23] G.M. Glenn, A.P. Klamczynski, D.F. Woods, B. Chiou, W.J. Orts, S.H. Imam, Encapsulation of plant oils in porous starch microspheres, J. Agric. Food Chem. 58 (7) (2010) 4180–4184.

[24] I. Mourtzinos, N. Kalogeropoulos, S. Papadakis, K. Konstantinou, V. Karathanos, Encapsulation of nutraceutical monoterpenes in β-cyclodextrin and modified starch, J. Food Sci. 73 (1) (2008) S89–S94.

[25] Z. Zhu, R. Zhuo, Slow release behavior of starch-g-poly (vinyl alcohol) matrix for 2,4,5-trichlorophenoxyacetic acid herbicide, Eur. Polym. J. 37 (9) (2001) 1913–1919.

[26] C. Yáñez, P. Cañete-Rosales, J.P. Castillo, N. Catalán, T. Undabeytia, E. Morillo, Cyclodextrin inclusion complex to improve physicochemical properties of herbicide bentazon: exploring better formulations, PLoS One 7 (8) (2012) e41072.

[27] B.W. Amoabeng, G.M. Gurr, C.W. Gitau, P.C. Stevenson, Cost: benefit analysis of botanical insecticide use in cabbage: implications for smallholder farmers in developing countries, Crop Prot. 57 (2014) 71–76.

[28] C.M. Boursier, D. Bosco, A. Coulibaly, M. Negre, Are traditional neem extract preparations as efficient as a commercial formulation of azadirachtin A? Crop Prot. 30 (3) (2011) 318–322.

[29] H.A. Fouad, L.R.D.A. Faroni, W. de Souza Tavares, R.C. Ribeiro, S. de Sousa Freitas, J.C. Zanuncio, Botanical extracts of plants from the Brazilian Cerrado for the integrated management of *Sitotroga cerealella* (Lepidoptera: Gelechiidae) in stored grain, J. Stored Prod. Res. 57 (2014) 6–11.

[30] B.-H. Feng, L.-F. Peng, Synthesis and characterization of carboxymethyl chitosan carrying ricinoleic functions as an emulsifier for azadirachtin, Carbohydr. Polym. 88 (2) (2012) 576–582.

[31] L. Higueras, G. López-Carballo, J.P. Cerisuelo, R. Gavara, P. Hernández-Muñoz, Preparation and characterization of chitosan/HP-β-cyclodextrins composites with high sorption capacity for carvacrol, Carbohydr. Polym. 97 (2) (2013) 262–268.

[32] L. Keawchaoon, R. Yoksan, Preparation, characterization and in vitro release study of carvacrol-loaded chitosan nanoparticles, Colloids Surf. B: Biointerfaces 84 (1) (2011) 163–171.

[33] M. Bhatia, S.B. Rajulapati, S. Sonawane, A. Girdhar, Synthesis and implication of novel poly (acrylic acid)/nanosorbent embedded hydrogel composite for lead ion removal, Sci. Rep. 7 (1) (2017) 16413.

[34] G.-O. Lim, S.-A. Jang, K.B. Song, Physical and antimicrobial properties of *Gelidium corneum*/nano-clay composite film containing grapefruit seed extract or thymol, J. Food Eng. 98 (4) (2010) 415–420.

[35] D. Bielska, A. Karewicz, K. Kamiński, I. Kiełkowicz, T. Lachowicz, K. Szczubiałka, et al., Self-organized thermo-responsive hydroxypropyl cellulose nanoparticles for curcumin delivery, Eur. Polym. J. 49 (9) (2013) 2485–2494.

[36] T.M. Taylor, J. Weiss, P.M. Davidson, B.D. Bruce, Liposomal nanocapsules in food science and agriculture, Crit. Rev. Food Sci. Nutr. 45 (7–8) (2005) 587–605.

[37] L.E. De-Bashan, J.-P. Hernandez, Y. Bashan, The potential contribution of plant growth-promoting bacteria to reduce environmental degradation—a comprehensive evaluation, Appl. Soil Ecol. 61 (2012) 171–189.

[38] M. Bhatia, R.S. Babu, S. Sonawane, P. Gogate, A. Girdhar, E. Reddy, et al., Application of nanoadsorbents for removal of lead from water, Int. J. Environ. Sci. Technol. 14 (5) (2017) 1–20.

# 9

# Nanofluids-based delivery system, encapsulation of nanoparticles for stability to make stable nanofluids

Parag Thakur[1], Shriram S. Sonawane[1], Shirish H. Sonawane[2], Bharat A. Bhanvase[3]

[1]DEPARTMENT OF CHEMICAL ENGINEERING, VISVESVARAYA NATIONAL INSTITUTE OF TECHNOLOGY, NAGPUR, INDIA [2]CHEMICAL ENGINEERING DEPARTMENT, NATIONAL INSTITUTE OF TECHNOLOGY, WARANGAL, INDIA [3]CHEMICAL ENGINEERING DEPARTMENT, LAXMINARAYAN INSTITUTE OF TECHNOLOGY, RTM NAGPUR UNIVERSITY, NAGPUR, INDIA

## Chapter outline

**Nomenclature** ..... **142**
**9.1 Introduction** ..... **142**
**9.2 Encapsulation of nanomaterials** ..... **143**
9.2.1 Types of nanocapsules ..... 143
9.2.2 Nanoparticles encapsulation techniques ..... 144
9.2.3 Encapsulation challenges ..... 145
**9.3 Nanofluid-based delivery system** ..... **145**
9.3.1 Preparation of nanofluids ..... 145
9.3.2 Nanofluid stability evaluation methods ..... 146
**9.4 Targeted drug delivery** ..... **148**
9.4.1 Passive targeting ..... 148
9.4.2 Active targeting ..... 148
**9.5 Applications of nanofluid-based delivery system** ..... **148**
9.5.1 Antibacterial activity of nanofluids ..... 148
9.5.2 Applications in polymerase chain reactions ..... 149
9.5.3 Applications in cancer treatment ..... 149
**9.6 Conclusion** ..... **149**
**References** ..... **149**

**Encapsulation of Active Molecules and their Delivery System. DOI: https://doi.org/10.1016/B978-0-12-819363-1.00009-0**

## Nomenclature

**DLS** dynamic light scattering
**ESM** electrostatic stabilization method
**IUPAC** International Union of Pure and Applied Chemistry
**PCR** polymerase chain reactions
**PDI** polydispersity index
$Z_D$ intensity weighted $Z$ average

## 9.1 Introduction

Nanotechnology is a branch of technology that deals with dimensions less than 100 nm. Long before the term Nanotechnology is used, physicist Richard Feynman described the term in which scientists would be able to manipulate and control individual atoms and molecules, while delivering a talk on the topic of "There is Plenty of Room at the Bottom" at American physical Society meeting on December 29, 1959. Nanofluids are one of the forms of nanotechnology. Previously, their applications as coolant are well explored by various researchers [1,2]. Now researchers are exploring the biomedical application of nanofluids. The target is to increase the efficiency of delivery to target site; this will increase the therapeutic index of compounds and minimize accumulation in healthy body sites to avoid toxicity [3,4].

Nanoparticles used in drug-delivery systems are of two types, namely, nanocapsules and nanospheres. Nanospheres and nanocapsules are schematically represented in Fig. 9–1. Nanospheres are systems in which drugs are uniformly dispersed, while nanocapsules are solid shelled, and the substance is trapped into this shell [5].

Usually, the size of nanocapsules is below 50 nm, but the upper size limit of nanocapsules may rise up to 1000 nm (i.e., 1 μm), although generally obtaining particle size is 100–500 nm. Thus this is the advantage of nanofluid-based delivery system. Because smallest capillaries in body are near to 5–6 μm. Thus nanofluids do not block micron-sized blood vessels; this property enables nanofluid as a great tool for drug delivery. The main mechanism used to transport nanofluid is peristaltic mechanism. Numeric evaluation of peristaltic transport in an asymmetric channel is available in literature [6]. Similarly, numeric methods suggested that increase in temperature profile and decrease in velocity profile can be achieved by increasing Brownian motion and thermophoresis parameters [7].

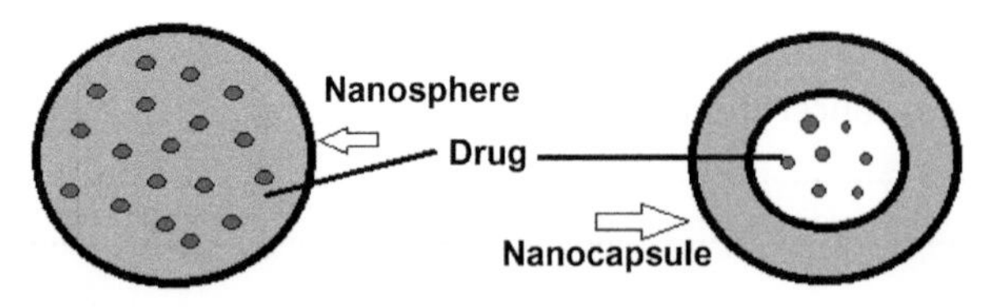


**FIGURE 9–1** Nanospheres and nanocapsules.

The numerical investigation of the nanofluid-based drug-delivery system is not attained by many researchers. Ghasemi et al. studied the effect of numeric parameters of Brownian motion and thermophoresis on nanoparticles fraction, Grashof number, and velocity values and found that nanoparticles fraction increases with increasing Brownian motion parameters and decreases with an increase in thermophoretic parameters, while velocity decreases with an increase in both parameters [8].

# 9.2 Encapsulation of nanomaterials

## 9.2.1 Types of nanocapsules

Nanocapsules can be made of polymeric nanoparticles, inorganic nanoparticles, biological nanoparticles, and liposome-based nanoparticles.

Advantages that encapsulated nanomaterials enjoy are

1. protection from adverse environment,
2. controlled release of drug, and
3. precise targeting [9].

### *9.2.1.1 Polymeric nanocapsules*

Polymeric nanoparticles are relatively stable and easy to modify than their counterparts. This makes them preferable for biomedical applications [10]. Poly-ε-caprolactone, poly(lactide), poly(lactide-*co*-glycolide), thiolated poly(methacrylic acid), and poly(*N*-vinylpyrrolidone) are common examples of synthetic polymers used in nanocapsules formulation. Apart from these, naturally occurring polymers such as chitosan, gelatine, sodium alginate, and albumin are also used for drug delivery.

Characteristic of polymeric nanocarriers are as follows [11]:

1. The polymer should be nontoxic and nonantigenic and should be biodegradable and biocompatible.
2. Drug particles should be preserved until they reach a site of action.
3. Nanoparticles should not have any potential harmful effect on body cells or tissue or they should not interact with each other.
4. Particles should be able to transverse the intervening membrane.
5. The particle should recognize the site of action and should get bound with it.
6. The release rate of the drug should be according to achievement of the continuous required healing effect.
7. After the release of the drug the shell should get degraded or eliminated from the body.

### *9.2.1.2 Inorganic nanocapsules*

Nowadays, inorganic nanocapsules are used in semiconductors quantum dots. Magnetic resonance imaging is an important tool in biomedical engineering. Iron oxide nanoparticles have important role in this application, as a contrast enhancer's [12] role of particle surface monolayer is highlighted by Rotello et al.—they reported a study on four cationic gold nanoparticles that have different structure and hydrophobicity. These gold nanoparticles possess different mechanisms for healthy and cancerous cells [13].

Metallic nanoparticles have charge on their surface. Nonspecific binding of biomolecule and nanoparticle takes place by electrostatic interactions but induces hydrophobic interactions. Nonspecific binding between nanoparticles and biomolecules is the main challenge. But this problem can be overcome by precise control over water content and operating parameters [14,15]. Even specific binding is done by derivating tiopronin in biotin; Au $(-S-EG_3)_n$/Tp-biotin particle is synthesized, which specifically binds to streptavidin with negligible nonspecific binding [16].

#### *9.2.1.3 Viral/virus-like nanocapsules*

Advantages of using virus-based nanocarriers are that they are available from 10 nm size to several micron sizes and have a variety of different shapes. CCMV (cowpea chlorotic mottle virus) has a central ribonucleic acid molecule which is surrounded by 180 identical coat protein subunits. These protein subunits are made of 20 hexamers and 12 pentamers. CPMV (cowpea mosaic virus) has a diameter of 30 nm. This virus can be stable up to 60°C and a pH range of 3–9. These characteristics make this virus a good candidate for nanocarriers. RCNMV (red clover necrotic mosaic virus) has thickness of 20 nm and an inner diameter of 17 nm [17].

### 9.2.2 Nanoparticles encapsulation techniques

The technique of nanoparticle encapsulation is chosen according to the requirement of the drug. Generally, the following methods of nanoparticle encapsulation are used extensively [18].

#### *9.2.2.1 Nano-precipitation method*

The solvent phase is generally organic in nature, while the other phase is aqueous phase. Generally, water is used in the nonsolvent phase. But it is also possible that both phases may be organic or both are aqueous phase. In this method, organic and aqueous phases are allowed to mix with each other. Organic to aqueous flow rate is kept low, generally below unity. Nanoparticle size is controlled by agitation rate mainly [19].

#### *9.2.2.2 Coacervation method*

In this method the phase separation of polyelectrolytes takes place from solution, and the newly formed coacervate phase gets deposited around active ingredients. pH and ionic strength are important parameters for nanoparticle size and shape. Apart from these factors including biopolymer type, concentration affects the nature of the complex formed [20].

#### *9.2.2.3 Emulsion diffusion method*

Emulsion is allowed to form by vigorous mixing of both phases. After forming an emulsion, water is added. The solvent gets diffused out to the external phase. Then nanocapsules are formed. Nanoparticle morphology can be influenced by parameters such as shear rates of the emulsification process, the chemical composition of organic phase, and polymer composition [21,22].

#### *9.2.2.4 Solvent evaporation method*

First, both phases are emulsified by vigorous mixing. Then, polymer–solvent is allowed to evaporate by various techniques, inducing polymer nanospheres precipitation. Size can be adjusted by varying control parameters [23].

### 9.2.3 Encapsulation challenges

The biological impact of nanomaterial is dependent on size, shape, chemical composition, solubility, and aggregation. Thus it is possible that these parameters can influence drug-delivery mechanism even if there is a possibility of causing tissue injury [24].

## 9.3 Nanofluid-based delivery system

Nanofluids are stable suspensions of base fluid such as water containing a particle of size up to 100 nm. Various researchers have tried to establish mathematical models to evaluate affecting parameters on drug-delivery capacity [25,26]. Nanofluid preparation method and stability evaluation are discussed in detail.

### 9.3.1 Preparation of nanofluids

The nanofluid preparation is a key parameter for better efficiency of nanofluid, as preparation method defines purity and particle size of nanoparticles. One and two-step methods are used to synthesize nanofluids [27]. In the one-step method, base fluid and nanomaterial are prepared simultaneously. While in the two-step method both are prepared differently and then nanomaterials are added in base fluid and stable dispersion is obtained. It is hard to remove the remaining residues of the incomplete reaction. But in the two-step method, nanomaterials are added in base fluid only after having pure nanomaterials. In one-step method, various methods such as, direct evaporation technique, chemical reduction, submerged arc nanoparticles synthesis system, laser ablation, microwave irradiation, and polyol process are involved. Nowadays, some researchers also developed microfluidics junctions to produce nanofluids. While in the two-step method, microwave-assisted synthesis and direct mixing techniques are classified. Jung et al. [28] prepared $Al_2O_3$/water nanofluid by using two-step method. Then to produce second nanofluid, nanoparticles of this nanofluid are broken down by using high-speed rotating procedure followed by electrostatic stabilization method (ESM); hydrochloric acid is then added to reduce pH to 7. Another nanofluid, nanofluid 3, is prepared by directly using ESM in above procedure. Here high-speed rotation stage is skipped; rotation speed is kept at 24,000 rpm in homogenizer for nanofluid 2. Chang and Chang [29] used modified plasma arc system. In this system, nanoparticles are directly added to water to form nanofluid. Graves et al. [30] used the two-step method to produce a copper–methanol nanofluid that is capped with short-chain molecules of (3-aminopropyl) trimethoxysilane.

## 9.3.2 Nanofluid stability evaluation methods

Nanomaterials have a tendency to get agglomerated due to van der Waals forces of attraction, and this agglomeration leads to sedimentation; thus the evaluation of nanofluid stability is a necessary and important task. Nanofluid stability is improved by certain methods; Li et al. [31] improved the nanofluids stability by using cationic Gemini surfactant. Nanoparticles used in this study were gold and silver nanoparticles dispersed in water. pH and spacer length are important parameters for stability. 1,3-bis(cetyltrimethylammonium) propane dibromide showed more stability for a wider range of pH of nanofluid and 1,8-bis (cetyltrimethylammonium) octane dibromide is used for more thermal stability applications. Lightweight negatively charged (2,2,6,6-tetramethylpiperidin-1-yl)oxyl-oxidized cellulose nanofibers are found to be a good stabilizing agent for the thermal application of water-based nanofluids [32].

Using various surfactants, stability of nanofluids can be increased; surfactants increase the wetting ability of nanoparticles. The stability of nanofluids is also highly dependent on pH value, for example, for stable CuO in water system optimum pH value is 9.5. While for alumina nanofluid value is 8.

### *9.3.2.1 Light-scattering techniques*

Dynamic light scattering technique is primarily used to obtain particle size dispersed in solution [33]. This is the fastest available technique to determine nanoparticles size in a nanofluid. The results obtained from this analyzer are shown in Fig. 9–2. Results obtained by this method can be verified using scanning electron microscopy and transmission electron microscopy.

### *9.3.2.2 Zeta potential measurements*

Zeta potential is the difference between the potential value at bulk of base fluid and the potential value at stationary layer of fluid around the nanoparticles. Nanoparticles with zeta potential more than $\pm 40-60$ mV are considered as electrically stable. The minimum concentration required for analysis depends on the relative refractive index and particle size. Bigger particle produces more scattered light; thus we will require less concentration to analyze. The laser beam has to pass through the sample; thus, we need to keep concentration accordingly. The same equipment used for size analysis can be used to analyze zeta potential of nanofluids. Systems with zeta potentials between $+30$ and $-30$ mV are unstable, thus value should be more than or less than of above value range.

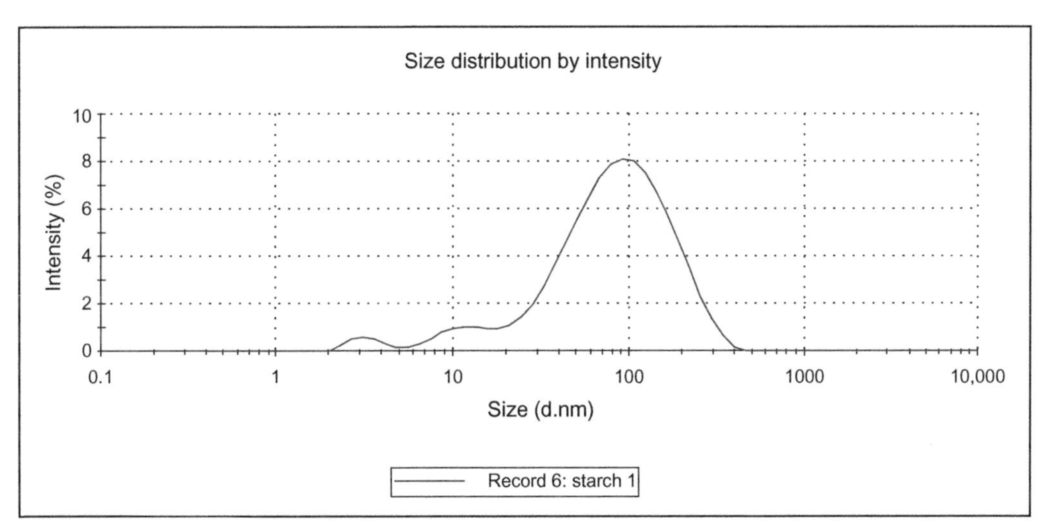


**FIGURE 9–2** Intensity versus size graph obtained from DLS analyzer. *DLS*, Dynamic light scattering.

pH is the most important parameter for zeta potential. For example, if we added acid in nanofluid, pH will decrease, which will increase positive charges on the particle surface. Zeta potential will increase. The point at which zero electrophoretic mobility occurs is called an isoelectric point [34].

### 9.3.2.3 Sedimentation and centrifugation method

Sedimentation is one of the simplest of all methods for stability. But it is a time-consuming method; thus, many times centrifugation is also added. Particles sedimentation velocity is given by the following equation [35]

$$V = \frac{2R^2}{9\mu}(\rho_P - \rho_L) \cdot g \tag{9.1}$$

where $V$ is the particle sedimentation velocity (m/s), $R$ is spherical particle radius (m), $\mu$ is liquid medium viscosity ($m^2/s$), $\rho_P$ and $\rho_L$ are particle and liquid densities, respectively ($kg/m^3$), and $g$ is acceleration due to gravity ($9.81\ m/s^2$). As per equation, to decrease sedimentation rate, we need to decrease particle size and increase medium fluid viscosity, and if possible, we should also decrease the difference between densities of particle and fluid. But as the square of particle size is present, particle size is the most dominant parameter. This reduced particle size will increase the stability of nanofluids. After having sufficient particle size, sedimentation effects will not work as Brownian motion will be a dominant parameter in nanometer range. But these small particles have tendency to agglomerate. Thus we need to prevent this agglomeration effect by using suitable method.

## 9.4 Targeted drug delivery

To have more uniform effects of the drug, reductions in side effects of the drug are the primary objective of efficient drug-delivery system. And these objectives are achieved by two methods, passive targeting and active targeting.

### 9.4.1 Passive targeting

Generally, polyethylene glycol (PEG) is added to the surface of nanoparticles. PEG is hydrophilic in nature, and thus water molecules form hydrogen bonds with oxygen molecule of PEG molecule. Due to this bond, the film of hydration around nanoparticles can be seen. This makes substance antiphagocytic. Thus ultimately nanoparticles remain in circulation for a longer period of time [36].

### 9.4.2 Active targeting

Active targeting can be done in many ways, for example, by using cell-specific ligands or by using magneto liposomes [37]. Targeting strategy is based on the requirement of the system. There is no ideal targeting method in drug delivery.

## 9.5 Applications of nanofluid-based delivery system

Research on micro-electro-mechanical systems has started in the early 1990s. The commercialization of many devices started in recent years, for example, porous silicon, microfabricated microneedles, and micro-reservoir drug-delivery systems [38]. Several studies regarding the immobilization of pepsin are also done using gold nanoparticles. The reusability of gold nanoparticles is also remarkable [39]. Organic, inorganic, and carbon nanotubes are also used for biomedical applications. These applications consist of nano-biosensors, nano–pharmacology, and efficient targeted drug delivery [40].

### 9.5.1 Antibacterial activity of nanofluids

Goharshadi et al. [41] synthesized zinc oxide nanoparticles by the microwave decomposition of acetate in glycerol using ionic liquid ([BMIM][$NTf_2$]). Ammonium citrate is used as a dispersant. This suspension is tested against *Escherichia Coli* DH5$\alpha$. And it is found that, at higher concentration of nanofluid, death of cell increases. Mahapatra et al. [42] used copper oxide nanoparticles of 80–160 nm size against four pathogenic bacterial strains, for example, *Pseudomonas*, *Shigella*, *Salmonella*, and *Klebsiella*. Nanoparticles showed antibacterial activity against these strains. These nanoparticles form complex inside the cell and result in the death of bacteria.

### 9.5.2 Applications in polymerase chain reactions

The polymerase is a common laboratory level technique. This method is used to make many copies of specific regions of deoxyribonucleic acid. Polymerase reaction occurs in three steps: denaturation, annealing, and extension. These steps occur at a temperature of 96°C, 55°C–65°C, and 80°C, respectively. This cycle repeats 25–35 times in a typical polymerase chain reactions (PCR). Due to high thermal conductivity, nanoparticles can affect PCR product. Gold nanoparticles can dramatically enhance PCR [43].

### 9.5.3 Applications in cancer treatment

Cancer is a general term; there are about 100 diseases that come under cancer. Cancer causes the growth of unwanted cells uncontrollably and damaging body tissues [44]. Graphene-based nanosensors can be utilized to the cancer diagnosis. These biosensors are highly sensitive and specific to the task [45]. Use of gold nanoparticle in cancer treatment is also available in literature [46]. Magnetite fluid hyperthermia is also used for cancer treatment. Biocompatible magnetite nanoparticles are used as a heat mediator. They have fewer side effects and more efficiency than other nanoparticles [47,48]. Ferrite-based magnetic particles are also used for hyperthermia treatment. Iron oxide ferrofluids are not used much because they are difficult to distinguish from hemoglobin. Thus they are attached to other metals such as zinc, nickel, and manganese [49]. SWCNT is also used as thermal enhancer, and some biocompatible molecules such as iron oxide are also used to directly inject into tissue [50].

## 9.6 Conclusion

Particle size, shape, and requirement of the drug are key parameters of efficient therapeutic effect. To achieve this, we need to choose the nanofluid preparation method and nanoparticles encapsulation method accordingly. Peristaltic flow is an important mechanism to deliver nanofluid to target location. There is scope to do more detailed numeric and experimentation study in this regard. The stability of nanofluid is an important parameter in efficient drug delivery. There are many methods to evaluate drug delivery. Nanofluids have a wide range of application in medical field, such as in PCR, to be used as antibacterial agents, and magnetite and gold nanoparticles can be used as a heat mediator.

## References

[1] R.S. Khedkar, S.S. Sonawane, K.L. Wasewar, Influence of CuO nanoparticles in enhancing the thermal conductivity of water and mono ethylene glycol based nanofluids, Int. Commun. Heat Mass Transf. 39 (5) (2012) 665–669.

[2] D.G. Subhedar, B.M. Ramani, A. Gupta, Experimental investigation of heat transfer potential of $Al_2O_3$/ water-mono ethylene glycol nanofluids as a car radiator coolant, Case Stud. Therm. Eng. 11 (2018) 26–34.

[3] D.K. Mishra, N. Balekar, P.K. Mishra, Nanoengineered strategies for siRNA delivery: from target assessment to cancer therapeutic efficacy, Drug Deliv. Transl. Res. 7 (2) (2017) 346–358.

[4] M. Talekar, T.H. Tran, M. Amiji, Translational nano-medicines: targeted therapeutic delivery for cancer and inflammatory diseases, AAPS J. 17 (4) (2015) 813–827.

[5] M. Vert, Y. Doi, K.H. Hellwich, M. Hess, P. Hodge, P. Kubisa, et al., Terminology for biorelated polymers and applications (IUPAC Recommendations 2012), Pure Appl. Chem. 84 (2) (2012) 377–410.

[6] F.M. Abbasi, T. Hayat, B. Ahmad, G.Q. Chen, Peristaltic motion of a non-Newtonian nanofluid in an asymmetric channel, Z. für Naturforschung A 69 (8-9) (2014) 451–461.

[7] S.E. Ghasemi, Thermophoresis and Brownian motion effects on peristaltic nanofluid flow for drug delivery applications, J. Mol. Liq. 238 (2017) 115–121.

[8] S.E. Ghasemi, M. Vatani, M. Hatami, D.D. Ganji, Analytical and numerical investigation of nanoparticle effect on peristaltic fluid flow in drug delivery systems, J. Mol. Liq. 215 (2016) 88–97.

[9] P.N. Ezhilarasi, P. Karthik, N. Chhanwal, C. Anandharamakrishnan, Nanoencapsulation techniques for food bioactive components: a review, Food Bioprocess Technol. 6 (3) (2013) 628–647.

[10] R. Singh, J.W. Lillard Jr, Nanoparticle-based targeted drug delivery, Exp. Mol. Pathol. 86 (3) (2009) 215–223.

[11] P.K. Ghosh, Hydrophilic polymeric nanoparticles as drug carriers, Indian J. Biochem. Biophys. 37 (2000) 273–282.

[12] J.J. Giner-Casares, M. Henriksen-Lacey, M. Coronado-Puchau, L.M. Liz-Marzan, Inorganic nanoparticles for biomedicine: where materials scientists meet medical research, Mater Today 19 (1) (2016) 19–28.

[13] K. Saha, S.T. Kim, B. Yan, O.R. Miranda, F.S. Alfonso, D. Shlosman, et al., Surface functionality of nanoparticles determines cellular uptake mechanisms in mammalian cells, Small 9 (2) (2013) 300–305.

[14] M. Zheng, F. Davidson, X. Huang, Ethylene glycol monolayer protected nanoparticles for eliminating nonspecific binding with biological molecules, J. Am. Chem. Soc. 125 (26) (2003) 7790–7791.

[15] M. Zheng, Z. Li, X. Huang, Ethylene glycol monolayer protected nanoparticles: synthesis, characterization, and interactions with biological molecules, Langmuir 20 (10) (2004) 4226–4235.

[16] M. Zheng, X. Huang, Nanoparticles comprising a mixed monolayer for specific bindings with biomolecules, J. Am. Chem. Soc. 126 (38) (2004) 12047–12054.

[17] Y. Ma, R.J. Nolte, J.J. Cornelissen, Virus-based nanocarriers for drug delivery, Adv. Drug Deliv. Rev. 64 (9) (2012) 811–825.

[18] D. Quintanar-Guerrero, E. Allémann, H. Fessi, E. Doelker, Preparation techniques and mechanisms of formation of biodegradable nanoparticles from preformed polymers, Drug Dev. Ind. Pharm. 24 (12) (1998) 1113–1128.

[19] C.E. Mora-Huertas, H. Fessi, A. Elaissari, Polymer-based nanocapsules for drug delivery, Int. J. Pharm. 385 (1-2) (2010) 113–142.

[20] P.N. Ezhilarasi, P. Karthik, N. Chhanwal, C. Anandharamakrishnan, Nanoencapsulation techniques for food bioactive components: a review, Food Bioprocess Technol. 6 (3) (2013) 628–647.

[21] D. Quintanar-Guerrero, E. Allémann, E. Doelker, H. Fessi, Preparation and characterization of nanocapsules from preformed polymers by a new process based on emulsification-diffusion technique, Pharm. Res. 15 (7) (1998) 1056–1062.

[22] D. Quintanar, H. Fessi, E. Doelker, E. Allemann, U.S. Patent No. 6,884,438, U.S. Patent and Trademark Office, Washington, DC, 2005.

[23] C.P. Reis, R.J. Neufeld, A.J. Ribeiro, F. Veiga, Nanoencapsulation I. Methods for preparation of drug-loaded polymeric nanoparticles, Nanomed.: Nanotechnol. Biol. Med. 2 (1) (2006) 8–21.

[24] A. Nel, T. Xia, L. Mädler, N. Li, Toxic potential of materials at the nanolevel, Science 311 (5761) (2006) 622–627.

[25] D. Tripathi, O.A. Bég, A study on peristaltic flow of nanofluids: application in drug delivery systems, Int. J. Heat Mass Transf. 70 (2014) 61–70.

[26] F.M. Abbasi, T. Hayat, A. Alsaedi, Peristaltic transport of magneto-nanoparticles submerged in water: model for drug delivery system, Phys. E: Low-Dimension. Syst. Nanostruct. 68 (2015) 123–132.

[27] I.M. Mahbubul, Preparation of nanofluid, Prep. Char. Prop. Appl. Nanofluid (2019) 15–45.

[28] H. Park, S.J. Lee, S.Y. Jung, Effect of nanofluid formation methods on behaviors of boiling bubbles, Int. J. Heat Mass Transf. 135 (2019) 1312–1318.

[29] H. Chang, Y.C. Chang, Fabrication of $Al_2O_3$ nanofluid by a plasma arc nanoparticles synthesis system, J. Mater. Process. Technol. 207 (1–3) (2008) 193–199.

[30] J.E. Graves, E. Latvytė, A. Greenwood, N.G. Emekwuru, Ultrasonic preparation, stability and thermal conductivity of a capped copper-methanol nanofluid, Ultrason. Sonochem. 55 (2019) 25–31.

[31] D. Li, W. Fang, Y. Feng, Q. Geng, M. Song, Stability properties of water-based gold and silver nanofluids stabilized by cationic gemini surfactants, J. Taiwan Inst. Chem. Eng. 97 (2019) 458–465.

[32] W.K. Hwang, S. Choy, S.L. Song, J. Lee, D.S. Hwang, K.Y. Lee, Enhancement of nanofluid stability and critical heat flux in pool boiling with nanocellulose, Carbohydr. Polym. 213 (2019) 393–402.

[33] F. Farahmandghavi, M. Imani, F. Hajiesmaeelian, Silicone matrices loaded with levonorgestrel particles: impact of the particle size on drug release, J. Drug Deliv. Sci. Technol. 49 (2019) 132–142.

[34] S. Bhattacharjee, DLS and zeta potential—what they are and what they are not? J. Control. Release 235 (2016) 337–351.

[35] P.C. Hiemenz, P.C. Hiemenz, Principles of Colloid and Surface Chemistry, vol. 188, M. Dekker, New York, 1986.

[36] L.E. van Vlerken, T.K. Vyas, M.M. Amiji, Poly(ethylene glycol)-modified nanocarriers for tumor-targeted and intracellular delivery, Pharm. Res. 24 (8) (2007) 1405–1414.

[37] P. Galvin, D. Thompson, K.B. Ryan, A. McCarthy, A.C. Moore, C.S. Burke, et al., Nanoparticle-based drug delivery: case studies for cancer and cardiovascular applications, Cell. Mol. Life Sci. 69 (3) (2012) 389–404.

[38] R.S. Shawgo, A.C.R. Grayson, Y. Li, M.J. Cima, BioMEMS for drug delivery, Curr. Opin. Solid State Mater. Sci. 6 (4) (2002) 329–334.

[39] K. Mukhopadhyay, S. Phadtare, V.P. Vinod, A. Kumar, M. Rao, R.V. Chaudhari, et al., Gold nanoparticles assembled on amine-functionalized Na – Y zeolite: a biocompatible surface for enzyme immobilization, Langmuir 19 (9) (2003) 3858–3863.

[40] Z. Aguilar, Nanomaterials for Medical Applications, Newnes, 2012.

[41] R. Jalal, E.K. Goharshadi, M. Abareshi, M. Moosavi, A. Yousefi, P. Nancarrow, ZnO nanofluids: green synthesis, characterization, and antibacterial activity, Mater. Chem. Phys. 121 (1–2) (2010) 198–201.

[42] O. Mahapatra, M. Bhagat, C. Gopalakrishnan, K.D. Arunachalam, Ultrafine dispersed CuO nanoparticles and their antibacterial activity, J. Exp. Nanosci. 3 (3) (2008) 185–193.

[43] M. Li, Y.C. Lin, C.C. Wu, H.S. Liu, Enhancing the efficiency of a PCR using gold nanoparticles, Nucl. Acids Res. 33 (21) (2005) e184.

[44] C. De Martel, J. Ferlay, S. Franceschi, J. Vignat, F. Bray, D. Forman, et al., Global burden of cancers attributable to infections in 2008: a review and synthetic analysis, Lancet Oncol. 13 (6) (2012) 607–615.

[45] G. Eskiizmir, Y. Baskın, K. Yapıcı, Graphene-based nanomaterials in cancer treatment and diagnosis, Fullerens, Graphenes Nanotubes, Elsevier Inc, 2018, pp. 331–374.

[46] K.S. Mekheimer, W.M. Hasona, R.E. Abo-Elkhair, A.Z. Zaher, Peristaltic blood flow with gold nanoparticles as a third grade nanofluid in catheter: application of cancer therapy, Phys. Lett. A 382 (2–3) (2018) 85–93.

[47] P. Das, M. Colombo, D. Prosperi, Recent advances in magnetic fluid hyperthermia for cancer therapy, Colloids Surf. B: Biointerfaces 174 (2019) 42–55.

[48] S. Laurent, S. Dutz, U.O. Häfeli, M. Mahmoudi, Magnetic fluid hyperthermia: focus on superparamagnetic iron oxide nanoparticles, Adv. Colloid Interface Sci. 166 (1–2) (2011) 8–23.

[49] I. Sharifi, H. Shokrollahi, S. Amiri, Ferrite-based magnetic nanofluids used in hyperthermia applications, J. Magn. Magn. Mater. 324 (6) (2012) 903–915.

[50] A. Sohail, Z. Ahmad, O.A. Beg, S. Arshad, L. Sherin, A review on hyperthermia via nanoparticle-mediated therapy, Bull. Cancer 104 (5) (2017) 452–461.

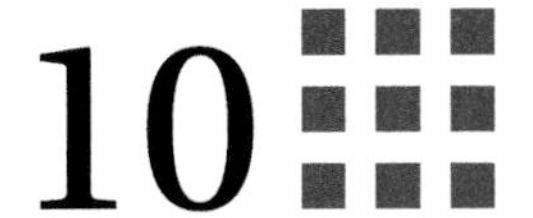

# Corrosion and nanocontainer-based delivery system

Uday Bagale[1], Dipak Pinjari[2], Shrikant Barkade[3], Irina Potoroko[4]

[1]DEPARTMENT OF FOOD AND BIOTECHNOLOGY, SOUTH URAL STATE UNIVERSITY, CHELYABINSK, RUSSIAN FEDERATION [2]NATIONAL CENTER FOR NANOSCIENCE AND NANOTECHNOLOGY, UNIVERSITY OF MUMBAI, MUMBAI, INDIA [3]CHEMICAL ENGINEERING DEPARTMENT, SINHGAD COLLEGE OF ENGINEERING, PUNE, INDIA [4]DEPARTMENT OF FOOD TECHNOLOGY AND BIOTECHNOLOGY, SCHOOL OF MEDICAL BIOLOGY, SUSU, CHELYABINSK, RUSSIA

## Chapter outline

**10.1 Introduction to corrosion problem** ..... **154**
**10.2 Container approach for corrosion prevention** ..... **154**
10.2.1 Micro- or nanocontainer approach ..... 154
10.2.2 Self-healing materials ..... 156
**10.3 Different types of container and their method preparation/fabrications** ..... **156**
10.3.1 Layer double hydroxide base micro- and nanocontainer ..... 156
10.3.2 Polymer shell and polyelectrolyte with ceramic core container ..... 157
10.3.3 Stimuli response with ceramic core ..... 158
10.3.4 Direct or inverse emulsion based container ..... 159
10.3.5 Container-based internal physical phenomena ..... 160
10.3.6 Based on chemical reaction nanocontainer ..... 161
**10.4 Distribution and performance of container for protective coating** ..... **164**
**10.5 Release of active compounds from container** ..... **165**
**10.6 Case studies** ..... **165**
10.6.1 Case study I: preparation of iron oxide nanocontainer by layer-by-layer method using ultrasound approach ..... 165
**10.7 Commercial applications and future prospectus** ..... **170**
**10.8 Conclusion/inference** ..... **171**
**References** ..... **171**

Encapsulation of Active Molecules and their Delivery System. DOI: https://doi.org/10.1016/B978-0-12-819363-1.00010-7

## 10.1 Introduction to corrosion problem

In the current scenario the global cost of corrosion is increasing every year for metallic structures.

The acceptable value of thickness lost for metallic structures due to corrosion is about $\sim$100 μm/year. Practically, the combination of various methods such as electrochemical methods (anodic and cathodic protections), metallurgical design, inhibitors, and coatings are used for the protection of metals. Due to advanced developments in surface science and technology, nanocontainers and hollow spheres have been emerging as the promising options for self-healing corrosion protection coatings of steel and aluminum alloys [1,2]. Various research groups are working on the synthesis and design of the nanocontainer shell, which should be stable, permeable to release/upload the active material, and must have the desired functionalities such as magnetic, catalytic, conductive, and targeting [3,4]. Design of polymer, inorganic, and layer-by-layer (LBL)-assembled nanocontainers are proposed to release the active and repairing material within short time after changes in the coating's integrity by the various researchers. Role of pH is one of the most important factors in the release of corrosion inhibitor from the nanocontainer shell on the metal parts along with other physical and chemical stimulants that actually prevents the drain of corrosion inhibitor out of the shell and enhances the coating stability [5]. Actually, this process is termed self-healing to seal the further corrosion activity. In this chapter the focus is placed upon the preparation methods of different types of container and their distribution and performance as protective coatings.

## 10.2 Container approach for corrosion prevention

For corrosion prevention, one of best ways to tackle is that the active corrosion inhibitor encapsulated in various coatings to make a self-repairing (self-healing) system. The human skins as well as other living organisms behave like the repairing mechanism of skin. Fig. 10–1 shows the healing mechanism of human skin. Generally, accidentally, if human skin is cut, it breaks the layer of epidermis and dermis, and blood starts to flow from the skin. At the same time, if the breaking of wall of larger blood cell takes place, it activates the platelet. These platelets change their structure from round to spiny and begin the join the breakage of layer to form new layer of epidermis. The newly formed layer of dermis and epidermis is the nature of wound healing of skin. This biochemical phenomenon has encouraged most of the renowned researchers to work on self-healing of living organisms [6–11].

Based on specific design of these active feedbacks, material in the form of micro- or nanocontainer was determined and classified as follows.

### 10.2.1 Micro- or nanocontainer approach

A perfect coating is the one which not only provides protection from the corrosive attacks but also gives good mechanical strength and better adhesion. This is achieved by embedding

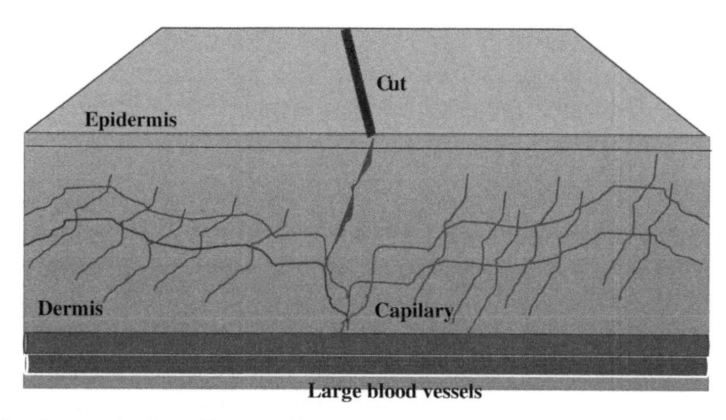


**FIGURE 10–1** Self-healing mechanism of human skin.

a corrosion inhibitor. [11] The main function of an inhibitor is to release itself at the time of any corrosive attack and terminate corrosion propagation at already damaged corrosion defects. Earlier, chromate and lead-based pigments were used as corrosion inhibitors, but now they are banned because of their toxic nature and bad effect on the environment. Chromium(VI) when released into the atmosphere causes health hazards such as cancer. So, nowadays 8-hydroxyquinoline (8-HQ), 1-*H*-benzotriazole-4-sulfonic acid (1-BSA), phosphate compounds, molybdates compounds, hydrotalcite pigments, lignosulfonate, dodecyl benzene sulfonic acid, etc., are used as inhibitors [5,12–16].

However, direct incorporation of inhibitors into the coatings can lead to several problems, including the significant shortcoming in the stability and self-repairing activity of the coating and leakage of the free inhibitor into the environment that again creates health problems [13]. Recent studies have shown that the encapsulation of nanocontainers into the coating matrix highly enhances the anticorrosive properties. One main advantage of nanocontainer with coating matrix is that it will distribute well in matrix and provide barrier for inhibitor to avoid its direct contact with atmosphere. Again, it provides barrier between the inhibitor and coating matrix without any sort of interaction and release the inhibitor when it requires for corrosion inhibition action [14]. Thus it provides an active protection [5,15]. Some disadvantages of nanocontainers are that they should be chosen in such a way that they are compatible with the coating matrix, they should not affect chemical and mechanical properties of coating, and should provide controlled release kinetics for a good trigger release. The addition of nanocontainer should be less for better control over the pigment volume concentration. So nanocontainers must be chosen in such a way that they provide all the benefits.

Some of the commonly used nanocontainers are halloysite nanotubes (HNTs). These are mainly used for industrial purpose because of their low-cost and self-healing properties. They are loaded with the corrosion inhibitor 2-mercaptobenzothiazole (MBT) for the protection of aluminum alloy [12]. Layer double hydroxides (LDHs) are mostly used in drug-delivery system, polymer stabilizers, heterogeneous catalysis, silica nanocapsules, and ceria nanocontainers [15]. Ceramic nanocontainers are loaded with MBT into hybrid epoxy coatings for protection of galvanized steels [5]. Cerium molybdate nanocontainers [$Ce_2(MoO_4)_3$] are loaded with 8-HQ or with 1-BSA as a corrosion inhibitor [16].

### 10.2.2 Self-healing materials

Based on self-healing chemistries, two part of self-healing system occurs, namely, intrinsic and extrinsic. In intrinsic self-healing, healing occurs within the material itself through its structure. The self-healing was achieved through the reversible reaction between the healing materials (polymer matrix). The process of healing can be done using several techniques/approaches such as hydrogen bonding, thermally reversible reaction, molecular diffusion and entanglement, and ionomeric arrangements [17]. Wool and O'Connor [18] proposed theory of intermolecular diffusion for chain entanglement for self-healing. On this assumption, they categorize two sections: (1) reversible covalent bonds [18] and (2) supramolecular interactions [19]. On other hand, extrinsic self-healing system and healing material are stored in container or reservoir to form microcapsules or vascular. Examples of vascular-based healing system are one-dimensional two-pack epoxy [20,21], 3D epoxy system with Grubbs catalyst [22], and two-pack epoxy with more number of healing cycles [23]. However, there is limitation due to the swelling for preserving the integrity resin matrix, lack of good compatibility to ensure adhesion, and time-consuming step of fiber dissolution and infusion of healing agents. Thus moving toward capsule-based system has more attraction. The main benefit of this capsule-based system is that it regains its original aesthetic look with better self-healing ability [24].

## 10.3 Different types of container and their method preparation/fabrications

There are several type of micro- or nanocontainer used for the encapsulation active agent for self-repairing coatings. These container always have their own technique for preparation, inhibitor encapsulation, and addition in coating formulation, which we will discuss in the later sections.

### 10.3.1 Layer double hydroxide base micro- and nanocontainer

LDH are clay-based nanostructure material, which naturally occur and are also called hydrotalcite. Their composition can be expressed by the following general formula ($M_{1-x}{}^{2+}$ $Mx^{3+}(OH)_2)Ax/n\ n\cdot\ mH_2O$, where the cations $M^{2+}$ can be represented by $Mg^{2+}$, $Zn^{2+}$, $Fe^{2+}$, $Co^{2+}$, $Cu^{2+}$, and so on, and $M^{3+}$ can be represented by $Al^{3+}$, $Cr^{3+}$, $Fe^{3+}$, $Ga^{3+}$, and others. The LDH material are holding brucite-like structure with octahedral hole where exchange of their cation ions are from trivalent to divalent. The excess of positive charge can be compensated by LDH in the form of guest–host complex. The exchange of anions with various species thus allows the synthesis of new material with specific properties application.

For example, as carrier and loading for biological compound and drugs in media separation and additive in polymer. The LDHs are also used as catalyst, polymer stabilizer, and corrosion inhibitor [25–27].

In some studies, in situ based prepared LDHs act as protective film when applied on metallic substrates. This is the simplest way for the protection of substrate when LDHs interact with corrosion inhibitor in the form of nanocontainer [28,29]. With LDHs provide better corrosion protection, some research modify interaction between LDHs and organic coating [30]. Another method for the use of these LDHs is inorganic carrier for the various types of corrosion inhibitor and their utilization in organic coatings. This type of nanocontainer not only impart active protection but also adsorb corrosive species from environment. Vanadates, nitrates, and chromate-free corrosive inhibitor are encapsulated in such nanocontainer [31–33]. Later, such Zn/Al LDH-nanocontainers are added to coating formulation to develop a self-healing mechanism as well as provide corrosion protection better than chromate-based coatings. LDHs with organic anions provide new class of family in LDHs. Williams and McMurray reported the successful use of benzotriazole (BTA), oxalate, and ethyl xanthate with LDHs [34,35]. The resulting layer system added in poly(vinylbutyral) coating for aluminum alloys. Kendig and Hon reported preparation of LDHs with 2,5-dimercapto-1,3,4-thiadiazolate and inhibiting action done by oxygen-reduction reaction for copper substrate [36]. Poznyak et al. [37] reported the nanocontainer preparation via anion-exchange reaction. They use the combination of Zn–Al and Mg–Al LDHs with MBT. This provide significant reduction of corrosion rate by releasing the organic anions from LDHs.

## 10.3.2 Polymer shell and polyelectrolyte with ceramic core container

Hallow or porous type material are naturally occurring count as container of this type. Several examples of micro- or nanocontainer were used for encapsulation purpose as shown in Fig. 10–2.

These containers based on their origin and chemical nature with extended porous nature can be served as robustic container where they can encapsulate corrosion inhibitor in it. The loading capacity varies for each nanocontainer depending on its total volume of cavities and pores. HNTs are cost-effective over carbon nanotube and other materials. It has two layers in the ratio of 1:1 aluminosilicate and has the general formula, $Al_2Si_2O_5(OH)_4$ $nH_2O$. It has two groups, one on the inside $Al_2O_3$ and the other is at the outside $SiO_2$. Halloysite has a length of 200–1000 nm with 10–25 nm as internal and 50 nm as external diameter. Recently, Feng and Cheng [38] reported the fabrication of halloysite nanocontainers and their compatibility with epoxy coating for anticorrosion. Based on soft template synthesis methods, these particles vary their size, morphology, as well as size distribution and pore structure. They also control the kinetics of hydrolysis and polycondensation reaction. There are two types of silica observed, first is the formation of spherical mesoporous silica, surfactant act as template method for this where cetyltrimethylammonium bromide (CTAB) dissolved in a sodium hydroxide solution. Tetraethoxysilanes (TEOS) were added to the above solution for hydrolysis and followed by polycondensation for the formation of silanol molecules on micellar templates, called

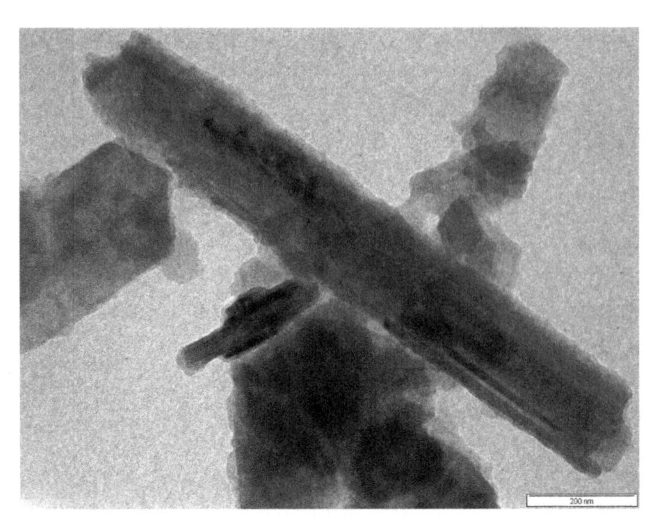


**FIGURE 10–2** Halloysite nanotube TEM image.

mesoporous silica. These particles have large surface area (1000 $m^2/g$) and pore volume (1 $cm^3/g$) with low particle size of diameter 3.5 nm. Second case is the formation of spherical hallow silica (SHS); in this synthesis, CTAB and TEOS disperse in water–ethanol medium with pH value raise to 11 by ammonium hydroxide to form SHS. Such silica nanoparticle does not show any disorder in there structure but has more thick shell with high porosity [39]. Another example of artificial porous ceramic nanoparticles is mesoporous titania ($TiO_2$), in this container formation where firstly mesoporous type nanoparticle will formed by hetrophase oxidation of titanium carbide by adding concentrated nitric acid. The titania nanoparticles have moderate polydispersity index (0.35) and average size of 400 nm with 0.35 $cm^3/g$ pore volume, which allows sufficient encapsulation of inhibitors. Next big step in this container preparation is encapsulation active corrosion inhibitor into the cavity of container. For this purpose the prepared hallow or porous nanoparticle dissolve in saturated solution of active agent with corresponding solvent. Then this mixture of solution is subject to vacuum filtration for multiple times for maximum loading. Such procedure leads to the decrease of the mobility of inhibitor inside narrow pore of container, which slows down the release of active agent from container [40]. The last step of the container is adsorption of layer of opposite charged polyelectrolyte on to the core. They are few polyelectrolyte material used such as polyethylene imine (PEI), polystyrene sulfonate (PSS), polymethacrylic acid. Depending on specific required properties on shell formation with various polyelectrolyte used.

## 10.3.3 Stimuli response with ceramic core

Basic component of this type of container will be similar to as describe in the above section. Only difference is the buildup of polyelectrolyte shell onto the small pores of container with blocking the opening ends. First, inner part of pores of hallow particle is essential to load a containers core, then dissolve with encapsulated material, and subsequently subject to vacuum filtration for maximum loading. In contrast to previous above section, here container with both open ends will be closed to prevent its early release of inhibitor from container in any surrounding environment. To achieve the complete closed nanocontainer, opening end of container is closed with impenetrable material (shell stoppers) which is made of physical or chemical interaction of active material. For example, shell stopper composed of ionic complexes or aggregates arising due to physical reaction of gelation, coacervation, or interfacial precipitation. Sometimes, surface modification by chemical is necessary of such pores openings for trigger release [41]. Such release of inhibitor from container can be triggered with change in pH, ionic strength, temperature, and concentrations of specific ions. When such scenario happened, these closed containers cause the swelling of stopper that causes the stimuli response of active agent release by opening the end of the containers. For example, as Abdullayev et al. [42,43] reported, the use of HNT with such type of nanocontainer with BTA as active agent encapsulated in it. For shell stopper formation, they used copper sulfate solution. This form insoluble compound with BTA on to the ends of nanocontainers. For stimuli response of active inhibitor agents, nanocontainer with ammonia solution mixed to form complex of ammonia–copper BTA, which leads to increase in the pH of solution. Such scenarios that make the stoppers were irreversible, removed, and released the inhibitor from container. If such container loaded in coating formulations makes them better coating compared to standard coating because of complex compounds (copper and ammonia) with BTA, better corrosion inhibition performances are seen. Such container not only gives corrosion inhibition but also performed improve bacterial and fouling resistance properties.

## 10.3.4 Direct or inverse emulsion based container

As can be seen from the previous section, container preparation use hollow micro- or nanoparticle as support or framework and invulnerable container core is a multistep process. Again, depending on the size and volume, porous nanoparticle played an important for maxing loading of active agent. This method is more complex for nanocontainer with respect to their encapsulation material and their container core formation with different polarity compounds. Furthermore, the different polarity among the core and dispersion solutant, where the containers were initially dislocated, can make the total procedure overcomplicated [44]. Subsequently, two more steps of multilayer shell to container make them not simple for fabrication, whereas this direct or inverse emulsion base nanocontainer method has many advantages. First one, this emulsion droplet not only played a dual role as core formation but also acted as soft template particles that make them reduce number of step for fabrication. Second big advantage is the formation liquid state dispersion phase and media in the initial stage of process. This minimized the requirement of energy of colloidal system preparation with the presence of surfactant, which reduced surface tension. Also this liquid state of disperse phase provide not only proper distribution of active agent content but also control the size of container. Another advantage is that this emulsion route is possible to prepare encapsulation approach core–shell morphology container.

## 10.3.5 Container-based internal physical phenomena

### *10.3.5.1 Interfacial solvent*

As seen in the earlier section, emulsion route nanocontainer, emulsion droplet act as soft template at which shell-formation precipitate takes place by physical or chemical process in the form dispersion media. The same droplets can act as carrier or core for the encapsulation of material in nanocontainer. In this nanocontainer formation, if initially the formation of homogenous solution takes place among the shell-forming material and emulsion droplet, it disturbs the solubility of shell-forming material and leads to start of its precipitation. In the easy way, precipitation happened due to physical change in reaction system means change in phase composition either removal of solvent or addition of solvent, which cause the lowering the solubility of shell-forming material with other material in one phase from other phase. This typical process leads to interfacial precipitation by the evaporation of solvent from dispersion media that makes the solvent evaporation-induced precipitation.

Grigoriev et al. [45] reported the synthesis of this type of nanocontainer, where they encapsulate sodium docusate as water repelling agent in O/W styrene emulsion through interfacial solvent route. They select ethyl acetate as dispersion media with water in the ratio of 5:9. Polystyrene micro-container of size of 1570 nm and polydispersity index (PDI) 0.8 is formed by the mean of high-energy method such as high-speed homogenizer. Despite this, the process is relatively easy and economical, but there are many challenges and limitation when they are adding to coating matrix. First, the application of such solvent interfacial precipitation to the coating matrix leads to volatile organic compounds (VOC) emission problems to environment. Another problem is that the selection of polymer for shell forming should have good polarity between the oil phase and aqueous phase by lowering the degree of cross-link polymer. This means restrict the usage of this container only in water base formulation.

### *10.3.5.2 Layer by layer method container*

From the last three decades, LBL method for container preparation became well known in the form plane solid and colloidal solid templates [46,47]. In the late 2000s, most of research already work on this method but found restricted due to its processing condition such chemical and physical [48–50]. Recently, liquid colloidal templates was develop for container [48,49]. Grigoriev et al. [45] reported the universal approach formation of micro- or nanocontainer on to emulsion LBL route. They reported that water immiscible substance dodecane is used as liquid core in emulsion droplets. For stabilizing the emulsion dispersion phase and providing a site for the deposition of polyelectrolyte layer, oil phase was doped with precursor (cationic surfactant) to rise to surface stability of this monolayer. The next step of LBL procedure is the adsorption of oppositely charge aqueous polyelectrolyte solution on to the monolayer. The second encapsulation step was done in an aqueous solution of cationic polyelectrolyte shell. Finally, with desired shell size of container, repetition of adsorption step was done. Such nanocontainer have average diameter of size 4.2 μm and was quite monodisperse (PDI = 0.32). The container prepared via such emulsion route LBL method provides support to protective coating when coating required corrosion resistance on demand trigger system [51–54]. Example of such nanocontainer was reported by several researchers with the use of different corrosion inhibitor. Bhanvase et al. [55] reported the nanocontainer-based delivery system for the corrosion protection. These authors found that 5 wt.% of nanocontainer loading in alkyd resins may be sufficient to make pH-controlled smart coatings. The release kinetics of corrosion inhibitors of LBL method could be the trigger pH-responsive for controlled release. The release studies for all the nanocontainers were carried out in an aqueous solution at acidic, basic, and neutral pH, which was obtained by adding HCl or NaOH. Sonawane et al. [56] reported the synthesis of LBL-assembled zinc oxide (ZnO) nanocontainer and its kinetic mechanism for controlled release of corrosion inhibitor from container. BTA as corrosion inhibitor was loaded in nanocontainer and later added in alkyd-based coating. Corrosion studies were evaluated using weight-loss method, Tafel plot. Result shows with increase in the concentration of nanocontainer in coating, corrosion rate decreases. Bhanvase et al. [57] reported that LBL assembled calcium zinc phosphate nanocontainer using sonochemical approach and its release kinetic mechanism. The prepared container due to the sonochemical approach is small and the uniform particle size are observed. Nanocontainer shows pH responsive release mechanism. Nanocontainer with increase in concentration from 0% to 4% decreases the corrosion rate from 2.2 to 0.15 mm/year.

#### *10.3.5.3 Pickering emulsion container*

Pickering emulsion container is emulsion route approach where hydrophilic or hydrophobic micro- or nanoparticle is attached to the interface of O/W or W/O emulsion. In this approach of nanocontainer, interfacial surface modification of micro- or nanoparticle is important so this can be done by using surfactant, which bonded to nanomaterial by covalent bond [58–60]. This modification to particle not only provides physical interaction between particle, surfactant, and surrounding solution but also provides interface to corrosion inhibitor in the core-forming emulsion droplets. However, care must take accounting, for surfactant selection, as it arise partial hydrophobic at surface modification when it contacts the corrosion inhibitors. Fabrication of such nanocontainer is a simple method by adopting interfacial method. There are two basic step of such system, first is the formation of building blocks particle by partial hydrophilization or hydrophobization in such a manner that use as container. Second is the addition of remaining component of container to dispersion phase of emulsion for nanocontainer form. Haase et al. [61] reported the multifunctional container in the form silica-coated polystyrene with different corrosion inhibitor of different concentration of 8-hydroxyquinoline (8-HQ) and MBT. They also reported that this type container shows better technique for ecofriendly waterborne coating, controls the size of nanocontainer, and increases the encapsulation percentage of inhibitor in container as compared to earlier reported container.

### 10.3.6 Based on chemical reaction nanocontainer

In this method of container preparation, chemical reaction take place either at interface between dispersion media and emulsion droplets or inside the droplets. Based on their location of chemical reaction, container formation is classified into two group, namely, in situ polymerization and polycondensation or polyaddition by interfacial method inside emulsion droplets.

#### *10.3.6.1 Interfacial polyaddition/interfacial polycondensation*

In this type of container preparation, one of the components was dispersed in continuous phase at outside droplets of emulsion, whereas others component are dissolved inside the droplets. This means they should have phase separation at the beginning of the chemical reaction. As reaction progress at emulsion droplets interface, both phase interact with each other and form product which neither is soluble in dispersion media nor in the droplet of emulsion. Such container form with core–shell morphology. Latnikova et al. [62] reported that the encapsulation of two different protective agents together was carried out by this method of container preparation. Also, Cho et al. [63] reported the novel siloxane-based self-healing system in the form of container for mild steel substrate. The main concept of these works is based on phase separation in which siloxane material does not encapsulate (phase separation form) but catalyst is encapsulated. It is reported that siloxane material with hydroxyl functional were economical, effective, easily available, and chemically stable under any environmental condition while performing the self-healing process. The catalyst such as dibutyltin dilaurate (DBTL) and HOPDMS and PDES blends with vinyl ester prepolymer. These types of self-healing system reported 88% recovery of fracture strength compared to dicyclopentadiene (DCPD)-based self-healing system. Keller et al. [64] reported the new elastomer self-healing system based on PDMS novel material. Two different core materials, namely, blend of vinyl based PDMS resin and platinum catalyst and other one activator for vinyl-PDMS, were encapsulated in UF shell material for the formation of microcapsules. Later these capsules incorporated in PDMS elastomer matrix and provided 70%–100% of recovery to original material. They also reported that with increase in the concentration of PDMS, the tea strength increased. Again in year of 2009, Cho et al. [65] reported the modified PDMS microcapsules self-healing system. In their work, healing agent is the combination of volume 96% HOPDMS and 4% polydiethoxysiloxane with dimethyldineodecanoate (DMDNT) in chlorobenzene as catalyst. These phase-separated microcapsules were incorporated in polyurethane polymeric shell and later encapsulated in PDMS resin for formation of self-healing corrosion inhibition coating. Cho et al. [66] reported to extend his previous work [65] by checking the effect of rheological behavior of healing agent and the concentration of capsules, catalyst, and catalyst activity on healing performance at room temperature. It was found that lowering the viscosity of healing agent with dual microcapsule in system provide healing performance of around 75% compared to original film. Again, these containers with cerium-based corrosion inhibitor are encapsulated in polyamide shell in the presence of organic solvent butyl phosphate. The resultant container has broad PDI with the average size of 6 μm. Such nanocontainer provide self-healing corrosion inhibition application when change in pH and trigger release of active inhibitor occurred.

### *10.3.6.2 In situ emulsion polymerization*

In this type of container fabrication, all reactant or component present inside the emulsion droplets from beginning of emulsion. Such droplet acts as reservoir or container for reactant and active agent. This plays an important role for forming liquid colloidal reactors as well as stabilized the reaction against premature polymerization. After the preparation of O/W or W/O emulsion with desired size distribution and stability, emulsion can be trigged to activate by the means of chemical or physical reaction such as catalyst, UV light to attain equilibrium state in polymerization. Grigoriev et al. [45] reported the preparation of 2-mercaptobenzothizole (2-MBT) encapsulated polyepoxy container for aluminum alloy substrates. In this nanocontainer preparation, different epoxy monomer is in 1:1 ratio, along with 20% MeBT of total mass. The last stage of emulsion is the incorporation of curing agent for in situ polymerization in excess quantity of epoxy monomer mass. Resulting nanocontainer of average size 2.3 μm with PDI close to 1 and have obviously the core/shell morphology. Later Pang and bond [67] reported that epoxy and amine microcapsules were incorporated in hallow fibers for self-repairing effect in plastic composite with healing efficiency around 40%–60%. However, it has some drawbacks due to limited healing resin availability. As time progress self-healing ability also vanishes. Yuan et al. [68] reported the dual microcapsule based epoxy self-healing system. In these systems, epoxy (base) and mercaptan (hardner) microcapsules prepared by in situ polymerization approach encapsulate in MF polymeric shell. Both microcapsules at different concentrations were incorporated in low viscous bisphenol–based epoxy matrix with diethylenetriamine (DETA). The 8:1 weight ratio was taken for epoxy and DETA. They reported that as the concentration of microcapsule increase from 1 to 5 wt.%, the healing efficiency also increases from 43.5% to 104.5%. Yuan et al. [69] reported the effect of alkality and activity of hardener microcapsule and viscosity of epoxy microcapsule on healing chemistries. They reported that high activity of mercaptan and highly basicity of amine and low viscous epoxy microcapsule provides maximum healing efficiency at rapid process. During self-healing mechanism, concentration of both microcapsules should be distributed equally to the damage site, which is highly dependent on the size and fraction of microcapsules used. Yuan et al. [70] reported to extend his previous work [69], by checking the effect of microcapsule on healing efficiency. Later they observed that the healing efficiency also depends on applied stress and fatigue crack range ($\Delta K_I$). Jin et al. [71] reported the use of dual capsule–based self-healing system. Low viscous bisphenol based epoxy and aliphatic amine microcapsule was prepared separately using in situ polymerization method. Later both capsules in ratio of 4:6 molar concentrations were incorporated in the epoxy thermoset–based host matrix for the preparation of a self-healing coating. Coating with these microcapsule shows 91% healing efficiency. Hart et al. [72] reported the novel 2-ethyl-4-methylimidazole (24-EMI) as latent catalyst for new self-healing system that can heal for multiple times with moderate temperature. 24-EMI along with liquid low viscous epoxy monomer and cycloaliphatic amine were encapsulated in host matrix. With the help of tapered double cantilever beam (TDCB) fracture test method, 90% recovery fracture obtained with addition of 10% 24-EMI in coating. They also reported that the effect of concentration of imidazole on epoxy toughens, glass transition temperature were evaluated. With increase in concentration of imidazole, there will be decrease in the $T_g$ of material, which is directly related to toughness fracture. Zhu et al. [73] reported the novel polypropylene material in the form in which container was used and in which epoxy and mercaptan-based healing agent were incorporated. This material along with defoaming agent improve the flow behavior of healing agent and healing efficiency.

## 10.4 Distribution and performance of container for protective coating

Distribution of container in coating matrix is an important aspect for not only providing effective barrier resistance but also strongly influencing their responsive ability. Traditionally, prepared container size should be optimal (10% of coating thickness). The coating and container content should be at proper ratio to ensure maintenance integrity of coating during coating processing [74]. Container location is an important part in coating matrix, which can influence the release of active inhibitor to repairing film. Recently, Borisova et al. [75] observed that, when distance between container location and substrate increases, there will be increasing in barrier performance, but this hampers its corrosion-resistance properties. Dispersion of container in coating matrix always recognizes how well is its distribution in matrix. For stability of container, they should provide electrostatic repulsion charge against aggregation of container. Nanocontainer is well distributed in coating matrix, which keeps away corrosion inhibitor from direct coating matrix. The corrosion inhibitor that releases from nanocontainer attacks onto the damage sites which provide superior corrosion resistance. Shchukin et al. [15] reported the first LBL approach for the preparation of nanocontainer. They have taken silica nanoparticle ($SiO_2$) with BTA as corrosion inhibitor, PEI/PSS as polyelectrolyte layer for formation of $SiO_2$/PEI/PSS/BTA/PSS/BTA nanocontainer with size 70 nm. These nanocontainers encapsulated in hybrid sol–gel coatings and applied on aluminum alloys. On pH responsive trigger mechanism, BTA is released from nanocontainer and protect the alloys by forming oxide layer. Zheludkevich et al. [5] reported the formation $SiO_2$/PEI/PSS/BTA/PSS/BTA nanocontainer of size 70 nm and their incorporation in hybrid silica-zirconia sol–gel coating for self-healing application. These coating is applied on metallic substrate and provide long-term corrosion inhibition and control release of inhibitor. Corrosion inhibition evaluation were carried out using electrochemical analyzer and scanning vibrating electrode technique (SVET) method. Result shows as the concentration of nanocontainer in coating increases, ohmic resistance of coating also increases as compared to conventional sol–gel coating. Skorb et al. [39] reported novel active smart material–based coating in the form of mesoporous silica nanocontainer. Unique 2-(benzothiazol-2-ylsulfanyl)-succinic acid corrosion inhibitor is encapsulated in silica nanoparticle by using LBL approach. These container-based coating systems mainly focus for industrial application due to its corrosion issue on metallic surfaces. Self-healing evaluation of such coating was carried out using SVET. The result shows that due to change in pH occur at corrosion damage, the container provide responsive release and makes the coating with self-healing ability.

Balaskas et al. [16] reported the formation of titanium nanocontainer with encapsulate 8-HQ as corrosion inhibitor in it.

## 10.5 Release of active compounds from container

Functional species such as corrosion inhibitors are often loaded in an active carrier, which can be added to the paint formulation for the maintenance and protective performances. These carriers are often called reservoirs because they store the functional species until they are in need. For the encapsulation of inhibitors, LBL method is one of the widely used technique in recent years. The LBL method allows us to synthesize multilayer coatings with controlled properties of permeability. It possesses a core material onto which a layer of polyelectrolyte is followed by an inhibitor layer and a polyelectrolyte layer such as PSS and poly(ethyl imine). This approach not only reduces the high solubility of inhibitor but also reduces the negative influence of corrosion inhibitor on the coating. There have been numerous investigations, which report on nanocontainer through LBL encapsulation [49–60]. The release kinetics of corrosion inhibitors of LBL method could be the trigger pH-responsive for controlled release. The released studies for all the nanocontainers carried out in an aqueous solution at acidic, basic, and neutral pH, which was obtained by adding HCl or NaOH. For the release of active compound, following section discusses the case studies of containers where performance and release was carried out.

## 10.6 Case studies

### 10.6.1 Case study I: preparation of iron oxide nanocontainer by layer-by-layer method using ultrasound approach

#### *10.6.1.1 Material and methods*

Sodium hydroxide (98%), sodium chloride (99%), ammonium persulfate, and xylene (99%) were procured from Molychem, Mumbai. Acetone, butanol, and sodium dodecyl sulfate (SDS) were purchased from SLR Chemicals, Mumbai, India. Hydrochloride acid (35%–38%) was procured from S.D. Fine Chemicals Ltd., Mumbai, India. BTA (98%) and acrylic acid (98%) were procured from Avra Chemicals, Hyderabad, India. Aniline (93%) and ferrous sulfate (98%) were procured from Zen Chemicals, Bangalore, India. Epoxy resin and hardener were procured from Aditya Birla Pvt. Ltd, Mumbai, India. Adhesion promoter, wetting, and dispersing agent were procured from Evonik Industries Mumbai, India. Demineralized water obtained from Millipore apparatus was used in all experiments.

#### *10.6.1.2 Preparation of iron oxide nanoparticle by ultrasound*

As reported by Bhanvase et al. [55], nanocontainer was prepared by using ultrasound approach by modification. Initially, 0.5 M ferrous sulfate and 0.1 M sodium hydroxide solutions were prepared using distilled water and sonicated for 10 min in pulsed mode. Later, NaOH solution was added drop by drop to ferrous sulfate solution, until a greenish brown precipitate was obtained. The solution was filtered and washed thoroughly with water and

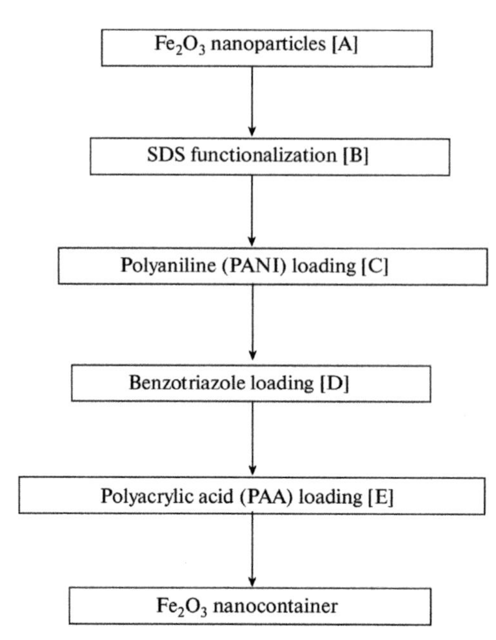


**FIGURE 10–3** Flow diagram for the layer-by-layer synthesis of $FeO_3$ nanocontainers. *Reprinted/adapted by permission from S. Shaik, U. Bagale, M. Ashokumar, S.H. Sonawane, Preparation of $Fe_2O_3$ nanoparticles by acoustic and hydrodynamic cavitation techniques and corrosion inhibition release studies using its nanocontainers, Prot. Metal Phys. Chem. Surf. 53 (5) (2017) 850–858. Springer Nature.*

acetone. On washing, greenish brown turned in to brownish yellow. The precipitate was then dried at 50°C for 3 h.

#### 10.6.1.2.1 Iron oxide nanocontainers preparation

Step-by-step procedure for the synthesis of iron oxide nanocontainers is shown in Fig. 10–3. Iron oxide nanoparticles obtained from above methods were used in the preparation of iron oxide nanocontainers.

Add 50 mL of SDS solution and 0.5 g of iron oxide nanoparticles to the above solution and sonicated for 15 min. This SDS functionalized iron oxide solution was further used in the preparation of nanocontainers. The adsorption of SDS on the nanoparticles changes the charge on the particles to negative.

Initially, 0.7 g of ammonium persulfate was dissolved in 10 mL of distilled water. To this solution, at temperature 10°C encapsulation of SDS–iron oxide nanoparticles were added under ultrasound irradiation followed by dropwise addition of 5 g of aniline under sonication for 45 min. Finally, a green colored positively charged PANI encapsulated SDS–iron oxide particles were obtained. The solid particles were filtered, washed, and dried at 50°C for 3 h.

The PANI-loaded iron oxide particles (0.3 g) was added into 4 M NaCl solution prepared in 50 mL distilled water. Further, encapsulation BTA layer onto the layer of PANI-loaded iron oxide nanoparticle was accomplished using 0.1 g of BTA solution in 10 mL water in ultrasound aided surrounds for 15 min. These nanoparticles is separated by centrifugation, and washed and dried for further use.

In order to maintain the stability of the nanocontainers, polyacrylic acid (PAA) was loaded onto the BTA–PANI–iron oxide nanoparticles. The loading of PAA onto BTA–PANI–iron oxide nanoparticles changes the surface charge of the particles. Two grams of PAA was added into 4 M NaCl solution followed by the addition of BTA–PANI–iron oxide nanoparticles. The mixture was then sonicated for 30 min to get uniform coatings of PAA onto the BTA–PANI–iron oxide nanoparticles. The loading of PAA created a positively charged polyelectrolyte layer on the BTA–PANI–iron oxide nanoparticles, which gives better stability of the particle in corrosive atmosphere. These nanocontainers were incorporate into solution of epoxy resin in order to formulate nanocontainer-based coatings.

The epoxy coating consists of different ingredients such as resin (Epoxy resin-70% by weight), wetting agent (0.2%), adhesion promoter (0.3%), butanol (3%), and xylene (7%). In order to prepare the nanocontainer-based resin coating, different weight percentages of iron oxide nanocontainers (1, 2, 3, 4 wt.%) were added into the coating solution. The prepared coating was applied on steel plates that were cut in required dimensions to carry out the electrochemical tests.

#### *10.6.1.3 Characterization of nanoparticle and nanocontainer*

The surface morphologies of the iron oxide nanoparticles and nanocontainers were characterized transmission electron microscopy (TEM) analysis at 200 kV. The particle size distribution of the nanoparticles and nanocontainers was analyzed by using particle size analyzer (Malvern Zetasizer, Nano S90-Version 7.02). With the help of UV spectrophotometer, we would be able to study the release mechanism of BTA at various pH range of solution. For corrosion inhibition efficiency the coated steel panels were dipped in an acidic solution (electrolyte). Tafel plots were drawn for the epoxy-coated plates using electrochemical workstation (Ivium Technologies, The Netherlands).

#### *10.6.1.4 Results and discussions*

##### **10.6.1.4.1 Morphological studies of iron oxide nanoparticles and nanocontainers**

The internal morphology that synthesizes nanocontainers and nanoparticles using ultrasound were evaluated by TEM analysis. This is shown in Fig. 10–4A and B, respectively. It can be observed that the particles show nanostructures with average dimensions of $81 \times 17$ nm (length to width) for particles synthesized by acoustic cavitation. The TEM image shows spherical shape of iron oxide nanocontainer containing iron oxide nanoparticles with the layer of PAA in Fig. 10–4B for ultrasound. Further, in TEM image, the presence of black dark spot observed over the structured iron oxide nanoparticles has confirmed the layer formation of PANI, BTA, and PAA. This results in increasing the particle size of iron oxide nanocontainer, the results of which show replication in particle size.

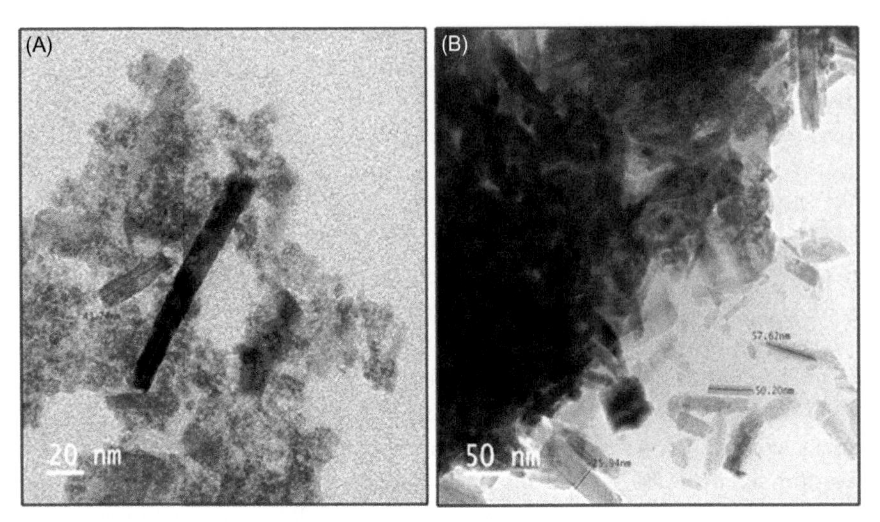


**FIGURE 10–4** TEM Images of $Fe_2O_3$ nanoparticles (A) and nanocontainers (B) produced by acoustic cavitation method. *Reprinted/adapted by permission from S. Shaik, U. Bagale, M. Ashokumar, S.H. Sonawane, Preparation of $Fe_2O_3$ nanoparticles by acoustic and hydrodynamic cavitation techniques and corrosion inhibition release studies using its nanocontainers, Prot. Metal Phys. Chem. Surf. 53 (5) (2017) 850–858. Springer Nature.*

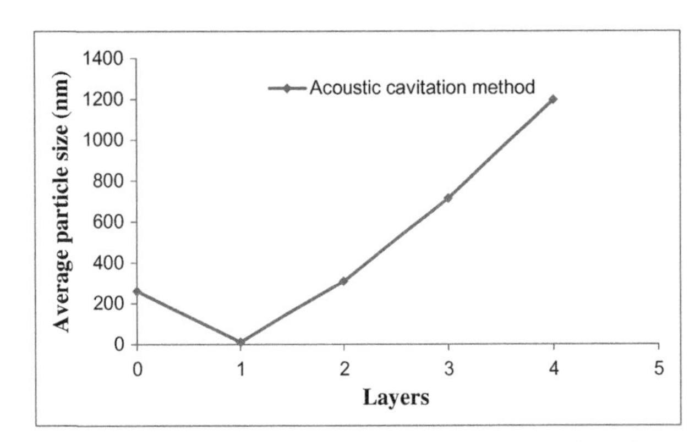


**FIGURE 10–5** Growth of $Fe_2O_3$ nanocontainer on layer-by-layer loading of polyelectrolyte and benzotriazole.

### 10.6.1.4.2 Layer-by-layer assembly

Fig. 10–5 shows growth of iron oxide nanocontainer after LBL loading of polyelectrolyte and BTA. The iron oxide nanoparticles average particle size of 260 nm with size distribution are shown in Fig. 10–5. With each layer, nanoparticles of iron oxide will tend to increase the average particle size to 1096 nm with the adsorption layer of PANI, BTA, and PAA. The reduction in particle size is due to the sonochemical approach which make the system in proper mixing and dispersion of particle with enhance solute transfer rate. The sonochemical

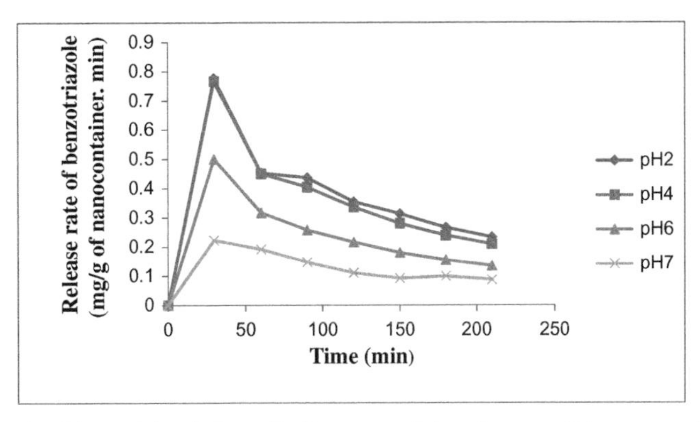


**FIGURE 10–6** Release rate of benzotriazole from $Fe_2O_3$ nanocontainer (nanoparticles-core material synthesized from acoustic technique). *Reprinted/adapted by permission from S. Shaik, U. Bagale, M. Ashokumar, S.H. Sonawane, Preparation of $Fe_2O_3$ nanoparticles by acoustic and hydrodynamic cavitation techniques and corrosion inhibition release studies using its nanocontainers, Prot. Metal Phys. Chem. Surf. 53 (5) (2017) 850–858. Springer Nature.*

technique improves the kinetic of reaction that provides better site formation of bubble growth and collapse of bubble to get reduction in size.

### *10.6.1.5 Release and release rate of benzotriazole*

The release rate and release concentration of BTA were studied to know the sensitivity of the nanocontainers at different pH (ranging from 2 to 7) conditions, and the results are reported in Fig. 10–6. The experiments were carried out at pH range 2–7 to validate responsive release BTA from iron oxide nanocontainers. From Fig. 10–6, it can be observed that the release of BTA is higher in acidic medium indicating that the iron oxide nanocontainers show better corrosion inhibition. The release concentration increases with respect to time until the concentration of BTA in nanocontainer and outside medium is equal. The maximum released amount of corrosion inhibitor was 47 mg/g of BTA for nanoparticles produced by acoustic at pH 2. Fig. 9 show the release rate of BTA from iron oxide nanocontainers produced using acoustic. As observed in TEM image (Fig. 10–4), iron oxide nanoparticle is clearly shown and less dimension (length × width) spherical structure nanoparticle observed. These spherical-shaped nanoparticles, prepared from ultrasound, adsorbed more BTA inside the iron oxide nanoparticle with less amount of BTA at outer surface of nanoparticle. This makes BTA release very slowly in aqueous media acoustic cavitation. This also means that slow release of BTA from iron oxide nanocontainer shows better corrosion control over long period. The initial BTA release rate is high and decreases with time, which might be due to a decrease in the concentration of BTA in iron oxide nanocontainers. It was found that the release rate is high in acidic conditions. When pH was increased from 2 to 7, the release rate decreased from 0.8 to 0.2 (mg/g of nanocontainers. min) for nanocontainers prepared using acoustic

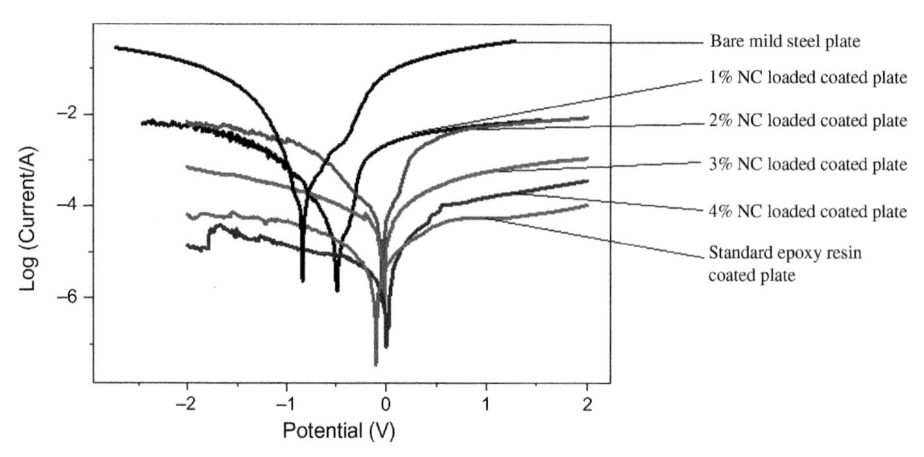


**FIGURE 10–7** Tafel plot of mild steel plate coated with different compositions of $Fe_2O_3$ nanocontainer (nanoparticles-core material synthesized from acoustic technique). *Reprinted/adapted by permission from S. Shaik, U. Bagale, M. Ashokumar, S.H. Sonawane, Preparation of $Fe_2O_3$ nanoparticles by acoustic and hydrodynamic cavitation techniques and corrosion inhibition release studies using its nanocontainers, Prot. Metal Phys. Chem. Surf. 53 (5) (2017) 850–858. Springer Nature.*

#### *10.6.1.6 Electrochemical analytical studies*

For electrochemical analysis, Tafel plot can be measured using electrochemical analyze. In this measurement counter, reference and working electrode are kept as graphite, Ag/AgcCl, and coated panel, respectively, in solution of HCl (0.5 M) with potential window of +2 to −2 V.

Tafel plots for different loading of iron oxide nanocontainers are shown in Fig. 10–7. It is observed that the current density ($I_{corr}$) decreases with an increase in percentage loading of iron oxide nanocontainers in epoxy coating. The rate of reaction of corrosion depends on the potential of the metal to transfer the electrons and ions from the solution. As the potential of the metal increases in positive direction (Fig. 10–7), the anodic reaction increases and the cathodic reaction decreases. Therefore the current density decreases; hence, the corrosion rate also decreases because of low transfer of ions and electrons. It is found that with an increase in iron oxide nanocontainer loading from 1 to 4 wt.%, the current density ($I_{corr}$) increase from $1.76 \times 10^{-5}$ to $4.04 \times 10^{-6}$ A/cm$^2$ for iron oxide nanocontainers. It is also observed that the corrosion rate decreases from 0.688 to 0.051 mm/year.

## 10.7 Commercial applications and future prospectus

Future applications of nanocontainers are expected to be much broader ranging from the area of catalyst, drug delivery systems, nanoreactors, gas storage, and smart coatings up to anticorrosion. Different types of active molecule/agents can be embedded inside the nano-container shell for the design and development of customized intelligent materials for the specific application in the actual environmental conditions. In commercial applications, internal and external stimulants play a very crucial role in the drain, release and repair mechanism of active molecule from the nanocontainer shell.

There is a need to focus on fundamental phenomena and mechanisms through which active molecules release from nanocontainers to provide the base in the precise selection of customized application prerequisite material. Nowadays, nanocontainers are the at entrance of the industrial application, and the main issues are in the development of commercial production facility for nanocontainer and carrying out required tests such as salt-spray test in the case of corrosion coatings. Also another issue is the encapsulation of more than one active agent for the generation of multifunctaional nanocontainers that can be used to tackle the multiple problems for the various commercial applications.

## 10.8 Conclusion/inference

The reported study concludes that there are various methods available for the fabrication of nanocontainer such as cost-effective halloysite nanocontainer, silica hallow nanocontainers, and titanium nanocontainer with various corrosion inhibitors. The nanocontainer is compatible with both water and solvent coating system. From the presented case study, we found out that the addition of nanocontainer in coating reduces the corrosion activity at alloy or mild steel surface. From EIS study in terms of Tafel plots, it is found that the corrosion rate decreases from 0.688 to 0.051 mm/year with the addition of nanocontainer. It has also found that there is a controlled release of corrosion inhibitor from such nanocontainer. These nanocontainers not only provide coating corrosion inhibition but also self-healing ability in some coating formulation that makes them functional or smart coatings.

## References

[1] M. Yeganeh, T.A. Nguyen, Methods for corrosion protection of metals at the nanoscale, Kenkyu J. Nanotechnol. Nanosci. 5 (2019) 37–44.

[2] D. Grigoriev, E. Shchukina, D.G. Shchukin, Nanocontainers for self-healing coatings, Adv. Mater. Interfaces 4 (2017) 1–11.

[3] D.V. Andreeva, D.G. Shchukin, Smart self-repairing protective coatings, Mater. Today 11 (2008) 24–30.

[4] S.R. White, N.R. Sottos, P.H. Geubelle, J.S. Moore, M.R. Kessler, S.R. Sriram, et al., Nature 409 (2001) 794; (b) A. Rahimi, S. Amiri, J. Polym. Res. 4 (2016) 1.

[5] M.L. Zheludkevich, D.G. Shchukin, K.A. Yasakau, H. Mohwald, M.G.S. Ferreira, Anticorrosion coatings with self-healing effect based on nanocontainers impregnated with corrosion inhibitor, Chemistry of Materials 19 (2007) 402–411.

[6] D.G. Shchukin, D.O. Grigoriev, H. Meohwald, Application of smart organic nanocontainers in feedback active coatings, Soft Matter 6 (2010) 720–725.

[7] I.A. Kartsonakis, I.L. Danilidis, G.S. Pappas, G.C. Kordas, Encapsulation and release of corrosion inhibitors into titania nanocontainers, J. Nanosci. Nanotechnol. 10 (2010) 1–9.

[8] S.K. Ghosh, Self-Healing Anticorrosion Coatings, Self-healing Materials: Fundamentals, Design Strategies, and Applications, Wiley-VCH Verlag GmbH & Co. KGaA, Weinheim, 2009, pp. 1–25.

[9] Y. Liang, M. Wang, C. Wang, J. Feng, J. Li, L. Wang, et al., Facile synthesis of smart nanocontainers as key components for construction of self-healing coating with superhydrophobic surfaces, Nanoscale Res. Lett. 11 (2016) 231–242.

[10] A. Stankiewicz, I. Szczygiel, B. Szczygiel, Self-healing coatings in anti-corrosion applications, J. Mater. Sci. 48 (2013) 8041–8051.

[11] M.F. Montemor, Functional and smart coatings for corrosion protection: a review of recent advances, Surf. Coat. Technol. 258 (2014) 17–37.

[12] D. Shchukin, H. Mohwald, Surface-engineered nanocontainers for entrapment of corrosion inhibitors, Adv. Funct. Mater. 17 (2007) 1451–1458.

[13] G. Gupta, N. Birbilis, A.S. Khanna, An epoxy based lignosulphonate doped polyaniline-poly(acrylamide co-acrylic acid) coating for corrosion protection of aluminium alloy 2024-T3, Int. J. Electrochem. Sci. 8 (3) (2013) 3132–3149.

[14] M. Evaggelos, K. Ioannis, P. George, K. George, Release studies of corrosion inhibitors from cerium titanium oxide nanocontainers, J. Nanopart. Res. 13 (2011) 541–554.

[15] D.G. Shchukin, M. Zheludkevich, K. Yasakau, S. Lamaka, M.G.S. Ferreira, H. Mohwald, Layer-by-layer assembled nanocontainers for self-healing corrosion protection, Adv. Mater. 18 (2006) 1672–1678.

[16] A.C. Balaskas, I.A. Kartsonakis, L.A. Tziveleka, G.C. Kordas, Improvement of anti-corrosive properties of epoxy-coated AA 2024-T3 with $TiO_2$ nanocontainers loaded with 8-hydroxyquinoline, Prog. Organ. Coat. 74 (2012) 418–426.

[17] G.H. Deng, C.M. Tang, F.Y. Li, H.F. Jiang, Y.M. Chen, Covalent cross-linked polymer gels with reversible sol–gel transition and self-healing properties, Macromolecules 43 (2010) 1191–1194.

[18] R.P. Wool, K.M. O'Connor, A theory crack healing in polymers, J. Appl. Phys. 52 (1981) 5953–5963.

[19] Y.X. Lu, Z.B. Guan, Olefin metathesis for effective polymer healing via dynamic exchange of strong carbon–carbon double bonds, J. Am. Chem. Soc. 134 (2012) 14226–14231.

[20] B.J. Blaiszik, S.L. Kramer, S.C. Olugebefola, J.S. Moore, N.R. Sottos, S.R. White, Self-healing polymers and composites, Annu. Rev. Mater. Res. 40 (2010) 179–211.

[21] V. Vahedi, P. Pasbakhsh, C.S. Piao, C.E. Seng, A facile method for preparation of self-healing epoxy composites: using electrospun nanofibers as microchannels, J. Mater. Chem. A 3 (2015) 16005–16012.

[22] K.S. Toohey, N.R. Sottos, J.A. Lewis, J.S. Moore, S.R. White, Self-healing materials with microvascular networks, Nat. Mater. 6 (8) (2007) 581–585.

[23] K.S. Toohey, C.S. Hansen, J.A. Lewis, S.R. White, N.R. Sottos, Delivery of two-part self-healing chemistry via microvascular networks, Adv. Funct. Mater. 19 (9) (2009) 1399–1405.

[24] H. Ullah, K.A. Azizli, Z.B. Man, M.C. Ismai, M.I. Khan, The potential of microencapsulated self-healing materials for microcracks recovery in self-healing composite systems: a review, Polym. Rev. (2016) 1–57.

[25] S. Albertazzi, F. Basile, A. Vaccari, Catalytic properties of hydrotalcite-type anionic clays, in: F. Wypych, K.G. Satyanarayana (Eds.), Clay Surfaces: Fundamentals and Applications., Elsevier, Amsterdam, 2004, pp. 496–546.

[26] A. Sorrentino, G. Gorrasi, M. Tortora, Incorporation of Mg-Al hydrotalcite into a biodegradable poly(3-caprolactone) by high energy ball milling, Polymer 46 (5) (2005) 1601–1608.

[27] S.J. Palmer, R.L. Frost, T. Nguyen, Hydrotalcites and their role in coordination of anions in Bayer liquors: anion binding in layered double hydroxides, Coord. Chem. Rev. 253 (1–2) (2009) 250–267.

[28] R.G. Buchheit, S.B. Mamidipally, P. Schmutz, Active corrosion protection in Ce-modified hydrotalcite conversion coatings, Corrosion 58 (1) (2002) 3–14.

[29] J. Tedim, M.L. Zheludkevich, A.N. Salak, Nanostructured LDH-container layer with active protection functionality, J. Mater. Chem. 21 (39) (2011) 15464–15470.

[30] R.B. Leggat, S.A. Taylor, S.R. Taylor, Adhesion of epoxy to hydrotalcite conversion coatings: II Surface modification with ionic surfactants, Colloids Surf., A 210 (1) (2002) 83–94.

[31] G. Williams, H.N. McMurray, Anion-exchange inhibition of filiform corrosion on organic coated AA2024-T3 aluminum alloy by hydrotalcite-like pigments, Electrochem. Solid-State Lett. 6 (3) (2003) B9–B11.

[32] P.V. Mahajanarn, R.G. Buchheit, Characterization of inhibitor release from Zn-Al hydrotalcite pigments and corrosion protection from hydrotalcite-pigmented epoxy coatings, Corrosion 64 (3) (2008) 230–240.

[33] M.L. Zheludkevich, S.K. Poznyak, L.M. Rodrigues, Active protection coatings with layered double hydroxide nanocontainers of corrosion inhibitor, Corros. Sci. 52 (2) (2010) 602–611.

[34] B.K.G. Theng, The Chemistry of Clay-Organic Reactions, Wiley, New York, 1974.

[35] G. Williams, H.N. McMurray, Inhibition of filiform corrosion on polymer coated AA2024-T3 by hydrotalcite- like pigments incorporating organic anions, Electrochem. Solid-State Lett. 7 (5) (2004) B13–B15.

[36] M.H. Kendig, M. Hon, A hydrotalcite-like pigment containing an organic anion corrosion inhibitor, Electrochem. Solid-State Lett. 8 (3) (2005) B10–B11.

[37] S.K. Poznyak, J. Tedim, L.M. Rodrigues, Novel inorganic host layered double hydroxides intercalated with guest organic inhibitors for anticorrosion applications, ACS Appl. Mater. Interfaces 1 (10) (2009) 2353–2362.

[38] Y.C. Feng, Y.F. Cheng, Fabrication of halloysite nanocontainers and their compatibility with epoxy coating for anti-corrosion performance, *Corros. Eng. Sci. Technol.* (2015) https://doi.org/10.1080/1478422X.2016.1142161.

[39] E.V. Skorb, D. Fix, D.V. Andreeva, H. Mohwald, D.G. Shchukin, Surface-modified mesoporous $SiO_2$ containers for corrosion protection, Adv. Funct. Mater. 19 (2009) 2373–2379.

[40] D. Borisova, D. Akcakayiran, M. Schenderlein, H. Mohwald, D.G. Shchukin, Nanocontainer-based anticorrosive coatings: effect of the container size on the self-healing performance, Adv. Funct. Mater. 23 (2013) 3799–3812.

[41] Z. Zheng, X. Huang, M. Schenderlein, Self-healing and antifouling multifunctional coatings based on pH and sulfide ion sensitive nanocontainers, Adv. Funct. Mater. 23 (26) (2013) 3307–3314.

[42] E. Abdullayev, R. Price, D. Shchukin, Halloysite tubes as nanocontainers for anticorrosion coating with benzotriazole, ACS Appl. Mater. Interfaces 1 (7) (2009) 1437–1443.

[43] E. Abdullayev, Y. Lvov, Clay nanotubes for corrosion inhibitor encapsulation: release control with end stoppers, J. Mater. Chem. 20 (32) (2010) 6681–6687.

[44] S. Moya, G.B. Sukhorukov, M. Auch, Microencapsulation of organic solvents in polyelectrolyte multilayer micrometer-sized shells, J. Colloid Interface Sci. 216 (2) (1999) 297–302.

[45] D.O. Grigoriev, M.F. Haase, N. Fandrich, Emulsion route in fabrication of micro and nanocontainers for biomimetic self-healing and self-protecting functional coatings, Bioinspir. Biomim. Nanobiomater. 1 (2) (2012) 101–116.

[46] G. Decher, J.B. Schlenoff, Multilayer Thin Films. Sequential Assembly of Nanocomposite Materials., Wiley, Weinheim, 2003.

[47] G.B. Sukhorukov, A. Fery, M. Brumen, Physical chemistry of encapsulation and release, Phys. Chem. Chem. Phys. 6 (16) (2004) 4078–4089.

[48] D. Guzey, D.J. McClements, Formation, stability and properties of multilayer emulsions for application in the food industry, Adv. Colloid Interface Sci. 128–130 (2006) 227–248.

[49] L. Nilsson, B.J. Bergenståhl, Adsorption of hydrophobically modified anionic starch at oppositely charged oil/water interfaces, J. Colloid Interface Sci. 308 (2) (2007) 508–513.

[50] E. Tjipto, K.D. Cadwell, J.F. Quinn, Tailoring the interfaces between nematic liquid crystal emulsions and aqueous phases via layer-by-layer assembly, Nano Lett. 6 (10) (2006) 2243–2248.

[51] K. Köhler, D.G. Shchukin, G.B. Sukhorukov, Drastic morphological modification of polyelectrolyte microcapsules induced by high temperature, Macromolecules 37 (25) (2004) 9546–9550.

[52] D.G. Shchukin, D.A. Gorin, H. Möhwald, Ultrasonically induced opening of polyelectrolyte microcontainers, Langmuir 22 (17) (2006) 7400–7404.

[53] T. Mauser, C. Déjugnat, H. Möhwald, Microcapsules made of weak polyelectrolytes: templating and stimuli-responsive properties, Langmuir 22 (13) (2006) 5888–5893.

[54] W. Ramsden, Separation of solids in the surface-layers of solutions and 'Suspensions' (observations on surface-membranes, bubbles, emulsions, and mechanical coagulation). Preliminary account, Proc. R. Soc. Lond. 72 (477–486) (1903) 156–164.

[55] B.A. Bhanvase, M.A. Patel, S.H. Sonawane, Kinetic properties of layer-by-layer assembled cerium zinc molybdate nanocontainer during corrosion inhibition, Corros. Sci. 88 (2014) 170–177.

[56] S.H. Sonawane, B.A. Bhanvase, A.A. Jamali, S.K. Dubey, S.S. Kale, D.V. Pinjari, et al., Improved active anticorrosion coatings using layer-by-layer assembled ZnO nanocontainers with benzotriazole, Chem. Eng. J. 189–190 (2012) 464–472.

[57] B.A. Bhanvase, Y. Kutbuddin, R.N. Borse, N.R. Selokar, D.V. Pinjari, P.R. Gogate, et al., Ultrasound assisted synthesis of calcium zinc phosphate pigment and its application in nanocontainer for active anticorrosion coatings, Chem. Eng. J. 231 (2013) 345–354.

[58] J. Frelichowska, M.A. Bolzinger, Y. Chevalier, Pickering emulsions with bare silica, Colloids Surf., A 343 (1–3) (2009) 70–74.

[59] U.T. Gonzenbach, A.R. Studart, E. Tervoort, Ultrastable particle-stabilized foams, Angew. Chem. Int. Ed. 45 (21) (2006) 3526–3530.

[60] M. Schmitt-Roziéres, J. Krägel, D.O. Grigoriev, From spherical to polymorphous dispersed phase transition in water/oil emulsions, Langmuir 25 (8) (2009) 4266–4270.

[61] M.F. Haase, D.O. Grigoriev, H. Mohwald, D.G. Shchukin, Development of nanoparticle stabilized polymer nanocontainers with high content of the encapsulated active agent and their application in water-borne anticorrosive coatings, Adv. Mater. 24 (2012) 2429–2435.

[62] A. Latnikova, D.O. Grigoriev, J. Hartmann, Polyfunctional active coatings with damage-triggered water-repelling effect, Soft Matter 7 (2) (2011) 369–372.

[63] S.H. Cho, H.M. Andersson, N.R. Sottos, S.R. White, P.V. Braun, Polydimethylsiloxane-based self-healing materials, Adv. Mater. 18 (8) (2006) 997–1000.

[64] M.W. Keller, S.R. White, N.R. Sottos, A self-healing poly(dimethyl siloxane) elastomer, Adv. Funct. Mater. 17 (2007) 2399–2404.

[65] S.H. Cho, S.R. White, P.V. Braun, Self-healing polymer coatings, Adv. Mater. 21 (2009) 645–649.

[66] S.H. Cho, S.R. White, P.V. Braun, Room-temperature polydimethylsiloxane-based self-healing polymers, Chem. Mater. 24 (2012) 4209–4214.

[67] J.W. Pang, I.P. Bond, A hollow fibre reinforced polymer composite encompassing self-healing and enhanced damage visibility, Compos. Sci. Technol. 65 (11) (2005) 1791–1799.

[68] Y.C. Yuan, M.Z. Rong, M.Q. Zhang, J. Chen, G.C. Yang, X. Li, Self-healing polymeric materials using epoxy/mercaptan as the healant, Macromolecules 41 (14) (2008) 5197–5202.

[69] Y.C. Yuan, M. Rong, M. Zhang, G.C. Yang, Study of factors related to performance improvement of self-healing epoxy based on dual encapsulated healant, Polymer 50 (2009) 5771–5781.

[70] Y.C. Yuan, M.Z. Rong, M.Q. Zhang, G.C. Yang, J.Q. Zhao, Self-healing of fatigue crack in epoxy materials with epoxy/mercaptan system, Polym. Lett. 5 (1) (2011) 47–59.

[71] H. Jin, G.M. Miller, S.J. Pety, A.S. Griffin, D.S. Stradley, D. Roach, et al., Fracture behavior of a self-healing, toughened epoxy adhesive, Int. J. Adhes. Adhes. 44 (2013) 157–165.

[72] K.R. Hart, N.R. Sottos, S.R. White, Repeatable self-healing of an epoxy matrix using imidazole initiated polymerization, Polymer 67 (2015) 174–184.

[73] Y. Zhu, X.J. Ye, M.Z. Rong, M.Q. Zhang, Self-healing glass fiber/epoxy composites with polypropylene tubes containing self-pressurized epoxy and mercaptan healing agents, Compos. Sci. Technol. 135 (2016) 146–152.

[74] D. Grigoriev, D. Akcakayiran, M. Schenderlein, Protective organic coatings with anticorrosive and other feedback active features: micro- and nanocontainers based approach, Corrosion 70 (5) (2014) 446–463.

[75] D. Borisova, H. Möhwald, D.G. Shchukin, Influence of embedded nanocontainers on the efficiency of active anticorrosive coatings for aluminum alloys part I: influence of nanocontainer concentration, ACS Appl. Mater. Interfaces 4 (6) (2012) 2931–2939.

# Encapsulation and delivery of active compounds using nanocontainers for industrial applications

Shailesh A. Ghodke[1], Shirish H. Sonawane[2], Bharat A. Bhanvase[3], Kalpana Joshi[4]

[1]DEPARTMENT OF CHEMICAL ENGINEERING, DR. D. Y. PATIL INSTITUTE OF ENGINEERING, MANAGEMENT AND RESEARCH, PUNE, INDIA [2]CHEMICAL ENGINEERING DEPARTMENT, NATIONAL INSTITUTE OF TECHNOLOGY, WARANGAL, INDIA [3]CHEMICAL ENGINEERING DEPARTMENT, LAXMINARAYAN INSTITUTE OF TECHNOLOGY, RTM NAGPUR UNIVERSITY, NAGPUR, INDIA [4]DEPARTMENT OF BIOTECHNOLOGY, SINHGAD COLLEGE OF ENGINEERING, PUNE, INDIA

## Chapter outline

**11.1 Introduction** .......... **178**
**11.2 Nanocontainer synthesis** .......... **179**
11.2.1 Polymeric nanocontainers .......... 179
11.2.2 Layer-by-layer assemblies .......... 180
11.2.3 Silica-based delivery system .......... 182
11.2.4 Halloysite nanocontainers .......... 183
**11.3 Control parameters for nanocontainer applications** .......... **185**
11.3.1 Size of container .......... 185
11.3.2 Surface charge on container .......... 185
11.3.3 pH-based response for release .......... 186
11.3.4 Encapsulation efficiency .......... 186
11.3.5 Release of active molecules .......... 186
**11.4 Active molecules to be delivered** .......... **187**
11.4.1 Corrosion inhibitor .......... 187
11.4.2 Drug .......... 188
11.4.3 Perfume .......... 189

Encapsulation of Active Molecules and their Delivery System. DOI: https://doi.org/10.1016/B978-0-12-819363-1.00011-9

**11.5 Conclusion and future prospective** ........ 190
**References** ........ 191

# 11.1 Introduction

Many studies have been reported in the last decade regarding successful attempts of nanotechnology in producing nanostructures of materials, organic, inorganic, and composites. Growing interest in encapsulation techniques to fabricate nanocontainers is observed in the last few years because of its ability capsulate active molecules. Nanoencapsulation can be defined as the technique of capturing core of active molecules/materials within a shell material so as to transport and create a sealed vessel in nanometric dimension. Nanocontainers synthesized with a typical size range from 0.1 nm to 1 μm (1000 nm) have variety of advantages, for example, in drug delivery long-lasting circulation in organism. The key feature of such nanocontainers is the capability to transport molecules with tunable permeability. Various facets of nanocontainers synthesis were reported successfully pertaining to delivery requirement in different fields [1–6]. All these studies present that the nanocontainers are designed in such a way that they not only retain the cargo but also preserve it until delivered at target sites.

Typically, there are three steps followed for synthesis of nanocontainers first being formation of shell around material to be encapsulated second ensuring no leakages during delivery and third ensuring the targeted delivery [7–9]. Primarily, encapsulation prohibits the contact of pH, oxygen, light, moisture, temperature, and other undesirable effects with active molecule. This leads to increase in shelf life of active material, thereby providing the formation of protective container. Nanometric range of nanocontainers implies to growth in specific surface area providing increased interfacial area with the surrounding environment [10]. The cases where prolonged release behavior is expected, the nanocontainers need adequate protective layer and tuneable behavior to trigger the release. Complexation and nanocontainer-conjugate approaches best describe the phenomenon of delivery system [11]. In complexation approach, nanocontainer shell entraps the active molecule that is further triggered by structural alteration within the container material. Techniques such as structural changes in nanocontainer wall through cleavage of shell, charging of functional groups, and carrier degradation are practiced for the complexation approach [12]. In nanocontainer-conjugate approach, excruciating of the link between the nanocontainer and the active molecule is responsible for the release of active molecules [13].

The present study reports encapsulation and delivery of active compounds using nanocontainers for industrial applications. Methods for fabricating nanocontainers are discussed in detail. Control parameters such as surface charge, pH, encapsulation efficiency, and release behavior required for obtaining targeted delivery applications are discussed in detail. Various types of nanocontainers used for delivery of common active molecules such as corrosion inhibitor, drug, and aroma are discussed in detail.

# 11.2 Nanocontainer synthesis

## 11.2.1 Polymeric nanocontainers

Many successful research attempts of polymeric nanocontainers have been reported because of versatility of polymeric compounds. Prospective use of polymeric nanocontainers are transport of drugs/genes/proteins [14], corrosion inhibitors [15], fragrance [16], etc. These nanocontainers are known for their uniform product size and ease of handling [17]. Polymeric nanotubes and hollow spherical nanocontainers are the two major types of polymeric nanocontainers. In order to synthesize spherical nanocontainers emulsion polymerization, suspension polymerization, core–shell precursor, self-assembly, and dendrimer are practiced [12]. The nanotubular polymeric assemblies are prepared through template-directed synthesis, emulsion polymerization, self-assembly macromolecules, and electrospinning. During synthesis by suspension polymerization, monomer(s) is dispersed in liquid phase (water) by using mechanical agitators so as to form polymerized droplets of monomer. Unlike suspension, the emulsion polymerization begins with an emulsion of monomer with a mixture of water and surfactant. A typical type of oil-in-water emulsion polymerization occurs when the oil (monomer) is emulsified in the presence of surfactant in a continuous water phase. Emulsion polymerization is a three-step process nucleation, growth, and stabilization of polymer particles, which itself is a complex process and is governed by free radical polymerization mechanisms in combination with various colloidal phenomena.

Amiri and Rahimi reported corrosion protection of hybrid nanocomposite coating wherein the formation and characterization nanocontainers were discussed in detail [18]. The nanocontainer synthesis was the result of creation of inclusion complexes of a corrosion inhibitor with β-cyclodextrin (β-CD) at room temperature and under sonication. Dual stimuli-responsive polymeric assembly using polyaniline was reported to deliver two corrosion inhibitors, namely, 3-NisA (3-nitrosalicylic acid) and 2-MBT (2-mercaptobenzothiazole) [19]. pH and redox responsiveness were perfectly utilized to deliver drugs with average diameter of 600 nm nanocontainers. They were also shown to have selective or combined delivery of active molecules either by chemical reduction or change in pH value. High content of the encapsulated corrosion inhibitor was reported by Haase et al. wherein MBT or benzotriazole (BTA) was successfully delivered in corrosion protection [20]. Polystyrene (PS) composite nanocontainers were prepared for the release purpose. The present nanocontainers were found to have advantages such as control over size of nanocontainers, and as high as 20% loading of inhibitor was obtained. Fig. 11–1 indicates the typical self-healing mechanism of self-healing coating. When a mechanical impact takes place onto the coating surface causing tearing off the coating, it leads to puncture of polymeric nanocontainers embedded in polymer matrix. This results to pouring of loaded corrosion inhibitor at the damaged sites. Layer of inhibitor then forms a protective barrier between metal surface and corrosive environment.

Nanoprecipitation was also found to be useful in producing nanocontainers to deliver fragrance on human skin through stabilized poly-L-lactic-acid-nanocapsules yielding 115 nm diameter [16]. Chlorobenzene and fluorescein were used as model fragrances, and their

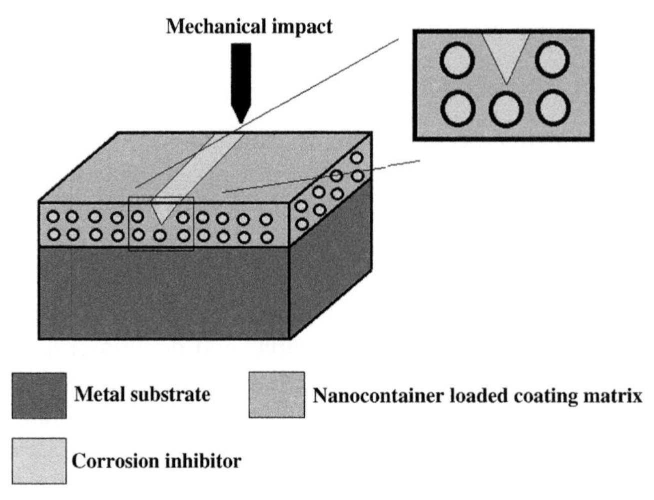


**FIGURE 11–1** Self-healing mechanism of self-healing coating.

delivery was tested on human skin. Release mechanism depicted the influence of chemical interaction between the compound and the polymeric matrix, molecular weight and volatility. Another study by Marturano et al. reports light-triggered release of basil and thyme essential oils by using polymeric nanocontainers in 100 nm [21]. The release behavior was function of irradiation time.

## 11.2.2 Layer-by-layer assemblies

Layer-by-layer (LbL) assemblies consist of class of polymers that possesses charged functionalities (polyelectrolyte) with opposite charges. Through complexation approach layers of polyelectrolyte with opposite charges are brought together to form a nanocontainer. The flexibility of spherical or substrate template can be availed through the alternate deposition of differently charged polyelectrolyte layers. The systems wherein variety of characteristics such as surface characteristics, permeability, elasticity, biocompatibility, morphology, and stability needs to be varied can be fulfilled by using LbL nanocontainers. The LbL assemblies offer various stimuli in order to provide controlled release. This controlled release is possible because of the balance of electrostatic interactions within the multilayer [22]. Active molecules can be loaded with two different approaches such as incorporation and adsorption or physical interaction. In the case of incorporation the active molecules are encapsulated during the formation of LbL assemblies, whereas in adsorption or physical interaction, the cargo is charged through adsorption or physical interaction between polyelectrolyte and active material [22,23]. Fig. 11–2 indicates the classic LbL approach for nanocontainer typically forms a boundary of surfactant to yield micelles further first polyelectrolyte forms a layer 1 similar strategy can be employed to increase more layers.

During the delivery of active molecule the process starts with the ionization of weak polyectrolyte of the functional groups that results into increase in repulsion between the

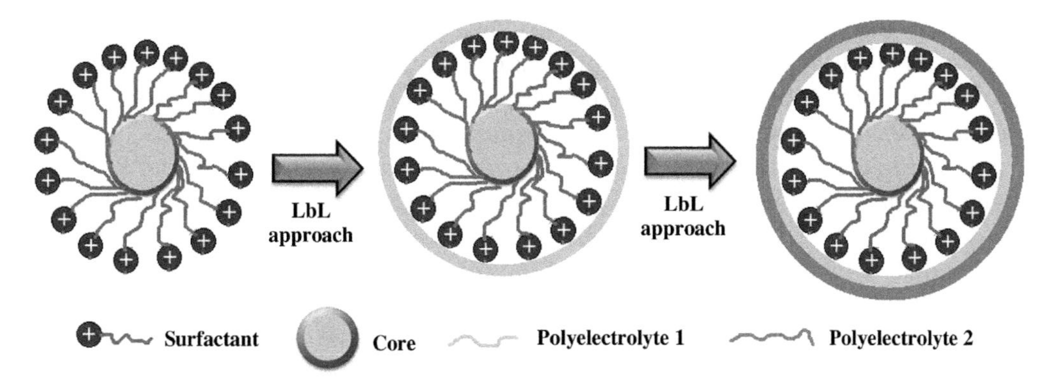


**FIGURE 11–2** Classic LbL approach for nanocontainer synthesis. *LbL*, Layer-by-layer.

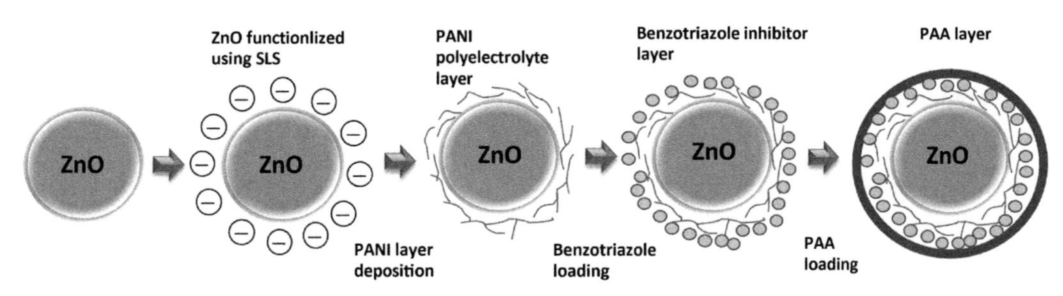


**FIGURE 11–3** Schematic illustration of the procedure for benzotriazole loading on ZnO nanocontainers [28].

uncompensated charges [24]. Then counter ions pierce the layered structure so as to balance the charge. Further osmotic pressure over the nanocontainers shell is created because of the higher ionic concentration of cargo within assembly [25]. Lastly, adjacent solute/water enters into the nanocontainer, which results into swelling in the shell wall followed by pore opening to release active material [26]. Poly(allylamine hydrochloride) and PS sulfonate (PSS) most widely used polyelectrolytes for LbL assembles because of their slow release rates in the delivery applications [27].

Sonawane et al. presented a detailed study on delivery of BTA by using LbL-assembled ZnO nanocontainers [28]. Alternate layers of polyaniline–BTA –polyacrylic acid were obtained on ZnO core to form LbL assembly of average size 950 nm. During release, maximum amount (0.86 mg/g of nanocontainers) was obtained at pH 3. Fig. 11–3 indicates the schematic illustration of the procedure for BTA loading on ZnO nanocontainers [28]. Similar study of magnesium zinc molybdate nanocontainer has been reported by Singh et al. [29]. Nanocontainers synthesized was with the help of magnesium zinc molybdate core with alternate layers of polypyrrole–BTA–polyacrylic acid The containers synthesized were of the size range 500–700 nm and were evaluated using epoxy-based coating. Another successful attempt of nanocontainer with core of oil phase and adsorbed polylectrolyte was used in self-healing epoxy coatings [30]. 2-Methylbenzothiazole and 2-MBT were used as corrosion inhibitors. Poly(diallyldimethylammonium chloride) and poly(sodium 4-styrenesulfonate) were used to form oppositely charged polyelectrolytes. Smaller nanocontainer size made it possible to uniformly mix into epoxy coating without altering the polymer matrix.

Novel drug-delivery application was developed to deliver lipophilic drug-loaded polyglutamate/polyelectrolyte nanocontainers [31]. Nanocontainer synthesis was through ultrasound emulsion followed by LbL technique. Soyabean oil was used to dissolve model drugs—rifampicin and lipophilic—followed by the layers of polyglutamate/polyethyleneimine/poly(acrylic acid). pH responsive nature of polyelectrolyte layers are useful in the delivery of active molecules. Another study reported use of biodegradable monodispersed polectrolyte nanocontainers for protein (bovine serum albumin) drug delivery [32]. Biocompatibility of synthesized nanocontainers also indicated better biocompatibility as compared to other higher molecular weight sulfates. Successful attempt of fragrance delivery using LbL structure from polyphenols (tannic acid) and proteins (bovine serum albumin) was reported by Sadovoy et al. [33]. The assembly was efficient enough to transport around eight different types of fragrances without leakages. Also, the shells were stable for a period of 2 months at a temperature of 4°C.

### 11.2.3 Silica-based delivery system

Mesoporous silica obtained from sodium silicate, alkoxides such as tetraethyl orthosilicate and tetramethylortosilicate is made to undergo polycondensation (in presence of surfactants) to use as form of cargo delivery vehicle in variety of delivery applications. In the last few years, one of the most versatile templates for the range of drug delivery has been silica-based delivery system. This is due to the variety of useful characteristics such as surface area and encapsulation efficiency. Two primary advantages of silica nanocontainers are large pore area of about 1 mL/g and surface area of 1000 $m^2/g$ [34]. Another essential advantage of silica-based nanocontainers is its ability to be functionalized with octyl groups so as to provide better dispersibility to applications wherein oil-based coating is required [35].

One of the recent studies depicted the delivery of sulfamethazine inhibitor using mesoporous silica nanocontainers (1 wt.%) so as to apply in epoxy coating [36]. The release mechanism revealed that the driving force for release was governed by the bond between N and O atoms of inhibitor ($sp^2$ electron pairs) and the iron atoms. Inhibitor loaded nanocontainers were found to be effective in corrosion protection wherein specimen was immersed in NaCl solution for 1 month. Another attempt to safeguard carbon steel substrate is by using silica nanocontainer employed MBT in epoxy coating [37]. The corrosion testing was conducted in NaCl (3.5 wt.%) environment. Three weight percentage of nanocontainer loading provided superior corrosion protection for the given condition, indicating sufficient amount of inhibitor loading in silica pores. Sodium molybdate as corrosion inhibitor was also employed in epoxy layer using silica nanocontainers [38]. Excellent corrosion protection was obtained in chlorine environment (8 weeks) immersion. Release rates calculated during entire process was found to be highest indicating the diffusion of inhibitor through nanocontainer pores.

In order to employ mesoporous nanocontainers for drug/gene delivery, attempt was made to immobilize the azobenzene incorporated DNA double strands in silica nanocontainers [39]. The strategy was to use the photoisomerization of azobenzene resulting in dehybridization/hybridization of DNA for the controlled release. Switching light from visible to UV triggered the cargo release. Further, the study was found to be useful for studies on cancer treatments. Another study reported intracellular drug release using silica nanocontainers with calcium phosphate [40]. Doxorubicin was used as model drug in release studies. Release of drug was found to be pH dependent. The overall performance of nanocontainer was found to provide advantages such as low drug leakages, high-structure stability, and tunable release.

Very recently, functionalized silica nanocontainers were engaged for the delivery of fragrance having antibacterial properties [41]. Four different model fragrances were tested for the delivery studies. Study found that the type of oil and surfactant was important in obtaining desirable properties such as encapsulation efficiency, surface area, and pore size. 3-Aminopropyl triethoxysilane was used to functionalize the nanocontainer wall. In order to provide further release delays, the functionalization was later grafted with hyaluronic acid. Other successful studies also reported various other techniques retarding the release rate of volatile perfume cores [42–44].

### 11.2.4 Halloysite nanocontainers

With a great advantage of dual surface charge, that is, negative on exterior surface and positive on the inner lumen, halloysite nanocontainers has been practiced to deliver the variety of active molecules in the recent past [45]. Basically, halloysites are aluminosilicate clays, tubular in nature and obtained from natural deposits [46]. Chemically, they are double layered aluminosilicates similar to kaoline clay [47]. Typically, halloysites have hollow tubular geometry with higher specific surface area. Advantages such as superior hydrodynamic and aerodynamic properties and better processability than spherical nanocontainers make halloysite a promising candidate for delivery applications [48].

Zahidah et al. presented an extensive review on the utility of halloysite nanocontainers for inhibitor delivery applications [49]. It was denoted the halloysite can indicate the encapsulation efficiency in the range of 5–20 wt.%. Joshi et al. reported the use of halloysite nanocontainers to deliver corrosion inhibitor in oil-based alkyd paints [50]. The release behavior was found to be as slow as 20–30 h for two corrosion inhibitors, namely, mercaptobenzimidazole (MBI) and BTA. To create the slow release effect the nanocontainer ends were closed with different approaches such as tube coating and stoppers formation. Stopper formation was found to be more effective than tube coating because of the fact that it provides additional sealing with the inhibitor restricting leakages. Detailed study on halloysite nanocontainers for loading drugs was presented by an extensive review article by Lvov et al. [51]. Very recently, clotrimazole drug was delivered using functionalized halloysite by Massaro et al. [52]. Bond formation between allyloxy β-CD and functional groups on halloysite nanocontainers (thiol groups) was successfully utilized to provide drug loading and delivery. Further in order to obtain mucoadhesion, nanocontainers were functionalized using ammonium groups. During release studies it was found the cyclodextrin was retained in its active form within halloysite lumen (restricting hydrolysis of imidazole ring). Another study reported cytocompatibility and uptake of polycations-coated nanotubes [53]. Poly(ethylenimine), poly (diallyldimethyl-ammonium), and poly(allylamine) were efficiently used to functionalize the halloysite nanocontainers.

Ghodke et al. reported the use of halloysite nanocontainers for the delivery of fragrance [54]. Fragrance-loaded nanocontainers were further coated with a layer of polymer so as to safeguard the volatile active material. The polymeric layer onto nanocontainer was found to be pH responsive. Release results indicated consistent fragrance release up to 5 h. Two different studies reported the delivery of herbicides using halloysite nanocontainers. In one case, biodegradable poly(vinyl alcohol)/starch film was opted to safeguard the active molecule release [55]. About 61% of atrazine loaded in the nanocontainers was released during 96 h of release profile. The same release studies without surface coating indicated 97% atrazine release. Matrix erosion and Fickian diffusion were the driving forces in delivering encapsulated fragrance. The second case of herbicide delivery mentioned the encapsulation of *Adenophora* (spreng.) [56]. Poly(vinyl alcohol)/starch composites were prepared using herbicide loaded halloysite nanocontainers. Almost 50% increase in release was observed when the biodegradable polymer layer was absent during the release studies.

Overall halloysite nanocontainers were found to be efficient in all the forms for delivery applications. Each application depicted the specific delivery requirement pertaining to targeted delivery of active molecules. Fig. 11–4 indicates the various nanocontainer preparations routes that have been used while designing a delivery system. Functionaliztion of

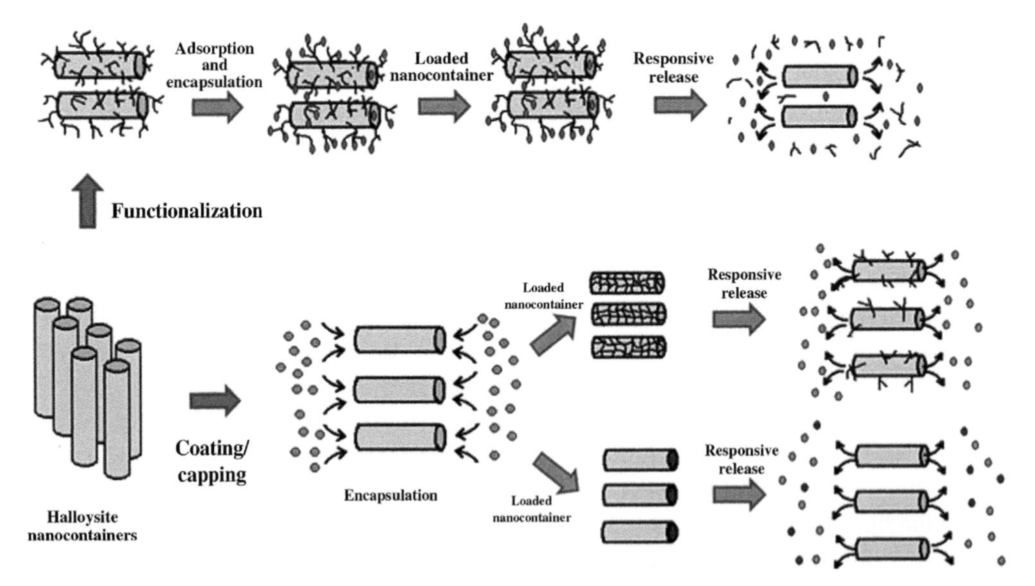


**FIGURE 11–4** Various nanocontainer preparations routes for designing a delivery system.

halloysite nanocontainer walls increased the adsorption sites in order to increase loading of active molecules. Capping/coating route describes loading of active molecules in the hollow lumen followed either by coating or capping so as to retard the molecule release behavior.

## 11.3 Control parameters for nanocontainer applications

Most significant parameter of nanocontainer-based delivery system is possible for using their properties to achieve targeted and responsive release. For selection of specific delivery system for the delivery of active material, various criteria such as size, surface charge, pH response, wall thickness, and encapsulation efficiency [57].

### 11.3.1 Size of container

More concern is shown in providing the nanometric dimensions of synthesized nanocontainers. This is because it is the only parameter that separates them from micrometer counterparts (microspheres/microcapsules). For a nanocontainer in the range of 20–30 nm, exceptional property changes can be observed because of exponential growth in the number of atoms being localized at nanocontainer surface [58]. Further, for a typical size decrease from 1 μm to 10 nm, nearly $10^2$ increase in surface area is observed for unit weight of nanocontainers. Therefore noteworthy properties can be observed from the nanocontainers as compared to their bulk state.

Shchukin et al. reported the general size ranges for nanocontainers used for corrosion inhibition purpose as a form of smart or active coating is in the range of 300–400 nm. This largest possible size is inevitable because of the fact that it can affect the integrity of coating matrix forming large hollow cavities that reduce the passive protective properties of the coatings [59]. The basic nature of in vivo applications of drug delivery prescribes that the use of nanometric dimensions only can be used for circulation [60]. According to Weiss et al. better drug absorption is observed in the case of smaller delivery containers [61]. This may be due to the fact that drug delivery is rate of a particular drug increases with corresponding surface area. One of the studies on controlled delivery of food delivery shows to have stable nanocontainers (in water) for 5 months if the particle size is less than 350 nm [62]. Parameters such as surfactants and degree of homogenization, sonication time, agitation speed, and nature and volume of phases decide the size of nanocontainers [63].

### 11.3.2 Surface charge on container

Factors such as nature of precursors, pH of synthesis medium, and charges on stabilizing agents are mainly responsible for the surface charge on nanocontainers. Subsequently, a nanocontainer bears positive charge if, for example, it is synthesized using cationic polymer using nonionic stabilizing agent. It is observed that the dispersion pH is primarily responsible for the magnitude of zeta-potential irrespective of the nature of stabilizing agent [63,64]. Therefore the surface charge of nanocontainer is independent of polymer concentration, stabilizer concentration, or the nature of active molecule. In the case of polyelectrolyte nanocontainers (LbL) the construction is dependent on two oppositely charged polyelectrolytes. Hence, the outermost layer of assemblies will decide the surface charge irrespective of number of layers used in LbL methodology.

### 11.3.3 pH-based response for release

The most common trigger used for responsive release is pH; the dispersion pH determines the release behavior through nanocontainers. For the case of polyelectrolyte nanocontainers, delivery of active molecules is observed due to protonation/deprotonation of charged groups. All nanocontainers except LbL assemblies release rates are observed due to repulsion between moieties. According to Delcea et al., protonation/deprotonation of charged groups takes place when the pH is altered relative to the p*K*a of the polyelectrolytes constituting the capsules walls [60].

In the case of oral drug consumption, when drug molecules pass through stomach to duodenum, it faces substantial pH changes causing them to deactivate drugs such as peptides, proteins, and nucleic acids. Therefore it is of utmost importance to safeguard the active molecules for delivery at targeted sites and design container shell accordingly. Another study of delivery of corrosion inhibitor through halloysite nanocontainers reported that the cargo release was triggered via pH change. The nanocontainers were sealed with metal stoppers by forming the metal complexes with the inhibitor at tube ends. Change in pH dissolved the complex resulting into inhibitor release [57].

### 11.3.4 Encapsulation efficiency

It is always desirable to have maximum encapsulation efficiency to have the desirable effect in less nanocontainer loading. Therefore encapsulation is one of the important characteristic. Some review studies related to nanocontainer synthesis and application reported that the emulsion-diffusion, coacervation, nanoprecipitation, and LbL are found to show as high as 80%–90% encapsulation efficiency [63,65]. According to Mora-Huertas et al., nanocontainer formation by using nanoprecipitation technique shows that regardless of technique and size of container, the type of oil governs drug-encapsulation efficiency [66]. Methods where good, and high concentration of ingredients is opted to synthesize nanocontainers and to show high encapsulation efficiency. This may be due to their contribution toward forming tightly sealed barrier structures with an inner aqueous phase capable of improving drug residence [63,67].

### 11.3.5 Release of active molecules

Aim of forming intact nanocontainers is to release the active materials only when the trigger is generated in the container wall, till that time there should be no cargo release. The trigger generated alters the shell properties initiating change in the affinity of active materials and nanocontainer wall. Primarily, the concentration of active material and their physical properties is important in generating specific release behaviors. For example, in the case of oily active materials used to deliver fragrance and flavor, release characteristics will depend on its solubility in the nearby medium and partition coefficient in terms of oil in water dissolutions [68]. Similarly, in the case of polymeric nanocontainers, the factors for release behavior may be molecular weight, concentration, size, and degradability.

## 11.4 Active molecules to be delivered

### 11.4.1 Corrosion inhibitor

All forms of metals being essential part of structures, vessels, equipment, pipes, etc., tend to undergo corrosion because of various reasons resulting into material loss [59,69]. One of the simplest tasks of corrosion protection is the provision of protective barrier between water and ion species ($O_2$ and $H^+$) with the metal surface being used [70]. Metal parts frequently undergoing degradation are to be refurbished with regular methods such as welding, recoating/repainting, or replacing small parts [71]. In addition, coatings (passive) are only effective when they are intact. These coatings are successful in corrosion protection up to certain level; however, the problem related to toxic pigments being used has created secondary pollutions. Even though conventional coatings are provided with corrosion inhibitors and mixed with organic resins before applications, their effectiveness is relatively for the short span [72]. Abovementioned disadvantages can be overcome by using "smart coating" with the nanocontainers with the ability to self-repair the damaged portion of the coating [73,74]. Various methodologies to synthesize nanocontainers are mentioned in preceding section.

Low molecular weight BTA is one of the most utilized corrosion inhibitor for metal parts. Because of its basic nature, it is soluble in water and hence may not be directly added to the coating material that may lead to the formation of empty voids. Therefore it is required to encapsulate BTA in nanocontainer [75]. Study on 2-MBT and 2-MBI was reported by Amiri and Rahimi [18]. The anticorrosive property of system was due to reversible bond formation between inhibitor and cyclodextrin. Container formation was the result of head-to-head channel-type structures and the formation of complex between cyclodextrin and corrosion inhibitor. Another approach to deliver bezotriazole was the use of silica nanocontainers as template followed by LbL depositions of polyectrolytes, wherein BTA was sandwiched between two layers [4]. Change in environment pH caused swelling in polyectrolytes layers resulting into inhibitor release. This behavior was observed at very high and very low pH values (pH 2 and pH 11). For environmental pH near neutral value corrosion inhibitor electrolyte shell remained closed and no purging of inhibitor was observed. Similar studies were reported by Tyagi et al. [76]. The pair of oppositely charged used in the approach was poly(diallyldimethylammonium chloride) and PSS. The study obtained inhibitor release of 0.1456 mg/L/g of silica nanocontainer at pH 11, resulting in development of passive layer of BTA on metal surface.

Polymeric nanocontainers with advantages such as high dispersibility in aqueous solution and compatibility with the coating matrix have been proposed by many material scientists. Delivery of 3-NisA and 2-MBT using polymeric nanocontainer was also reported for anticorrosive coating [19]. pH change or reduction were applied onto nanocontainer for cargo release through polyaniline nanocontainers. The strategy was to encapsulate 3-NisA within polymeric shell and adsorption of 2-MBT onto gold nanoparticle–loaded polymeric nanocontainers. On the introduction of reduction, 2-MBT start releasing form the shell surface, and on pH change, encapsulated 3-NisA starts dissolving. The system was found to be advantageous in providing selective delivery targets.

### 11.4.2 Drug

Many of drug/proteins are used for medicinal purposes in their therapeutic ranges. These ranges are such above which they are toxic and below which they are ineffective. In order to attain the optimal effects, it is necessary to deliver proteins in their therapeutic ranges for longer period of time [77]. Therefore it is necessary not only to deliver these drugs and proteins but to maintain their concentration by supplying them at targeted sites for longer period of time [78].

With the advent of nanotechnology, many researchers have attempted the use of nanomaterials as therapeutic and diagnostic tools [79]. Out of all nanomaterials, nanocontainers were practiced for variety of drug-delivery applications [80–82]. Nanocontainers provide the facility of safeguarding the active material and provide stability to delivery formulations. However, it is utmost important to look for specific type of nanocontainer which would provide the targeted drug delivery without leakages and required trigger. Recently poly(acrylic acid) and poly(ethyleneimine) were used to prepare the LbL nanocontainer for the delivery of indomethacin [1]. The basic nature of indomethacin is poorly soluble in water and causes decrease bioavailability. Hence, the delivery system required high amount of drug to make available at targeted site. Therefore the LbL approaches to fabricate nanocontainers of about 90–180 nm with loading efficiency with drug content of 0.25%.

Those conditions, wherein hydrophilic drug from water-in-oil is to be delivered in aqueous medium, need higher colloidal stability and strength to sustain osmotic pressure. Considering above requirement, efforts were made to prepare delivery system with (1) response to redox and pH—polyvinylferrocene-*b*-poly(2-vinylpyridine) and (2) response to pH an temperature—PS-*b*-poly(*N*,*N*-dimethylaminoethyl methacrylate) [83]. The system not only showed higher stability but also demonstrated selective release profile. Another study of doxorubicin release was indicated by mesoporous silica nanocontainers with the ability to respond to three different triggers such as pH, light, and reduction [84]. The study conferred multiresponsive nanocontainers for sophisticated environment of blood is necessary and important. Atom transferred radical polymerization of poly(2-(diethylamino)ethyl methacrylate) onto mesoporous silica nanocontainers was employed to create the responsive behavior in the shell. Clotrimazole is used for the treatment of candidiasis. The drug is found to have low stability in water (approximately 0.5 mg/L) and short plasma life of 3–6 h. Covalent bond between outer surface of halloysite and cyclodextrin was formed during drug loading. The participants for these covalent bonding were β-CD and thiol groups present onto halloysite nanocontainers.

## 11.4.3 Perfume

Many of the attempts to encapsulate perfume or aroma were for various purposes so as to protect the taste, aroma, or texture of food, mask off-flavors, and enhance the shelf-life and stability of the ingredient and the finished food product [7,54,85]. Nanocontainers for perfume delivery as compared to their micrometer level counterparts are found to possess antimicrobial property because of their subcellular size results into passive cellular absorption [86]. Most of the aromatic perfumes are found to be liquid at room temperature; however, they are very much sensitive to physical phenomenon such as temperature, air, light, and irradiations [87]. These reasons have made the determination of suitable delivery system without altering the active molecule quality and quantity.

Many of aromatherapy products are related to delivery of fragrances or perfume molecules that leads to killing pathogenic bacteria and well-being of the user. When such oils are to be used in day-to-day life, they tend to evaporate in presence of air, light, and temperature; hence, preference is given to prepare fragrant textile [88,89]. Alginate nanocontainer prepared through miniemulsion technique was grafted onto cotton by using microwave curing method [90]. Peppermint oil was employed as model fragrance. Microwave curing was able to produce two desirable effects, namely, higher microbial activity and higher degree of cross-linking. This is resulted into substantial stability of fragrance for about 25 washing cycles. Handling one or more perfume molecules has also been a problem because of requirement of keeping the perfume oils in closed conduit. In this regard, LbL assembly was suggested by Sadovoy et al. [33]. A perfume mixture of 10 ingredients was made to transport using LbL assembly. Bovine serum albumin and tannic acid were used for the container shell. Sunflower oil was used as core for all 10 perfume ingredient. All the 10 constituents were successfully stable for 3 days. To study the prolonged fragrance release, Ghodke et al. utilized halloysite nanocontainers to transport rose fragrance [54]. Halloysite nanocontainers were coated with a thin layer of polyelectrolyte to produce the pH-based release. Release results indicated higher release rates in the acidic region for first 5 h. Fickian diffusion was observed during the experimental results.

Essential oils are composed of phenylpropenes, terpenes, and terpenoids. Essential oils have a variety of useful characteristics such as antioxidant, antibacterial, antiviral, antifungal, and antiinflammatory. However, they must be protected during applications as they are extremely volatile and have a tendency to oxidize very quickly. Hence, it is required to maintain their compositions for delivery applications. An attempted have been made to encapsulate the essential oil—peppermint oil—onto mesoporous silica nanocontainers followed by coating with poly(vinyl alcohol) [41]. Hyaluronic acid was used for functionalization of the shell and was able to produce prolonged release rates. To provide fragrance in textile products a detailed study of pH responsive silica nanocontainers were reported by Yilmaz [88]. LbL approach was used to form alternate layers of two different polysaccharides—hyaluronic acid and chitosan. Encapusalation efficiency as well as pH responsiveness properties were excellent for the prescribed application. Another silica-based nanocontainer was shown to have excellent nonsticky nature [43,89,91]. Model fragrance citronella was delivered from triethoxysilane monomers using sol–gel process. The delivery system was found to produce prolonged citronellal release.

## 11.5 Conclusion and future prospective

Various types of nanocontainers, their synthesis methodologies, and functionalization have been discussed in the previous sections. Control parameters necessary for the fine-tuning of delivery applications were also discussed in detail. Numbers of industrial applications need the provision of smart delivery applications depending on type of active materials required. Various types of nanocontainers and desirable triggers are available in order to encapsulate active molecules and regulate the release rates. It is necessary to select the type of modality and type of response useful for the respective applications. Unstable and low-solubility drugs are susceptible to harmful environment such as pH, light, or enzymatic attack, while their journey through digestive tract and efforts is made through provision of various types of biocompatible nanocontainers. In the case of drug-delivery systems, all the payloads are biodistributed by using nanocontainers, hence their physicochemical properties are not responsible for therapeutic concentrations [92]. Many drug-delivery architecture are in less than 50 nm, making them for the possible for their use in blood circulation without blood clogging [93].

Polymeric nanocontainers are most versatile of all nanocontainers as they provide flexibility, stability, and structural integrity. This is mainly possible because of the degree of cross-linking. However, biocompatibility and biodegradability have been the greatest hurdle at all times for the drug and perfume-delivery applications. Another useful and versatile nanocontainer has been the LbL assemblies with the ability to activate with physical as well as chemical stimuli. However, very few studies report the in vivo release applications. Employing LbL approach, template-based assemblies have been able to provide wide delivery application. Only shortcoming of this technique is that the final nanocontainer sometimes reaches to micrometer range. Halloysite nanocontainers have provided different means of producing stimuli for payload release. Natural surface opposite surface charge on inner and outer surface have made them to be able to delivery positive and negative charge cargo. Because of good dispersibility and colloidal characteristics, they have been found to be useful for a variety of release environment. Mesoporous silica nanocontainers with large surface area and smaller nanocontainer size have the proved their applications in biological delivery systems without any significant cytotoxicity.

Overall, all the studies reported in literature seem favorable and inspiring; however, plenty of scope is in fundamental and systematic investigations. Complexity in designing nanocontainers makes it difficult to scale up which ends up with limited practical use. Consistent efforts in more sophisticated delivery system will definitely introduce more and more practical solutions for different delivery systems.

# References

[1] A.B. Mirgorodskaya, R.A. Kushnazarova, A.V. Nikitina, I.I. Semina, I.R. Nizameev, M.K. Kadirov, et al., Polyelectrolyte nanocontainers: controlled binding and release of Indomethacin, J. Mol. Liq. 272 (2018) 982–989.

[2] J. Li, Z. Yu, H. Jiang, G. Zou, Q. Zhang, Photo and pH dual-responsive polydiacetylene smart nanocontainer, Mater. Chem. Phys. 136 (2012) 219–224.

[3] M. Shah, M.I. Naseer, M.H. Choi, M.O. Kim, S.C. Yoon, Amphiphilic PHA–mPEG copolymeric nanocontainers for drug delivery: Preparation, characterization and in-vitro evaluation, Int. J. Pharm. 400 (2010) 165–175.

[4] Y. Feng, Y.F. Cheng, An intelligent coating doped with inhibitor-encapsulated nanocontainers for corrosion protection of pipeline steel, Chem. Eng. J. 315 (2017) 537–551.

[5] B. Qian, M. Michailidis, M. Bilton, T. Hobson, Z. Zheng, D. Shchukin, Tannic complexes coated nanocontainers for controlled release of corrosion inhibitors in self-healing coatings, Electrochim. Acta 297 (2019) 1035–1041.

[6] S.A. Ghodke, S.H. Sonawane, B.A. Bhanvase, S. Mishra, K.S. Joshi, I. Potoroko, Halloysite nanocontainers for controlled delivery of gibberellic acid, Russ. J. Appl. Chem. 90 (1) (2017) 120–128.

[7] Z. Fang, B. Bhandari, Encapsulation of polyphenols – a review, Trends Food Sci. Technol. 21 (2010) 510–523.

[8] B.F. Gibbs, S. Kermasha, I. Alli, C.N. Mulligen, Encapsulation in food industry: a review, Int. J. Food Sci. Nutr. 50 (1999) 213–224.

[9] M.R. Mozafari, K. Khosravi-Darani, G.G. Borazan, J. Cui, A. Pardakhty, S. Yurdugul, Encapsulation of food ingredients using nanoliposome technology, Int. J. Food Prop. 11 (2008) 833–844.

[10] V. Marturano, P. Cerruti, C. Carfagn, M. Giamberini, B. Tylkowski, V. Ambrog, Photo-responsive polymer nanocapsules, Polymer 70 (2015) 222–230.

[11] M.A.C. Stuart, W.T.S. Huck, J. Genzer, M. Müller, C. Ober, M. Stamm, et al., Emerging applications of stimuli-responsive polymer materials, Nat. Mater. 9 (2010) 101–113.

[12] S. Ghodke, B. Bhanvase, S. Sonawane, S. Mishra, K. Joshi, Nanoencapsulation and nanocontainer based delivery systems for drugs, flavors, and aromas. July in: A.M. Grumezescu (Ed.), Nanotechnology in Food Industry (I-X) Multi-vol. 2, Encapsulations, Elsevier, Amsterdam, The Netherlands, 2016, pp. 673–715.

[13] E. Fleige, M.A. Quadir, R. Haag, Stimuli-responsive polymeric nanocarriers for the controlled transport of active compounds: concepts and applications, Adv. Drug Deliv. Rev. 64 (2012) 866–884.

[14] C. Tapeinos, E.K. Efthimiadoua, N. Boukos, G. Kordas, Sustained release profile of quatro stimuli nanocontainers as a multi sensitive vehicle exploiting cancer characteristics, Colloids Surf., B 148 (2016) 95–103.

[15] N.P. Tavandashti, M. Ghorbani, A. Shojaei, J.M.C. Mol, H. Terryn, K. Baert, et al., Inhibitor-loaded conducting polymer capsules for active corrosion protection of coating defects, Corros. Sci. 112 (2016) 138–149.

[16] B. Hosseinkhani, C. Callewaert, N. Vanbeveren, N. Boon, Novel biocompatible nanocapsules for slow release of fragrances on the human skin, New Biotechnol. 32 (1) (2015) 40–46.

[17] M.S. Islam, J.H. Yeum, A.K. Das, Synthesis of poly(vinyl acetate–methyl methacrylate) copolymer microspheres using suspension polymerization, J. Colloid Interface Sci. 368 (1) (2002) 400–405.

[18] S. Amiri, A. Rahimi, Synthesis and characterization of supramolecular corrosion inhibitor nanocontainers for anticorrosion hybrid nanocomposite coatings, J. Polym. Res. 22 (2015) 66–72.

[19] L.P. Lv, K. Landfester, D. Crespy, Stimuli-selective delivery of two payloads from dual responsive nanocontainers, Chem. Mater. 26 (2014) 3351–3353.

[20] M.F. Haase, D.O. Grigoriev, H. Möhwald, D.G. Shchukin, Development of nanoparticle stabilized polymer nanocontainers with high content of the encapsulated active agent and their application in water-borne anticorrosive coatings, Adv. Mater. 24 (2012) 2429–2435.

[21] V. Marturano, V. Bizzarro, A. De Luise, A. Calarco, V. Ambrogi, M. Giamberini, et al., Essential oils as solvents and core materials for the preparation of photo-responsive polymer nanocapsules, Nano Res. 11 (5) (2018) 2783–2795.

[22] A. Pomorska, K. Yliniemi, B.P. Wilson, D. Shchukin, D. Johannsmann, G. Grundmeier, QCM study of the adsorption of polyelectrolyte covered mesoporous $TiO_2$ nanocontainers on SAM modified Au surfaces, J. Colloid Interface Sci. 362 (2011) 180–187.

[23] D.G. Shchukin, M. Zheludkevich, K. Yasakau, S. Lamaka, M.G.S. Ferreira, H. Möhwald, Layer-by-layer assembled nanocontainers for self-healing corrosion protection, Adv. Mater. 18 (2006) 1672–1678.

[24] D.V. Andreeva, E.V. Skorb, D.G. Shchukin, Layer-by-layer polyelectrolyte/inhibitor nanostructures for metal corrosion protection, Appl. Mater. Interface 2 (7) (2010) 1954–1962.

[25] B.M. Wohl, J.F.J. Engbersen, Responsive layer-by-layer materials for drug delivery, J. Control. Release. 158 (2012) 2–14.

[26] V. Kozlovskaya, E. Kharlampieva, M.L. Mansfield, S.A. Sukhishvili, Poly(methacrylic acid) hydrogel films and capsules: response to pH and ionic strength, and encapsulation of macromolecules, Chem. Mater. 18 (2006) 328–336.

[27] S.A. Sukhishvili, Responsive polymer films and capsules via layer-by-layer assembly, Curr. Opin. Colloid Interface Sci. 10 (2005) 37–44.

[28] S.H. Sonawane, B.A. Bhanvase, A.A. Jamali, S.K. Dubey, S.S. Kale, D.V. Pinjari, et al., Improved active anticorrosion coatings using layer-by-layer assembled ZnO nanocontainers with benzotriazole, Chem. Eng. J 189–190 (2012) 464–472.

[29] H.K. Singh, K.V. Yeole, S.T. Mhaske, Synthesis and characterization of layer-by-layer assembled magnesium zinc molybdate nanocontainer for anticorrosive application, Chem. Eng. J 295 (2016) 414–426.

[30] M. Kopec, K. Szczepanowicz, G. Mordarski, K. Podgórna, R.P. Socha, P. Warszynski, et al., Self-healing epoxy coatings loaded with inhibitor-containing polyelectrolyte nanocapsules, Prog. Org. Coat 84 (2015) 97–106.

[31] X.R. Teng, D.G. Shchukin, H. Möhwald, A novel drug carrier: lipophilic drug-loaded polyglutamate/polyelectrolyte nanocontainers, Langmuir 24 (2008) 383–389.

[32] S. Shu, C. Sun, X. Zhang, Z. Wu, Z. Wang, C. Li, Hollow and degradable polyelectrolyte nanocapsules for protein drug delivery, Acta Biomater. 6 (2010) 210–217.

[33] A.V. Sadovoy, M.V. Lomova, M.N. Antipina, N.A. Braun, G.B. Sukhorukov, Layer-by-layer assembled multilayer shells for encapsulation and release of fragrance, Appl. Mater. Interfaces 5 (18) (2013) 8948–8954.

[34] D. Borisova, H. Möhwald, D.G. Shchukin, Influence of embedded nanocontainers on the efficiency of active anticorrosive coatings for aluminum alloys Part I: Influence of nanocontainer concentration, Appl. Mater. Interfaces 4 (6) (2012) 2931–2939.

[35] M.J. Hollamby, D. Fix, I. Dönch, D. Borisova, H. Möhwald, D. Shchukin, Hybrid polyester coating incorporating functionalized mesoporous carriers for the holistic protection of steel surfaces, Adv. Mater. 23 (2011) 1361.

[36] M. Yeganeh, N. Asadi, M. Omidi, M. Mahdavian, An investigation on the corrosion behavior of the epoxy coating embedded with mesoporous silica nanocontainer loaded by sulfamethazine inhibitor, Prog. Org. Coat. 128 (2019) 75–81.

[37] M. Rahsepar, F. Mohebbi, H. Hayatdavoudi, Synthesis and characterization of inhibitor-loaded silica nanospheres for active corrosion protection of carbon steel substrate, J. Alloy Compd. 709 (2017) 519–530.

[38] M. Yeganeh, A. Keyvani, The effect of mesoporous silica nanocontainers incorporation on the corrosion behaviour of scratched polymer coatings, Prog. Org. Coat. 90 (2016) 296–303.

[39] Q. Yuan, Y. Zhang, T. Chen, D. Lu, Z. Zhao, X. Zhang, et al., Photon-manipulated drug release from a mesoporous nanocontainer controlled by azobenzene-modified nucleic acid, Nano 6 (7) (2012) 6337–6344.

[40] H.P. Rim, K.H. Min, H.J. Lee, S.Y. Jeong, S.C. Lee, pH-tunable calcium phosphate covered mesoporous silica nanocontainers for intracellular controlled release of guest drugs, Angew. Chem. Int. Ed. 50 (2011) 8853–8857.

[41] A. Jobdeedamrong, R. Jenjob, D. Crespy, Encapsulation and release of essential oils in functional silica nanocontainers, Langmuir 34 (44) (2018) 13235–13243.

[42] Z. Cao, C. Xu, X. Ding, S. Zhu, H. Chen, D. Qi, Synthesis of fragrance/silica nanocapsules through a sol–gel process in miniemulsions and their application as aromatic finishing agents, Colloid Polym. Sci. 293 (2015) 1129–1139.

[43] P. Kidsaneepoiboon, S.P. Wanichwecharungruang, T. Chooppawa, R. Deephum, T. Panyathanmaporn, Organic–inorganic hybrid polysilsesquioxane nanospheres as UVA/UVB absorber and fragrance carrier, J. Mater. Chem. 21 (2011) 7922–7930.

[44] S. Wang, M. Zhang, D. Wang, W. Zhang, S. Liu, Synthesis of hollow mesoporous silica microspheres through surface sol–gel process on polystyrene-*co*-poly(4-vinylpyridine) core–shell microspheres, Micropor. Mesopor. Mater. 139 (2011) 1–7.

[45] V. Vergaro, E. Abdullayev, R. Cingolani, Y. Lvov, S. Leporatti, Cytocompatibility and uptake of halloysite clay nanotubes, Biomacromolecules 11 (2010) 820–828.

[46] W.N. Xing, L. Ni, P.W. Huo, Z.Y. Lu, X.L. Liu, Y.Y. Luo, et al., Preparation high photocatalytic activity of CdS/halloysite nanotubes (HNTs) nanocomposites with hydrothermal method, Appl. Surf. Sci. 259 (2012) 698–704.

[47] S.A. Ghodke, S.H. Sonawane, B.A. Bhanvase, S. Mishra, K.S. Joshi, I. Potoroko, Functionalization, uptake and release studies of active molecules using halloysite nanocontainers, J. Inst. Eng. E 100 (1) (2019) 59–70.

[48] Y.M. Lvov, D.G. Shchukin, H. Mohwald, R.R. Price, Halloysite clay nanotubes for controlled release of protective agents, Appl. Mater. Interfaces 2 (2008) 814–820.

[49] K.A. Zahidah, S. Kakooei, M.C. Ismail, P.B. Raja, Halloysite nanotubes as nanocontainer for smart coating application: a review, Prog. Org. Coat. 111 (2017) 175–185.

[50] A. Joshi, E. Abdullayev, A. Vasiliev, O. Volkova, Y. Lvov, Interfacial modification of clay nanotubes for the sustained release of corrosion inhibitors, Langmuir 29 (24) (2013) 7439–7448.

[51] Y.M. Lvov, M.M. DeVilliers, R.F. Fakhrullin, The application of halloysite tubule nanoclay in drug delivery, Exp. Opin. Drug Deliv. 13 (7) (2016) 977–986.

[52] M. Massaro, A. Campofelice, C.G. Colletti, G. Lazzara, R. Noto, S. Riela, Functionalized halloysite nanotubes: efficient carrier systems for antifungine drugs, Appl. Clay Sci. 160 (2018) 186–192.

[53] E. Tarasova, E. Naumenko, E. Rozhina, F. Akhatova, R. Fakhrullin, Cytocompatibility and uptake of polycations-modified halloysite clay nanotubes, Appl. Clay Sci. 169 (2019) 21–30.

[54] S.A. Ghodke, B.A. Bhanvase, K.S. Joshi, S.H. Sonawane, S. Mishra, Studies on fragrance delivery from inorganic nanocontainers: encapsulation, release and modeling studies, J. Inst. Eng. India Ser., E 96 (1) (2015) 45–53.

[55] B. Zhong, S. Wang, H. Dong, Y. Luo, Z. Jia, X. Zhou, Halloysite tubes as nanocontainers for herbicide and its controlled release in biodegradable poly(vinyl alcohol)/starch film, J. Agric. Food Chem. 65 (2017) 10445–10451.

[56] X. Zeng, B. Zhong, Z. Jia, Q. Zhang, Y. Chen, D. Jia, Halloysite nanotubes as nanocarriers for plant herbicide and its controlled release in biodegradable polymers composite film, Appl. Clay Sci. 171 (2019) 20–28.

[57] D.V. Andreeva, D.G. Shchukin, Smart self-repairing protective coatings, Mater. Today 11 (10) (2008) 24–30.

[58] A. Mihranyan, N. Ferraz, M. Strømme, Current status and future prospects of nanotechnology in cosmetics, Prog. Mater. Sci. 57 (2012) 875–910.

[59] D.G. Shchukin, H. Mohwald, Self-repairing coatings containing active nanoreservoirs, ACS Small 3 (6) (2007) 926–943.

[60] M. Delcea, H. Möhwald, A.G. Skirtach, Stimuli-responsive LbL capsules and nanoshells for drug delivery, Adv. Drug Deliv. Rev. 63 (2011) 730–747.

[61] J. Weiss, Takhistov, D.J. Mcclements, Functional materials in food nanotechnology, J. Food Sci. 71 (9) (2006) 107–116.

[62] M.J. Fabra, A. Hambleton, P. Talens, F. Debeaufort, A. Chiralt, A. Voilley, Influence of interactions on water and aroma permeabilities of i-carrageenan–oleic acid–beeswax films used for flavour encapsulation, Carbohydr. Polym. 76 (2009) 325–332.

[63] C.E. Mora-Huertas, H. Fessi, A. Elaissari, Polymer-based nanocapsules for drug delivery, Int. J. Pharm. 385 (2010) 113–142.

[64] H.H. Joo, H.Y. Lee, Y.S. Guan, J.C. Kim, Colloidal stability and in vitro permeation study of poly(e-caprolactone) nanocapsules containing hinokitiol, J. Ind. Eng. Chem. 14 (2008) 608–613.

[65] M. Kakran, M.N. Antipina, Emulsion-based techniques for encapsulation in biomedicine, food and personal care, Curr. Opin., Colloid Interface Sci. 18 (2014) 47–55.

[66] C.E. Mora-Huertas, O. Garrigues, H. Fessi, A. Elaissari, Nanocapsules prepared via nanoprecipitation and emulsification–diffusion methods: comparative study, Eur. J. Pharm. Biopharm. 80 (2012) 235–239.

[67] S. Khoee, M. Yaghoobian, An investigation into the role of surfactants in controlling particle size of polymeric nanocapsules containing penicillin-G in double emulsion, Eur. J. Med. Chem. 44 (6) (2008) 2392–2399.

[68] A. Madene, M. Jacquot, J. Scher, S. Desobry, Flavour encapsulation and controlled release – a review, Int. J. Pharm. 41 (2006) 1–21.

[69] D.Y. Wu, S. Meure, D. Solomon, Self-healing polymeric materials: a review of recent developments, Prog. Polym. Sci. 33 (2008) 479–522.

[70] B.J. Blaiszik, S.L.B. Kramer, S.C. Olugebefola, J.S. Moore, N.R. Sottos, S.R. White, Self-healing polymers and composites, Annu. Rev. Mater. Res. 40 (2010) 179–211.

[71] H. Pulikkalparambil, S. Siengchin, J. Parameswaranpillai, Corrosion protective self-healing epoxy resin coatings based on inhibitor and polymeric healing agents encapsulated in organic and inorganic micro and nanocontainers, Nanostruct. Nano Obj. 16 (2018) 381–395.

[72] M. Zheludkevich, J. Tedim, M. Ferreira, "Smart" coatings for active corrosion protection based on multifunctional micro and nanocontainers, Electrochim. Acta 82 (2012) 314–323.

[73] A. Stankiewicz, I. Szczygiel, B. Szczygiel, Self-healing coatings in anti-corrosion applications, J. Mater. Sci. 48 (2013) 8041–8051.

[74] S.R. White, N.R. Sottos, P.H. Geubelle, J.S. Moore, M.R. Kessler, S.R. Sriram, et al., Autonomic healing of polymer composites, Nature 409 (2001) 794–797.

[75] K. Kamburova, N. Boshkova, N. Boshkov, G. Atanassova, T. Radeva, Hybrid zinc coatings for corrosion protection of steel using polyelectrolyte nanocontainers loaded with benzotriazole, Colloids Surf., A 559 (2018) 243–250.

[76] M. Tyagi, B.A. Bhanvase, S.L. Pandharipande, Computational studies on release of corrosion inhibitor from layer-by-layer assembled silica nanocontainer, Ind. Eng. Chem. Res. 53 (2014) 9764–9771.

[77] M. DeVilliers, Y. Lvov, Layer-by-layer self-assembled nanoshells for drug delivery, Adv. Drug Deliv. Rev. 63 (2011) 699–701.

[78] S.R. Levis, P.B. Deasy, Use of coated microtubular halloysite for the sustained release of diltiazem hydrochloride and propranolol hydrochloride, Int. J. Pharm. 253 (2003) 145–157.

[79] M. Jahanshahi, Z. Babaei, Protein nanoparticle: a unique system as drug delivery vehicles, Afr. J. Biotech. 7 (2008) 4926–4934.

[80] J. Panyam, V. Labhasetwar, Biodegradable nanoparticles for drug and gene delivery to cells and tissue, Adv. Drug Deliv. Rev. 55 (2003) 329–347.

[81] I.M. Rodriguez-Cruz, V. Merino, M. Merino, O. Deiz, A. Nacher, D. Quintanar-Guerrero, Polymeric nanospheres as strategy to increase the amount of triclosan retained in the skin: passive diffusion vs. iontophoresis, J. Microencap. 30 (1) (2013) 72–80.

[82] K. Cho, X. Wang, S. Nie, Z. Chen, D.M. Shin, Therapeutic nanoparticles for drug delivery in cancer, Clin. Cancer Res. 14 (2008) 1310–1316.

[83] R.H. Staff, M. Gallei, K. Landfester, D. Crespy, Hydrophobic nanocontainers for stimulus-selective release in aqueous environments, Macromole 47 (2014) 4876–4883.

[84] Y. Zhang, Y. Zhao, C.Y. Ang, M. Li, S.Y. Tan, Q. Qu, et al., Polymer coated hollow mesoporous silica nanoparticles for triple-responsive drug delivery, Appl. Mater. Interface 7 (32) (2015) 18179–18187.

[85] M.A. Augustin, Y. Hemar, Nano- and micro-structured assemblies for encapsulation of food ingredients, Chem. Soc. Rev. 38 (2009) 902–912.

[86] F. Donsi, M. Annunziata, M. Sessa, G. Ferrari, Nanoencapsulation of essential oils to enhance their antimicrobial activity in foods, LWT—Food Sci. Technol. 44 (2011) 1908–1914.

[87] M.X. Quintanilla-Carvajal, B.H. Camacho-Diaz, L.S. Meraz-Torres, J.J. ChanonaPerez, L. Alamilla-Beltran, A. Jimenez-Aparicio, et al., Nanoencapsulation: a new trend in food engineering processing, Food Eng. Rev. 2 (2010) 39–50.

[88] M.D. Yilmaz, Layer-by-layer hyaluronic acid/chitosan polyelectrolyte coated mesoporous silica nanoparticles as pH-responsive nanocontainers for optical bleaching of cellulose fabrics, Carbohydr. Polym. 146 (2016) 174–180.

[89] C.I. Beristain, H.S. Garcia, E.J. Vernon-Carter, Spray-dried encapsulation of cardamom (*Elettaria cardamomum*) essential oil with mesquite (*Prosopis juliflora*) gum, LWT Food Sci. Technol. 34 (2001) 398–401.

[90] S. Ghayempour, S.M. Mortazavi, Microwave curing for applying polymeric nanocapsules containing essential oils on cotton fabric to produce antimicrobial and fragrant textiles, Cellulose 22 (6) (2015) 4065–4075.

[91] E.D. Cristina, Understanding true aromatherapy: understanding essential oils, Home Health Care Manage. Pract. 16 (2004) 474–479.

[92] G.M. Barratt, Therapeutic applications of colloidal drug carriers, Pharm. Sci. Technol. Today 3 (2000) 163–171.

[93] K. Letchford, H. Burt, A review of the formation and classification of amphiphilic block copolymer nanoparticulate structures: micelles, nanospheres, nanocapsules and polymersomes, Eur. J. Pharm. Biopharm. 65 (2007) 259–269.

# 12

# Virus-like particles: nano-carriers in targeted therapeutics

Gundappa Saha[1], Prakash Saudagar[2], Vikash Kumar Dubey[3]

[1]DEPARTMENT OF BIOSCIENCES AND BIOENGINEERING, INDIAN INSTITUTE OF TECHNOLOGY GUWAHATI, GUWAHATI, INDIA [2]DEPARTMENT OF BIOTECHNOLOGY, NATIONAL INSTITUTE OF TECHNOLOGY WARANGAL, HANAMKONDA, INDIA [3]SCHOOL OF BIOCHEMICAL ENGINEERING, INDIAN INSTITUTE OF TECHNOLOGY BHU, VARANASI, INDIA

**Chapter outline**

**12.1 Introduction** ..... 197
**12.2 Role of virus-like particles as good drug delivery vectors** ..... 199
**12.3 Virus-like particles: overcoming limitations of other therapeutic approaches** ..... 200
**12.4 Prerequisite factors in designing of virus-like particles as therapeutics** ..... 201
**12.5 Immune responses induced by virus-like particles** ..... 203
**12.6 Current applications of virus-like particles as targeted therapeutics** ..... 204
**12.7 Conclusion** ..... 206
**References** ..... 206

## 12.1 Introduction

Recent advancement in the field of nanotechnology has the potential to make strategic targeted therapeutic tools to treat various diseases with minimal side effects and higher efficacy. Targeted drug delivery has been in the long list of advanced medical treatment approaches but with challenges and proper opportunities. It offers primarily three main benefits over other conventional therapies such as (1) drugs can be targeted at the specific site of action, minimizing the off-target effects; (2) higher local drug concentrations can be achieved at the diseased site increasing the efficacy; and (3) ease of drug transportation to the challenging sites such as brain through blood–brain barrier (BBB) [1]. However, structural heterogeneity, instability, and limited solubility are the major concerns related to targeted drug deliveries [2].

Encapsulation of Active Molecules and their Delivery System. DOI: https://doi.org/10.1016/B978-0-12-819363-1.00012-0

To cope with these challenges, various nanoparticle (NP)-based delivery cargos are being used with much safer and effective deliveries at the site of action, including liposomal, polymer-based, metal-based, and protein-based NPs. The protein-based NPs include virus-like particles (VLPs), which are formed spontaneously by self-assembly of viral proteins by in vitro protein expressions in a suitable expression system (usually eukaryotic host) or during infections [3]. They closely resemble the structures of wild-type virions but are replication and infection incompetent due to the absence of their genetic materials. In contrast to other synthetic delivery vehicles, VLPs are uniform in their size distribution ranging from 20 to 100 nm with higher stability, solubility, symmetric spatial organization, ease of production, safety, and negligible toxicity issues [4]. Moreover, these VLPs allow site-selective modifications and conjugations through mutagenesis of amino acid in the protein backbone and thus have greater control over a site-specific target of the drugs with desired amount of drug doses to be delivered efficiently [5,6].

The classification of VLPs depends on many parameters such as virus taxonomy (e.g., adenovirus and retrovirus) or synthesis methods (e.g., expression in plant, yeast, or animal cell) or architecture (e.g., enveloped or nonenveloped). The most common expression system for VLPs is animal cells owing to their ability to process the posttranslational modifications required for proper protein folding [7]. Despite many advantages of animal cell expression system, there are few limitations such as high production costs and difficulties with scaling up. These issues can be resolved using insect cell expression system with simultaneous expression of many viral proteins aiding in the assembly of VLPs [8]. However, many studies were reported for the successful self-assembly of VLPs in yeast expression system such as *Pichia pastoris*. The two common vaccines against human papillomavirus (HPV) and hepatitis B virus (HBV) are also based on yeast expression system [9,10]. Few common expression systems used recently for diverse VLPs are listed in Table 12–1.

**Table 12–1** Different expression systems used for different virus-like particles (VLPs).

| VLPs | Expression system | References |
|---|---|---|
| *Plasmodium falciparum* HBcAg | *Bacteria—Escherichia coli* | Sällberg et al. [11] |
| HPV | *Bacteria—Lactobacillus casei* | Aires et al. [12] |
| HBV | *Yeast—Pichia pastoris* | Lünsdorf et al. [9] |
| DENV | | Tang et al. [13] |
| HIV/HBV (HBsAg) | *Plant—Arabidopsis thaliana* | Greco et al. [14] |
| HBV (HBcAg) | *Plant—Nicotiana tabacum* | Tomasz [15] |
| HIV | *Insect—Spodoptera frugiperda* | Gheysen et al. [16] |
| Influenza | | Pushko et al. [17] |
| NV | | El-Kamary et al. [18] |
| Influenza | *Mammalian*—Vero | Barrett et al. [19] |
| Ebola | *Mammalian*—HEK | Kawaoka and Yamayoshi [20] |
| Chikungunya | | Akahata et al. [21] |

*DENV*, Dengue virus; *HBV*, hepatitis B virus; *HIV*, human immunodeficiency virus; *HPV*, human papillomavirus; *NV*, Norwalk virus.

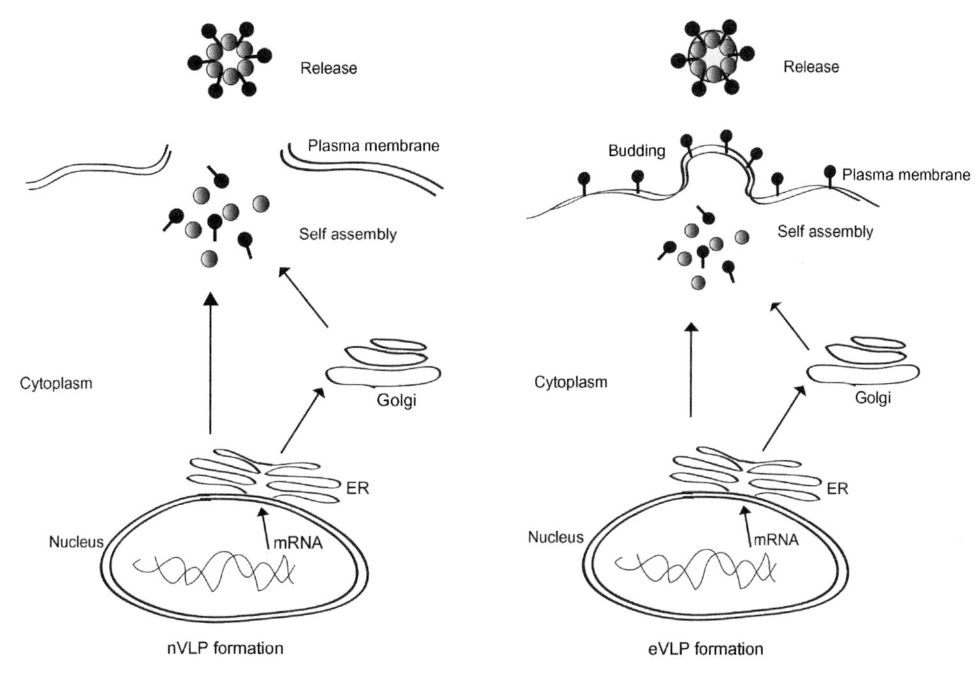


**FIGURE 12–1** A schematic representation of the formation of eVLPs and nVLPs. *eVLPs*, Enveloped virus-like particles; *nVLPs*, nonenveloped virus-like particles.

Based on the structure of viruses, VLPs can be enveloped and nonenveloped like parental viruses. Nonenveloped VLPs (nVLPs) include expression of one or more structural viral proteins, whereas enveloped VLPs (eVLPs) include host cell membrane with viral proteins displayed on the surface. These eVLPs are more complex in contrast to nVLPs and provide additional possibilities to integrate heterogeneous antigens and adjuvants as per requirements [22]. A schematic representation of the formation of eVLPs and nVLPs has been shown in Fig. 12−1.

The efficiency of VLPs expression is high and can be easily purified owing to their higher molecular weight using various purification techniques such as sucrose gradient centrifugation and size exclusion chromatography [23].

## 12.2 Role of virus-like particles as good drug delivery vectors

Targeted drug delivery to site-specific cells, tissues, or organs was a challenging task in terms of efficient drug uptake, endosomal entrapment, and restricted drug delivery to certain cell types along with damage to cells in the delivery process [24]. On the other hand, VLPs offer a promising adaptable platform to deliver drugs or peptides to cells liberating the encapsulated therapeutic cargo inside the cell. The internalization of VLPs is accomplished via receptor-mediated endocytosis [25]. Cellular endocytosis is a multistep process initiated from engulfing of molecules from the plasma membrane of the cell to intracellular vesicle formation, which finally traffics through the cell. Dynamin-like GTPase plays a role in the release of intracellular cargos which gets traveled along the cytoskeleton to fuse with early endosomes and maturing in the late endosome. Finally, late endosome fuses with the prelysosomal compartment to make lysosome where cargo degradation occurs [26]. However, this endosomal uptake pathway is also known to be a rate-limiting barrier for various drug delivery mechanisms where bioactive molecules get entrapped and degraded in the lysosomal compartment of the cell [27]. Like parental viruses, VLPs overcome this limitation by escaping the endosomal uptake and execute successful drug delivery [28]. Due to the lack of virus genome, VLPs do not pose any risk associated with viral replication and infection. Thus the limitation associated with whole virus administration can be alleviated using VLPs [29].

As an emerging and promising nano-carrier drug delivery platform, VLPs can be modified genetically and chemically on their outer surface and inner cavities to facilitate the formulation of new material that could meet higher solubility, biocompatibility, and higher efficiency to uptake drug and deliver accordingly. Protein fusion can be done on the surface of VLPs to introduce heterologous peptides/proteins or chemical modifications of the capsid forming protein can also be done to functionalize the surface of VLPs with new specificity, tropism and physical properties [30]. For example, specific recognition of human hepatocellular carcinoma (HCC) was accomplished due to VLP modification with a specific HCC-binding peptide, SP94 and enabled efficient targeting [31]. Direct drug targeting using VLPs increases the interaction of the transported drug molecules with the specific cells, tissues, or organs. Few VLPs show natural tropism like their parental origin. For example, HBV infects the liver, and thus HBV-VLPs are used to target drugs in hepatocytes [32]. A more specific targeting can be achieved by attaching a receptor recognizing domain to the VLPs by conjugation or by genetic modifications. For example, surface antigen of HBV was coexpressed with maltose-binding protein (MBP) as a fusion with VLP and upon receptor binding, anti-MBP antibody was used to detect successful delivery using flow cytometry [33].

Currently, VLPs are increasing their popularity in the field of therapeutics as an efficient drug delivery vehicle due to its versatility in targeting specific site of action, escaping endosomal uptake and degradation of drug cargos, and facilitating chemical and genetic modifications to increase the efficiency of drug delivery.

## 12.3 Virus-like particles: overcoming limitations of other therapeutic approaches

Since last decade, NP-based delivery vehicles are being used by the scientific community for targeted therapeutic approaches. Recently, many challenges came across the usage of lipid- and polymer-based NP delivery vehicles, which include the lack of stability, lower solubility, structural heterogeneity, endosomal degradation of encapsulated therapeutic cargo and spontaneous off-target effects. Polymer-based NPs show slow and nonuniform drug release, whereas liposomal NPs are restricted by particle instability, off-target drug effects, and rapid clearance [34,35]. High toxicity and lack of specificity are the common issues of metal-based NPs [36]. Although PEGylation of NPs help to avoid phagocyte-mediated clearances, they reduce their ability to be uptaken by the targeted cells [37].

In order to tackle the above limitations, protein-based NPs were launched in the form of VLPs as an emerging class of targeted delivery vehicles. VLPs have a unique feature to resemble structurally and functionally like their parental viruses and form an outstanding vehicle to pack and deliver therapeutic cargos such as drugs, siRNA, aptamers, peptides, and proteins to their desired location [1,31,38]. VLP surface modification is another unique feature in order to efficiently extravasate from the vasculature of blood vessels by interaction with specific ligands displayed on the surface of epithelial cells and avoiding off-target effects and organ accumulation and toxicity. Overall, a combination of surface functionalization of VLPs and their capability to load and deliver therapeutics to their intended cellular location opens up many avenues for all the unmet medical needs.

## 12.4 Prerequisite factors in designing of virus-like particles as therapeutics

Current progresses in the field of targeted drug delivery in the form of VLPs show a great deal of promising results. Although VLPs have many advantages over other NP-based delivery vehicles, there are still rooms for various modifications in terms of more specificity, higher solubility, more biocompatibility, and many more. Few important factors are discussed later which should be taken care of while designing VLPs as delivery vehicles.

1. *Stability and uniformity*: VLPs should be highly uniform and should degrade and release all the therapeutic cargo at once which is in contrast to other NP-based vehicles. Although slow release is sometimes beneficial, but in most cases rapid/immediate release is most effective like in cancer therapy. The stability of VLPs poses poor performance in comparison to other NP-based agents. This problem should be addressed efficiently to make it a better therapeutic vehicle. Studies were done on MS2, HBVc, and Qβ VLPs. Like Qβ, the mutant MS2 VLP was engineered to form disulfide bonds between pentamers and hexamers that helped to increase the dissociation temperature and thereby increasing the stability [39,40]. Additional 240 disulfide bonds were incorporated successfully in mutant HBVc VLP that helps in linking every coat protein covalently and shown to be highly stable in PBS [41]. These incorporations of new disulfide bonds through engineering of VLPs will help in stabilizing during modification, formulation, storage, and administration of vaccines and targeted therapeutics.
2. *Cargo loading and release ability:* One of the most important characteristics of VLPs is its capability to load therapeutic cargo, transport to its desired target of action, and release the therapeutic for quick action. There are ranges of molecules that can be loaded inside these VLPs as per requirements such as drugs, proteins, peptides, nucleic acids, and fluorescent probes. For efficient encapsidated of cargos inside the VLPs, covalent method of attachments is more advantageous in comparison to noncovalent methods. Since viral

genomes are encapsidated inside for natural viruses, VLPs have also evolved in a similar manner to encapsidated nucleic acid very efficiently than that of other smaller molecules such as drugs and probes. For example, MS2 VLPs are suited for loading RNA molecules and require a stem loop RNA hairpin to get assembled inside the capsids [35], whereas CCMV, P22, HBVc, and Qβ use the electrostatic adsorption to load nucleic acids effectively [42,43]. On the other hand, loading peptides or proteins into VLPs can be done using various ways such as fusion of amino acid sequence of coat protein with the peptides or proteins or conjugating the peptides/proteins with the genome or passive encapsidation. HBVc VLPs were shown to encapsidate peptides/proteins via genetic fusion, whereas Qβ and MS2 use conjugation strategy to load peptides/proteins [31,44,45]. Smaller molecules are assembled inside the capsid by bacteriophages efficiently; for example, MS2 has been reported to load fluorescent probes (such as Fluorescein and Alexa Fluor 488), cancer drugs [such as taxols and doxorubicin (DOX)], and few other molecules such as porphyrin [31,46,47]. Plant virus–based VLPs are also being used for loading smaller molecules due to their unique characteristics of bonding through noncovalent attachments. For example, CPMV VLPs have been reported to load proflavin and fluorescent probes noncovalently along with small molecules adsorbed electrostatically by CPMV's RNA genome [48].

3. *Surface functionalization:* One of the crucial parameters to be considered while designing VLPs is their surface modifications to achieve different functions. This has been in practice recently owing to VLP's capacity to get modified genetically or chemically via various methods. The ligands/biomolecules that are used for attaching on VLP surfaces can be helpful for specific cellular targeting or reducing the immune response or facilitating extravasations. On the basis of various reactive amino acids or genetic methods, various modifications can be considered during VLP designing.
    a. *Cysteine-based modifications*: The presence of free sulfhydryl group makes cysteine as one of the most commonly used reactive amino acids to be presented on VLP surfaces. Under oxidative conditions, they spontaneously form disulfide bonds with ligands containing sulfhydryl groups. Various cell-penetrating peptides and fluorescent probes are being displayed on the surface of VLPs by conjugation chemistries such as maleimide forming irreversible thioether bonds with cysteine residues between pH 6.5 and 7.5 [49–51].
    b. *Lysine-based modifications*: After cysteine, lysine is another reactive amino acid, which can be modified and exploited to make VLPs more efficient. Amide bonds are formed at the surface-exposed lysine residues using *n*-hydroxysuccinimide (NHS) ester reactions. This method can be well demonstrated by displaying transferrin on MS2 VLPs, which allows it to cross the BBB [52]. This technique opened up new avenues in the field of neurological disorders.
    c. *Aspartate-based modifications*: Unlike cysteine- and lysine-based modifications, this is not a commonly used method and requires multiple steps. The chemistry is mainly based on NHS to form NHS ester to form a stable amide bond with ligands. For example, fluorescent probes or peptides were displayed on CCMV VLPs using this strategy [50].

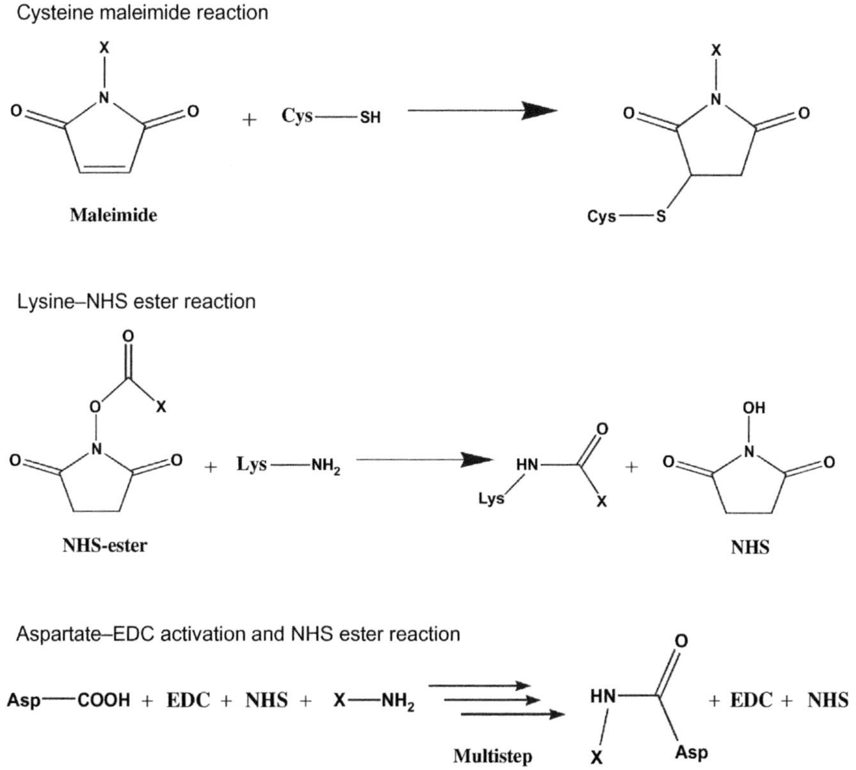

Cysteine maleimide reaction

Maleimide + Cys—SH → Cys—S (thioether adduct)

Lysine–NHS ester reaction

NHS-ester + Lys—$NH_2$ → Lys—HN—C(=O)—X + NHS

Aspartate–EDC activation and NHS ester reaction

Asp—COOH + EDC + NHS + X—$NH_2$ → (Multistep) X—HN—C(=O)—Asp + EDC + NHS

**FIGURE 12–2** Few common conjugation chemistries required for the functionalization of VLPs. *VLPs*, Virus-like particles.

- **d.** *Genetic modifications*: This modification is another covalent attachment method where the surface ligand's gene is being fused with the gene of VLPs coat protein. This has been worked for most of the VLPs with proper peptide folding and assembly except CCMV [53–55]. Various heterogeneous proteins were displayed using a genetic fusion method including antibody fragments for specific targeting.
- **e.** *Affinity-based modifications*: The main concept of this type of modification is that it does not require any alteration in the coat proteins of VLPs. Most importantly, the higher affinity of protein—ligand complex serves the whole purpose of targeting specific cellular sites. For example, decoration (dec) proteins have a high affinity for the surface of P22, and by fusing ligands with "dec" protein, P22 VLPs increase their efficiency to address their specificity [56] (Fig. 12–2).

## 12.5 Immune responses induced by virus-like particles

The geometric structural pattern of VLPs helps in facilitating their engagement with the innate and adaptive immune system. Since their final shape resembles the shape of their parental virus and owing to their lack of original viral genome, they have shown fruitful results in the field of vaccinology. The outer and inner surfaces of VLPs can be functionalized with heterologous antigens by genetic fusion or chemical cross-linking to enhance their immunogenicity by engagement of various pathogen recognition receptors (PRRs) [57]. The properties of VLPs in terms of size, structure, charge distribution, hydrophobicity, or hydrophilicity, etc. makes it a most favorable NP for the uptake by antigen-presenting cells (APCs) [58]. The antigens are processed and finally presented as peptides on the surface of APCs via MHC-I and MHC-II molecules. When the antigens are presented by MHC-II molecules, stimulation of $CD4^+$ T cells takes place, whereas on the other hand, induction of $CD8^+$ immune response occurs when presented via MHC-II molecules.

VLPs induce high levels of long-lasting humoral immune responses. Unlike other soluble antigens, the highly ordered, complex, and repetitive structure of VLPs actually promotes B-cell receptor (BCR) cross-linking that bypasses the prior need of T-cell help. The initiation of isotype switching is another feature of humoral immunity elicited by VLPs and is dependent on TLR7 or TLR9 signaling on B cells [59,60]. The VLPs undergo BCR-mediated endocytosis, where the nucleic acid packaged within reaches the endocytic compartment to activate endosomal TLRs. The packaged ssRNA, CpGs, or dsRNA in VLPs are mainly recognized by endosomal TLRs which results in the activation of signaling pathways, initiating the expression of important genes to induce inflammatory responses. The major transcription factors that get activated are NFκβ, AP-1, IRF3, and IRF7, and they stimulate the expression of many inflammatory molecules such as cytokines, chemokines, and adhesion molecules. There are other PRRs and nuclear sensors apart from TLRs, which get elicited by VLPs and help to initiate immune responses such as RIG-I and STING which stimulate type-I IFN induction upon viral infection [61]. Thus it highlights how VLPs are used to package nucleic acids in order to enhance the immunogenicity of the adjuvants and the immune responses. Various successful applications were recently been reported to induce protective immune responses against infectious and other noninfectious diseases in preclinical and clinical models. This accelerated the use of VLPs as vaccines in many diseases ranging from infection to asthma to hypertension. The highly immunogenic properties of VLPs are the main parameters for its enormous use in the field of medical science owing to its viral fingerprints retained from the parental viruses.

## 12.6 Current applications of virus-like particles as targeted therapeutics

VLPs have a vast range of applications in the field of medical science due to their diversity, specificity, and multigenicity of viruses. Due to their repetitive structural pattern that resembles their parent viruses, they interact in a similar fashion like their parental pathogens inducing both humoral- and cell-mediated immune responses. Numerous VLPs have been used as vaccines owing to their role in eliciting a potential immune response through MHC-I and MHC-II molecules. These VLP vaccine candidates are considered as "self-adjuvant" possessing molecules which favors uptake by professional APCs [58].

Among the four VLP vaccines commercially available in the market, HBV and HPV are the most potential VLPs in terms of efficacy and safety. The virus surface antigen (HBsAg)-based HBV-VLPs were the first VLPs vaccine to be successfully produced during the 1980s in different expression systems [62,63]. The HPV VLPs vaccine, overexpressed in *P. pastoris* [64] or plant cells, showed almost 98% of efficiency in protecting individuals from infection and considered to be one of the greatest milestones in cancer prevention [65]. Few VLPs vaccine candidates are found to be immunogenic in animal models, and under preclinical stage for human consumption includes dengue, rabies, rotavirus (RV), hepatitis C virus, and human immunodeficiency virus-1 [3,7].

Chimeric VLPs are also designed to enhance humoral immunity by inducing antibodies against self-antigens and are mostly used for noninfectious diseases such as cancer (melanoma), Alzheimer's disease, asthma, and other allergic diseases [66–68]. Targeted delivery of drugs is of prime importance to prevent off-target effects of systemic administrations. Thus VLPs are used as specific carriers to address the issue of specific drug delivery. For example, anticancer drugs such as DOX and bleomycin (BLM) were coupled to VLPs (DOX to RV VLPs and BLM to adenoviral VLPs) and showed high drug action specificity with pronounced growth inhibition of targeted cancer cells and improved bioavailability [69,70].

The ability of packaging oligonucleotides and plasmids into VLPs makes them a promising carrier of nucleic acid. Many successful gene therapies were achieved by expressing a specific gene encoded by plasmids and delivered by different viral VLPs. For example, the plasmid encoding β-galactosidase gene was efficiently delivered to the site of action by a polyomavirus VP1 VLP [71]. In lieu of advanced VLP-mediated gene therapy, antigens, cytokines, antibodies, reporter proteins such as green fluorescent protein or MBP are tethered on the outer surface of VLPs using genetic fusion or by natural interactions (such as ligand–receptor interaction and antigen–antibody interaction) to address directed immunotherapy. For example, Herceptin antibody was conjugated to the polyoma VLP to direct it against tyrosine kinase receptor HER2 in order to achieve directed cancer immunotherapy [72]. In case of glioblastoma treatment, VLPs have shown promising outcomes with improved rate of drug delivery and reduced growth of tumor size. The major hurdle for other therapeutics to cross the BBB has been significantly surpassed by the use of VLPs. Three VLPs were designed and modified with DOX and showed higher drug delivery and in vitro cellular uptake and increased survival rates in glioma-bearing mice [73]. Recently, an interesting application of VLPs in diagnostic was reported where HBcAg-VLPs loaded with $Fe_3O_4$ were used for magnetic resonance imaging due to high paramagnetic contrast agent [74].

Although VLPs are considered as efficient delivery platforms, there are few limitations in terms of their use as gene or drug carriers. The preexisting immunity in most individuals for common viruses is a major concern in the usage of VLPs as drug carriers. In this case, polyethylene glycol is used as immune masking agents to suppress the unwanted immunogenicity of VLPs. Another demerit of VLPs-based treatment is repetitive administrations that are avoided except cancer immunotherapy where tumor recognition is stimulated in an escalating manner. More in-depth studies are required to overcome the large-scale production of

VLPs vaccine and also to translate preclinical research to licensed products. Their biodistribution, toxicity, stability, and clearance are also to be considered before its usage in the medical field.

## 12.7 Conclusion

The concept of VLP-mediated targeted drug delivery is by far a giant step in the field of nanotechnology where many disease treatment strategies were highly promising. The characteristic features of VLPs in terms of size, structural pattern, stability, and inducing immune response with surface functionalization and modifications made it possible to encapsulate therapeutic cargos for specific drug targeting. The induction of both humoral- and cell-mediated immunity is one of the major advantages of using VLPs as vaccines. In fact, chemical and genetic modifications of VLPs make it more useful for therapeutic application with increased site-directed delivery, enhanced immunogenicity, and reduced side effects. The most efficient and promising VLPs are HBVc, MS2, and Qβ considering both merits and demerits. On the other hand, CCMV, CPMV, and P22 need more improvement in terms of technology for future applications. There are few unexplored areas of research in VLP-mediated therapeutics where additional progress are required to improve the technology such as (1) reducing off-target effects of organ accumulation in liver and spleen, (2) improving extravasation from the blood vessels, and (3) increasing the local concentration around targeted cells.

Recently, in 2016, the world's first influenza vaccine, Cadiflu-S, based on VLPs, was commercialized by CPL Biologicals Private Limited, a joint venture of Cadila Pharmaceuticals Limited, India and Novovax Inc., United States. This is one of the major breakthroughs as Cadiflu-S does not contain any virus and eliminates the possibility of infection to the vaccinated person. But there are still many challenges faced during the commercialization of VLPs-based vaccines. Further researches are essential to understand the behavioral characteristics of VLPs in in vivo systems and how to translate the results from the laboratory to clinical trials is still under investigation. All these suggestions will finally indicate a bright future for VLP-mediated targeted therapeutics.

## References

[1] P. Anand, A. O'neil, E. Lin, T. Douglas, M. Holford, Tailored delivery of analgesic ziconotide across a blood brain barrier model using viral nanocontainers, Sci. Rep. 5 (2015) 12497.

[2] G. Casi, D. Neri, Antibody–drug conjugates and small molecule–drug conjugates: opportunities and challenges for the development of selective anticancer cytotoxic agents, J. Med. Chem. 58 (2015) 8751–8761.

[3] A. Roldão, M.C.M. Mellado, L.R. Castilho, M.J.T. Carrondo, P.M. Alves, Virus-like particles in vaccine development, Expert Rev. Vaccines 9 (2010) 1149–1176.

[4] A.E. Czapar, N.F. Steinmetz, Plant viruses and bacteriophages for drug delivery in medicine and biotechnology, Curr. Opin. Chem. Biol. 38 (2017) 108–116.

[5] A.C. Obermeyer, S.L. Capehart, J.B. Jarman, M.B. Francis, Multivalent viral capsids with internal cargo for fibrin imaging, PLoS One 9 (2014) e100678.

[6] W. Wu, S.C. Hsiao, Z.M. Carrico, M.B. Francis, Genome-free viral capsids as multivalent carriers for taxol delivery, Angew. Chem. (Int. Ed. Engl.) 48 (2009) 9493–9497.

[7] N. Kushnir, S.J. Streatfield, V. Yusibov, Virus-like particles as a highly efficient vaccine platform: diversity of targets and production systems and advances in clinical development, Vaccine 31 (2012) 58–83.

[8] F. Liu, X. Wu, L. Li, Z. Liu, Z. Wang, Use of baculovirus expression system for generation of virus-like particles: successes and challenges, Protein Expr. Purif. 90 (2013) 104–116.

[9] H. Lünsdorf, C. Gurramkonda, A. Adnan, N. Khanna, U. Rinas, Virus-like particle production with yeast: ultrastructural and immunocytochemical insights into *Pichia pastoris* producing high levels of the Hepatitis B surface antigen, Microb. Cell Fact. 10 (2011) 48.

[10] M.-K. Woo, J.-M. An, J.-D. Kim, S.-N. Park, H.-J. Kim, Expression and purification of human papillomavirus 18 L1 virus-like particle from *Saccharomyces cerevisiae*, Arch. Pharmacal Res. 31 (2008) 205–209.

[11] M. Sällberg, J. Hughes, J. Jones, T.R. Phillips, D.R. Milich, A malaria vaccine candidate based on a hepatitis B virus core platform, Intervirology 45 (2002) 350–361.

[12] K.A. Aires, A.M. Cianciarullo, S.M. Carneiro, L.L. Villa, E. Boccardo, G. Pérez-Martinez, et al., Production of human papillomavirus type 16 L1 virus-like particles by recombinant *Lactobacillus casei* cells, Appl. Environ. Microbiol. 72 (2006) 745.

[13] Y.-X. Tang, L.-F. Jiang, J.-M. Zhou, Y. Yin, X.-M. Yang, W.-Q. Liu, et al., Induction of virus-neutralizing antibodies and T cell responses by dengue virus type 1 virus-like particles prepared from *Pichia pastoris*, Chin. Med. J. 125 (2012) 1986–1992.

[14] R. Greco, M. Michel, D. Guetard, M. Cervantes-Gonzalez, N. Pelucchi, S. Wain-Hobson, et al., Production of recombinant HIV-1/HBV virus-like particles in *Nicotiana tabacum* and *Arabidopsis thaliana* plants for a bivalent plant-based vaccine, Vaccine 25 (2007) 8228–8240.

[15] P. Tomasz, Is an oral plant-based vaccine against hepatitis B virus possible? Curr. Pharm. Biotechnol. 13 (2012) 2692–2704.

[16] D. Gheysen, E. Jacobs, F. De Foresta, C. Thiriart, M. Francotte, D. Thines, et al., Assembly and release of HIV-1 precursor Pr55gag virus-like particles from recombinant baculovirus-infected insect cells, Cell 59 (1989) 103–112.

[17] P. Pushko, T.M. Tumpey, F. Bu, J. Knell, R. Robinson, G. Smith, Influenza virus-like particles comprised of the HA, NA, and M1 proteins of H9N2 influenza virus induce protective immune responses in BALB/c mice, Vaccine 23 (2005) 5751–5759.

[18] S.S. El-Kamary, M.F. Pasetti, P.M. Mendelman, S.E. Frey, D.I. Bernstein, J.J. Treanor, et al., Adjuvanted intranasal Norwalk virus-like particle vaccine elicits antibodies and antibody-secreting cells that express homing receptors for mucosal and peripheral lymphoid tissues, J. Infect. Dis. 202 (2010) 1649–1658.

[19] P.N. Barrett, W. Mundt, O. Kistner, M.K. Howard, Vero cell platform in vaccine production: moving towards cell culture-based viral vaccines, Expert Rev. Vaccines 8 (2009) 607–618.

[20] Y. Kawaoka, S. Yamayoshi, Mapping of a region of *Ebola virus* VP40 that is important in the production of virus-like particles, J. Infect. Dis. 196 (2007) S291–S295.

[21] W. Akahata, Z.-Y. Yang, H. Andersen, S. Sun, H.A. Holdaway, W.-P. Kong, et al., A virus-like particle vaccine for epidemic chikungunya virus protects nonhuman primates against infection, Nat. Med., 16, 2010, p. 334.

[22] F. Cheng, S. Mukhopadhyay, Generating enveloped virus-like particles with in vitro assembled cores, Virology 413 (2011) 153–160.

[23] M.T. Arevalo, T.M. Wong, T.M. Ross, Expression and purification of virus-like particles for vaccination, J. Vis. Exp. (2016) 54041.

[24] S.J. Kaczmarczyk, K. Sitaraman, H.A. Young, S.H. Hughes, D.K. Chatterjee, Protein delivery using engineered virus-like particles, Proc. Natl. Acad. Sci. U.S.A. 108 (2011) 16998–17003.

[25] L.H.L. Lua, N.K. Connors, F. Sainsbury, Y.P. Chuan, N. Wibowo, A.P.J. Middelberg, Bioengineering virus-like particles as vaccines, Biotechnol. Bioeng. 111 (2014) 425–440.

[26] L.M. Bareford, P.W. Swaan, Endocytic mechanisms for targeted drug delivery, Adv. Drug Deliv. Rev. 59 (2007) 748–758.

[27] H.K. Shete, R.H. Prabhu, V.B. Patravale, Endosomal escape: a bottleneck in intracellular delivery, J. Nanosci. Nanotechnol. 14 (2014) 460–474.

[28] P. Fender, R.W.H. Ruigrok, E. Gout, S. Buffet, J. Chroboczek, Adenovirus dodecahedron, a new vector for human gene transfer, Nat. Biotechnol. 15 (1997) 52–56.

[29] E. Crisci, J. Bárcena, M. Montoya, Virus-like particles: the new frontier of vaccines for animal viral infections, Vet. Immunol. Immunopathol. 148 (2012) 211–225.

[30] Y. Ma, R.J.M. Nolte, J.J.L.M. Cornelissen, Virus-based nanocarriers for drug delivery, Adv. Drug Deliv. Rev. 64 (2012) 811–825.

[31] C.E. Ashley, E.C. Carnes, G.K. Phillips, P.N. Durfee, M.D. Buley, C.A. Lino, et al., Cell-specific delivery of diverse cargos by bacteriophage MS2 virus-like particles, ACS Nano 5 (2011) 5729–5745.

[32] A. Shlomai, Y. Lubelsky, O. Har-Noy, Y. Shaul, The "Trojan horse" model-delivery of anti-HBV small interfering RNAs by a recombinant HBV vector, Biochem. Biophys. Res. Commun. 390 (2009) 619–623.

[33] J.-H. Park, E.-A. Choi, E.-W. Cho, Y.-J. Lee, J.-M. Park, S.-Y. Na, et al., Detection of cellular receptors specific for the hepatitis B virus preS surface protein on cell lines of extrahepatic origin, Biochem. Biophys. Res. Commun. 277 (2000) 246–254.

[34] M. Elsabahy, K.L. Wooley, Design of polymeric nanoparticles for biomedical delivery applications, Chem. Soc. Rev. 41 (2012) 2545–2561.

[35] T.M. Allen, P.R. Cullis, Liposomal drug delivery systems: from concept to clinical applications, Adv. Drug Deliv. Rev. 65 (2013) 36–48.

[36] A.Z. Wang, R. Langer, O.C. Farokhzad, Nanoparticle delivery of cancer drugs, Annu. Rev. Med. 63 (2012) 185–198.

[37] J.K. Armstrong, G. Hempel, S. Koling, L.S. Chan, T. Fisher, H.J. Meiselman, et al., Antibody against poly (ethylene glycol) adversely affects PEG-asparaginase therapy in acute lymphoblastic leukemia patients, Cancer 110 (2007) 103–111.

[38] P. Kelly, P. Anand, A. Uvaydov, S. Chakravartula, C. Sherpa, E. Pires, et al., Developing a dissociative nanocontainer for peptide drug delivery, Int. J. Environ. Res. Public Health 12 (2015) 12543–12555.

[39] B.C. Bundy, J.R. Swartz, Efficient disulfide bond formation in virus-like particles, J. Biotechnol. 154 (2011) 230–239.

[40] A.E. Ashcroft, H. Lago, J.M.B. Macedo, W.T. Horn, N.J. Stonehouse, P.G. Stockley, Engineering thermal stability in RNA phage capsids via disulphide bonds, J. Nanosci. Nanotechnol. 5 (2005) 2034–2041.

[41] Y. Lu, W. Chan, B.Y. Ko, C.C. Vanlang, J.R. Swartz, Assessing sequence plasticity of a virus-like nanoparticle by evolution toward a versatile scaffold for vaccines and drug delivery, Proc. Natl. Acad. Sci. U.S.A. 112 (2015) 12360–12365.

[42] S. Qazi, H.M. Miettinen, R.A. Wilkinson, K. Mccoy, T. Douglas, B. Wiedenheft, Programmed self-assembly of an active P22-Cas9 nanocarrier system, Mol. Pharm. 13 (2016) 1191–1196.

[43] A. Strods, V. Ose, J. Bogans, I. Cielens, G. Kalnins, I. Radovica, et al., Preparation by alkaline treatment and detailed characterisation of empty hepatitis B virus core particles for vaccine and gene therapy applications, Sci. Rep. 5 (2015) 11639.

[44] G. Beterams, B. Böttcher, M. Nassal, Packaging of up to 240 subunits of a 17 kDa nuclease into the interior of recombinant hepatitis B virus capsids, FEBS Lett. 481 (2000) 169–176.

[45] J.-K. Rhee, M. Hovlid, J.D. Fiedler, S.D. Brown, F. Manzenrieder, H. Kitagishi, et al., Colorful virus-like particles: fluorescent protein packaging by the Qβ capsid, Biomacromolecules 12 (2011) 3977–3981.

[46] N. Stephanopoulos, G.J. Tong, S.C. Hsiao, M.B. Francis, Dual-surface modified virus capsids for targeted delivery of photodynamic agents to cancer cells, ACS Nano 4 (2010) 6014–6020.

[47] G.J. Tong, S.C. Hsiao, Z.M. Carrico, M.B. Francis, Viral capsid DNA aptamer conjugates as multivalent cell-targeting vehicles, J. Am. Chem. Soc. 131 (2009) 11174–11178.

[48] I. Yildiz, K.L. Lee, K. Chen, S. Shukla, N.F. Steinmetz, Infusion of imaging and therapeutic molecules into the plant virus-based carrier cowpea mosaic virus: cargo-loading and delivery, J. Control. Release 172 (2013) 568–578.

[49] A. Chatterji, W.F. Ochoa, M. Paine, B.R. Ratna, J.E. Johnson, T. Lin, New addresses on an addressable virus nanoblock: uniquely reactive Lys residues on cowpea mosaic virus, Chem. Biol. 11 (2004) 855–863.

[50] E. Gillitzer, D. Willits, M. Young, T. Douglas, Chemical modification of a viral cage for multivalent presentation, Chem. Commun. (2002) 2390–2391.

[51] S. Kang, G.C. Lander, J.E. Johnson, P.E. Prevelige, Development of bacteriophage P22 as a platform for molecular display: genetic and chemical modifications of the procapsid exterior surface, ChemBioChem 9 (2008) 514–518.

[52] F.A. Galaway, P.G. Stockley, MS2 viruslike particles: a robust, semisynthetic targeted drug delivery platform, Mol. Pharm. 10 (2013) 59–68.

[53] H. Peyret, A. Gehin, E.C. Thuenemann, D. Blond, A. El Turabi, L. Beales, et al., Tandem fusion of hepatitis B core antigen allows assembly of virus-like particles in bacteria and plants with enhanced capacity to accommodate foreign proteins, PLoS One 10 (2015) e0120751.

[54] J.P. Phelps, N. Dang, L. Rasochova, Inactivation and purification of cowpea mosaic virus-like particles displaying peptide antigens from *Bacillus anthracis*, J. Virol. Methods 141 (2007) 146–153.

[55] A. Servid, P. Jordan, A. O'neil, P. Prevelige, T. Douglas, Location of the bacteriophage P22 coat protein C-terminus provides opportunities for the design of capsid-based materials, Biomacromolecules 14 (2013) 2989–2995.

[56] B. Schwarz, P. Madden, J. Avera, B. Gordon, K. Larson, H.M. Miettinen, et al., Symmetry controlled, genetic presentation of bioactive proteins on the P22 virus-like particle using an external decoration protein, ACS Nano 9 (2015) 9134–9147.

[57] M.B. Heo, S.-Y. Kim, W.S. Yun, Y.T. Lim, Sequential delivery of an anticancer drug and combined immunomodulatory nanoparticles for efficient chemoimmunotherapy, Int. J. Nanomed. 10 (2015) 5981–5992.

[58] V. Manolova, A. Flace, M. Bauer, K. Schwarz, P. Saudan, M.F. Bachmann, Nanoparticles target distinct dendritic cell populations according to their size, Eur. J. Immunol. 38 (2008) 1404–1413.

[59] J.M. Clingan, M. Matloubian, B cell-intrinsic TLR7 signaling is required for optimal B cell responses during chronic viral infection, J. Immunol. (Baltimore, MD: 1950) 191 (2013) 810–818.

[60] A. Jegerlehner, P. Maurer, J. Bessa, H.J. Hinton, M. Kopf, M.F. Bachmann, TLR9 signaling in B cells determines class switch recombination to IgG2a, J. Immunol. 178 (2007) 2415.

[61] P. Nakhaei, J. Hiscott, R. Lin, STING-ing the antiviral pathway, J. Mol. Cell Biol. 2 (2009) 110–112.

[62] D.P. Aden, A. Fogel, S. Plotkin, I. Damjanov, B.B. Knowles, Controlled synthesis of HBsAg in a differentiated human liver carcinoma-derived cell line, Nature 282 (1979) 615–616.

[63] P. Valenzuela, A. Medina, W.J. Rutter, G. Ammerer, B.D. Hall, Synthesis and assembly of hepatitis B virus surface antigen particles in yeast, Nature 298 (1982) 347–350.

[64] N. Hanumantha Rao, P. Baji Babu, L. Rajendra, R. Sriraman, Y.-Y.S. Pang, J.T. Schiller, et al., Expression of codon optimized major capsid protein (L1) of human papillomavirus type 16 and 18 in *Pichia pastoris*; purification and characterization of the virus-like particles, Vaccine 29 (2011) 7326–7334.

[65] Y. Deleré, O. Wichmann, S.J. Klug, M. Van Der Sande, M. Terhardt, F. Zepp, et al., The efficacy and duration of vaccine protection against human papillomavirus: a systematic review and meta-analysis, Dtsch. Arzteblatt Int. 111 (2014) 584–591.

[66] M. Braun, C. Jandus, P. Maurer, A. Hammann-Haenni, K. Schwarz, M.F. Bachmann, et al., Virus-like particles induce robust human T-helper cell responses, Eur. J. Immunol. 42 (2012) 330–340.

[67] B. Chackerian, M. Rangel, Z. Hunter, D.S. Peabody, Virus and virus-like particle-based immunogens for Alzheimer's disease induce antibody responses against amyloid-β without concomitant T cell responses, Vaccine 24 (2006) 6321–6331.

[68] G. Senti, P. Johansen, S. Haug, C. Bull, C. Gottschaller, P. Müller, et al., Use of A-type CpG oligodeoxynucleotides as an adjuvant in allergen-specific immunotherapy in humans: a phase I/IIa clinical trial, Clin. Exp. Allergy 39 (2009) 562–570.

[69] Q. Zhao, W. Chen, Y. Chen, L. Zhang, J. Zhang, Z. Zhang, Self-assembled virus-like particles from rotavirus structural protein VP6 for targeted drug delivery, Bioconjugate Chem. 22 (2011) 346–352.

[70] M. Zochowska, A. Paca, G. Schoehn, J.-P. Andrieu, J. Chroboczek, B. Dublet, et al., Adenovirus dodecahedron, as a drug delivery vector, PLoS One 4 (2009) e5569.

[71] T. May, S. Gleiter, H. Lilie, Assessment of cell type specific gene transfer of polyoma virus like particles presenting a tumor specific antibody Fv fragment, J. Virol. Methods 105 (2002) 147–157.

[72] S. Gleiter, H. Lilie, Cell-type specific targeting and gene expression using a variant of polyoma VP1 virus-like particles, Biol. Chem. 384 (2003) 247–255.

[73] A.J. Finbloom, L.I. Aanei, M.J. Bernard, H.S. Klass, K.S. Elledge, K. Han, et al., Evaluation of three morphologically distinct virus-like particles as nanocarriers for convection-enhanced drug delivery to glioblastoma, Nanomaterials 8 (2018).

[74] L. Shen, J. Zhou, Y. Wang, N. Kang, X. Ke, S. Bi, et al., Efficient encapsulation of $Fe_3O_4$ nanoparticles into genetically engineered hepatitis B core virus-like particles through a specific interaction for potential bioapplications, Small 11 (2015) 1190–1196.

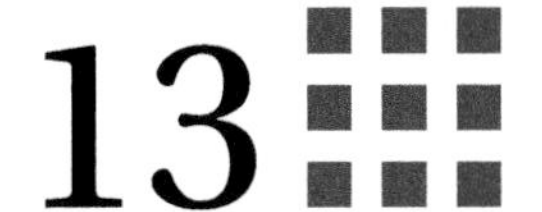

# Formulation development and in vitro multimedia drug release study of solid self-microemulsifying drug delivery system of ketoconazole for enhanced solubility and pH-independent dissolution profile

Vinod Mokale[1], Shivraj Naik[1], Trupti Khatal[1], Shirish H. Sonawane[2], Irina Potoroko[3]

[1]DEPARTMENT OF PHARMACEUTICAL TECHNOLOGY, UNIVERSITY INSTITUTE OF CHEMICAL TECHNOLOGY, NORTH MAHARASHTRA UNIVERSITY, JALGAON, INDIA [2]CHEMICAL ENGINEERING DEPARTMENT, NATIONAL INSTITUTE OF TECHNOLOGY, WARANGAL, INDIA [3]DEPARTMENT OF FOOD TECHNOLOGY AND BIOTECHNOLOGY, SCHOOL OF MEDICAL BIOLOGY, SUSU, CHELYABINSK, RUSSIA

## Chapter outline

**13.1 Introduction** .... **212**
**13.2 Material and method** .... **212**
13.2.1 Material .... 212
13.2.2 Methods .... 212
**13.3 Results and discussion** .... **216**
13.3.1 Solubility study .... 216
13.3.2 Physicochemical compatibility of the drug to excipients .... 218
**13.4 Conclusion** .... **229**
**Acknowledgment** .... **230**
**Conflict of interest** .... **230**
**References** .... **230**

Encapsulation of Active Molecules and their Delivery System. DOI: https://doi.org/10.1016/B978-0-12-819363-1.00013-2

## 13.1 Introduction

Effective treatment of many diseases relies on the oral route of drug delivery. For new chemical entities (NCE), oral drug delivery development is becoming a challenging issue because of poor bioavailability of the NCE due to the inadequate drug dissolution and absorption [1]. Approximately 40% of new drugs are lipophilic in nature leading toward severe consequences such as poor oral bioavailability, high variability among individuals, and deficient toward dose proportionality. Thus dissolution governs the absorption ratio or percentage from the gastrointestinal tract. Various approaches have been implemented to improve the dissolution rate of the drugs, and among them self-microemulsifying drug delivery systems (SMEDDSs) improve its oral bioavailability of poorly soluble drugs [2].

It is a system of particular class of emulsions that contains oil and surfactant with/without cosurfactant that instantly forms emulsion upon mixing with water, which requires little or no energy input [3]. Chemically ketoconazole is $C_{26}H_{28}C_{12}N_4O_4$ is 1-acetyl-4-[4-[[(2RS, 4SR)-2-(2,4-dichlorophenyl)-2-(1*H*-imidazol-1-ylmethyl) 1,3-dioxolan-4-yl] methoxy] phenyl] piperazine. Ketoconazole is an antifungal agent. Oral ketoconazole is mostly beneficial for the treatment of various medical conditions such as blastomycosis, candida infections, chromomycosis and coccidioidomycosis [4]. Ketoconazole is a poorly water-soluble drug having $\log P$ value of 4.4. Peak plasma levels occur within 2 h, after which the therapeutic drug plasma concentration falls abruptly [5]. In past years lots of new approaches have been already applied to enhance the solubility of ketoconazole such as solid inclusion complexes [6], solid dispersion [7], and nanostructured crystals of salt of ketoconazole [8]. Ketoconazole exhibits pH-dependent drug dissolution that increases with decreasing pH [9]. The present investigation was aimed at developing solid SMEDDS of ketoconazole to enhance its solubility by oral route and to overcome pH-dependent dissolution.

## 13.2 Material and method

### 13.2.1 Material

Ketoconazole got as a gift sample for research work from Aarti Drugs Ltd; Mumbai, India. Gift samples of various grades of Capmul were obtained from Abitec Corporation. BASF Chemicals, Mumbai, India gifted Cremophor RH 40 and Cremophor EL. Capriyol 90, Transcutol P and Labrasol were gifted by Gattefosse, France. Acrysol EL was gifted by Corel Pharma Chem. Ahmedabad, India. PEG 400 and Propylene glycol purchased from S.D. Fine chemicals, Mumbai, India. Aerosil 200 was purchased from Sigma Life Science, USA. The materials used in this study were GRAS approved and within the limits as per IIG. All other chemicals were of analytical grade.

### 13.2.2 Methods

#### *13.2.2.1 Equilibrium solubility studies of ketoconazole*

Excipient selection in the formulation development of SMEDDS was done on the basis of the solubility of the ketoconazole in it. The solubility of ketoconazole in different oils, surfactants, and cosurfactants (S/CoS) was determined as demonstrated by Mandawgade et al. with some modifications [10]. 3 g of vehicle and excess amount of ketoconazole were added in the vials followed by heating at 40°C to facilitate the drug solubilization and then mixed homogeneously using mechanical stirrer. This mixture was sonicated for 15 min and kept for about 48 h at 30°C. After reaching equilibrium the mixture then centrifuges for 10 min at 5000 RPM (Remi centrifuge) and the supernatant was filtered through a 0.45 μm membrane filter to remove the excess amounts of the drug. The solubility of the ketoconazole in different vehicles was quantified by UV spectrometer (Hitachi U-2900) at 269.5 nm using methanol as diluent.

#### *13.2.2.2 Physicochemical compatibility of ketoconazole with excipients*

##### 13.2.2.2.1 Visual inspection method

This study was performed to get the physicochemical compatibility study of the different excipients used. The study was performed by placing the ketoconazole in combination with different excipient for 3 months at two different temperatures (room temperature and 40°C). The samples were virtually inspected after every 15 days to determine any color change.

##### 13.2.2.2.2 Fourier transform-infrared spectroscopy study

After 3 months of placing the ketoconazole in combination with different excipients, Fourier transform-infrared (FTIR) of all samples was done to determine any incompatibility between the excipients and ketoconazole. KBr pellet method (1 mg sample mixed with around 40 mg KBr and compressed into a disc using a manual press) was used to scan the FTIR to study the chemical structures of the ketoconazole and other excipients using FTIR spectrophotometer (FTIR-8400; Shimadzu). Scanning range was at 4000–400 $cm^{-1}$.

##### 13.2.2.2.3 Differential scanning calorimetry

Differential scanning calorimetry (DSC) characterization was performed using Shimadzu DSC-60 thermal analyzer. The required quantity of sample was placed in an aluminum pan with another similar reference pan. The analysis was done at 10°C–300°C with heating flow of 10°C/min. The nitrogen purge gas flow was maintained at 50 mL/min.

#### *13.2.2.3 Self-emulsification study*

This study was performed by mixing various oils and surfactants at a fixed ratio (2:3) for 10 min using vortex mixer. Then, 0.5 mL of this mixture was added into 500 mL water maintaining room temperature and mixed properly by using magnetic stirrer. The tendency of spontaneous emulsification and dispersibility was observed visually and assigned grades as per the grading system (Table 13–1). For study the effect of cosurfactants on emulsification and dispersibility, ratio of surfactant to cosurfactant was maintained at constant (3:1) and mix to oil at a ratio of 3:1 based on requirements as stated by Pouton [3].

#### *13.2.2.4 Apparent partition coefficient studies*

Equal volumes of water and 1-octanol were mixed in a shaker for 3 h to attain the saturation and then two phases were left for separation it overnight. A known concentration of

**Table 13–1** Grading system for visual assessment of self-microemulsification efficiency.

| Sr. no. | Dispersibility and appearance | Time of self-emulsification (min) | Grade | Inference |
|---|---|---|---|---|
| 1 | A clear and transparent microemulsion | <1 | $A^+$ | Rapid microemulsion |
| 2 | Transparent, gel-like microemulsion | 3–5 | A | Microemulsion |
| 3 | A turbid emulsion | >5 | B | Emulsion |
| 4 | Poor emulsification with coalescence of oil droplets | No emulsion | C | No emulsion |

ketoconazole was dissolved into the 1-octanol phase along with gentle shaking then further mixed with the aqueous phase. The obtained mixture was agitated for 6 h at 37°C and two phases were separated again after centrifugation done. The drug concentrations in both the phases were calculated spectrophotometrically to determine the partition coefficient [11].

#### *13.2.2.5 Construction of pseudo-ternary phase diagrams*

A formulation component was screened by constructing pseudo-ternary phase diagrams using water titration method. To plot the ternary plots oil and S/CoS mixtures were used at certain mass ratios. S/CoS were used at specific concentration ratios as 1:1, 2:1, 3:1, and 4:1. Oils and S/CoS form a transparent and homogenous mixture upon vortexing for 5 min. Then these mixtures were titrated sequentially with water at specific ratios of 9:1, 8:2, 7:3, 6:4, 5:5, 4:6, 3:7, 2:8, and 1:9 until the solution or mixture get turbidity or phase separation observed. At last at end point, the amounts of all the individual components added were noted [12].

#### *13.2.2.6 Effect of ketoconazole on phase diagram*

To investigate effect of ketoconazole on the ternary phase diagram, there were added to the boundary formulations of microemulsion region of phase diagram at predetermined concentration. Self-microemulsifying ability was assessed visually upon infinite dilution [13].

#### *13.2.2.7 Preparation of ketoconazole liquid self-microemulsifying drug delivery system*

By varying the S/CoS ratios such as 1:1, 2:1, 3:1, and 4:1, four different pseudo-ternary phase diagrams were obtained. Out of these the phase diagram with the largest microemulsion region was selected for further studies. Different points in the microemulsion region were selected and the concentration values related to those points were noted and formulations were prepared, respectively. Accurately weighed ketoconazole was dispersed into the oil phase and heated to 50°C under vortex. S/CoS was mixed separately and heated to 50°C. Ketoconazole-containing oil phase was transferred into S/CoS mixture with continuous mixing with heated at 50°C in a sonicator till the ketoconazole completely dissolved. The resultant liquid formulation was kept at room temperature for further processing.

### *13.2.2.8 Characterization of liquid self-microemulsifying drug delivery system*

#### 13.2.2.8.1 Transmittance test

Test was done to investigate effect of the dilution on the stability of the optimized formulation wherein the transmittance was measured at 650 nm at UV spectrophotometer (Hitachi U-2900). The analysis was performed in three replicates [14].

#### 13.2.2.8.2 Globule size and zeta potential analysis

Optimized SMEDDS formulation was studied for the globule size measurement and zeta potential analysis using Malvern Zetasizer (Nano ZS90) with a 50 mV laser. The particle size and polydispersity indices measurements are based upon the photon correlation spectroscopy, whereas the zeta potential measurements are based upon the electrophoretic mobility. The reconstituted SMEDDS formulation (0.1 mL) was diluted to 20 mL with distilled water and analyzed at 25°C with a fixed angle of 90 degrees [15]. Each globule size measurement value represents the average of at least three values. Similarly, the zeta potential analysis was performed on the same sample at the abovementioned instrument and testing conditions [16].

### *13.2.2.9 Preparation of solid ketoconazole self-microemulsifying drug delivery system*

The spray drying technique was used for the transformation of the liquid SMEDDS formulation to the solid SMEDDS by using Aerosil 200 as a suitable carrier. Spraymate spray dryer (Version 1.0, Jay instruments) was used in this study. The spray drying process consisted of the Aerosil 200: liquid SMEDDS in the ratio of 500 mg:1 mL. 2.5 g of Aerosil 200 was suspended in 250 mL methanol and 5 mL liquid SMEDDS formulation was added to this suspension with constant stirring. Obtained methanolic suspension was passed to nozzle keeping flow rate of 5 mL/min by using peristaltic pump. Spray drying was carried out at the inlet and outlet temperatures of 120°C and 70°C respectively under constant supply of compressed air with a pressure of 3 kg/cm$^2$ [17]. The yield of the powder obtained after spray drying was calculated from the following formula:

$$\text{Process Yield} = \text{weight of products obtained from spray dryer/weight of total dissolved solids} \times 100$$

### *13.2.2.10 Characterization of solid self-microemulsifying drug delivery system*

#### 13.2.2.10.1 Differential scanning calorimetry

DSC analysis performed on Shimadzu DSC-60 thermal analyzer (Shimadzu, Japan). The required quantity of samples was placed in an aluminum pan along with the similar reference pan and scanned at range of 10°C–300°C at a heating rate of 10°C/min, nitrogen purge gas flow was of 50 mL/min.

#### 13.2.2.10.2 X-ray diffraction study

To study the crystallographic analysis or the physical status characterization, that is, amorphous or crystalline nature, X-ray diffraction (XRD) patterns of (1) pure ketoconazole,

(2) Aerosil 200 (carrier), (3) physical mixture, and (4) SMEDDS were obtained using Bruker X-ray diffractometer over the angle range ($2\theta$) of 20–80 degrees.

#### 13.2.2.10.3 Surface morphological study

To study surface morphology of the solid particles, field emission scanning electron microscopy (FE-SEM S 4800, Hitachi, Japan) was used to know the surface morphology of (1) pure ketoconazole, (2) Aerosil 200, and (3) solid SMEDDS.

### *13.2.2.11 In vitro multimedia drug release study*

For in vitro drug release studies of the optimized formulation, we used USP type II dissolution test apparatus (Eletrolab, TDT-06T India). The analysis was performed at 37°C ± 0.5°C with 50 RPM. An accurately weighed amount of SMEDDS was put into the 900 mL of 0.1 N HCL as a dissolution medium. The sampling was done at specific time intervals and 5 mL samples were pulled out from the dissolution media, which were replaced by same amount of fresh media to maintain the sink condition. Pulled out samples were get filtered and analyzed at 269.5 nm using UV spectrophotometer. Ketoconazole concentration in each sample was determined using a previously plotted ketoconazole calibration curve [18]. Ketoconazole exhibits pH-dependent dissolution profile that increases with decreasing pH and to overcome this issue, it was formulated as a self-emulsifying drug delivery system. Thus to study the effect of pH on the drug release, multimedia dissolution testing was carried out at different pH conditions such as gastric condition (0.1 N HCl), gastric feeding state (pH 4.5 phosphate buffer) and intestinal conditions (pH 6.8 phosphate buffer) and the results were compared.

### *13.2.2.12 Drug content determination*

The drug was extracted from the optimized SMEDDS formulation using methanol after sonication and determined by UV spectrophotometer at 269.5 nm after suitable dilutions using the previously plotted calibration curve.

### *13.2.2.13 Freeze-thawing*

Thermodynamic stability of the final formulation was evaluated by using freeze-thawing study that was performed at different temperatures and subjected to freeze-thaw cycles at varying temperatures from −20°C, 1°C–4°C to 40°C for 1, 2, and 3 days, then they were serially diluted and centrifuged and observed its stability and phase separation [3].

# 13.3 Results and discussion

## 13.3.1 Solubility study

The solubility of ketoconazole in the water at 25°C was found to be 14.6 μg/mL. Apart from this aqueous solubility determination, the solubility study was also carried out in various oils, surfactants, and cosurfactants used in the formulation development. All the studies were performed in triplicate as shown in Fig. 13–1 and Table 13–2.

#### *13.3.1.1 Screening of oils*

Oil having the highest solubility for ketoconazole was selected for the formulation development purpose. The solubility study of ketoconazole in various oils was shown in Fig. 13–1.

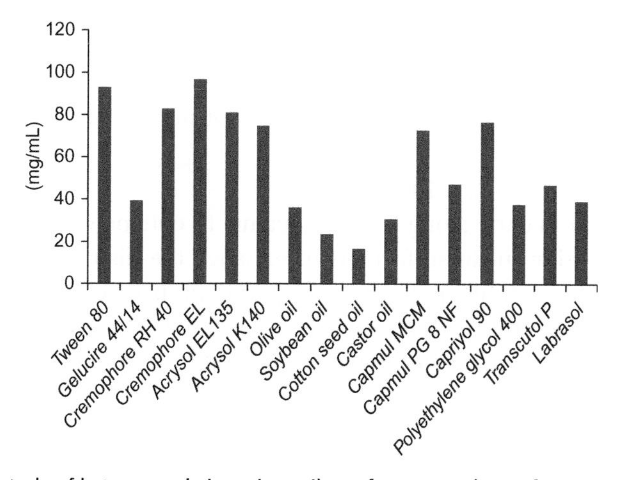


**FIGURE 13–1** Solubility study of ketoconazole in various oils, surfactants and cosurfactants.

**Table 13–2** Solubility study of ketoconazole in different vehicles.

| Sr. no. | Vehicle | Solubility (mg/mL) | SD |
|---|---|---|---|
| 1 | Tween 80 | 93.05 | 1.29 |
| 2 | Gelucire 44/14 | 39.34 | 1.27 |
| 3 | Cremophor RH 40 | 83.18 | 1.57 |
| 4 | Cremophor EL | 97.57 | 1.84 |
| 5 | Acrysol EL 135 | 81.90 | 1.40 |
| 6 | Acrysol K 140 | 75.23 | 2.36 |
| 7 | Olive oil | 36.71 | 1.85 |
| 8 | Soybean oil | 24.20 | 2.62 |
| 9 | Cotton seed oil | 17.41 | 1.54 |
| 10 | Castor oil | 31.38 | 1.09 |
| 11 | Capmul MCM | 73.19 | 1.41 |
| 12 | Capmul PG 8 NF | 47.20 | 1.36 |
| 13 | Capriyol 90 | 76.71 | 1.42 |
| 14 | Polyethylene glycol 400 | 38.27 | 1.37 |
| 15 | Transcutol P | 47.06 | 1.29 |
| 16 | Labrasol | 39.83 | 1.07 |

Among all the different oils screened, Capryol 90 (76.71 mg/mL) and Capmul MCM (73.19 mg/mL) showed the highest solubility for ketoconazole and due to this, they were selected for further studies.

#### 13.3.1.2 Screening of surfactants

Nonionic surfactants are most commonly used for the formulation of the SMEDDS due to its less toxic nature as compared with the ionic surfactants. In the present study, various surfactants (Acrysol EL 135, Acrysol K 140, Cremophor EL, Cremophor RH 40, and Tween 80) were screened to determine the solubility of the ketoconazole. From the different surfactants studied, Cremophor EL (97.57 mg/mL), Tween 80 (93.05 mg/mL), and Acrysol K 140 (75.23 mg/mL) showed highest solubility. Thus they were selected for formulation development.

#### 13.3.1.3 Screening of cosurfactants

Cosurfactants are used majorly along with surfactants to enhance the microemulsion region. They are used as an adjuvant to surfactants to enhance the dispersibility and dissolution of the drug. In the present study polyethylene glycol, propylene glycol, Transcutol P, and Labrasol were used to determine the solubility of ketoconazole. Labrasol (39.83 mg/mL) and Transcutol P (47.06 mg/mL) showed the highest solubility for ketoconazole, hence they were selected for further study.

### 13.3.2 Physicochemical compatibility of the drug to excipients

#### 13.3.2.1 Visual inspection

The visual inspection of ketoconazole in combination with different excipients was done for 3 months. The visual inspection shows that there was no change in color or odor indicating physical compatibility of ketoconazole with the excipients.

#### 13.3.2.2 Fourier transform-infrared spectroscopy

Results of the FTIR analysis are shown in Fig. 13–2. The IR spectrum of pure ketoconazole shows the characteristic peaks at 1643 $cm^{-1}$ indicate C = O stretching, 2997.48 $cm^{-1}$ indicate olefinic C–H stretch, 2962.76 $cm^{-1}$ show methyl C–H asymmetric stretching, 2880 $cm^{-1}$ for methyl and methylene C–H asymmetric stretching, 1451.48 and 1366.61 $cm^{-1}$ that indicate methyl asymmetric and symmetric bend, respectively. From the results, it is evident that the physical mixing of ketoconazole with the excipients used in the study, that is, S/CoS showed no major changes in the position of the characteristic peaks of ketoconazole indicating the ketoconazole–excipients compatibility (Fig. 13–3).

#### 13.3.2.3 Differential scanning calorimetry

DSC thermograms of the pure ketoconazole API, Aerosil 200, physical mixture and SMEDDS formulation were shown in Fig. 13–4. Pure ketoconazole showed a sharp endothermic peak at 153.12°C owing to its melting point while physical mixture also exhibited an endothermic peak at 152.64°C indicating that the drug is in crystalline form. Optimized SMEDDS

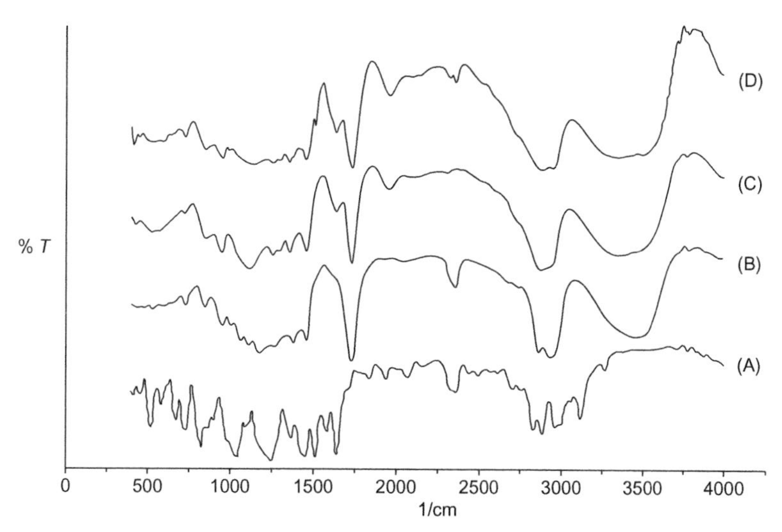


**FIGURE 13–2** FTIR spectra of (A) pure ketoconazole, (B) Aerosil 200, (C) physical mixture, and (D) ketoconazole SMEDDS. *FTIR*, Fourier transform-infrared; *SMEDDS*, self-microemulsifying drug delivery system.

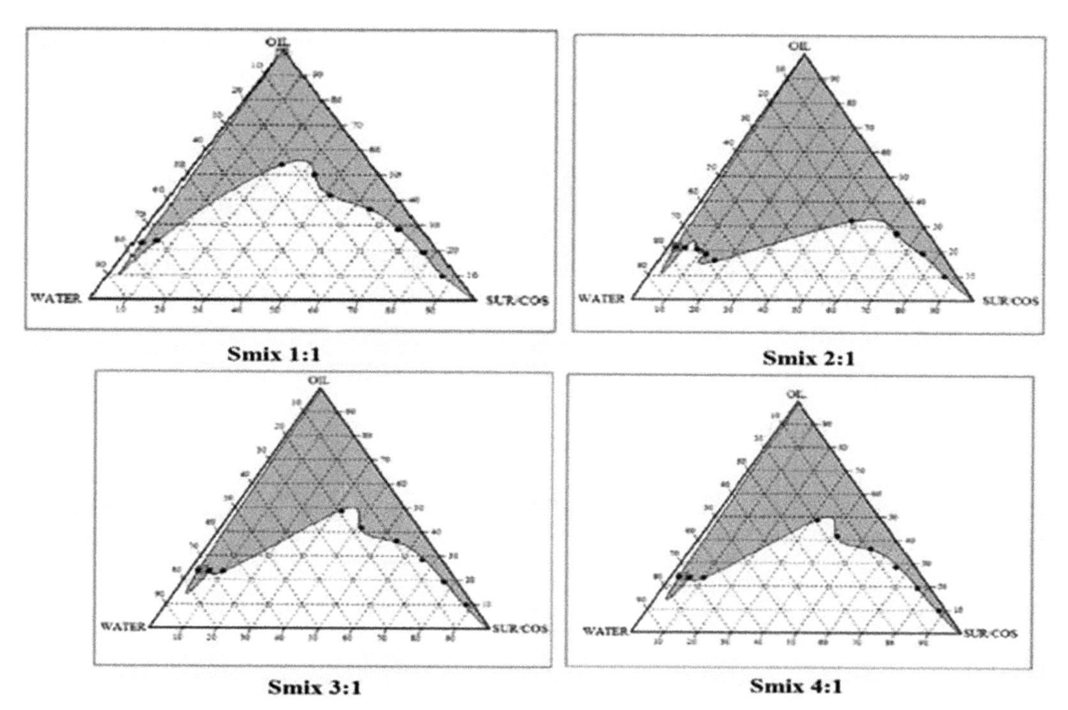


**FIGURE 13–3** Ternary phase diagrams with different SUR/COS ratio indicating microemulsion region. *SUR/COS*, Surfactant:cosurfactant.

formulation showed a smaller endothermic peak at 152.50°C indicating that the drug is transformed from crystalline to the amorphous form. There was no significant change in the positions of the endothermic peaks indicating no drug–excipients interaction that validates the ketoconazole–excipients compatibility.

#### 13.3.2.4 Self-emulsification study

Emulsification time is the crucial parameter in the optimization process of the SMEDDS and microemulsion formulation development. If the formulation is microemulsion, then this emulsification time must be less than 1 min. In the present study the emulsification time was determined as per the USP XXII using USP dissolution apparatus test apparatus II, that is, paddle apparatus. Each formulation (200 mg) was added dropwise into 500 mL of purified water at 37°C with gentle agitation at 50 RPM. The results are shown in Tables 13–3 and 13–6.

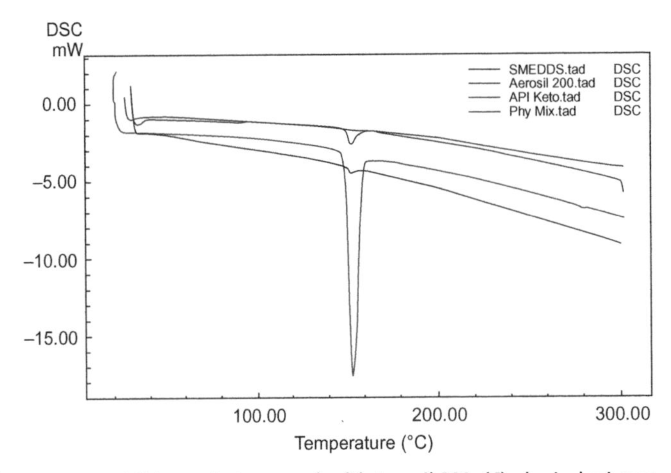


**FIGURE 13–4** DSC thermograms of (A) pure ketoconazole, (B) Aerosil 200, (C) physical mixture, and (D) Ketoconazole SMEDDS. *DSC*, Differential scanning calorimetry; *SMEDDS*, self-microemulsifying drug delivery system.

**Table 13–3** Emulsification efficiency of various surfactants.

| | Oils | | | | | | |
|---|---|---|---|---|---|---|---|
| **Surfactants** | **Olive oil** | **Soybean oil** | **Cotton seed oil** | **Castor oil** | **Capmul MCM C8** | **Capmul PG 8 NF** | **Capriyol 90** |
| Tween 80 | B | C | C | B | B | B | A |
| Gelucire 44/14 | B | C | D | B | C | C | B |
| Cremophor RH 40 | B | B | B | A | A | A | A |
| Cremophor EL | B | B | C | A | A | A | $A^{+}$ |
| Acrysol EL 135 | B | C | C | B | A | A | A |

From the results, it was observed that the formulations F1, F2, F3, and F4 were formed within the required time period so as to be declared as SMEDDS (Table 13–4).

### 13.3.2.5 Apparent partition coefficient studies

From the calculations, it was observed that the octanol–water partition coefficient, that is, log $P$ value for ketoconazole was 4.4, which implies that the ketoconazole will be having good solubility in lipophilic solvents (Table 13–5).

### 13.3.2.6 Construction of pseudo-ternary phase diagrams

The selection of oil and S/CoS and also the oil:S/CoS plays a crucial role in the formation of microemulsion [19]. In the present study the components of the optimized formulation were Capryol 90 as an oil phase, Chremophor EL as surfactant and Labrasol:Transcutol P in the

**Table 13–4** Emulsification study of surfactant and cosurfactant combinations with oil [surfactant/cosurfactant (Smix) 3:1 and Smix/Oil 3:1].

| | Oils | | | | | | |
|---|---|---|---|---|---|---|---|
| **Surfactant + cosurfactant (3:1)** | **Olive oil** | **Soybean oil** | **Cotton seed oil** | **Castor oil** | **Capmul MCM C8** | **Capmul PG 8 NF** | **Capriyol 90** |
| Tween 80 + PEG | B | C | C | B | B | B | A |
| Tween 80 + Transcutol P | B | C | C | B | B | B | A |
| Tween 80 + Labrasol | B | B | C | B | B | B | A |
| Tween 80 + (Transcutol P + Labrasol, 1:1) | B | B | B | B | A | B | A |
| Gelucire 44/14 + PEG | B | C | C | B | C | C | B |
| Gelucire 44/14 + Tanscutol P | B | C | C | B | B | C | B |
| Gelucire 44/14 + Labrasol | B | C | C | B | B | B | A |
| Gelucire 44/14 + (Transcutol P + Labrasol, 1:1) | B | B | B | B | A | A | A |
| Cremophor RH 40 + PEG | B | B | B | B | A | A | A |
| Cremophor RH 40 + Transcutol P | B | B | B | B | A | A | A |
| Cremophor RH 40 + Labrasol | B | B | B | B | A | A | A |
| Cremophor RH 40 + (Transcutol P + Labrasol, 1:1) | B | B | B | B | A | A | A |
| Cremophor EL + PEG | B | B | C | A | A | A | A |
| Cremophor EL + Transcutol P | B | B | B | A | A | A | A |
| Cremophor EL + Labrasol | B | B | B | A | A | A | A |
| Cremophor EL + (Transcutol P + Labrasol, 1:1) | B | B | B | A | A | A | $A^+$ |
| Acrysol EL 135 + PEG | B | C | C | B | A | A | A |
| Acrysol EL 135 + Transcutol P | B | B | B | A | A | A | A |
| Acrysol EL 135 + Labrasol | B | B | B | A | A | A | A |
| Acrysol EL 135 + (Transcutol P + Labrasol, 1:1) | B | B | B | A | A | A | A |

ratio of 1:1 as cosurfactants. Four different surfactants to cosurfactants ratios were screened as 1:1, 2:1, 3:1, and 4:1 and the phase diagrams were constructed after the titration. From the results and the ternary plots it was observed that the oil Capryol 90 gave a wider microemulsion region with 2:1 ratio of the surfactant to cosurfactant than remaining three ratios of 1:1, 3:1, and 4:1. When these mixtures were added into the aqueous media, they formed spontaneous self-emulsifying systems containing fine emulsion upon gentle agitation. From the various phase diagrams, it was observed that the microemulsion region was highest when the surfactant to cosurfactant ratio was 2:1 and oil to S/CoS ratio was 1:3. Hence, the microemulsion components were selected at these specified concentrations (Table 13–6).

#### *13.3.2.7 Effect of drug on the phase diagram*

The effect of ketoconazole on the phase diagram is shown in Table 13–7. Initially, the placebo phase diagram was optimized and in that ketoconazole was added in a stepwise manner and the resulting formulations in the form of fine emulsion were observed visually. The results showed that even after the addition of 200 mg of ketoconazole, the final formulation formed was transparent indicating that there was no significant effect of ketoconazole addition on the ternary plot (Fig. 13–5).

#### *13.3.2.8 Preparation of liquid self-microemulsifying drug delivery system*

Various formulations are prepared by following various combinations of oils as Capryol 90, Capmul MCM C8 and Capmul PG 8 NF; surfactants as Cremophor EL, Cremophor RH 40, Acrysol K 140; and cosurfactant as PEG 400, Propylene glycol, Labrasol, Transcutol

**Table 13–5** Composition of selected ketoconazole self-microemulsifying drug delivery system formulations.

| | | | Fraction in formulation (%) | | | |
|---|---|---|---|---|---|---|
| **Component** | **Selected component** | **Ketoconazole (mg)** | **F1** | **F2** | **F3** | **F4** |
| Oil | Capryol 90 | 200 | 9.83 | 12.43 | 15.16 | 17.90 |
| Surfactant | Cremophor EL | 200 | 48.33 | 46.93 | 45.47 | 44.00 |
| Cosurfactant | Labrasol + Transcutol P (1:1) | 200 | 41.84 | 40.64 | 39.37 | 38.10 |

**Table 13–6** Visual assessment of various formulations.

| Formulation code | Self-emulsification time (s) | Grade |
|---|---|---|
| F1 | 52 | $A^+$ |
| F2 | 30 | $A^+$ |
| F3 | 38 | $A^+$ |
| F4 | 47 | $A^+$ |

**Table 13–7** Effect of drug on the phase diagram.

| Amount of drug (mg/mL) | Visual inspection | Amount of drug (mg/2 mL) | Visual inspection |
|---|---|---|---|
| 20 | Transparent | 20 | Transparent |
| 40 | Transparent | 40 | Transparent |
| 60 | Transparent | 60 | Transparent |
| 80 | Transparent | 80 | Transparent |
| 100 | Transparent | 100 | Transparent |
| 120 | Turbid | 120 | Transparent |
| 140 | Turbid | 140 | Transparent |
| 160 | Turbid | 160 | Transparent |
| 180 | Turbid | 180 | Transparent |
| 200 | Turbid | 200 | Transparent |
| 220 | Turbid | 220 | Turbid |
| 240 | Turbid | 240 | Turbid |
| 260 | Turbid | 260 | Turbid |
| 280 | Turbid | 280 | Turbid |
| 300 | Turbid | 300 | Turbid |

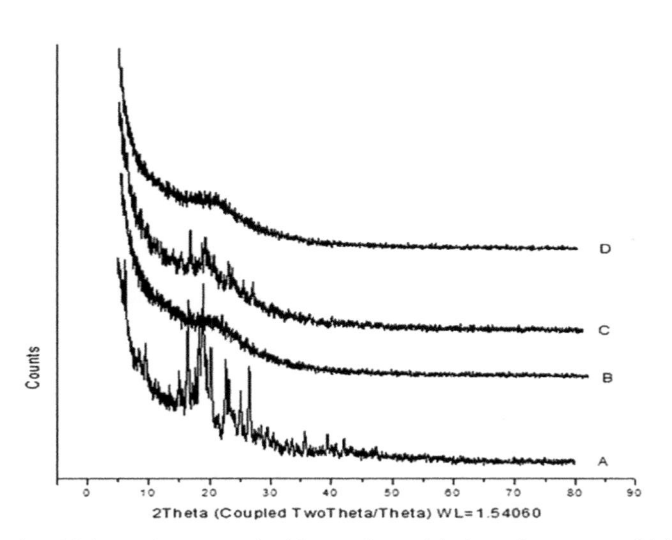


**FIGURE 13–5** XRD graphs of (A) pure ketoconazole, (B) Aerosil 200, (C) physical mixture, and (D) ketoconazole SMEDDS. *SMEDDS*, Self-microemulsifying drug delivery system; *XRD*, X-ray diffraction.

P. Ketoconazole loading in all the formulations was found to be more than 0.1% in the solubilized form. Developed liquid SMEDDS formulations were transparent and homogenous as shown in Table 13–5.

### *13.3.2.9 Characterization of liquid self-microemulsifying drug delivery system*

#### 13.3.2.9.1 Transmission test

This test investigates the clarity and transparency of the optimized microemulsion formulation wherein the transmittance of light from the formulation was checked spectrophotometrically at 650 nm. In this study the transmittance of the light was checked in the optimized formulations along with its serial dilutions as 50 times and 100 times dilution with water and with 0.1 N HCl. The results of this test are specified in Table 13–8, which clearly shows that the formulations F1, F2, F3, and F4 are clear and transparent. Also, from Figs. 13–6 and 13–7 it is evident that the results are similar even after diluting the formulation with 0.1 N HCl, which mimics the gastric environment indicating the stability of the formulation.

**Table 13–8** Percentage (%) Transmittance results of various ketoconazole self-microemulsifying drug delivery system formulations upon dilution with water as well as with 0.1 mol/L HCl.

| | % Transmittance | | | |
|---|---|---|---|---|
| **Formulation code** | **50 times dilution with water** | **100 times dilution with water** | **50 times dilution with 0.1 mol/L HCl** | **100 times dilution with 0.1 mol/L HCl** |
| F1 | 96.23 | 96.76 | 97.13 | 97.79 |
| F2 | 99.17 | 99.53 | 99.35 | 99.87 |
| F3 | 97.56 | 97.92 | 98.21 | 98.72 |
| F4 | 96.08 | 96.39 | 96.43 | 96.89 |

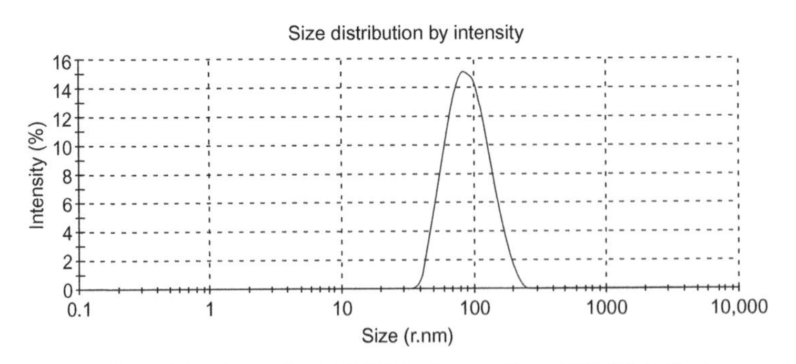


**FIGURE 13–6** Droplet size analysis of developed liquid SMEDDS formulation. *SMEDDS*, Self-microemulsifying drug delivery system.

Results

| | | | Size (r.nm): | % Intensity | Width (r.nm): |
|---|---|---|---|---|---|
| **Z-Average (r.nm):** | 81.36 | **Peak 1:** | 112.6 | 79.6 | 47.59 |
| **PdI:** | 0.492 | **Peak 2:** | 1864 | 11.4 | 587.3 |
| **Intercept:** | 0.952 | **Peak 3:** | 10.95 | 9.0 | 2.347 |
| **Result quality:** | **Refer to quality report** | | | | |

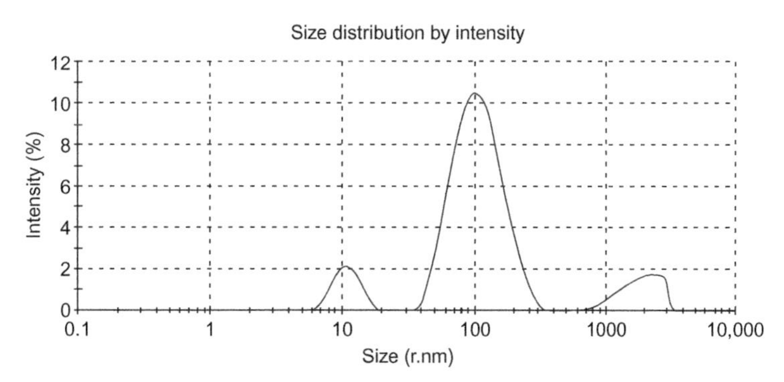


**FIGURE 13–7** Droplet size analysis of developed solid SMEDDS formulation after reconstitution. *SMEDDS*, Self-microemulsifying drug delivery system.

#### 13.3.2.9.2 Globule size and zeta potential analysis

Average globule size of optimized liquid SMEDDS formulation of ketoconazole was found to be 83.53 nm and the polydispersity index was 0.175 indicating monodisperse system. Results of the zeta potential analysis are shown in Table 13–8. As the surfactants used for the formulation development were nonionic in nature, the zeta potential values of the developed formulations after 100 times dilution were found to be in between 0.01 and 0.1 mV (Figs. 13–8 and 13–9; Table 13–9).

### *13.3.2.10 Preparation of solid self-microemulsifying drug delivery system*

Liquid SMEDDS formulation of ketoconazole was mix into the solid SMEDDS with the help of spray drying technology and Aerosil 200 as an inert carrier. The process yield of spray drying was found to be in the range of 65%–70%. Results of spray drying are given in Table 13–10.

### *13.3.2.11 Characterization of solid self-microemulsifying drug delivery system*

#### 13.3.2.11.1 Differential scanning calorimetry

Fig. 13–4 shows the DSC thermograms of the pure ketoconazole, Aerosil 200, physical mixture of ketoconazole and Aerosil 200 and the optimized SMEDDS formulation. Thermogram of ketoconazole showed a sharp endothermic peak at 153.12°C owing to its melting point and it also confirmed the crystalline nature. Aerosil 200 did not exhibit any peak at all indicating highly amorphous nature. Physical mixture exhibited a sharp endothermic peak at 152.64°C owing to the melting point of the ketoconazole and its crystalline nature. The optimized SMEDDS formulation of ketoconazole showed a small endothermic peak at 152.50°C

**Results**

| | | Mean (mV) | Area (%) | Width (mV) |
|---|---|---|---|---|
| **Zeta potential (mV):** 0.0190 | **Peak 1:** | 0.0190 | 100.0 | 3.49 |
| **Zeta deviation (mV):** 3.49 | **Peak 2:** | 0.00 | 0.0 | 0.00 |
| **Conductivity (mS/cm):** 2.59e–4 | **Peak 3:** | 0.00 | 0.0 | 0.00 |

**Result quality: See result quality report**

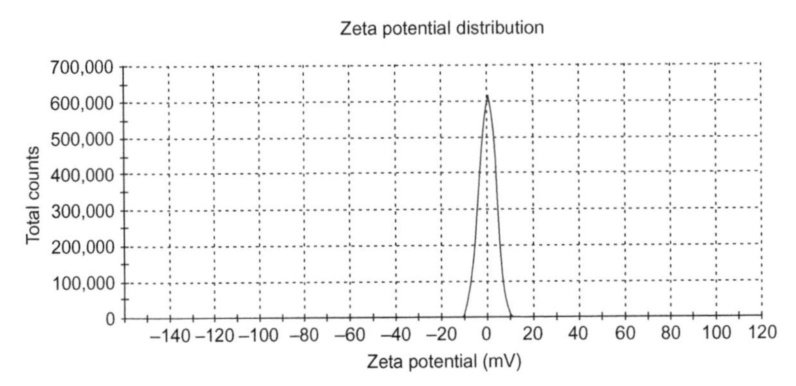


**FIGURE 13–8** Zeta potential analysis of developed liquid SMEDDS formulation. *SMEDDS*, Self-microemulsifying drug delivery system.

**Results**

| | | Mean (mV) | Area (%) | Width (mV) |
|---|---|---|---|---|
| **Zeta potential (mV):** 0.0632 | **Peak 1:** | 0.0632 | 100.0 | 3.73 |
| **Zeta deviation (mV):** 3.73 | **Peak 2:** | 0.00 | 0.0 | 0.00 |
| **Conductivity (mS/cm):** 1.69e–4 | **Peak 3:** | 0.00 | 0.0 | 0.00 |

**Result quality: See result quality report**

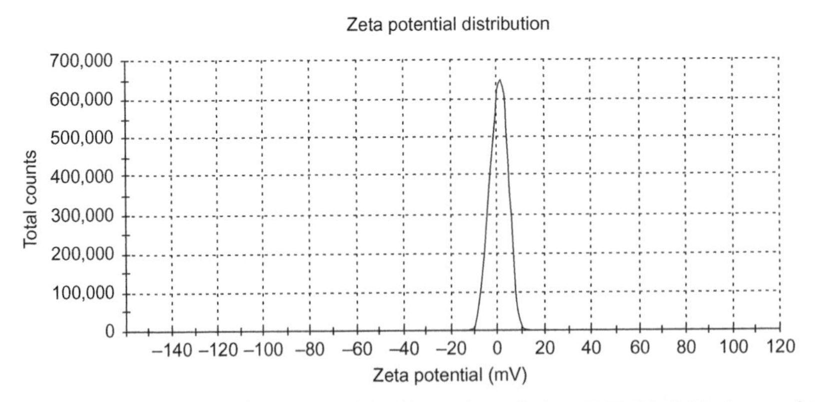


**FIGURE 13–9** Zeta potential analysis of developed solid SMEDDS formulation. *SMEDDS*, Self-microemulsifying drug delivery system.

**Table 13–9** Droplet size, polydispersity index, and zeta potential of self-microemulsifying drug delivery system formulations.

| Formulation code | Droplet size (nm) | PDI | Zeta potential |
|---|---|---|---|
| F1 | 81.36 | 0.492 | 0.0632 |
| F2 | 83.53 | 0.175 | 0.0190 |
| F3 | 156.23 | 0.649 | 0.0781 |
| F4 | 949.4 | 1 | 0.126 |

*PDI*, Polydispersity index.

**Table 13–10** Spray drying yield.

| Formulation code | % Yield | Drug content (mean) ± SD (mg/g) |
|---|---|---|
| F1 | 74.25 | 69.24 |
| F2 | 77.50 | 73.89 |
| F3 | 68.25 | 63.51 |
| F4 | 70.00 | 67.39 |

indicating inhibition of the crystalline nature of ketoconazole. From these results, it was evident that maximum drug was either became amorphous or was present in the solubilized form.

#### 13.3.2.11.2 X-ray diffraction study

XRD graphs of the pure ketoconazole, Aerosil 200, physical mixture, and SMEDDS formulation were shown in Fig. 13–5, which showed 76.7%, 39.5%, 77.1%, and 41.9% crystallinity, respectively. The fall in the crystallinity from 76.7% to 41.9% was attributed to either the amorphization or the solubilization of the ketoconazole, which shows the enhanced efficacy of the novel developed formulation.

#### 13.3.2.11.3 Surface morphological study

Scanning electron micrographs of ketoconazole, Aerosil 200, and final optimized formulation of ketoconazole SMEDDS were shown in Fig. 13–10. FE-SEM image of ketoconazole reflects its crystalline nature. Aerosil 200 also showed rough surfaces. SMEDDS showed well-separated particles without agglomerations. Rough surfaces of Aerosil 200 and sharp edges of ketoconazole were disappeared showing that the drug was solubilized in lipid vehicle, which was adsorbed evenly on the carrier.

### *13.3.2.12 In vitro multimedia drug release study*

0.1 N HCl was selected as dissolution media for in vitro drug release pattern based on the solubility of the maximum dose of ketoconazole (which is 200 mg) in 1 L of media [18]. Further, the drug release from novel optimized solid SMEDDS formulation was compared

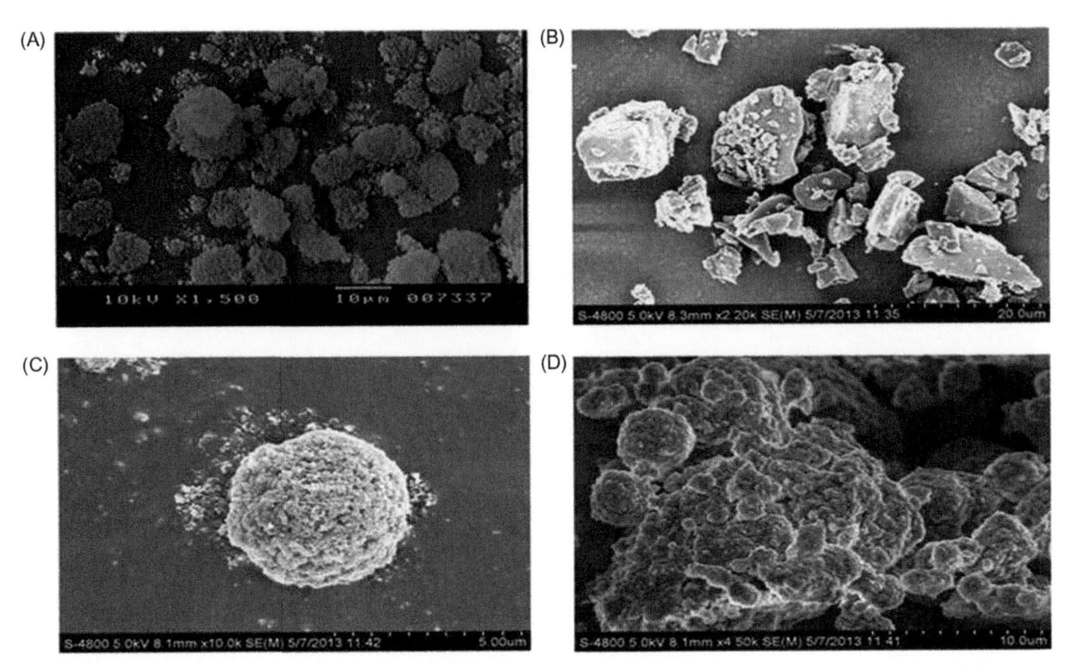


**FIGURE 13–10** Images of field emission microscopy of (A) Aerosil 200, (B) ketoconazole API, (C) single particle of developed spray dried solid SMEDDS formulation of ketoconazole, and (D) developed spray dried solid SMEDDS formulation of ketoconazole. *SMEDDS*, Self-microemulsifying drug delivery system.

with pure ketoconazole (i.e., intrinsic dissolution) and marketed formulation as shown in Fig. 13–11. These results clearly showed that the significantly higher drug release was observed from novel SMEDDS formulation than the pure ketoconazole and marketed formulation. It was also observed that ketoconazole showed pH-dependent release. As pH increases the dissolution of ketoconazole decreases, whereas the developed SMEDDS formulation showed pH-independent drug dissolution, which overcomes the problem associated with native ketoconazole and portraying the effectiveness of the novel formulation.

### *13.3.2.13 Drug content determination*

The solid SMEDDS formulation was reconstituted with water and the drug content was measured spectrophotometrically. It was observed that the ketoconazole content was in the range of 65–70 mg/g. All the studies were performed in triplicate.

### *13.3.2.14 Freeze-thawing*

There was no phase separation observed in the optimized formulation after the freeze-thaw cycling with respect to the specified temperature conditions and time indicating the overall stability and the robustness of the novel ketoconazole formulation.

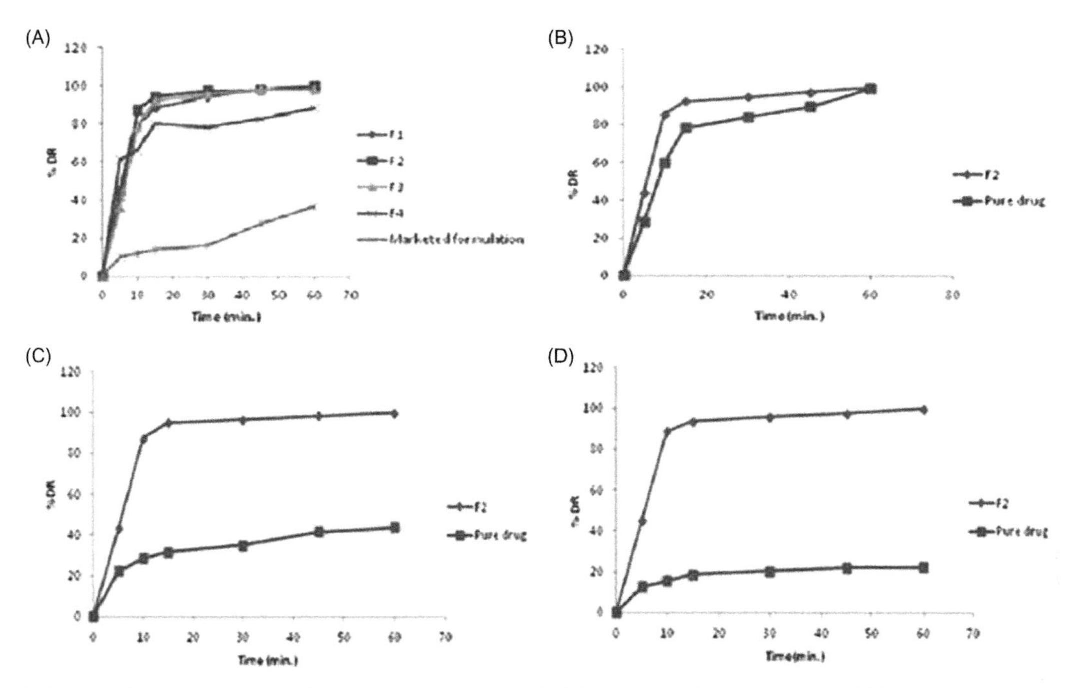


**FIGURE 13–11** Percentage cumulative drug release (% DR) of (A) developed ketoconazole SMEDDS formulations with marketed formulation, (B) optimized ketoconazole SMEDDS formulation with pure ketoconazole at 0.1 N HCl, (C) optimized ketoconazole SMEDDS formulation with pure ketoconazole at pH 4.5 phosphate buffer, and (D) optimized ketoconazole SMEDDS formulation with pure ketoconazole at pH 6.8 phosphate buffer. *SMEDDS*, Self-microemulsifying drug delivery system.

## 13.4 Conclusion

The solid SMEDDS of ketoconazole was developed using Capryol 90 as an oil phase, Cremophor EL as a surfactant, Labrasol:Transcutol P in the ratio of 1:1 as cosurfactants, and Aerosil 200 as an inert carrier for spray drying. Pure ketoconazole is a highly water-insoluble drug with a log $P$ value of 4.4. The optimized ketoconazole formulation showed enhanced solubility as the drug was present in the solubilized form, which was shown by the DSC and XRD analysis. FTIR and DSC analysis exhibited physicochemical compatibility among ketoconazole and other excipients used for the formulation development. In vitro drug dissolution studies indicated that SMEDDS has higher drug release rate as compared to pure ketoconazole and marketed product. Also, from these results it is evident that the problem of pH-dependent drug dissolution of ketoconazole was solved successfully. Droplet size analysis exhibited the uniform dispersion of globules throughout the medium. Zeta potential measurements showed that the formulated SMEDDS was physically stable. FE-SEM images exhibited that the crystalline nature of ketoconazole was transformed into a solubilized amorphous form. Thus the present study illustrated the potential use of SMEDDS to effectively deliver the poorly water-soluble drugs by oral route. It should be noted that maximum of 100 mg/mL drug loading was obtained in the present studies. Therapeutic dose of ketoconazole is 200 mg. So, 2 mL of vehicle (placebo formulation) was used. For transforming 200 mg/2 mL into solid form quantity of Aerosil 200 required was nearly 1 g. Final bulk of formulation became nearly 3 g. This implies that the final dose of ketoconazole SMEDDS was 200 mg/3 g, which was too bulky to incorporate into the capsule. Therapeutic dose of ketoconazole can be administered via capsule in multiple doses. Hence, further studies need to be carried out to increase the ketoconazole loading into SMEDDS using different excipient combinations.

# Acknowledgment

Authors are thankful to the Director of the University Institute of Chemical Technology (UICT), North Maharashtra University (NMU) Jalgaon, Maharashtra, India for providing a platform to carry out the research work.

# Conflict of interest

Authors have no conflict of interest regarding this article.

# References

[1] V. Jannin, et al., Approaches for the development of solid and semi-solid lipid-based formulations, Adv. Drug Deliv. Rev. 60 (6) (2008) 734–746.

[2] M. Shui, J. Andrew, J. Christopher, Int. J. Pharm. 167 (1998) 155–164.

[3] C.W. Pouton, Lipid formulations for oral administration of drugs: non-emulsifying, self-emulsifying and self-microemulsifying drug delivery systems, Eur. J. Pharm. Sci. 11 (2000) S93–S98.

[4] A.J. Canilo Mufloz, C. Tur, J. Torres, J. Antimicrob. Chemother. 37 (1996) 815.

[5] E.J. Elder, J.C. Evans, B.D. Scherzer, J.E. Hitt, G.B. Kupperblatt, S.A. Saghir, et al., Drug Dev. Ind. Pharm. 33 (7) (2007) 755.

[6] T. Virmanil, N. Parvez, S. Yadav, K. Pathak, Solid inclusion complexes of class II imidazole derivative with β-cyclodextrin, Continental J. Pharm. Sci. 1 (2007) 1–8.

[7] G. Balata, M. Mahdi, R.A. Bakera, Improvement of solubility and dissolution properties of ketoconazole by solid dispersions and inclusion complexes, Asian J. Pharm. Sci. 5 (1) (2010) 1–12.

[8] A.H. Hosmani, Y.S. Thorat, Synthesis and evaluation of nanostructured particles of salt of ketoconazole for solubility enhancement, Dig. J. Nanomater. Biostruct. 6 (3) (2011) 1411–1418.

[9] R. Zhou, P. Moench, C. Heran, X. Lu, N. Mathias, T.N. Faria, et al., pH-dependent dissolution *in vitro* and absorption *in vivo* of weakly basic drugs: development of a canine model, Pharm. Res. 22 (2) (2005).

[10] S.D. Mandawgade, S. Sharma, S. Pathak, V.B. Patravale, Int. J. Pharm. 362 (2008) 179–183.

[11] A.A. Kassem, M.A. Marzouk, A.A. Ammar, G.H. Elosaily, Preparation and in vitro evaluation of self-nanoemulsifying drug delivery systems (SNEDDS) containing clotrimazole, Drug Discov. Ther. 4 (5) (2010) 373–379.

[12] A.K. Meena, et al., Formulation development of an albendazole self-emulsifying drug delivery system (SEDDS) with enhanced systemic exposure, Acta Pharm. 62 (2012) 563–580.

[13] L. Wei, J. Li, L. Guo, et al., Preparation and evaluation of SEDDS and SMEDDS containing Carvedilol, Drug Dev. Ind. Pharm. 31 (2005) 785–794.

[14] H. Shen, M. Zhong, et al., Preparation and evaluation of self-microemulsifying drug delivery systems (SMEDDS) containing atorvastatin, J. Pharm. Pharmacol. 58 (2006) 1183–1191.

[15] M.J. Lawrence, G.D. Rees, Emulsion based media as a novel delivery system, Adv. Drug Deliv. Rev. 45 (2000) 89–121.

[16] Y. Zhaoa, C. Wanga, A.H.L. Chowb, K. Renc, T. Gongc, Z. Zhangc, et al., Self-nanoemulsifying drug delivery system (SNEDDS) for oral delivery of Zedoary essential oil: Formulation and bioavailability studies, Int. J. Pharm. 383 (2010) 170–177.

[17] D.H. Oh, et al., Int. J. Pharm. 420 (2011) 412–418.

[18] USP 30-NF25, Centre for Drug Evaluation and Research Package for Application No 21-204, 2007.

[19] M.J. Groves, The self-emulsifying action of mixed surfactants in oil, Acta Pharm. Suec. 13 (1976) 361–372.

# Molecular recognition, selective targeting, and overcoming gastrointestinal digestion by folic acid–functionalized oral delivery systems in colon cancer

Pallab Kumar Borah, Raj Kumar Duary

*DEPARTMENT OF FOOD ENGINEERING AND TECHNOLOGY, SCHOOL OF ENGINEERING, TEZPUR UNIVERSITY, TEZPUR, INDIA*

## Chapter outline

**Nomenclature** ..... **234**
**14.1 Introduction** ..... **234**
**14.2 Structure and function of the folate receptor** ..... **235**
**14.3 Expression of folate receptor in normal and malignant tissues** ..... **236**
**14.4 Folic acid–functionalized uptake of oral delivery systems via folate receptor–mediated endocytosis** ..... **237**
**14.5 Functionalization of folic acid on oral delivery systems** ..... **238**
14.5.1 Effect of folic acid functionalization on hydrodynamic size of systems ..... 239
14.5.2 Effect of folic acid functionalization on surface charge of systems ..... 240
**14.6 Folic acid–functionalized systems for colon cancer** ..... **240**
14.6.1 Colon cancer cell selectivity of folic acid–functionalized systems in vitro ..... 241
14.6.2 Colon tumor selectivity of folic acid–functionalized systems in vivo ..... 241
**14.7 Oral delivery systems and gastrointestinal digestion** ..... **245**
**14.8 Conclusion** ..... **249**
**Acknowledgments** ..... **250**
**References** ..... **250**

Encapsulation of Active Molecules and their Delivery System. DOI: https://doi.org/10.1016/B978-0-12-819363-1.00014-4

# Nomenclature

| | |
|---|---|
| **FR** | folate receptor |
| $K_a$ | binding affinity |
| **PDB** | protein data bank |
| **FOLR** | human folate receptor |
| **DNA** | deoxyribonucleic acid |
| **RNA** | ribonucleic acid |
| **DCC** | 1,3-dicyclohexylcarbodiimide |
| **NHS** | 1-hydroxysuccinimide |
| **DMAP** | 4-dimethylaminopyridine |
| **EDC** | 1-ethyl-3-(3-dimethylaminopropyl) carbodiimide |
| **PLGA** | poly(lactic-*co*-glycolic acid) |
| **PEG** | polyethylene glycol |
| **FITC** | fluorescein isothiocyanate |
| **c.** | approximately |
| **cf.** | compare |
| **nm** | nanometer |
| **mV** | millivolt |
| **M** | molar |
| **mM** | millimolar |

# 14.1 Introduction

Colon cancer very often referred to as colorectal or large bowel cancer makes reference to cancerous growth in the colon, rectum, or cecum. A total of 862,000 deaths related to colon cancer are reported worldwide each year, making it currently the second most causal form of cancer death globally [1]. Colon cancer–related deaths are clearly on the rise as evidenced by the World Health Organization reports of 639,000 deaths in the year 2009 [2]. This is roughly a 20% increment in the cases of deaths in a period of 5 years. Current therapeutic strategies rely on anticancer formulations that are unable to distinguish amidst normal versus cancerous cells accurately. This leads to adverse toxicity on the one hand and, on the other, dramatically limits sufficient anticancer formulation buildup in the cancerous tumor. As such, a considerable effort has been dedicated to the development of targeted therapies using different targeted nanomaterials.

A promising targeted therapeutic avenue may be the utilization of the water-soluble vitamin B9, folic acid (an oxidized form of folate), which can achieve selective targeting and molecular recognition of colon cancer cells. Such selective targeting and molecular recognition exist because the survival and proliferation in colon cancer cells are dependent on a cell's ability to acquire the folic acid. The vitamin is indispensable for the biosynthesis of nucleotide bases, therefore, is ingested by proliferating cells in elevated quantities [3]. The higher uptake of folic acid is achieved by specialized cell surface receptors known as folate receptors (FRs), which colon cancer cells overexpress [4]. This effectively discerns the receptor as a prospective target for folic acid–functionalized oral delivery systems. FRs may be further competent as an "orally" delivered tumor-specific target because the receptors are expressed particularly on the apical (facing externally) surfaces of the colonic epithelial membrane, more so than the accessibility to FR-directed delivery systems that are blood-borne and transferred along with the plasma. It is of note that, with the transformation of the epithelia and subsequent loss in the polarity of the cells, targeted systems in plasma can also readily approach the FRs [5].

In this chapter, we have summarized the folic acid–functionalized oral delivery systems as a receptor-targeted system. Further, we have identified areas of noteworthy success in folic acid–assisted targeting of colon cancer both in vitro and in vivo and also pointed out the gastrointestinal obstacles that must be triumph over before the full potential of tumor targeting implementations of folic acid can be realized. First, we will discuss the structural and functional aspects of the FR, techniques of folic acid functionalization on oral delivery systems, their effects on the colloidal aspects (hydrodynamic size and charge) of the systems, and their cancer cell targeting capability, as it sets the stage to further the discussion on the gastrointestinal hurdles (such as gastrointestinal digestion and corona formation) and ways to surmount it using folic acid.

## 14.2 Structure and function of the folate receptor

The FR-$\alpha$ (Fig. 14–1A), FR-$\beta$ (Fig. 14–1B), and FR-$\gamma/\gamma'$ comprise the characterized family of human FRs with a molar mass of c. 38 kDa. These receptors are c. 80% identical in their amino acid sequence similarity (Fig. 14–1C) yet comprise varied expression patterns [8]. A glycophospholipid (glycosylphosphatidylinositol) anchor is responsible for the attachment of FR-$\alpha$ and FR-$\beta$ to the cellular membrane surface [8]. A distinctive feature of distinction between FR-$\alpha$ ($K_d \sim 0.1$ nM) and FR-$\beta$ ($K_d \sim 1$ nM) lay in the high-affinity binding of the former to folic acid [9–11]. In addition, FR-$\alpha$ has contrasting stereospecificity for the reduced folate [12]. On the other hand, FR-$\gamma/\gamma'$ (truncated protein) is a soluble form of the receptor and is devoid of the glycophospholipid anchor for membrane anchoring. FR-$\gamma/\gamma'$ is believed to be lowly expressed as it is scarcely detectable [8]. It is reported to have a binding affinity of $K_d \sim 0.4$ nM for folic acid [13]. Ref. [14] has reported of an FR-$\delta$ variant gene in humans (on chromosome 11q14). The authors predicted a 27.7 kDa protein comprising 241 amino acids; however, gene expression was not observed in 59 adults and embryonic tissue sources. This, according to them, should only suggest a spatial and temporal pattern of expression, which is highly restricted. The authors also suggest that it could be a spliced variant or a pseudogene.

Other receptors involved in intestinal folate absorption, such as the reduced folate carrier and the proton-coupled folate transporter, have also been reported [15,16]. The reduced folate carrier protein is expressed in intestinal epithelial cells (apical membrane domains) and functions optimally at pH c. 7.4 [15], whereas the proton-coupled folate transporter is expressed primarily in the proximal regions of the small intestine (with very limited expression in the distal regions of the small intestine and colon) and functions optimally at acidic

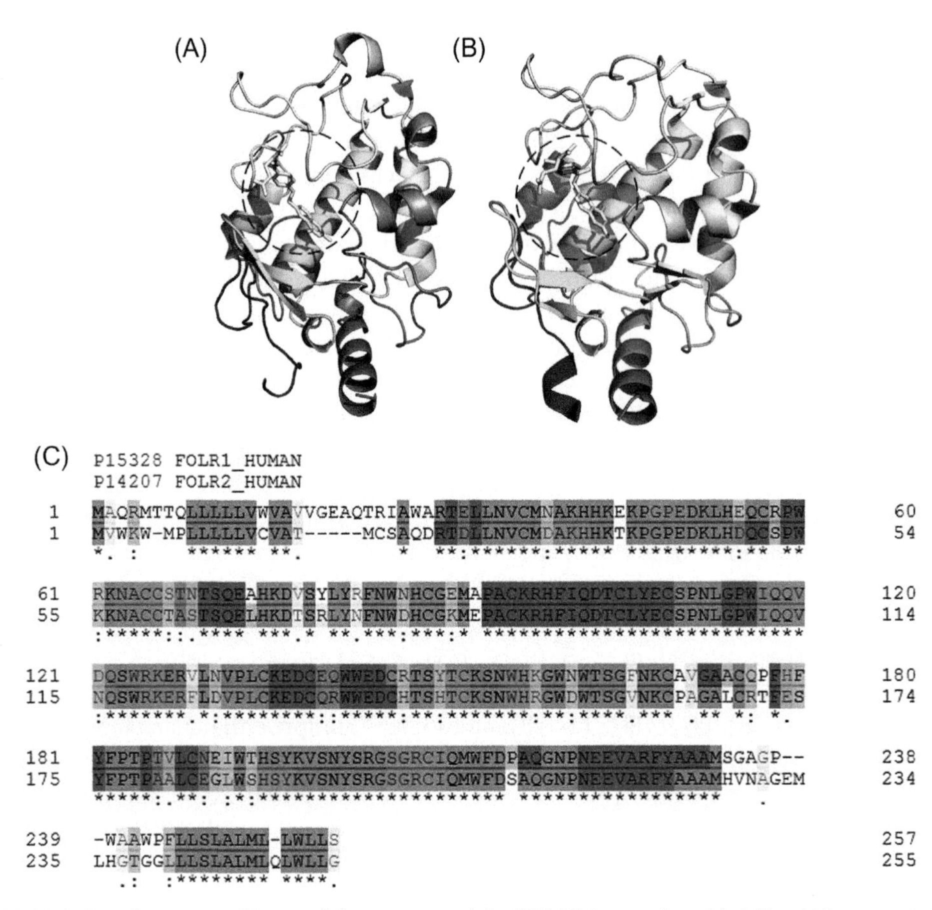


**FIGURE 14–1** Crystal structure of human folate receptor alpha (FOLR1) in complex with folic acid (represented as a stick model), RCSB PDB ID: 4LRH [6] (A), human folate receptor beta (FOLR2) in complex with the antifolate methotrexate (represented as a stick model), RCSB PDB ID: 4KN0 [7] (B) and amino acid sequence alignment between FOLR1 and FOLR2 performed in UniProt (https://www.uniprot.org/align). The asterisk indicates positions that consist of single and fully conserved residues, the colon indicates conserved residues with strongly similar properties, and the period indicates conserved residues with weakly similar properties (C).

pH [16]. Since they are particularly unrelated to the colonic tissue, we will only discuss the FRs in this chapter. Readers seeking more information on the reduced folate carrier and the proton-coupled folate transporter are directed toward excellent reviews by Said [17,18].

## 14.3 Expression of folate receptor in normal and malignant tissues

The expressions of the various isoforms of the FRs are reported by various researchers [8,19] to be dependent on (1) differentiation and (2) tissue specification. In comparison to noncancerous cells, the expression of FR-α is quite strikingly elevated (100–300 times) in cancerous cells of epithelial origin, particularly the breast, lung, kidney, colon, and brain [19–21]. However, some exception includes normal epithelia in placenta, kidneys, and the choroid plexus, where FRs are overexpressed. Folic acid is indispensable for the one-carbon transfer reactions, which are necessitated in the synthesis and replication of DNA (an essential step for cell division, growth, and survival) and the biosynthesis of *S*-adenosylmethione that is indispensable for methylation of RNA, DNA, proteins, and phospholipids [22]. An elevated requirement of folic acid, particularly in rapidly dividing cancer cells, has a direct correlation with the expression of FRs. The latter are incrementally overexpressed as colon cancer stages advance [4,23]. It is of note that researchers [24,25] have reported that this overexpression correlated with failure to respond to chemotherapy in ovarian and breast cancers. Further exploration is necessary to establish this for colon cancers if indeed, there exists a relationship. However, for colon cancers that are at an advanced stage and/or resistant to standard chemotherapy, folic acid–functionalized oral delivery systems may serve as an effective alternate strategy.

On the other hand, FR-β is known to be overexpressed in malignancies associated with nonepithelial origin (leukemias and sarcomas) [8,26] and in activated (but not resting) macrophages [27]. de Boer et al. [28] also reported that the expression of FR-β was negligible in colorectal tumor samples. Also, FR-γ and FR-γ′ are expressed at shallow levels in hematopoietic tissues (such as lymphoid cells) [8]. In addition, Yamaguchi et al. [29] reported that FR-γ and FR-γ′ receptors are not expressed in colorectal adenocarcinoma cells. Hence, in this chapter, further discussion will surround the FR-α. The FR-α will be referred to as the folate receptor from here onward if otherwise stated.

## 14.4 Folic acid–functionalized uptake of oral delivery systems via folate receptor–mediated endocytosis

Folic acid–functionalized oral delivery systems are known to be nondestructively taken into mammalian cells by process of FR-mediated endocytosis (Fig. 14–2) [30,31], although the exact mechanism is quite nebulous. Notwithstanding, disagreement exists in the pathways involved in the endocytosis of FRs anchored to glycophospholipid [32,33]. Researchers have claimed that these glycophospholipid are arranged on lipid rafts in the cell membrane (from the foundational concepts of the fluid mosaic model) that are FR affluent and measure c. 70 nm in diameter [34,35]. Researchers have also therefore explored avenues in patchy “folic acid–presenting” nanoparticles. For example, Poon et al. [36,37] describes nanoparticles and micelles formed of linear dendritic polymers conjugated with folic acid using 1,3-dicyclohexylcarbodiimide/1-hydroxysuccinimide (DCC/NHS) chemistries. The system carried the anticancer drug paclitaxel to in vitro KB (FR -positive subline of the ubiquitous KERATIN-forming tumor cell line HeLa) and in vivo to nude mice induced with a KB xenograft. Negligible uptake was observed in A375 (FR-negative human melanoma cell line) cells and nude mice induced with an A375 xenograft. The authors reported that the pattern of the

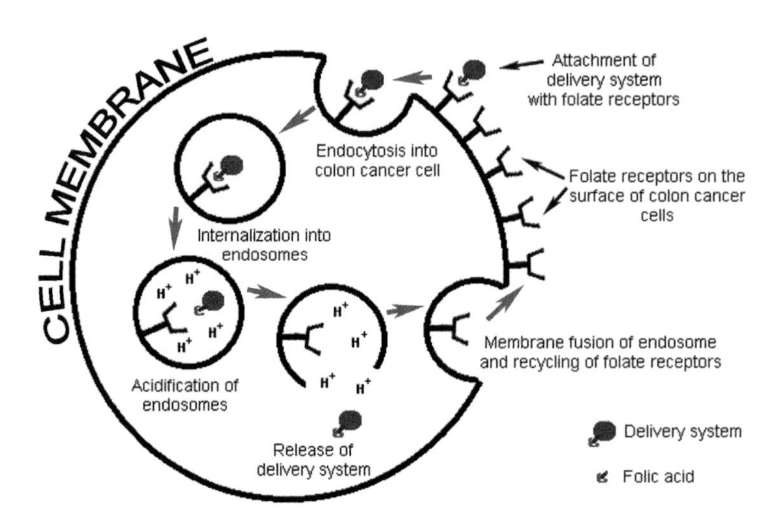


**FIGURE 14–2** Delivery mechanism of folic acid–functionalized oral delivery systems to folate receptor–positive colon cancer cells. The figure demonstrates the attachment of folic acid–functionalized oral delivery system with folate receptors, endocytosis into colon cancer cells, internalization into endosomes, acidification of the endosomes leading to the dissociation of folic acid from the folate receptor, release of the delivery systems inside the cell, and membrane fusion of endosomes to the cell membrane leading to subsequent recycling of the folate receptors back to the apical cell surface.

surfaces of nanoparticles with variable sized folic acid presenting clusters in different spatial arrangements impacted the intracellular uptake both in vitro and in vivo, although this hypothesis needs further exploration and establishment in colon cancer cells and xenografts. Nevertheless, folic acid conjugates transport beyond the cell membrane of mammalian cells into the cytoplasm via the endocytosis pathway aided by FRs as described more in detail by Ref. [21]. The mechanism of FR-mediated endocytosis is often described as a continuous process, with folic acid conjugates (regardless of size) are internalized at a rate of c. $1-2 \times 10^5$ molecules/cell/h [5]. The optimal binding pH of folic acid to FRs for such internalization is reported to be c. $\sim 7.6$, and the binding is known to decrease substantially at pH c. $<5$ and $>9$ [10]. Continuing from here, folic acid conjugates are internalized into endosomes that have a degenerating acidic pH $\sim 4.3-5.0$ [38]. Since folic acid-binding to FRs dramatically decreases at pH c. $<5$, it seems quite natural that the dissociation of folic acid conjugates (inside the endosomes) from the folate receptors should now proceed. The FRs are hypothesized to recycle back to the cell surface soon after this. This is illustrated in Fig. 14–2.

## 14.5 Functionalization of folic acid on oral delivery systems

The conjugation of folic acid to oral delivery systems (macromolecular systems comprising proteins, polysaccharides, silica, etc.) is usually achieved by chemical modification reactions such as DCC, NHS, 4-dimethylaminopyridine (DMAP), and 1-ethyl-3-(3-dimethylaminopropyl) carbodiimide (EDC)–mediated activation of folic acid. A wide variety of oral delivery systems have been prepared using these conjugation chemistries. For example, Ref. [39] developed cationic amphiphilic starch nanoparticles that were surface-functionalized with bovine serum albumin–conjugated folic acid (using EDC/NHS conjugation chemistries); Ref. [40] developed folic acid–polyethylene glycol-surface-functionalized starch nanoparticles (using DCC/NHS conjugation chemistries); Ref. [41] developed folic acid–conjugated pullulan nanoparticles (using DCC/DMAP conjugation chemistries, also known as Steglich esterification); Ref. [42] developed folic acid–functionalized hydroxyethyl starch nanocapsules (using EDC conjugation chemistries); etc.

### 14.5.1 Effect of folic acid functionalization on hydrodynamic size of systems

The hydrodynamic size of oral delivery systems is one of the paramount parameters to estimate their interaction with mammalian cells. A consensus exists that smaller nano-sized delivery systems are more readily absorbed or endocytosed by mammalian cells [43]. Ref. [44] also reported that systems c. <50 nm demonstrates a higher quantification of cellular uptake in comparison to larger systems. Also, solid lipid nanoparticles of c. 100 nm demonstrated a 2.5-fold higher uptake in comparison to c. 1000 nm systems by Caco-2 (human epithelial colorectal adenocarcinoma) cells [44]. It is reported that the conjugation with folic acid results in modification of the colloidal hydrodynamic size of delivery systems. For example, Ref. [39] reported the synthesis of c. <100 nm cationic amphiphilic starch nanoparticles surface-functionalized with bovine serum albumin–conjugated folic acid. It is interesting to note that the basal cationic amphiphilic starch nanoparticles had a hydrodynamic diameter of c. 83 nm. Similar observations were made by Ref. [40]. The authors developed folic acid surface–functionalized starch nanoparticles via a water-in-oil microemulsion templating method. It was interesting to note that upon esterification of the nanoparticles with folic acid, the authors reported an increase in hydrodynamic diameter (from c. 50 to 130 nm). A similar observation was made by Ref. [41], when folic acid was conjugated to pullulan nanoparticles. The authors reported that the hydrodynamic diameter of the nanoparticles increased from c. 185 to 261 nm upon folic acid conjugation. In another instance, Ref. [42] reported an increase of hydrodynamic diameter from c. 275 to 307 nm in hydroxyethyl starch nanocapsules upon folic acid conjugation. Ref. [45] developed folic acid surface–functionalized guar gum nanoparticles (using DCC/DMAP conjugation chemistry) charged with methotrexate. The authors reported that folic acid functionalization was seen to increase the colloidal hydrodynamic diameter of the system from c. 307 to 325 nm. From the previous observations, it is quite evident that the surface functionalization leads to altered hydrodynamic diameters of the systems, and such colloidal structure rearrangements might lead to alterations in the physical properties of the system and therefore affect its stability and/or release kinetics. Our group has recently demonstrated that conjugation of folic acid results in hierarchical structural rearrangements (at multiple length scales) in starch and such rearrangements can have profound impacts on the colloidal hydrodynamic diameter [11]. We advise careful control of folic acid–functionalized oral delivery systems, as smaller nanosystems should also be able to diffuse and migrate easily through mucus layers (existing on the apical surface of intestinal epithelia) in the colon and also avoid being cleared by the latter, in addition to ease of endocytosis as discussed earlier. It is widely accepted that the mucus mesh consists of c. 50–1800 nm pores, as such, nanosystems c. <200 nm hydrodynamic diameter are deemed favorable to overcome the mucus barrier [43].

### 14.5.2 Effect of folic acid functionalization on surface charge of systems

Surface charge is a profoundly important parameter for determining the absorption site and absorption rate of oral delivery systems. Folic acid is an anionic molecule (ζ-potential, c. −34 mV) [11], as such surface functionalization of oral delivery systems with folic acid results in the gain of a net negative charge. For example, Ref. [39] reported the synthesis of amphiphilic starch nanoparticles surfaces functionalized with bovine serum albumin–conjugated folic acid that had a net ζ-potential of c. −22 mV. It is interesting to note that although the amphiphilic starch was cationic (ζ-potential, c. 28 mV), surface functionalization with folic acid was able to deliver a net negative charge to the nanocapsules at pH 7.4. Similar observations were made by Ref. [40]. The authors developed folic acid surface–functionalized starch nanoparticles with a ζ-potential of −12 mV. Also, Ref. [42] reported the development of hydroxyethyl starch nanocapsules that acquired a ζ-potential of −12 mV upon folic acid functionalization. From the abovementioned examples, the disparity between the ζ-potential of folic acid and that of the systems are quite evident. It may be safe to state that, despite the surface functionalization of the systems, the coverage of folic acid over the system's surface might not have been completed. Our group reported a comparison between the degree of substitution (esterification/functionalization) of the starch polymer by folic acid and the corresponding ζ-potential [11]. This may give future researchers clues to controlling the amount of folic acid conjugation to various macromolecules, in order to achieve the desired level of anionic surface charge. Researchers have demonstrated that systems that present anionic surfaces have a c. 30 times higher diffusion rate through the colonic mucus mesh, in comparison to cationic counterparts [46]. The authors proposed that the similarly charged mucus mesh repels the anionic systems; therefore, in the absence of interactive forces, the anionic systems can diffuse rapidly [46].

## 14.6 Folic acid–functionalized systems for colon cancer

The landmark studies carried out by Leamon et al. [30,47,48] for the first time established that folic acid–conjugated systems (folic acid–conjugated horseradish peroxidase, ferritin, ribonuclease, serum albumin, and momordin) can be successfully targeted at FR presenting cancer cells via the mechanism of folic acid–mediated endocytosis. In particular, the $IC_{50}$ of momordin and *Pseudonomous* exotoxins was reduced to c. $<10^{-9}$ M after folic acid conjugation, in comparison to the unconjugated form (c. $>10^{-5}$ M). This was, however, not true when experimented on cancer cells that do not express FRs, confirming the importance of folic acid conjugation. The authors also demonstrated that folic acid–conjugated toxins deliver a highly specific targeting, molecular recognition, and assassination of FR presenting Caco-2 cells, cultured alongside with FR-negative WI-38 fibroblast cells derived from lung tissue and HS-67 lymphoblastoid cells [48,49]. Additional experiments on HeLa cervical cancer, XC sarcoma, SKOV3 ovarian carcinoma, Caov-3 ovarian carcinoma, and KB cells confirmed the studies. In the following sections, we discuss the various in vitro and in vivo attempts that have been made so far concerning folic acid–functionalized systems targeted at colon cancer.

### 14.6.1 Colon cancer cell selectivity of folic acid–functionalized systems in vitro

As discussed earlier, systems that are functionalized with folic acid govern significant interests as a variety of colon cancer cells (Caco-2, HT-29, CT-26, etc.) overexpress FRs on their apical cellular surfaces. Ref. [50] described the synthesis and biological evaluation of folic acid–functionalized β-cyclodextrin-based complexes (average size of the supramolecule, 1.5–2.5 nm) targeted at the HT-29 (human colon adenocarcinoma) cells. The authors reported a c. twofold higher cellular uptake and a c. sixfold higher toxicity in comparison to nonfunctionalized systems. Among larger systems, Ref. [45] developed folic acid surface–functionalized guar gum nanoparticles charged with methotrexate to target Caco-2 cells. The authors reported that the folic acid–functionalized systems resulted in a c. 5%–10% higher inhibition of the cancer cells than their nonfunctionalized counterparts. Table 14–1 summarizes some of the significant attempts to develop various folic acid–functionalized systems against colon cancer cells in vitro. For readers who are interested outside this concise realm of folic acid targeting of colon cancer are directed toward other excellent reviews [5,71–74].

### 14.6.2 Colon tumor selectivity of folic acid–functionalized systems in vivo

The reports of folic acid–functionalized oral drug delivery systems evaluated in vivo are scare. Ref. [45] showed that Wistar rats orally fed with FITC-loaded folic acid surface–functionalized guar gum nanoparticles loaded with methotrexate were localized and taken up in the small intestine and colon. Ref. [61] demonstrated that mesoporous silica nanoparticles (loaded with γ-secretase inhibitors) that are functionalized with amalgamation of poly(ethylene glycol), poly(ethylene imine), and folic acid proved to be useful for oral gavage in 18 FVB/N adult male mice and resulted in considerably higher Notch inhibition compared to free doses of γ-secretase inhibitors, notably in the colon. The authors reported that nanoparticle penetration was affected by the variation in the pore size and charge of mucin molecules in the mucus layers of the small intestine and the colon. In another

**Table 14–1** Experimental folic acid–functionalized oral delivery systems evaluated against colon cancer cells in vitro.

| Delivery system | Payload | Evaluation model | Major findings | References |
|---|---|---|---|---|
| Silica nanoparticles | Camptothecin | SW480 | Drug-loaded fluorescent folic acid–functionalized nanoparticles reduced the proliferation and survival of colorectal adenocarcinoma cells | [51] |
| Chitosan nanoparticles | 5-Aminolevulinic acid | HT-29 | Bioassay results demonstrated that folic acid–chitosan nanoparticles were readily taken up by cancer cells compared to regular chitosan nanoparticles. In addition, a novel photodynamic detection system to enhance the accuracy of endoscopic diagnosis for early colorectal cancer was reported | [52] |
| Chitosan nanoparticles | 5-Aminolaevulinic acid | HT-29 and Caco-2 | Nanoparticles were reported to be taken up more readily by HT-29 and Caco-2 cells after short-term uptake periods, most likely via folate receptor–mediated endocytosis. In addition, protoporphyrin IX was observed to accumulate (5-aminolevulinic acid induces accumulation of protoporphyrin IX) in cancer cells, which was correlated with folate receptor expression in cells and the folic acid functionalization of the nanoparticles | [53] |
| Chitosan nanoparticles | Calcein (fluorescent marker) | HT-29 | Folic acid–conjugated nanoparticles exhibited improved uptake in HT-29 cells. The authors suggest that it can, therefore, be a potential targeted drug delivery system for ameliorating colorectal cancer | [54] |
| Folate-modified self-microemulsifying drug delivery system | Curcumin | HT-29 | Cellular uptake studies analyzed with fluorescence microscopy and flow cytometry indicated that the system could efficiently bind with the folate receptors on the surface of colon cancer cells. Besides, folic acid–functionalized systems showed significant cytotoxicity than the nonfunctionalized systems | [55] |
| PLGA nanoparticles | Paclitaxel | Caco-2 | Functionalization of nanoparticles with folic acid increased the transport into cancer cells by up to c. > eightfold compared to free paclitaxel | [56] |
| Guar gum nanoparticles | Curcumin | Caco-2 | Nanoparticles functionalized with folic acid demonstrated higher cytotoxicity in Caco-2 cells. Nanoparticles loaded with curcumin dosage of up to 2000 mg/kg of body weight were found to be nontoxic. The curcumin payload was additionally reported to be protected from release in the upper gastrointestinal tract | [57] |

| | | | | |
|---|---|---|---|---|
| PLGA nanoparticles | 5-Fluorouracil | HT-29 | Nanoparticles with folic acid functionalization demonstrated higher uptake by HT-29 cancer cells compared to the pure drugs and nonfunctionalized nanoparticles | [58] |
| Chitosan nanoparticles | 5-Fluourouacil and leucovorin | – | Dual drug–loaded microcapsules functionalized with folic acid measuring 15–35 μm were developed. Although no in vitro studies were conducted to evaluate the folate receptor–mediated uptake theory, other studies revealed that the microcapsules demonstrated excellent pH-dependent release profile, and no drug release was observed in simulated gastric and intestinal fluids | [59] |
| Chitosan nanoparticles | 5-Fluorouracil | HT-29 | Folic acid–functionalized microspheres inhibited the proliferation of tumor cells and induced apoptosis over an extended time. The authors also reported that the evaluation of the tissue distribution of the drug demonstrated a high concentration in colon cells | [60] |
| Mesoporous silica nanoparticles | γ-Secretase inhibitors | HT-29 and Caco-2 | Inhibitor-loaded nanoparticles functionalized with poly(ethylene glycol), poly(ethylene imine), and folic acid demonstrated enhanced intestinal goblet cell differentiation as compared to free drug | [61] |
| Liposomes | Recombinant IL15 plasmid | CT-26 | The authors reported that the folic acid–functionalized liposomes incremented the transfection efficacy of the reporter gene, alongside enhancing the expression and secretion of the IL15 gene in colon cancer cells. Such was not observed in the nontargeted liposomes | [62] |
| Cationic amphiphilic starch nanoparticles | 2-Deoxy-D-glucose, and α-tocopheryl succinate | HT-29 | Folic acid–functionalized nanoparticles significantly promoted the anticancer efficiency compared with administration of free drugs | [39] |
| *N*-Benzyl-*N*,*O*-succinyl-chitosan nanoparticles | Curcumin | HT-29 and Caco-2 | The folic acid–functionalized nanoparticles demonstrated low cytotoxicity in Caco-2 and HT-29 cells. In addition, the drug-loaded folic acid–functionalized nanoparticles presented more significant cell inhibition of HT-29 cells | [63] |
| Liposomes | 5-Fluorouracil | HT-29, Caco-2, and CT26 | Folic acid–functionalized liposomal drug exhibited higher cytotoxicity than free drug and liposomal drug. In addition, the targeted liposomes were observed to trigger necrosis in HT-29 cells | [64] |
| Polyamidoamine dendrimers G4 | Oxaliplatin | SW 480 | Nano-complex demonstrated the significant controlled release of oxaliplatin at different pH. In addition, the folic acid tagging to the nano-complexes resulted in enhanced cellular uptake in the cancer cells | [65] |

(Continued)

**Table 14–1 (Continued)**

| Delivery system | Payload | Evaluation model | Major findings | References |
|---|---|---|---|---|
| Solid lipid nanoparticles | Oxaliplatin | HT-29 | The authors reported that anticancer activity on HT-29 cells revealed the higher potency of the oxaliplatin–lipid–folic acid system as compared to oxaliplatin-lipid and oxaliplatin solution | [66] |
| PLGA-PEG$_{2k}$-triphenylphosphonium/PLGA-hyd-PEG$_{4k}$-folic acid and PT/PLGA-SS-PEG$_{4k}$-folic acid nanoparticles | 10-Hydroxycamptothecin | SW620 | The authors reported that folic acid–functionalized nanoparticles were able to deliver a higher concentration of drugs to the nucleus and mitochondria of the tumor cell, which resulted in the enhancement of cytotoxicity | [67] |
| Solid lipid nanoparticle | Doxorubicin and superparamagnetic iron oxide nanoparticles | CT26 | Results strongly confirmed that the dextran shells on folic acid–functionalized nanoparticles surface retarded the cellular transport of the folic acid–coated nanoparticles by the proton-coupled folic acid transporter proteins and enhanced the particle residence in the colon by specific association with dextranase. The study highlights the importance of both folic acid and dextran functionalization of oral delivery vehicles for greater efficacy in combating colon cancer | [68] |
| Liposomes | 5-Fluorouracil | HT-29 and Caco-2 | Folic acid–functionalized liposomes in HT-29 cells were observed to trigger the mitochondrial apoptotic pathway by decreasing the mitochondrial membrane potential, releasing of cytochrome C and promoting the substantial activity of caspase 3 and 7 | [69] |
| Fibrous nano-silica | Doxorubicin | HT-29 | Folic acid–functionalized nano-silica was observed to both quantitative and qualitative attach to the cancer cells, compared to nonfunctionalized counterparts | [70] |

SW480, SW620, HT-29, Caco-2, and CT-26 are human colorectal adenocarcinoma cell lines. Studies are presented in chronological order of publication. *IL15*, Interleukin-15; *PEG*, polyethylene glycol; *PLGA*, poly(lactic-co-glycolic acid); *PT*, poly(lactic-co-glycolic acid)–polyethylene glycol$_{2k}$-3-carboxypropyltriphenylphosphonium bromide.

excellent study, Ref. [68] demonstrated that orally delivered folic acid-D-α-tocopheryl polyethylene glycol 1000 succinate–functionalized solid lipid nanoparticles (FSLN) facilitate cellular level targeting and FR-mediated internalization by colon cancer cells. The authors developed dextran-coated FSLN (abbreviated as DFSLN) with the specific purposes to (1) avoid the recognition by proton-coupled folic acid transporters localized on brush border membrane surfaces of the small intestine and (2) extend the localization in the colon by the bracketing of dextran by dextranase enzymes exclusive to the colon. The authors hypothesized that upon the enzymatic degradation of the dextran shell by dextranase, FSLN could molecularly recognize and selectively target the CT-26 cell–induced colon tumors in BALB/C mice. The authors reported that (1) doxorubicin-loaded DFSLN and (2) superparamagnetic iron oxide nanoparticle–loaded DFSLN in combination with a high-frequency magnetic field were able to inhibit primary colon tumors (c. 15-fold decrease in tumor node mass compared to phosphate-buffered saline control). In addition, the authors reported that the ascites volume of peritoneal metastasis and amount of tumor nodules were both significantly reduced in mice receiving DFSLNs with or without high-frequency magnetic field. The authors reported no systemic side effects based on the levels of blood alkaline phosphatase and aspartate aminotransferase.

However, other studies in female C57BL/6 mice with 24JK-FBP (mouse sarcoma cell line transfected with the human FR gene) xenografts have demonstrated that both folic acid–functionalized and –nonfunctionalized systems display enhanced uptake by tumor tissues without any difference [75]. The authors assumed that nanoparticle accumulation in tumors was governed by passive targeting mechanisms such as the enhanced permeability and retention effect, in the case of both functionalized and nonfunctionalized systems. Therefore any advantage that a folic acid–functionalized system would sustain in an in vivo environment may only be limited to selective targeting, molecular recognition, and internalization within individual cancer cells residing in the tumor. This statement can be partly connected to the results of Ref. [68], described earlier. A mere folic acid surface functionalization of solid lipid nanocarriers was not able to dramatically reduce the tumor volumes, without the aid of the dextran coating (Fig. 14–3A). Therefore, for a tumor environment, there seems to be no "one-method-does-it-all" folic acid surface functionalization strategy. The option of approach depends considerably on the specific function required. Lessons learned from other colon cancer–targeted oral delivery systems have to be compounded alongside the molecular recognition and selective targeting capability of folic acid for FRs. It is of note that other in vivo evaluations on colon and colorectal cell xenograft studies do exist but were excluded from this chapter as the studies were not "oral" in nature, but the administration of the delivery systems were peritoneal.

## 14.7 Oral delivery systems and gastrointestinal digestion

The oral avenue of drug administration is the most preferred, with c. 70% of systems administered orally. However, current oral "macroscopic" delivery systems (tablets and capsules)

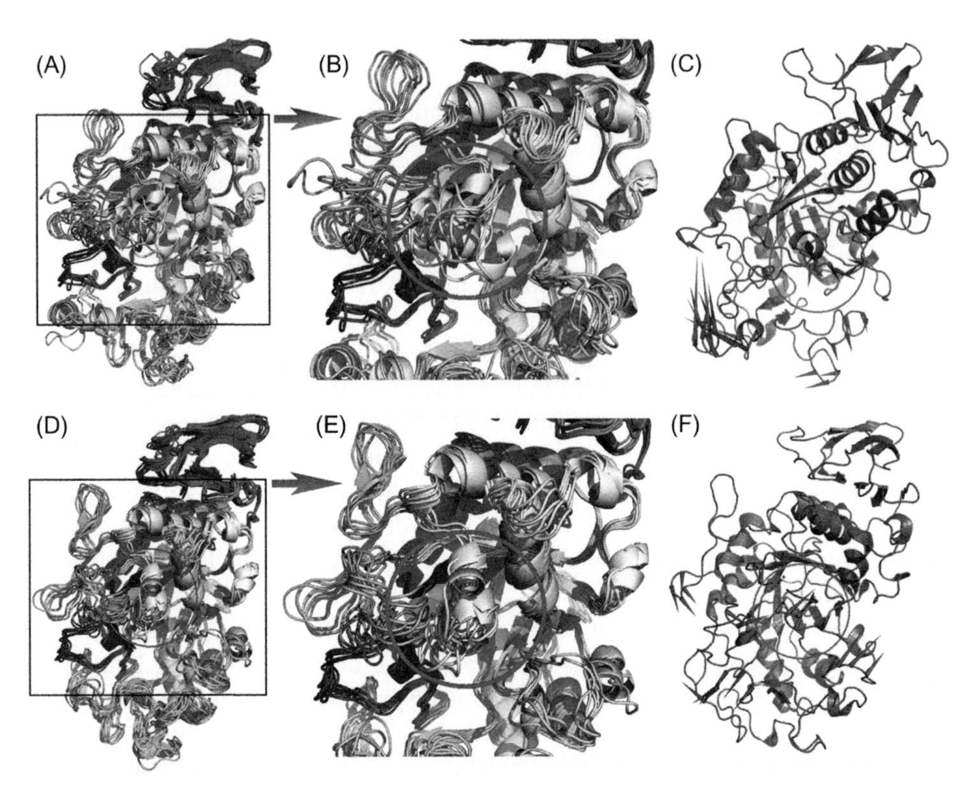


**FIGURE 14–3** Representative molecular structures of the major metastable states of human pancreatic α-amylase, RCSB PDB ID: 1HNY as obtained from clustering of molecular dynamics trajectory (A), enlarged view of boxed region to the left (B), structural modes for the first eigenvector (*red porcupines* depicting the direction and magnitude of prominent motions) for 1HNY (C), representative molecular structures of the major metastable states of the folic acid bound 1HNY complex as obtained from clustering of molecular dynamics trajectory (D), an enlarged view of boxed region to the left (E), structural modes for the first eigenvector (*red porcupines* depicting the direction and magnitude of prominent motions) for folic acid bound 1HNY complex (F). The red circle encloses the polypeptide loop 7.

are known to not perform optimally in the multiplex gastrointestinal conditions. This is partly because the gastrointestinal environment is a highly complex physiology [76] and has mostly been oversimplified and underrated during the evaluation of oral delivery systems. Therefore, not to duplicate the same limitations, it is of profound importance that we direct prime mindfulness to the design and evaluation of "nanoscopic" and "microscopic" oral delivery systems. As discussed previously in this chapter, researchers have developed numerous promising folic acid–functionalized colon cancer–targeted oral delivery systems; however, their evaluation in the complex gastrointestinal physiology has been largely underestimated. We know very limited about the complex infrastructural relationship that exists amidst these oral delivery systems and components of the gastrointestinal environment. One of the prime gastrointestinal barriers to the functioning and sustenance of oral delivery systems is the harsh gastrointestinal passage compounding with variable viscosity, variable ionic concentrations (c. 10–200 mM), variable pH (c. 1.2–7.5), and swirling with surface-active bile components, digestive enzymes (salivary α-amylase, pancreatic α-amylase, α-glycosidase, trypsin, chymotrypsin, other proteases, lipases, etc.), lipids, food proteins, peptides, and their debris [77]. In addition, the gastrointestinal scene is exceedingly dynamic, which constantly transforms in reciprocation to internal secretions and external food/liquid ingestions [78]. An excellent review by Ref. [76] discusses these challenges. It is natural that during the journey of an oral delivery system through the gastrointestinal passage, it will interact with the resident enzymes before reaching the colon for any molecular recognition and specific targeting of cancer cells and tumors to take place. Ref. [77] describes such interaction of the complex gastrointestinal fluids with nanoparticles to result in macromolecular crowding and formation of a "protein corona" around the surface coverage of the nanoparticles. This, in many cases, may cause colloidal instability of the oral delivery vehicle. Recently, researchers [79,80] have demonstrated that silver nanoparticles in simulated gastric digesta containing the enzyme pepsin, results in the formation of a protein envelop around the nanoparticles. This was seen to induce nanoparticle aggregation. Such effects are not just limited to metal nanoparticles but extend to polymeric nanoparticles exposed to digestive enzymes, namely, pepsin, α-amylase, and trypsin [81]. The authors reported that the thicknesses of the protein corona formed by α-amylase and trypsin were c. 25–100 and c. 50–100 nm, respectively. Such is accompanied by a significant increase in the ζ-potential of the nanoparticles. Food components may also tend to get attached to the surfaces of nanomaterials. Ref. [82] reported that poly(acrylic acid)–coated silver nanoparticles subjected to in vitro digestion in the presence of food components, such as carbohydrates, proteins, and lipids to simulate realistic conditions, displayed hydrodynamic size distributions that were larger (resulting from formation of aggregates via clustering of 120–150 primary particles) than original nanoparticles. It was interesting that the authors reported comparable internalization of the original nanoparticles and the nanoparticles that were digested in the presence of food components by Caco-2 cell. The authors hypothesized that orally ingested poly (acrylic acid)–coated silver nanoparticles could reach the intestine in a nanoparticulate form even if enclosed within food components. Ref. [83] also demonstrated that although lipid nanoparticles remain stable when exposed to acidic pH (pH c. 1.2); however, high pepsin and bile salt concentrations result in increased hydrodynamic diameter and polydispersity of lipid nanoparticles due to aggregation. This was observed to not prevent cellular uptake in Caco-2 cells, or the small intestine and colon cells of orally fed female C57BL/6 mice. Aggregation, however, prevented gene silencing by the siRNA drug payload. However, at times, the opposite was observed. Ref. [84] reported that in vitro digestion (in the presence of α-amylase, porcine pepsin, bile salts, and pancreatin from porcine pancreas) of carboxylated iron oxide nanoparticles [coated with poly(maleic)-alt-1-octadecene] in the presence of bread resulted in an increase in the hydrodynamic diameter and changed the ζ-potential; however, uptake by Caco-2 cells was enhanced compared to the original nanoparticles.

Although the studies are not conclusive yet, and further research is required to comprehend the scenario, it can be fair to propose that the formation of a protein corona comprising enzymes will negatively impact the structural stability of oral delivery systems owing to enzymatic digestion. Ref. [85] reported that nanoparticles consisting of proteins, starches, or lipid might be digested by gastrointestinal enzymes, altering their properties before reaching specific regions within the gastrointestinal tract. For example, Ref. [86] reported c. 100% hydrolysis of starch nanocrystals (prepared by acid hydrolysis of native starch) in the presence of pancreatic α-amylase (290 U/mL) and amyloglucosidase (500 U/mL). Therefore digestion is likely to affect the structural integrity and thereby affect the surface characteristics, such as folic acid functionalization and ζ-potential. The latter two are profoundly crucial for molecular recognition and specific targeting of colon cancer, and interaction with mucus and enterocyte proteins, as discussed earlier. Such digestion could also eventually lead to a compromise of the systems colloidal structure and result in premature drug release. In such scenarios, all the predictions made by researchers about "capability of molecular recognition and specific targeting of colon cancer" stand null and void.

However, researchers have proposed some remedial avenues that folic acid may seemingly provide to overcome enzyme-induced hydrolysis and structural disassembly of oral delivery systems. In the seminal work done by Ref. [87], folic acid was reported to reduce the enzymatic activity of porcine pancreatic α-amylase. The authors reported that 0.0004 mol/L folic acid could reduce the relative activity of porcine pancreatic α-amylase by c. 70%. Our group [88] reported a detailed description (including inhibition mechanisms, detailed kinetics, and binding regime) of the inhibition of human pancreatic α-amylase by folic acid. We observed that folic acid inhibited α-amylase activity by binding competitively to the active site cavity of α-amylase and decreased the disorder in the neighboring loops 3 and 7, the latter being profoundly crucial mobile loops in α-amylase for starch digestion. Readers are advised to take note of the more significant displacement of loop 7 (enclosed within *circles*) in the unbound state (Fig. 14–3A and B). After substrate binding to the active site cavity of α-amylase, this displacement measuring c. 0.6 nm is essential for starch hydrolysis [89]. The displacement is comparatively restricted in the folic acid bound α-amylase complex (Fig. 14–3D and E) and therefore may have negative implications for optimal starch hydrolysis. In addition, Fig. 14–3C and F shows the structural modes calculated from eigenvectors (as *red porcupines*) to demonstrate alterations in the direction and magnitude of prominent motions in unbound and folic acid bound α-amylase complex. Folic acid-binding to α-amylase significantly alters the direction of the prominent motions in the polypeptide loop 7 region (cf. Fig. 14–3C and F, enclosed within *circles*), further asserting the loss of optimal starch hydrolysis. The structural modes were calculated using principal components analysis from 200 ns of molecular dynamics simulation trajectories [88,90,91]. In addition, our group also demonstrated that folic acid, when esterified to starches, can induce a hierarchical restructuring of starch polymers that can lead to inhibition of digestion by pancreatic α-amylase; Ref. [92,93] however reported that folic acid does not inhibit the maltose (hydrolysis product of starches produced by α-amylase) digesting enzyme α-glucosidase. Therefore utilization of folic acid functionalization on oral delivery systems for achieving molecular recognition and selective targeting of colon cancer, alongside evading gastrointestinal digestion, may be enzyme and/or polymer specific. Ref. [87] also reported that the residual activity of pepsin decreases to about 35% of the original, upon incubation with folic acid. However, in comparison with pepsin, trypsin had a smaller activity loss. Using computational studies, the authors additionally reported that folic acid-binding to the substrate-binding pocket of pepsin directed the inhibition mechanism. The binding site of folic acid was however not near to the active site of trypsin. Although gastrointestinal digestion is just one facet of the hurdles faced by oral delivery systems, a broader comprehension of the evolution of oral delivery systems in terms of macromolecular crowding, protein corona formation, and colloidal stability during their voyage through the gastrointestinal tract is firmly necessary for the development of effective systems. For further reading into protein corona formations in oral nanoparticle systems, readers are directed toward the excellent review by Refs. [77,85], which also discusses the factors impacting the gastrointestinal fate of food-grade nanoparticle systems. The authors additionally discussed that ingested oral delivery systems might also be subjugated to a variety of mechanical forces as they pass through the gastrointestinal tract (such as mouth mastication, esophageal and small intestinal peristalsis, and stomach churning), which may alter the properties of the oral delivery systems. Very recently it has been demonstrated that coronae may also be formed by certain nonnutritive compounds (flavonoids, polyphenols, etc.) in foods, which can tightly bind to the surface of nano- and microparticles [94–96]. Therefore further research is needed to explore the interaction of folic acid and folic acid–functionalized oral delivery systems with various gastrointestinal and food components. Evaluation of their in vitro and in vivo fate is of profound importance. It seems that we have just scratched the surface.

## 14.8 Conclusion

The days of unmethodological oral drug delivery into healthy and pathologic tissues of the colon alike are possibly nearing an end. Beyond the passive tumor targeting possibilities widely practiced till date via the enhanced permeation and retention effects of tumor tissues, molecular recognition and selective targeting of cancer cell via folic acid surface–functionalized oral delivery systems seem a tremendous improvement. We caution though that the nutritional factor (folic acid), which is intricately involved in the genesis and propagation of colon cancer, should be clearly distinguished from its ability to aid oral delivery systems to deliver anticancer drugs effectively. The two processes are quite different. As seen from the discussion in this chapter, it is seemingly impossible to make extensive generalizations from the effects obtained by one type of delivery system over another, as the gastrointestinal environment is highly complicated and most systems discussed have not been tested to such rigor. It is, therefore, exceedingly difficult at present to recommend a particular type of oral delivery system or a particular design strategy. It is, nevertheless, reasonable to believe that folic acid functionalization of oral delivery systems is a promising future technology to battle colon cancer. Further attention needs to be paid to evaluate the gastrointestinal fate of these systems. The necessity to systematically evaluate folic acid–functionalized oral drug delivery systems for the stability of nanoformulation in biologically relevant gastrointestinal environments containing food components is of utmost necessity. Overall, a discernment of the structural (surface, hydrodynamic size, charge, and erosion) evolution of the oral delivery systems in terms of corona formations, gastrointestinal enzymatic digestion, and colloidal stability will significantly supplement the development of effective nanotechnology systems for ameliorating colon cancer.

## Acknowledgments

PKB acknowledges an Institutional Scholarship from Tezpur University, India. PKB additionally thanks Mr. Zaved Hazarika (Department of Molecular Biology and Biotechnology, Tezpur University, India) for technical discussion related to the analysis of principal components.

## References

[1] World Health Organization, Fact sheets on cancer. <https://www.who.int/news-room/fact-sheets/detail/cancer/>, 2015 (accessed 15.03.19).

[2] T.W. Wong, G. Colombo, F. Sonvico, Pectin matrix as oral drug delivery vehicle for colon cancer treatment, AAPS PharmSciTech 12 (2011) 201–214.

[3] J.T. Fox, P.J. Stover, Folate-mediated one-carbon metabolism, Vitam. Horm. 79 (2008) 1–44.

[4] J. Shia, D.S. Klimstra, J.R. Nitzkorski, P.S. Low, M. Gonen, R. Landmann, et al., Immunohistochemical expression of folate receptor α in colorectal carcinoma: patterns and biological significance, Hum. Pathol. 39 (2008) 498–505.

[5] Y. Lu, P.S. Low, Folate-mediated delivery of macromolecular anticancer therapeutic agents, Adv. Drug. Deliv. Rev. 64 (2012) 342–352.

[6] C. Chen, J. Ke, X.E. Zhou, W. Yi, J.S. Brunzelle, J. Li, et al., Structural basis for molecular recognition of folic acid by folate receptors, Nature 500 (2013) 486.

[7] A.S. Wibowo, M. Singh, K.M. Reeder, J.J. Carter, A.R. Kovach, W. Meng, et al., Structures of human folate receptors reveal biological trafficking states and diversity in folate and antifolate recognition, Proc. Natl Acad. Sci. U.S.A. 110 (2013) 15180–15188.

[8] F. Shen, J. Ross, X. Wang, M. Ratnam, Identification of a novel folate receptor, a truncated receptor, and receptor type. beta. in hematopoietic cells: cDNA cloning, expression, immunoreactivity, and tissue specificity, Biochemistry 33 (1994) 1209–1215.

[9] M. Da Costa, S.P. Rothenberg, Purification and characterization of folate binding proteins from rat placenta, Biochim. Biophys. Acta 1292 (1996) 23–30.

[10] B.A. Kamen, J.D. Caston, Properties of a folate binding protein (FBP) isolated from porcine kidney, Biochem. Pharmacol. 35 (1986) 2323–2329.

[11] P.K. Borah, M. Rappolt, R.K. Duary, A. Sarkar, Effects of folic acid esterification on the hierarchical structure of amylopectin corn starch, Food Hydrocoll. 86 (2019) 162–171.

[12] W. Xin, S. Feng, J.H. Freisheim, L.E. Gentry, M. Ratnam, Differential stereospecificities and affinities of folate receptor isoforms for folate compounds and antifolates, Biochem. Pharmacol. 44 (1992) 1898–1901.

[13] F. Shen, M. Wu, J.F. Ross, D. Miller, M. Ratnam, Folate receptor type. Gamma. Is primarily a secretory protein due to lack of an efficient signal for glycosylphosphatidylinositol modification: protein characterization and cell type specificity, Biochemistry 34 (1995) 5660–5665.

[14] O. Spiegelstein, J.D. Eudy, R.H. Finnell, Identification of two putative novel folate receptor genes in humans and mouse, Gene 258 (2000) 117–125.

[15] L.H. Matherly, Z. Hou, Structure and function of the reduced folate carrier: a paradigm of a major facilitator superfamily mammalian nutrient transporter, Vitam. Horm. 79 (2008) 145–184.

[16] A. Qiu, M. Jansen, A. Sakaris, S.H. Min, S. Chattopadhyay, E. Tsai, et al., Identification of an intestinal folate transporter and the molecular basis for hereditary folate malabsorption, Cell 127 (2006) 917–928.

[17] H.M. Said, Recent advances in carrier-mediated intestinal absorption of water-soluble vitamins, Annu. Rev. Physiol. 66 (2004) 419–446.

[18] H.M. Said, Intestinal absorption of water-soluble vitamins in health and disease, Biochem. J. 437 (2011) 357–372.

[19] J.F. Ross, P.K. Chaudhuri, M. Ratnam, Differential regulation of folate receptor isoforms in normal and malignant tissues in vivo and in established cell lines. Physiologic and clinical implications, Cancer 73 (1994) 2432–2443.

[20] A.C. Antony, Folate receptors, Annu. Rev. Nutr. 16 (1996) 501–521.

[21] B.A. Kamen, A. Capdevila, Receptor-mediated folate accumulation is regulated by the cellular folate content, Proc. Natl Acad. Sci. U.S.A. 83 (1986) 5983–5987.

[22] S.-W. Choi, J.B. Mason, Folate and carcinogenesis: an integrated scheme, J. Nutr. 130 (2000) 129–132.

[23] M. D'angelica, J. Ammori, M. Gonen, D.S. Klimstra, P.S. Low, L. Murphy, et al., Folate receptor-α expression in resectable hepatic colorectal cancer metastases: patterns and significance, Mod. Pathol. 24 (2011) 1221.

[24] L.C. Hartmann, G.L. Keeney, W.L. Lingle, T.J. Christianson, B. Varghese, D. Hillman, et al., Folate receptor overexpression is associated with poor outcome in breast cancer, Int. J. Cancer 121 (2007) 938–942.

[25] G. Toffoli, A. Russo, A. Gallo, C. Cernigoi, S. Miotti, R. Sorio, et al., Expression of folate binding protein as a prognostic factor for response to platinum-containing chemotherapy and survival in human ovarian cancer, Int. J. Cancer 79 (1998) 121–126.

[26] J.F. Ross, H. Wang, F.G. Behm, P. Mathew, M. Wu, R. Booth, et al., Folate receptor type β is a neutrophilic lineage marker and is differentially expressed in myeloid leukemia, Cancer 85 (1999) 348–357.

[27] M.J. Turk, G.J. Breur, W.R. Widmer, C.M. Paulos, L.C. Xu, L.A. Grote, et al., Folate-targeted imaging of activated macrophages in rats with adjuvant-induced arthritis, Arthritis Rheum. 46 (2002) 1947–1955.

[28] E. de Boer, L.M. Crane, M. Van Oosten, B. Van der Vegt, T. van der Sluis, P. Kooijman, et al., Folate receptor-beta has limited value for fluorescent imaging in ovarian, breast and colorectal cancer, PLoS One 10 (2015) e0135012.

[29] T. Yamaguchi, K. Hirota, K. Nagahama, K. Ohkawa, T. Takahashi, T. Nomura, et al., Control of immune responses by antigen-specific regulatory T cells expressing the folate receptor, Immunity 27 (2007) 145–159.

[30] C.P. Leamon, P.S. Low, Delivery of macromolecules into living cells: a method that exploits folate receptor endocytosis, Proc. Natl Acad. Sci. U.S.A. 88 (1991) 5572.

[31] J.J. Turek, C.P. Leamon, P.S. Low, Endocytosis of folate-protein conjugates: ultrastructural localization in KB cells, J. Cell Sci. 106 (1993) 423–430.

[32] S. Mayor, K.G. Rothberg, F.R. Maxfield, Sequestration of GPI-anchored proteins in caveolae triggered by cross-linking, Science 264 (1994) 1948–1951.

[33] K.G. Rothberg, Y.-S. Ying, B.A. Kamen, R. Anderson, Cholesterol controls the clustering of the glycophospholipid-anchored membrane receptor for 5-methyltetrahydrofolate, J. Cell Biol. 111 (1990) 2931–2938.

[34] R. Varma, S. Mayor, GPI-anchored proteins are organized in submicron domains at the cell surface, Nature 394 (1998) 798.

[35] M. Wu, J. Fan, W. Gunning, M. Ratnam, Clustering of GPI-anchored folate receptor independent of both cross-linking and association with caveolin, J. Membr. Biol. 159 (1997) 137–147.

[36] Z. Poon, S. Chen, A.C. Engler, H.I. Lee, E. Atas, G. Von Maltzahn, et al., Ligand-clustered "patchy" nanoparticles for modulated cellular uptake and in vivo tumor targeting, Angew. Chem. Int. Ed. 49 (2010) 7266–7270.

[37] Z. Poon, J.A. Lee, S. Huang, R.J. Prevost, P.T. Hammond, Highly stable, ligand-clustered "patchy" micelle nanocarriers for systemic tumor targeting, Nanomed. Nanotechnol. Biol. Med. 7 (2011) 201–209.

[38] R.J. Lee, S. Wang, P.S. Low, Measurement of endosome pH following folate receptor-mediated endocytosis, Biochim. Biophys. Acta 1312 (1996) 237–242.

[39] X. Lei, K. Li, Y. Liu, Z.Y. Wang, B.J. Ruan, L. Wang, et al., Co-delivery nanocarriers targeting folate receptor and encapsulating 2-deoxyglucose and α-tocopheryl succinate enhance anti-tumor effect in vivo, Int. J. Nanomed. 12 (2017) 5701–5715.

[40] S. Xiao, C. Tong, X. Liu, D. Yu, Q. Liu, C. Xue, et al., Preparation of folate-conjugated starch nanoparticles and its application to tumor-targeted drug delivery vector, Chin. Sci. Bull. 51 (2006) 1693–1697.

[41] H.-Z. Zhang, X.-M. Li, F.-P. Gao, L.-R. Liu, Z.-M. Zhou, Q.-Q. Zhang, Preparation of folate-modified pullulan acetate nanoparticles for tumor-targeted drug delivery, Drug. Deliv. 17 (2010) 48–57.

[42] G. Baier, D. Baumann, J.R.M. Siebert, A. Musyanovych, V. Mailänder, K. Landfester, Suppressing unspecific cell uptake for targeted delivery using hydroxyethyl starch nanocapsules, Biomacromolecules 13 (2012) 2704–2715.

[43] B. Mukherjee, B.S. Satapathy, S. Bhattacharya, R. Chakraborty, V.P. Mishra, Chapter 19—Pharmacokinetic and pharmacodynamic modulations of therapeutically active constituents from orally administered nanocarriers along with a glimpse of their advantages and limitations, in: A.M. Grumezescu (Ed.), Nano- and Microscale Drug Delivery Systems, Elsevier, 2017, pp. 357–375.

[44] R. Solaro, F. Chiellini, A. Battisti, Targeted delivery of protein drugs by nanocarriers, Materials 3 (2010) 1928–1980.

[45] M. Sharma, R. Malik, A. Verma, P. Dwivedi, G.S. Banoth, N. Pandey, et al., Folic acid conjugated guar gum nanoparticles for targeting methotrexate to colon cancer, J. Biomed. Nanotechnol. 9 (2013) 96–106.

[46] J.S. Crater, R.L. Carrier, Barrier properties of gastrointestinal mucus to nanoparticle transport, Macromol. Biosci. 10 (2010) 1473–1483.

[47] C.P. Leamon, P. Low, Cytotoxicity of momordin-folate conjugates in cultured human cells, J. Biol. Chem. 267 (1992) 24966–24971.

[48] C.P. Leamon, I. Pastan, P. Low, Cytotoxicity of folate-Pseudomonas exotoxin conjugates toward tumor cells. Contribution of translocation domain, J. Biol. Chem. 268 (1993) 24847–24854.

[49] C. Leamon, P. Low, Selective targeting of malignant cells with cytotoxin-folate conjugates, J. Drug. Target. 2 (1994) 101–112.

[50] J.-J. Yin, S. Sharma, S.P. Shumyak, Z.-X. Wang, Z.-W. Zhou, Y. Zhang, et al., Synthesis and biological evaluation of novel folic acid receptor-targeted, β-cyclodextrin-based drug complexes for cancer treatment, PLoS One 8 (2013) e62289.

[51] J. Lu, M. Liong, J.I. Zink, F. Tamanoi, Mesoporous silica nanoparticles as a delivery system for hydrophobic anticancer drugs, Small 3 (2007) 1341–1346.

[52] S. Yang, F. Lin, K. Tsai, M. Shieh, A novel detection of early colorectal cancer by chitosan nanoparticles conjugated with folic acid, in: Technical Proceedings of the 2007 NSTI Nanotechnology Conference and Trade Show, NSTI-Nanotech, 2007, pp. 242–245.

[53] S.-J. Yang, F.-H. Lin, K.-C. Tsai, M.-F. Wei, H.-M. Tsai, J.-M. Wong, et al., Folic acid-conjugated chitosan nanoparticles enhanced protoporphyrin IX accumulation in colorectal cancer cells, Bioconjugate Chem. 21 (2010) 679–689.

[54] P. Li, Y. Wang, F. Zeng, L. Chen, Z. Peng, L.X. Kong, Synthesis and characterization of folate conjugated chitosan and cellular uptake of its nanoparticles in HT-29 cells, Carbohydr. Res. 346 (2011) 801–806.

[55] L. Zhang, W. Zhu, C. Yang, H. Guo, A. Yu, J. Ji, et al., A novel folate-modified self-microemulsifying drug delivery system of curcumin for colon targeting, Int. J. Nanomed. 7 (2012) 151.

[56] E. Roger, S. Kalscheuer, A. Kirtane, B.R. Guru, A.E. Grill, J. Whittum-Hudson, et al., Folic acid functionalized nanoparticles for enhanced oral drug delivery, Mol. Pharm. 9 (2012) 2103–2110.

[57] R. Khatik, P. Dwivedi, M. Upadhyay, V.K. Patel, S.K. Paliwal, A.K. Dwivedi, Toxicological evaluation and targeting tumor cells through folic acid modified guar gum nanoparticles of curcumin, J. Biomater. Tissue Eng. 4 (2014) 143–149.

[58] Y. Wang, P. Li, L. Chen, W. Gao, F. Zeng, L.X. Kong, Targeted delivery of 5-fluorouracil to HT-29 cells using high efficient folic acid-conjugated nanoparticles, Drug. Deliv. 22 (2015) 191–198.

[59] P. Li, Z. Yang, Y. Wang, Z. Peng, S. Li, L. Kong, et al., Microencapsulation of coupled folate and chitosan nanoparticles for targeted delivery of combination drugs to colon, J. Microencapsul. 32 (2015) 40–45.

[60] K. Ganguly, A.R. Kulkarni, T.M. Aminabhavi, In vitro cytotoxicity and in vivo efficacy of 5-fluorouracil-loaded enteric-coated PEG-crosslinked chitosan microspheres in colorectal cancer therapy in rats, Drug. Deliv. (2015) 1–14. Available from: https://doi.org/10.3109/10717544.2015.1089955.

[61] D. Desai, N. Prabhakar, V. Mamaeva, D.Ş. Karaman, I.A. Lähdeniemi, C. Sahlgren, et al., Targeted modulation of cell differentiation in distinct regions of the gastrointestinal tract via oral administration of differently PEG-PEI functionalized mesoporous silica nanoparticles, Int. J. Nanomed. 11 (2016) 299.

[62] X. Liang, M. Luo, X.-W. Wei, C.-C. Ma, Y.-H. Yang, B. Shao, et al., A folate receptor-targeted lipoplex delivering interleukin-15 gene for colon cancer immunotherapy, Oncotarget 7 (2016) 52207.

[63] K. Tansathien, T. Woraphatphadung, W. Sajomsang, T. Rojanarata, T. Ngawhirunpat, P. Opanasopit, Development of folic-BSCS polymeric micelles containing curcumin for targeted delivery to colorectal cancer, Thai J. Pharm. Sci. 42 (2018) 168–171.

[64] S. Handali, E. Moghimipour, M. Rezaei, Z. Ramezani, M. Kouchak, M. Amini, et al., A novel 5-Fluorouracil targeted delivery to colon cancer using folic acid conjugated liposomes, Biomed. Pharm. 108 (2018) 1259–1273.

[65] A. Narmani, M. Kamali, B. Amini, A. Salimi, Y. Panahi, Targeting delivery of oxaliplatin with smart PEG-modified PAMAM G4 to colorectal cell line: in vitro studies, Process. Biochem. 69 (2018) 178–187.

[66] K. Rajpoot, S.K. Jain, Colorectal cancer-targeted delivery of oxaliplatin via folic acid-grafted solid lipid nanoparticles: preparation, optimization, and in vitro evaluation, Artif. Cells Nanomed. Biotechnol. 46 (2018) 1236–1247.

[67] H.-Q. Li, W.-L. Ye, M.-L. Huan, Y. Cheng, D.-Z. Liu, H. Cui, et al., Mitochondria and nucleus delivery of active form of 10-hydroxycamptothecin with dual shell to precisely treat colorectal cancer, Nanomedicine (2019) 14.

[68] M.-Y. Shen, T.-I. Liu, T.-W. Yu, R. Kv, W.-H. Chiang, Y.-C. Tsai, et al., Hierarchically targetable polysaccharide-coated solid lipid nanoparticles as an oral chemo/thermotherapy delivery system for local treatment of colon cancer, Biomaterials 197 (2019) 86–100.

[69] S. Handali, E. Moghimipour, M. Kouchak, Z. Ramezani, M. Amini, K.A. Angali, et al., New folate receptor targeted nano liposomes for delivery of 5-fluorouracil to cancer cells: strong implication for enhanced potency and safety, Life Sci. 227 (2019) 39–50.

[70] J. Soleymani, M. Hasanzadeh, M.H. Somi, N. Shadjou, A. Jouyban, Highly sensitive and specific cytosensing of HT 29 colorectal cancer cells using folic acid functionalized-KCC-1 nanoparticles, Biosens. Bioelectron. 132 (2019) 122–131.

[71] A. Garcia-Bennett, M. Nees, B. Fadeel, In search of the Holy Grail: folate-targeted nanoparticles for cancer therapy, Biochem. Pharmacol. 81 (2011) 976–984.

[72] A.R. Hilgenbrink, P.S. Low, Folate receptor-mediated drug targeting: from therapeutics to diagnostics, J. Pharm. Sci. 94 (2005) 2135–2146.

[73] Y.G. Assaraf, C.P. Leamon, J.A. Reddy, The folate receptor as a rational therapeutic target for personalized cancer treatment, Drug Resist. Updates 17 (2014) 89–95.

[74] L. Xu, Q. Bai, X. Zhang, H. Yang, Folate-mediated chemotherapy and diagnostics: an updated review and outlook, J. Control. Release 252 (2017) 73–82.

[75] W. Guo, T. Lee, J. Sudimack, R.J. Lee, Receptor-specific delivery of liposomes via folate-PEG-chol, J. Liposome Res. 10 (2000) 179–195.

[76] M. Koziolek, M. Grimm, F. Schneider, P. Jedamzik, M. Sager, J.-P. Kuehn, et al., Navigating the human gastrointestinal tract for oral drug delivery: uncharted waters and new frontiers, Adv. Drug. Deliv. Rev. 101 (2016) 75–88.

[77] A. Berardi, F. Baldelli Bombelli, Oral delivery of nanoparticles-let's not forget about the protein corona, Expert. Opin. Drug Deliv. 16 (2019) 563–566.

[78] E.L. Mcconnell, H.M. Fadda, A.W. Basit, Gut instincts: explorations in intestinal physiology and drug delivery, Int. J. Pharm. 364 (2008) 213–226.

[79] A.P. Ault, D.I. Stark, J.L. Axson, J.N. Keeney, A.D. Maynard, I.L. Bergin, et al., Protein corona-induced modification of silver nanoparticle aggregation in simulated gastric fluid, Environ. Sci.: Nano 3 (2016) 1510–1520.

[80] L. Pindǎková, V. Kašpárková, K. Kejlová, M. Dvořáková, D. Krsek, D. Jírová, et al., Behaviour of silver nanoparticles in simulated saliva and gastrointestinal fluids, Int. J. Pharm. 527 (2017) 12–20.

[81] Y. Wang, M. Li, X. Xu, W. Tang, L. Xiong, Q. Sun, Formation of protein corona on nanoparticles with digestive enzymes in simulated gastrointestinal fluids, J. Agric. Food Chem. 67 (2019) 2296–2306.

[82] D. Lichtenstein, J. Ebmeyer, P. Knappe, S. Juling, L. Böhmert, S. Selve, et al., Impact of food components during in vitro digestion of silver nanoparticles on cellular uptake and cytotoxicity in intestinal cells, Biol. Chem. 396 (2015) 1255–1264.

[83] R.L. Ball, P. Bajaj, K.A. Whitehead, Oral delivery of siRNA lipid nanoparticles: fate in the GI tract, Sci. Rep. 8 (2018) 2178.

[84] D. Di Silvio, N. Rigby, B. Bajka, A. Mackie, F. Baldelli Bombelli, Effect of protein corona magnetite nanoparticles derived from bread in vitro digestion on Caco-2 cells morphology and uptake, Int. J. Biochem. Cell Biol. 75 (2016) 212–222.

[85] D.J. Mcclements, H. Xiao, Is nano safe in foods? Establishing the factors impacting the gastrointestinal fate and toxicity of organic and inorganic food-grade nanoparticles, Sci. Food 1 (2017) 6.

[86] C. Liu, S. Jiang, Z. Han, L. Xiong, Q. Sun, In vitro digestion of nanoscale starch particles and evolution of thermal, morphological, and structural characteristics, Food Hydrocoll. 61 (2016) 344–350.

[87] W. Shi, Y. Wang, H. Zhang, Z. Liu, Z. Fei, Probing deep into the binding mechanisms of folic acid with α-amylase, pepsin and trypsin: an experimental and computational study, Food Chem. 226 (2017) 128–134.

[88] P.K. Borah, A. Sarkar, R.K. Duary, Water-soluble vitamins for controlling starch digestion: conformational scrambling and inhibition mechanism of human pancreatic α-amylase by ascorbic acid and folic acid, Food Chem. 288 (2019) 395–404.

[89] G. Andre, V. Tran, Putative implication of α-amylase loop 7 in the mechanism of substrate binding and reaction products release, Biopolymers 75 (2004) 95–108.

[90] N. Desdouits, M. Nilges, A. Blondel, Principal component analysis reveals correlation of cavities evolution and functional motions in proteins, J. Mol. Graph. Model. 55 (2015) 13–24.

[91] P.K. Borah, S. Chakraborty, A.N. Jha, S. Rajkhowa, R.K. Duary, In silico approaches and proportional odds model towards identifying selective ADAM17 inhibitors from anti-inflammatory natural molecules, J. Mol. Graph. Model. 70 (2016) 129–139.

[92] P.K. Borah, M. Rappolt, R.K. Duary, A. Sarkar, Structurally induced modulation of in vitro digestibility of amylopectin corn starch upon esterification with folic acid, Int. J. Biol. Macromol. 129 (2019) 361–369.

[93] E.D. Wahengbam, A.J. Das, B.D. Green, J. Shooter, M.K. Hazarika, Effect of iron and folic acid fortification on in vitro bioavailability and starch hydrolysis in ready-to-eat parboiled rice, Food Chem. 292 (2019) 39–46.

[94] X. Cao, C. Ma, Z. Gao, J. Zheng, L. He, D.J. Mcclements, et al., Characterization of the interactions between titanium dioxide nanoparticles and polymethoxyflavones using surface-enhanced Raman spectroscopy, J. Agric. Food Chem. 64 (2016) 9436–9441.

[95] M. Zembyla, B.S. Murray, A. Sarkar, Water-in-oil Pickering emulsions stabilized by water-insoluble polyphenol crystals, Langmuir 34 (2018) 10001–10011.

[96] M. Zembyla, B.S. Murray, S.J. Radford, A. Sarkar, Water-in-oil Pickering emulsions stabilized by an interfacial complex of water-insoluble polyphenol crystals and protein, J. Colloid Interface Sci. 548 (2019) 88–89.

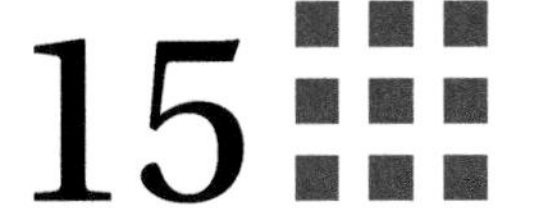

# Mathematical modeling and simulation of the release of active agents from nanocontainers/microspheres

Ashish P. Pradhane[1], Divya P. Barai[1], Bharat A. Bhanvase[2], Shirish H. Sonawane[3]

[1]CHEMICAL ENGINEERING DEPARTMENT, LAXMINARAYAN INSTITUTE OF TECHNOLOGY, RASHTRASANT TUKADOJI MAHARAJ NAGPUR UNIVERSITY, NAGPUR, INDIA [2]CHEMICAL ENGINEERING DEPARTMENT, LAXMINARAYAN INSTITUTE OF TECHNOLOGY, RTM NAGPUR UNIVERSITY, NAGPUR, INDIA [3]CHEMICAL ENGINEERING DEPARTMENT, NATIONAL INSTITUTE OF TECHNOLOGY, WARANGAL, INDIA

## Chapter outline

**15.1 Introduction** ..... **258**
**15.2 Mechanism of release in nanocontainers** ..... **258**
15.2.1 Drug delivery ..... 259
15.2.2 Anticorrosive self-healing coatings ..... 261
15.2.3 Wastewater treatment ..... 262
15.2.4 Fertilizers ..... 264
15.2.5 Catalysis ..... 264
15.2.6 Encapsulants in food industry ..... 266
15.2.7 Gas storage ..... 267
**15.3 Modeling of release of active agents** ..... **268**
15.3.1 Diffusion model ..... 269
**15.4 Simulation of release of active agents** ..... **281**
**15.5 Summary** ..... **284**
**References** ..... **284**

Encapsulation of Active Molecules and their Delivery System. DOI: https://doi.org/10.1016/B978-0-12-819363-1.00015-6

## 15.1 Introduction

Nanocontainers, as the name suggests, are a type of nano-sized containers that contain some active substances in their interior, which depend upon its final application. These active agents are needed to be released from the nanocontainer in order to serve the purpose at the point and time of application. Thus, nanocontainers help to save the loaded active agent from getting unnecessarily exposed to the environment other than that at its prime requirement location. The release of the active agents from the nanocontainers depends solely on its mechanism that is governed by either internal or external factors that stimulate the release process. These factors can be physical, chemical, or biochemical in nature and are considered to be related to either within or to the surrounding of the nanocontainer. There are various terminologies that alternatively represent nanocontainers, depending upon its application, such as microspheres, nanoshells, nanocarriers, and nanocapsule [1–6].

Recently, nanocontainers have found numerous applications where controlled release of active agents at a particular location and time is required [7]. According to the application and requirement of release rate, a nanocontainer can be synthesized using various techniques [8–11]. For understanding the release profile of various active agents, a particular set of assumptions with proper formation of equations are required that can exhibit results similar to what actually happens. A mathematical model translates the observed phenomenon in a mathematical expression, though it is only the simplified version of reality. One may predict the outcome of a particular process without even conducting it, or even simulate the process using the available technology. Hence, the number of necessary experimentations can be minimized. The usefulness of models can be seen through their applications in various fields such as biology, economy, genetics, psychology, medicine, engineering, and technology. So, there has been development of various model equations by various researchers in this field that can be used to predict the release behavior of different types of active agents in different applications. Basically, the release of the active agent from the nanocontainer is nothing but a process of diffusion, which depends on various properties of the nanocontainer, the embedded agent, and the presence and intensity of internal and external stimulating factors. Models developed for such systems serve more when applied for simulation purposes. Simulation of the release of active agents is nothing but a way of ideally analyzing the process of diffusion taking place through the nanocontainer considering some assumptions suitable for a particular application. There are various tools of simulating the release behavior such as the molecular dynamics, Monte Carlo, and the dissipative particles dynamics simulation. This chapter deals with the review of the investigations done on modeling and simulation of the release of the agents from the nanocontainer due to various stimulus mechanisms in different applications.

## 15.2 Mechanism of release in nanocontainers

Nanocontainers are able to carry agents that can be useful at distinct applications. Since their discovery, there have been many applications explored where they provide promising performance. The nanocontainer has been found to be smartly working for a typical application for which

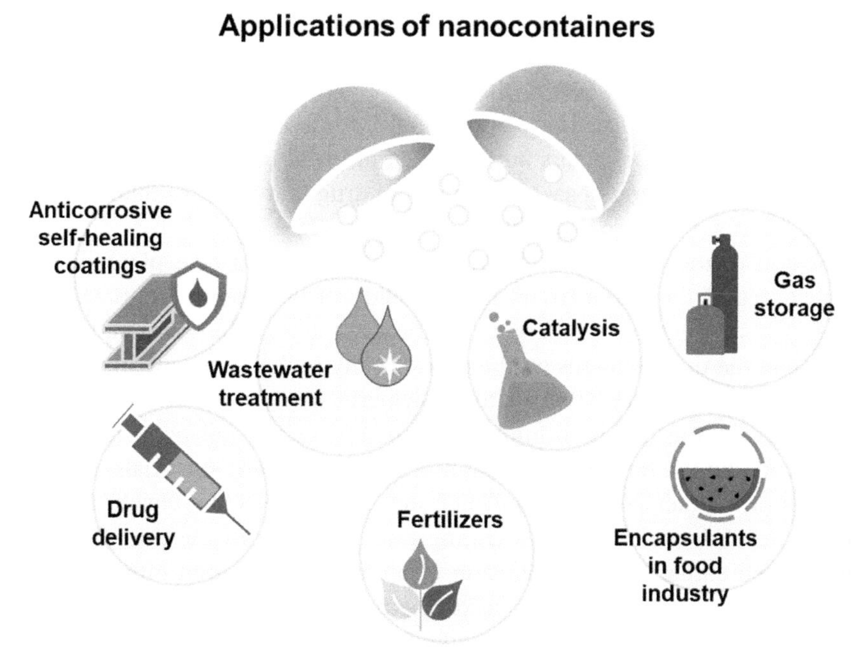


**FIGURE 15–1** Applications of nanocontainers.

it has been employed. Fig. 15–1 shows various applications of nanocontainers. For its performance the main two processes that can be manipulated during its synthesis are loading of the active agent into the nanocontainer and release pattern of the agents from the nanocontainer. While loading may depend and vary according to the type of nanocontainer, agent, and application. The release from the nanocontainer is far more complex because of typical requirements of the application. A brief discussion of nanocontainers for various applications and their release mechanism has been presented in this section of the chapter.

## 15.2.1 Drug delivery

Advances in nanotechnology in the field of medicine have allowed transport of drugs encapsulated into a nanocarrier through a pathway to the precise location in the human body for targeted treatment. The various types of nanocarriers that carry the drug are dendrimer, liposomes, micelle, polymer, nanosphere, nanocage, nanotube, nanorod, etc. [7]. Several types of drugs can be encapsulated inside nanocontainers made of various types of materials. This helps to improve the bioavailability of the drug thus protecting it from getting released at unnecessary site. Along with all these advantages, it is also possible to regulate its release rate according to the need by choosing and altering the nanocontainer material. Various types of drugs can be loaded into the nanocontainers as reported earlier [12–15].

As the drug is expected to affect a particular area at a particular concentration, its release is essential. On the other hand, it must not release above a level of toxicity. All this can be made sure by a controlled release of the drugs so as to provide the minimum amount of drug for the treatment purpose. There are several factors on which this controlled release of drugs depends. In fact, these factors are exploited in order to ensure the release of the drug from the nanocontainer. The release mechanism of drug that includes diffusion, degradation, swelling, and internal or external stimuli is depicted in Fig. 15–2.

The diffusion of the drug from inside the nanocarrier can take place due to diffusion through the pores of the nanocarriers mostly made of polymeric material [16]. This takes place due to the concentration difference of the drug across the membrane [17]. Thus the rate of release of the drug depends on the permeability of the nanocontainer. Another mechanism is based on the degradation of the polymeric nanocontainers due to enzymatic or hydrolytic attack on the bonds forming the structure of the polymer [18]. The release of drugs is governed by the rate of degradation of these polymeric materials. Altering the various properties of the nanocontainer polymer such as its molecular weight and composition can alter its rate of degradation, thus altering rate of release of the drug from it. The third mechanism stated earlier involves diffusion of an aqueous solution into the nanocontainer thus making it to swell. Thus the drug inside the nanocontainer gets released due to the liquid uptake and the rate of release is dependent on the rate of diffusion of the liquid [19]. It has been claimed that such release of drugs mostly follows zero-order kinetics, which depends on drug distribution inside the nanocontainer and the composition of polymer [20]. Other mechanisms of drug release are the stimuli-controlled release mechanisms. The

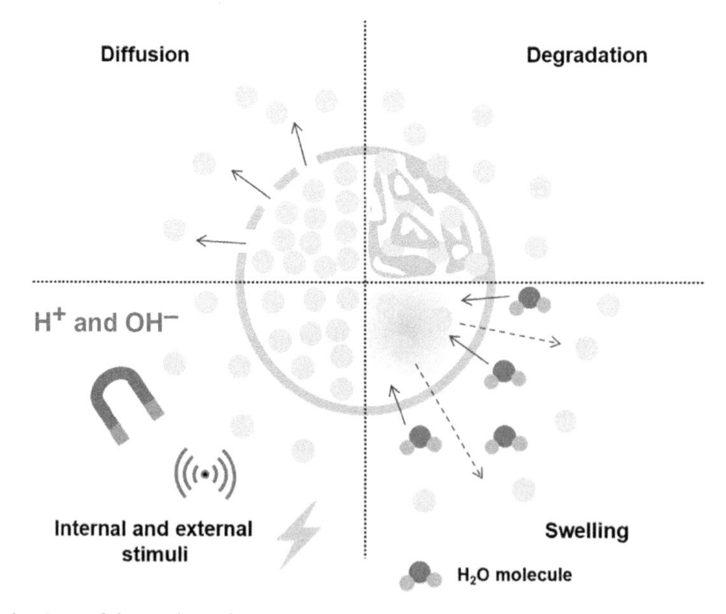


**FIGURE 15–2** Mechanisms of drug release from a nanocontainer.

stimuli can be due to changes in the environmental or internal conditions of the nanocarrier to which the response is the release of the drug from the nanocontainer. It can be changes in temperature, pH, ionic strength, ultrasound, and electric or magnetic field [21]. For this purpose the nanocarriers are designed in such a way that they respond to any of the above or a combination of stimuli. Ultrasound is also one of the usual stimuli used in drug delivery applications [22,23].

### 15.2.2 Anticorrosive self-healing coatings

One of the common types of degradation of materials exposed to natural environment is corrosion. Along with deteriorating the esthetics, it leads to degradation and devaluation of the material. This is most commonly overcome by applying one or more types of coating to the corrosion-prone surfaces. Coatings can work as active or passive protectors of surfaces from the environment. Passively protecting coatings are coatings that separate the surroundings from the surface by forming a barrier of oxides between it, whereas active coatings involve use of an inhibitor that is added to the environment so as to react with the corroding element. This then gets deposited over the surface and in turn protects it from getting corroded [24]. Nanocontainers can act as nanoreservoirs containing anticorrosive agents dispersed in the coating, which get released due to one or more external stimuli so as to protect the material from corrosion as shown in Fig. 15–3 [25].

It was first published by White et al. [26] that a healing agent can be microencapsulated in a composite matrix containing a catalyst that polymerizes the healing agent. Thus the corrosion inhibitors can be encapsulated into a nanocontainer contained by a matrix to form a "self-healing" surface. The autonomous or stimulant ability of the self-healing coating is to restore its tendency of protecting the metal surface against corrosion [27].

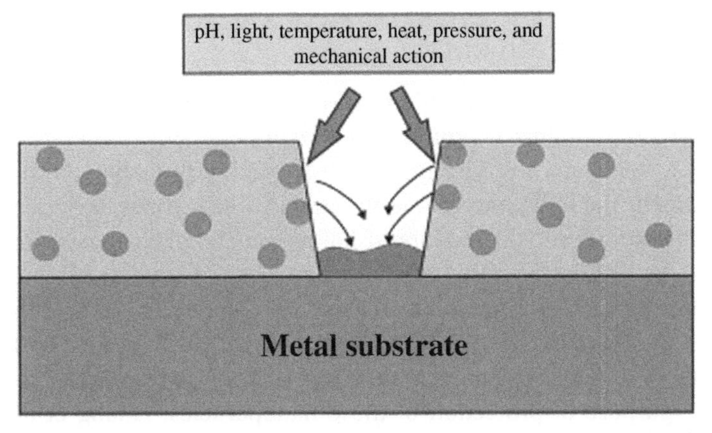


**FIGURE 15–3** Illustration of release of inhibitor in a self-healing coating from a nanocontainer [25]. *[Reprinted with permission from K.A. Zahidah, S. Kakooei, M.C. Ismail, P.B. Raja, Halloysite nanotubes as nanocontainer for smart coating application: a review, Prog. Org. Coat. 111 (2017) 175–185. Copyright (2017) Elsevier.]*

There are several external stimuli for the active agents in a nanocontainer for self-healing coating applications. The basic external stimulus is mechanical impact [26]. Due to mechanical impact over the nanocontainer, a crack develops in its structure thus instantly releasing the embedded active agent. This requires the nanocontainer material to be strong enough to carry the agent and at the same time it must be brittle enough to ease crack formation and discharge of the active agent. Further, it is known that corrosion in anodic areas leads to decrease in local pH and that in cathodic areas leads to an increase in the pH. One of the external stimuli to nanocontainers for self-healing anticorrosion applications is the change in pH at the local area of corroded surface. One can synthesize pH-sensitive nanocontainers, thus making them responsive to the pH changes in their surroundings. Reduction of corrosion up to zero has been achieved by employing such pH-sensitive nanocontainers to sol–gel coatings [28]. The mechanisms of a pH-responsive nanocontainer shell for controlled release of corrosion inhibitor has been studied by Refs. [8,29].

The ionic strength of the coating is found to be affecting the release rate of the agents from the nanocontainer [27,30]. Electromagnetic irradiation is also one of the external stimuli, which employs ultraviolet (UV) rays to irradiate over the nanocontainer shell. For exploiting this type of stimulus, the nanocontainer is designed in such a way that it becomes responsive to the typical wavelength of radiation. The irradiation incident on the radiation-sensitive component of the nanocontainer shell leads to release of the agent inside it [31]. This is also possible for some radiation-sensitive components of the nanocontainer shell by using infrared light as well as UV light [32]. Another stimulus to the nanocontainers for release of anticorrosion agents through them is difference in electrochemical potentials across the nanocontainer shell as studied by Vimalanandan et al. [33]. It was found that due to the corrosion metal, its electrode potential decreased which resulted in the reduction of the shell of the nanocontainer which in turn facilitated release of the encapsulated anticorrosion agent.

### 15.2.3 Wastewater treatment

Water being the most essential and important substance to all life forms on the earth must be protected. Water from industries containing a lot of pollutants is a threat to life, thus compelling us to treat such water in order to make it reusable or potable. Since this fact, several wastewater treatment methods have come up among which many are based on metal ions or dye removal. Photocatalytic degradation using nanomaterials has been found to be one of the promising advanced oxidative methods for removal of organic compounds from wastewater [34,35]. Among such nanomaterials, $TiO_2$ is very common [36,37]. These metal oxide particles are susceptible to dissolution when exposed to acidic conditions. Thus their protection becomes important. Nanocontainers can thus help in photocatalytic degradation along with serving the purpose of protection of these nanoparticles. Huang et al. [38] developed Fe-filled carbon nanocapsules onto which $TiO_2$ nanoparticles were immobilized for photocatalytic degradation of nitrogen oxide gas. For this they synthesized carbon nanocapsules made of concentric layers of graphitic sheets with a cavity inside for accommodation of the

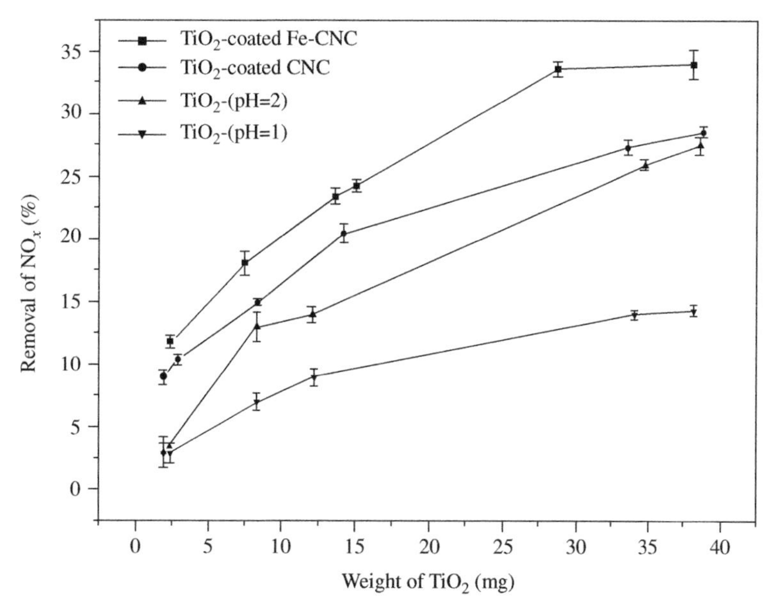


**FIGURE 15–4** Photocatalytic efficiency of different nanomaterials as given by Huang et al. [38]. *[Reprinted with permission from H. Huang, G. Huang, H. Chen and Y. Lee, Immobilization of $TiO_2$ nanoparticles on Fe-filled carbon nanocapsules for photocatalytic applications, Thin Solid Films 515 (2006) 1033–1037. Copyright (2006) Elsevier.]*

iron particles. Over this, $TiO_2$ nanoparticles were immobilized. Thus the synthesis of magnetic $TiO_2$ photocatalyst was possible along with protection of the Fe core by the resistant carbon shell, which could sustain acidic conditions and thus prevent dissolution of iron. The plot for percent removal of $NO_x$ against amount of $TiO_2$ nanoparticles in different types of nanomaterials synthesized was as depicted in Fig. 15–4.

Tian et al. [39] synthesized a fluorescence sensor-releasing nanocontainer, which can detect cadmium ions present in wastewater. For this, they coupled 2-(2-formylquinolin-8-yloxy)-*N,N*-diisopropylacetamide, which is a fluorescence sensor to silica nanoparticles. This sensor was adhered over the silica shell with the assistance of Schiff base bonds. These bonds become weak at acidic conditions thus letting the sensor to get released from the nanocontainer into the water containing cadmium ions for their detection. This process is claimed to be very fast and easily reproducible. Due to the fact that $Fe_3O_4$ nanoparticles cannot sustain acidic conditions, therefore its encapsulation inside a protecting core become vital. Thus $Fe_3O_4$@carbon nanocapsules were synthesized by Cheng et al. [40] for treatment of wastewater by removal of heavy metals. This process of protecting the magnetite inside the carbon core improved magnetic stability of the magnetite, acid resistance, and hydrophilicity along with enhancement in the treatment efficiency and easy regeneration due to porous pH-sensitive adsorption and desorption of the heavy metal ions. Further, removal of 12 heavy metal ions was studied by Dib et al. [41] using mesoporous nanocontainer made of silica shell and iron oxide core. Iron oxide present in the nanocontainer facilitates easy separation using magnets. Mesoporous structure of the silica shell provides efficient adsorption of the metal ions. It was also claimed that the nanocontainers could be easily regenerated for reuse. Hydrogels encapsulating various kinds of photocatalytically active nanoparticles have also been studied by various researchers for wastewater treatment [42].

### 15.2.4 Fertilizers

Attack of insects on the crops is a serious problem of farmers. Usage of fertilizer, pesticides, herbicides, etc. thus becomes important and unavoidable. Fertilizers and pesticides formulations used in agriculture are prone to loss of their activity by soil infiltration, runoff, leaching, and photo- or biodegradation. Therefore, it becomes compelling to have controlled release of the fertilizer or pesticides into the soil. This can be achieved if it is encapsulated in a carrier that will allow its regulated release as required according to the target site, time, or triggering stimulus. Such nanocarriers can provide a wide range of agrochemicals to the plants. At the same time, the active agents inside such nanocarriers are less likely to degrade in the soil. Polymer-based nanocarriers employed for controlled and slow release of fertilizers have been studied by many researchers [43,44]. Later, it was found that starch being a natural polymer, is totally biodegradable and has potential to be used as encapsulating material for fertilizers with controlled release ability [45]. Alongside, halloysite nanotubes (HNTs) developed a lot of interest among researchers as wonderful nanocarriers of active agents [46–49]. Thus embedment of fertilizers or herbicides into the lumen of HNTs also got into practice for controlled release but was found to retard the release as it was a single nanoclay delivery system [50]. Thus double delivery system was further developed so as to have effective controlled release of herbicides employing HNTs into poly(vinyl alcohol)–starch blends [51,52]. It has been reported that the use of biodegradable polymer such as the poly(vinyl alcohol) gets adhered to the openings at both sides of the HNTs and acts as a stopper for controlling the release of the herbicide. This decreased leaching of the herbicides from the soil thus providing effective biodegradable mulch for agricultural pesticide application. Further, starch reinforced cellulose nanofibers (CNF) were developed for insecticide encapsulation by Patil et al. [53]. They encapsulated dimethyl phthalate, which is an insect repellant, into the starch/CNF granules and studied the mechanism of release, which is reportedly due to swelling of the granules because of water being diffused into it. On the other hand, the nanocellulose network creates resistance to the agent release due to its barrier effect. Another natural polymer, lignin, has been also found to be applicable as one of the components of nanocontainer by Yiamsawas et al. [54]. They synthesized lignin–polyurea/polyurethane nanocontainers for loading of hydrophilic substances that can be released by enzyme stimulus. The enzymes also have an optimum set of pH and temperature for optimum working at which their rate of delignification leading to the nanocarrier degradation will proceed.

### 15.2.5 Catalysis

We already know how a catalyst helps to increase the rate of a reaction by providing the low-energy pathway for the reaction to proceed. Nanocontainers are found to have the capacity to carry molecules and to provide the necessary environment for reactions inside them and so acting as nanoreactors and thus they find applications in catalysis [55]. Being aware of the fact that nanocontainers can be loaded as well as emptied as they allow passage of molecules through the shell becomes very useful in such applications. There have been studies on a typical kind of nanocontainers mainly designated as core@void@shell [56]. They are also known as yolk–shell nanoparticles and nanorattles. These nanobodies differ from the core–shell nanoparticles in the way that they contain a cavity between the shell and the core, which acts as a lumen that may carry the foreign materials. This is possible only because of the presence of pores in the structure of the shell that connect the inner cavity to the outer surrounding of the nanocontainer [57,58]. These pores can prove to be very significant in the kinetics of the reaction taking place inside the shell. These nanobodies have various applications in catalysis, which have been discovered lately [56].

Silica is found to be very promising shell, which protects the nanoparticles encapsulated within it, which would otherwise sinter at high reaction temperature [59]. A temperature-sensitive corking and uncorking of the nanocontainers for oxidation of carbon monoxide based on capillary condensation of water vapor was proposed by Lin et al. as shown in Fig. 15–5 [59].

Priebe and Fromm [60] have developed a technique for synthesis of such nanocontainers and have studied their catalytic applications. They have prepared silica nancontainers that encapsulate silver nanoparticles within them and have managed to maintain a void between the nanoparticles and the shell along with pores in the structure of the silica shell that provide pathway for migration of molecules. This is confirmed by transporting nitric acid through these pores which in turn resulted into disappearance of the silver nanoparticles due to their dissolution. Rubio-Cervilla et al. [61] has explained the synthesis of functional single chain nanoparticles by folding of individual polymer chains. This provides a dense local environment resembling to that in a nanocontainer for catalytic activity to take place. This type of

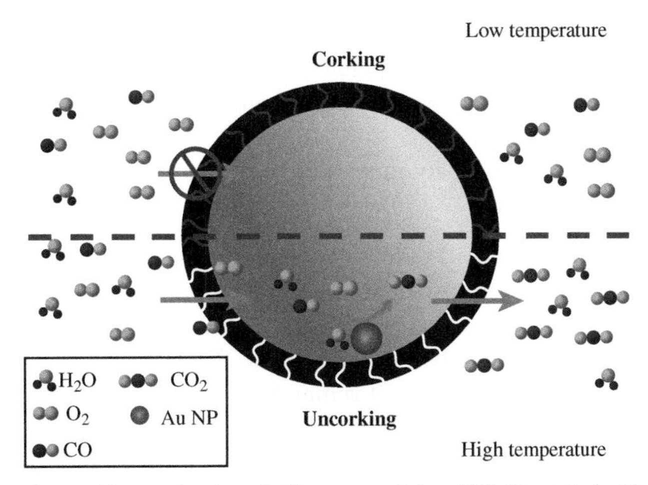


**FIGURE 15–5** Corking and uncorking mechanism of silica nanocontainer [59]. *[Reprinted with permission from C. Lin, X. Liu, S. Wu, K. Liu and C. Mou, Corking and uncorking a catalytic yolk-shell nanoreactor: stable gold catalyst in hollow nanosphere, J. Phys. Chem. Lett. 2 (2011) 2984–2988. Copyright (2011) American Chemical Society.]*

nanocontainers can also be used for $CO_2$ capture. Lin et al. [59] has explained how the CO oxidation taking place in the silica nanocontainer with gold nanoparticles can be switched on and off by changing the temperature. Do et al. [62] studied the degradation of acetaminophen, a pharmaceutical, using magnetic yolk–shell structured $Fe_3O_4@SiO_2$ nanoparticles in a hydrogen peroxide–initiated Fenton-like reaction. They found that the catalytic activity of the $Fe_3O_4@SiO_2$ yolk–shell nanoparticles increased with increase in the void space in the nanocontainer and decrease in the silica shell thickness. Also, due to degradation of the acetaminophen inside the shell, its concentration decreases thus producing a concentration gradient for diffusion to take place through the silica shell and enhancing the degradation process. Tin nanoparticles embedded in carbon nanoboxes have been synthesized by Yang et al. [63] in an attempt to produce anode for next-generation lithium-ion batteries. They claimed that the void in the structure is about 60%, which assists in volume expansion the $Li^+$ ion insertion/deinsertion reaction. Similarly, for application in lithium-sulfur batteries, Zhou et al. [64] synthesized polyaniline–sulfur nanocontainers that served the same purpose of volume variation accommodation for sulfur during lithiation. These yolk–shell nanocontainers found to be performing better than the core–shell nanoparticles of the same nanocomposite material.

### 15.2.6 Encapsulants in food industry

Micronutrients form a major source of bioactive ingredients that are essential in a healthy food diet. It has been known since then that their consumption is beneficial as far as human health awareness is concerned. But, there are numerous limitations to their utilization due to complexities in their incorporation in daily food products which are as follows [65]:

1. They are less soluble in oil and water.
2. They can degrade during food processing, storage, and transport due to changes in physical and chemical conditions.
3. Their off-flavors may disturb the acceptability of the main food product by the consumer.
4. They are susceptible to reaction with other micronutrients or food ingredients.
5. They have less bioavailability.
6. They are required in very small amounts, which cannot be easily distributed uniformly throughout the food product.

In order to overcome all these limitations, protect these essential micronutrients and to increase their bioavailability, encapsulation has emerged as an interesting remedy for nanodelivery of such micronutrients. This involves use of a nanocarrier that will carry the micronutrient inside it and release when required. As these materials are themselves also consumed along with the micronutrient, it is mandatory for them to be safe, edible, and nontoxic. Fig. 15–6 depicts the advantages of this method.

Use of such nanocarriers is a common practice for encapsulating flavors or aromas into food [66]. Natural biopolymers are identified as suitable materials for synthesizing nanocontainers for such systems [67,68]. It has been studied that the controlled and targeted release of micronutrients ensures optimal dosage thereby increasing its biological and economical

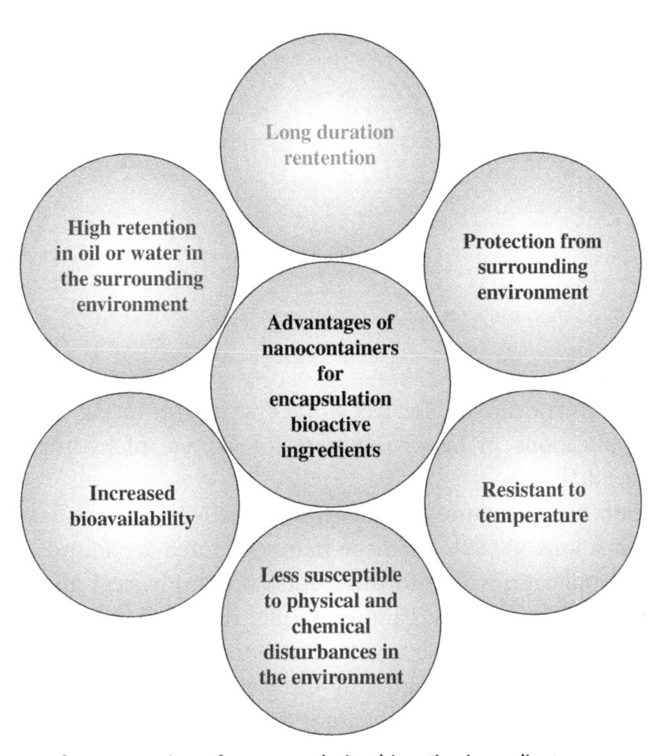


**FIGURE 15–6** Advantages of nanocontainers for encapsulating bioactive ingredients.

effectiveness [69]. Essential oils have been encapsulated by several researchers and have been thus studied [70]. The release of the encapsulated agent from the nanocarriers depends upon its solubility, diffusion, and degradation of the shell by biological activities [71]. Also, the loading capacity of the nanocarrier is determined by the release rate of the active agent [72]. Mechanisms involved in release of these essential oils from the nanocarriers include dissolution of the matrix, diffusion of the oils through the matrix, desorption of the oil, and erosion of matrix by enzymatic action. [73]. There are several other applications of nanotechnology in food industry [74].

## 15.2.7 Gas storage

Safe and economical storage of gas for various reasons and applications has been a need of science. Encapsulating gas into pores of a material involves use of high pressure so as to force the gas molecules into the pores, which otherwise, that is, at ambient conditions cannot be accessed by the gas. Nanocapsules made out of carbon nanostructures have found applications in storage of gas [75–77]. Molecular-dynamic modeling has been used by Volkova et al. [78] to study the absorption, storage, and desorption of methane gas in such structures that respond to high temperature for the release to take place. They demonstrated

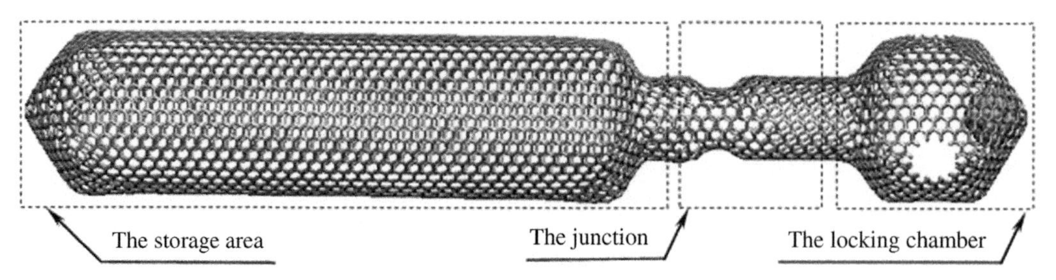


**FIGURE 15–7** Schematic of a nanocapsule for natural gas storage [78].

a typical structure of a nanocapsule for gas storage as shown in Fig. 15–7. Thus there have been studies on applications of nanocontainers for storage of hydrogen [79,80], methane [76], and noble gases [81].

Apart from carbon nanostructures, other types of nanocontainers have also been studied for gas storage applications [80,82]. All these nanostructures are found to have great potentials for gas storage applications due to their outstanding physical and chemical properties [83]. A computational study done by Tang et al. [84] demonstrated the use of single-walled carbon nanotubes (SWCNTs) as efficient gas containers for hydrogen. It was claimed that up to 5 MPa of gas pressure can be sustained by these nanotubes without any gas leakage and that release of gas can be stimulated by heating the nanotube. Also, density function theory has been used to study the properties of nitrogen gas contained in carbon fullerene-like structures by Barajas-Barraza and Guirado-Lopez [85]. Shang et al. [86] studied the use of zeolite potassium chabazite as a nanocontainer for storage of nitrogen and methane having temperature-controlled value-mechanism for release and tunable properties for storage of a variety of gases. Also, a vast study of gas storage capacity of boron oxide ($B_{20}O_{30}$) nanocapsules for various gases such as hydrogen, nitrogen, carbon monoxide, carbon dioxide, ammonia, methane, and chlorine has been done by Zamani et al. [87] using Monte Carlo simulation, density function theory, and Moller–Plesset second-order perturbation theory.

## 15.3 Modeling of release of active agents

Based on the discussed need of encapsulation and the possible mechanisms of release of active agents from nanocontainers/microspheres, the focus on something of great importance, that is, the mathematical modeling for the given system is essential. Many mathematical models have been formulated in the past describing the release of active agents from containers, specifically in the field of drug release. The necessity of predetermined release profile of active agents lies in (1) knowing the exact release mechanism of the release process and (2) quantitatively predicting the resulting release kinetics. Comprehension of all factors contributing in the release of active agents is required while formulating a mathematical model for a given system.

As stated earlier, the release of active agents must occur in response to some external stimuli to have some sort of control over the release process. One must also have some understanding about the release mechanism of these active agents through the walls of nanocontainers/microsphere. The easiest mechanism to understand is rupturing of shell material in response to applied force or shear. This is most commonly seen in the case of fragrance release and inkless papers. Shells synthesized must be brittle enough to undergo rupture. The force that must be applied to rupture the shell depends upon the material used and properties such as thickness of shell, which are controllable parameters in the synthesis process.

Another simple mechanism, resulting in the release of active agents, is dissolution of shell material, as a result of melting, interactions with solvent, hydrolysis, enzymatic degradation, chemical and/or photochemical reactions, etc. Release of flavors from cake mixture, upon addition of water, is an example of such release mechanism. The active agents are completely released as the shell material is totally dissolved. If one wishes to have an immediate or one-time release of the entire encapsulated active agents, the previous two release mechanisms must be preferred.

What if a prolonged release of active agents is required? Permeation and diffusion of active agents through the walls of permeable shell can result in such release. The diffusion may be Fickian or non-Fickian. The rate of release is dependent on the permeability of shell materials and the structure (size and shape) of active agents. Shell material, capable of swelling, can also be employed as an alternative to the permeable shells. Such shells can be fabricated with the use of responsive materials, which have the capability of undergoing reversible changes in response to stimuli or triggers such as pH, temperature, light, and ionic strength. Swelling of shell results in release of active agents and thus the release profile can be controlled and maintained with the help of triggering mechanism. There are many other phenomena that contribute to the release process and it is beyond the bounds of possibility to enlist all the possible mechanisms contributing in the release of active agents from these nanocontainers/microspheres.

## 15.3.1 Diffusion model

Diffusion, erosion, and swelling were found to be the most important rate-controlling mechanisms in the release of active agents. Fick's second law can be used to describe the phenomenon of diffusion. For one dimension, Fick's second law can be stated as given in the following equation:

$$\frac{\partial C}{\partial t} = D\frac{\partial^2 C}{\partial x^2} \tag{15.1}$$

For spherical particles (microspheres), Fick's second law, for one-dimensional drug release, takes the form as given in the following equation:

$$\frac{\partial C}{\partial t} = \frac{1}{r^2}\frac{\partial}{\partial r}\left(Dr^2\frac{\partial C}{\partial r}\right) \tag{15.2}$$

where $C$ represents local concentration of the active agents at any time $t$ and at a distance $r$ from the center of the particle and $D$ represents the diffusion coefficient of active agents in the matrix. To obtain an analytic solution of the equation, one must consider the boundary conditions, which are dependent on the mechanism of mass transfer and volume of the surrounding system. These boundary conditions categorize the mass transfer into two main cases:

1. Perfect sink conditions, which may exist either for surface of a microsphere or for the surrounding medium. Having a perfect sink condition at the surface can be understood as having a constant concentration of the active agents on the surface. This may happen in the scenario when the surface mass transfer resistance is negligible or medium surrounding the microspheres is infinitely large. Second possibility is having a perfect sink condition in the surrounding medium considering the mass transfer resistance at the surface to be finite.
2. Perfectly mixed finite volume of the surrounding, in which the resistance to mass transfer at the surface may be negligible or nonnegligible. The surface concentration of the surrounding medium varies with time.

Depending on the region where the diffusion of active agents occurs, one may categorize a diffusion-controlled system into reservoir and matrix system as given in Fig. 15–8 [88].

One may express the release of active agents through solid polymer matrix through Eq. (15.3), which is Fick's second law for spherical particles with the assumption of constant diffusion coefficient.

$$\frac{\partial C}{\partial t} = D\left(\frac{\partial^2 C}{\partial r^2} + \frac{2}{r}\frac{\partial C}{\partial r}\right) \tag{15.3}$$

Application of boundary conditions and mass balance yield following analytical solution upon integration as given in the following equation [89]:

$$\frac{C_l}{C_{l\infty}} = 1 - \sum_{n=1}^{\infty}\frac{6\alpha(1+\alpha)}{9+9\alpha+\left(\alpha q_n^2\right)}\exp\left(-\frac{Dq_n^2 t}{R^2}\right) \tag{15.4}$$

where $C_l$ is concentration of active agents in the bulk liquid, $\alpha$ is ratio of release to retained active agents at equilibrium, the transcendent equation, written as Eq. (15.5), with positive nonzero roots are represented by $q_n$ in the earlier equation.

$$\tan q_n = \frac{3q_n}{3+\alpha q_n^2} \tag{15.5}$$

The geometry of the container is also an important parameter to consider as complex 3D geometries, which are more relevant, often are difficult to treat. One more important parameter while employing Fick's law is diffusivity or diffusion coefficient. Assumption of constant

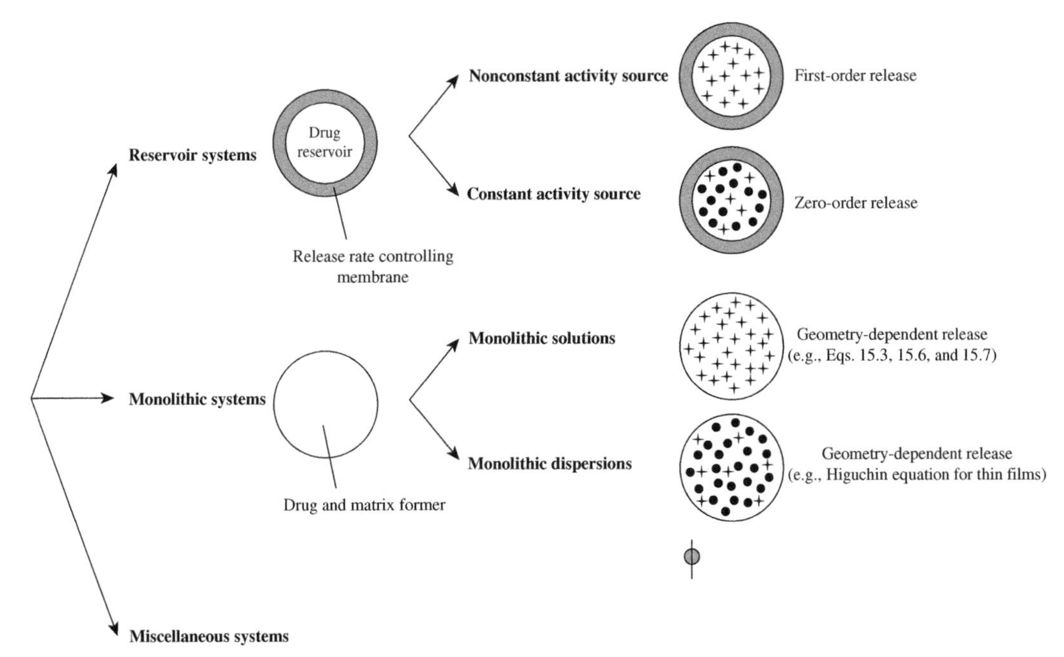


**FIGURE 15–8** Classification system for primarily diffusion-controlled drug delivery systems. Stars represent individual active agent molecules and black circles active agent crystals and/or amorphous aggregates (applicable to any type of geometry) [88]. *[Reprinted with permission from J. Siepmann and F. Siepmann, Mathematical modeling of drug delivery, Int. J. Pharm. 364 (2008) 328–343. Copyright (2008) Elsevier.]*

diffusivity relatively simplifies the treatment of model, but it is only applicable for containers that do not swell significantly while they are in contact with water. Thus assuming constant diffusivity will result in development of unrealistic models. Along with the geometry, polymer dissolution and drug solubility are also the factors that must be taken into consideration. Polymer dissolution will result in complicating Fick's second law as it will result in moving boundary conditions [90]. For poorly soluble drugs, drug dissolution is also an important parameter to consider as it can be rate limiting.

#### *15.3.1.1 Models for reservoir system*

The reservoir model is said to be the simplest model, based on Fick's law of diffusion, describing the release of active agents from a reservoir or depot. The depot is considered as a spherical particle and is assumed to be surrounded by a spherical shell barrier of inner radius $R_i$ and outer radius $R_o$. The active agents must diffuse through the barrier thickness $(R_o - R_i)$. Further assumptions that are considered to obtain distribution of active agents within the barrier $(R_i \leq r \leq R_o)$ are as follows:

1. The surface of depot or reservoir $(r = R_i)$ is assumed to have a constant concentration.
2. The outer surface of shell $(r = R_o)$ is assumed to have zero concentration of active agents as all the mass transfer resistances are assumed to be confined within the barrier and

there is no resistance to mass transfer at the outer surface, surrounded by an infinitely large volume of surrounding medium.

Upon solving the unsteady state diffusion equation to obtain concentration profile and cumulative amount of active agents released ($M_t$) with respect to time, following solutions as given in Eqs. (15.6) and (15.7) are obtained [91].

$$C(r,t) = \frac{C_i R_i}{r}\frac{(R_o - r)}{(R_o - R_i)} - \frac{2}{\pi}\sum_{n=1}^{\infty}\frac{C_i R_i}{n}\sin\left(n\pi\frac{r - R_i}{R_o - R_i}\right) \times \exp\left(-\frac{n^2\pi^2}{(R_o - R_i)^2}Dt\right) \tag{15.6}$$

$$\frac{M_t}{4\pi R_i (R_o - R_i) C_i} = \frac{Dt}{(R_o - R_i)^2} - \frac{1}{6} - \frac{2}{\pi}\sum_{n=1}^{\infty}\frac{(-1)^n}{n^2}\exp\left(-\frac{n^2\pi^2}{(R_o - R_i)^2}Dt\right) \tag{15.7}$$

The abovementioned equation, when considered for infinite observation time ($t \to \infty$), can be simplified as Eq. (15.8). The diffusion rate of active agents through the containers becomes constant as the series term approaches zero.

$$\frac{M_t}{t} = \frac{4\pi R_o R_i}{(R_o - R_i)} D C_i \tag{15.8}$$

One must be careful while using the solutions obtained earlier as it appears to suggest the cumulative amount of active agents released $M_t$ to increase with time. Thus, the validity of the solution must be limited for a certain period of time which may change with a changed system.

The reservoir systems can be further classified based upon the solubility of active agents in the container material (usually a polymer), as systems having initial concentration of active agents lesser than or more than its solubility in the container material.

If the initial concentration of the active agents is less than its solubility in the surrounding polymeric material, the released active agents will not be replaced in the barrier membrane, resulting in a drop in the active agent concentration with time. A first-order kinetics prevails as specified in Eq. (15.9), irrespective of the geometry, for the following conditions [88,92]:

1. No swelling or dissolution of the barrier membrane
2. Perfect sink conditions for the entire release period
3. Constant permeability of the active agents through the barrier

$$\frac{dM_t}{dt} = \frac{ADK_p C_t}{(R_o - R_i)} = \frac{ADK_p}{(R_o - R_i)}\frac{(M_0 - M_t)}{V} \tag{15.9}$$

where, $M_0$ is the initial amount of active agent present inside the containers, and $A$ is the surface area of the containers. Other symbols have meaning as defined in the earlier sections.

Contrary to the abovementioned case, if the initial concentration of the active agents is more than its solubility in the container material, the released active agents will be replaced by dissolution in the container materials and thus its concentration will remain constant in the barrier membrane. Zero-order release kinetics prevails as specified in Eq. (15.10), again irrespective of geometry for the following conditions:

1. Physical properties of barrier membrane, such as its thickness and permeability, remain unchanged with time.
2. Perfect sink conditions for the entire release period.

$$\frac{dM_t}{dt} = \frac{AJ_{\text{lim}}}{R_o - R_i} = \frac{ADK_pC_s}{R_o - R_i} \tag{15.10}$$

where, $J_{\text{lim}}$ is limiting flux in barrier membrane and $C_s$ is active agent's solubility of in the reservoir.

### *15.3.1.2 Models for matrix (monolithic) systems*

Release of active agents through nanocontainers/microspheres, made of nonbiodegradable polymers, could be described by matrix models. An assumption, while considering matrix, is made that there is uniform distribution of active agents inside the container. Like reservoir model, these models can be further categorized into two subclasses:

1. Initial loading of active agents is less than its solubility in the container material (dissolution model).
2. Initial loading of active agents is more than its solubility in the container material.

#### 15.3.1.2.1 Dissolution model

Though dissolution of active agents is not a primary mechanism of active agent transport out of polymeric shell, it can be considered as a contributed factor with subsequent diffusion. As stated earlier, this model is applicable when the lesser amount of active agents is loaded in containers, resulting in uniform dissolution of active agents in the container material. Quantitative assessment of effect of dissolution of active agents can be done by modifying the diffusion model by incorporation of a term to consider the effects of dissolution. Eqs. (15.11) and (15.12) can be used for expressing the concentration profile and fraction of active agent released, respectively, with negligible mass transfer resistance at the surface [17].

$$\frac{C - C_0}{KC_b - C_0} = 1 + \frac{2R_o}{\pi r}\sum_{n=1}^{\infty}\frac{(-1)^n}{n}\sin\left(\frac{n\pi r}{R_o}\right) \times e^{\left(-\left(Dn^2\pi^2 t/R_o^2\right)\right)} \tag{15.11}$$

$$\frac{M_t}{M_\infty} = 1 - \frac{6}{\pi^2}\sum_{n=1}^{\infty}\frac{1}{n^2}e^{\left(-\left(Dn^2\pi^2 t/R_o^2\right)\right)} \tag{15.12}$$

If there exists a finite mass transfer by convection, the concentration profile and the fraction of active agent released can be given as Eqs. (15.13) and (15.14), respectively [93].

$$\frac{C - KC_0}{C_o - KC_b} = \frac{2ShR_o}{r}\sum_{n=1}^{\infty}\frac{1}{\beta_n^2 + Sh^2 - Sh}\frac{\sin\beta_n(r/R)}{\sin\beta_n} \times e^{\left(-(\beta_n^2/R^2)Dt\right)} \tag{15.13}$$

$$\frac{M_t}{M_\infty} = 1 - \sum_{n=1}^{\infty}\frac{6Sh^2}{\beta_n^2\left(\beta_n^2 + Sh^2 - Sh\right)}e^{\left(-(\beta_n^2/R^2)Dt\right)} \tag{15.14}$$

where, the values $\beta_n$ are roots of another transcendent equation, written as follows and $Sh$ is the well-known Sherwood number. The equations written with the assumption of finite mass transfer with convection become equivalent to equations with assumption of negligible mass transfer resistance, at $Sh \gg 1$.

Polakovič et al. [89] formulated a similar dissolution model as given in Eq. (15.15) on the basis of following assumptions:

1. Dissolution of the active agent is rate-controlling step of the release process
2. Negligible mass of active agent is dissolved inside a particle in comparison to the actual mass of active agent
3. Driving force of transport is the difference between concentration of solid active agent and its equilibrium concentration.

$$r_d = -\frac{dc}{dt} = k\left(C - K_p C_l\right) \tag{15.15}$$

where, $r_d$ is rate of dissolution of active agents, $C$ is local concentration of active agents, $k$ is apparent dissolution constant, $K_p$ is coefficient of partition, and $C_l$ is the concentration of active agent in the bulk liquid. After application of mass balance and integration of previous equation, Eq. (15.16) was obtained.

$$C_l = \frac{C_0}{K_p(\alpha + 1)}\left[1 - \exp\left(-\frac{\alpha + 1}{\alpha}kt\right)\right] \tag{15.16}$$

where, the symbols used represents same parameters as in diffusion models.

For the abovementioned case of $Sh \gg 1$, relatively simplified solutions have been proposed by Baker and Lonsdale [92,94], with following approximations for both early time as well as late time period as given in Eqs. (15.17) and (15.18), respectively.

$$\frac{M_t}{M_\infty} = 6\left(\frac{Dt}{\pi r^2}\right)^{1/2} - \frac{3Dt}{r^2}; \quad \frac{M_t}{M_\infty} < 0.4 \tag{15.17}$$

$$\frac{M_t}{M_\infty} = 1 - \frac{6}{\pi^2}\exp\left(-\frac{\pi^2 Dt}{r^2}\right)^{\frac{1}{2}}; \quad \frac{M_t}{M_\infty} > 0.6 \tag{15.18}$$

For thin films with the assumption of edge effects to be negligible, the equation can be written as given in the following equation [17]:

$$\frac{M_t}{M_\infty} = 1 - \frac{8}{\pi^2}\sum_{n=0}^{\infty}\frac{1}{(2n+1)^2}e^{-(\pi^2 Dt(2n+1)^2/L^2)} \tag{15.19}$$

For cylinders, with the consideration of mass transfer in axial as well as radial direction, as in the case of HNTs, the same can be written as the following equation [93]:

$$\frac{M_t}{M_\infty} = 1 - \frac{32}{\pi^2}\sum_{n=1}^{\infty}\frac{1}{q_n^2}e^{\left(-(Dtq_n^2/R_o^2)\right)} \times \sum_{p=0}^{\infty}\frac{1}{(2p+1)^2}e^{\left(-(\pi^2 Dt(2p+1)^2/H^2)\right)} \tag{15.20}$$

#### 15.3.1.2.2 Dispersion model

The second case of initial loading of active agents more than its solubility in the containers can be described by dispersion models. Like a reservoir system, the dispersion system can be visualized to be consisted of two regions:

1. The non-diffusing core
2. The diffusing region surrounding the core

The core is different from the reservoir or depot, as the former consists of a polymer matrix and its thickness and it shrinks as the active agents diffuse through the diffusion region [91]. Hence, there is a necessity to incorporate moving boundary condition, which results in complicating the governing equations.

The dispersion model, wherein the transfer of active agents is diffusion controlled, for geometries such as planar sheets and spheres was introduced by Higuchi, assuming the system to be in a pseudo steady state. The well-known and the simplest form of Higuchi's equation for a thin film with negligible edge effect is given as in the following equation [95,96]:

$$M_t = S\sqrt{C_s Dt(2C_0 - C_s)} \tag{15.21}$$

where $S$ is the surface area available for release of active agents to the surrounding medium. The pseudo steady-state concentration profile for a sphere in the diffusing region can be given as the following equation [91]:

$$\frac{C}{C_s} = \frac{r'(R_o - r)}{r(R_o - r')} \tag{15.22}$$

where $r'$ is interface position of moving boundary. The fractional cumulative release of active agents can be given as the following equation [91]:

$$\frac{M_t}{M_\infty} = 1 - \left(\frac{r'}{R_o}\right)^3 \tag{15.23}$$

Other geometries and the dispersion models describing the release process have been widely studied by researchers all around the world, which are broadly described in available literature [97–101].

#### 15.3.1.3 Models for swelling controlled release systems

Incorporation of swelling polymers is done in release system to have a controlled release of active agents, specifically when it has a low diffusivity value in the polymer. The polymer employed is generally hydrophilic as it is susceptible to swelling; allowing water to get absorbed into the polymer, making it less entangled. The degree on entanglement varies with the concentration of polymer. Absorption of water into polymer changes the concentration of the polymers and thus changes its degree of entanglement. It results in swelling of matrix and formation of gel layer wherein enhanced diffusion occurs as a result of increased mobility of active agents. Hence, a deviation from Fickian model can be observed for such systems as the release depends upon many other factors such as disentanglement and dissolution.

Fig. 15–9 depicts the detailed polymer swelling mechanism, as developed by Lee and Peppas [102], considering the motion of swelling front. If there is negligible water absorption into the polymer, the release of active agents through the glassy region is Fickian diffusion-controlled. Contrary to this, if water mobility is significantly dominant, the drug release is described by "non-Fickian Case-II transport," wherein a sharp interface, moving at a constant velocity, and zero-order release of active agents exist. In between these two extremities, there exists another case of "anomalous transport" with intermediate transport characteristics.

Demant'eva et al. [103] studied the release kinetics of Myramistin that is a bactericidal drug from silica nanocontainers. They found that the release of this drug follows zero-order kinetics and that it is pH-sensitive. From this, they summarized that the water-penetration stage is the main rate-controlling stage and that a decrease in pH from 6.8 to 5, the release increases, and the time required for releasing the drug is less.

Korsmeyer et al. [104,105] developed a simple, semiempirical, power law–based equation describing the release of active agents from a swelling-controlled system as given in the following equation:

$$\frac{M_t}{M_\infty} = kt^n \tag{15.24}$$

where $k$ is release constant while $n$ is release exponent. It is observed as a superposition of Fickian diffusion and non-Fickian Case-II transport, with a feature that enables us to identify the release mechanisms from the value of release exponent. The values of $n$, as seen to be influenced by geometry, also vary with particles size distribution as per the observation of Ritger and Peppas [106]. Korsmeyer model was modified by Kim and Fassihi [107], with the inclusion of lag time and initial burst effect; the equation is given as follows:

$$\frac{M_t}{M_\infty} = k\left(t - t_{\text{lag}}\right)^n + b \tag{15.25}$$

Another modified form of Korsmeyer model was given by Peppas and Sahlin [108] as stated in Eq. (15.26), with the decoupling of diffusion and non-Fickian Case-II transport. In the following equation, $k_1$, $k_2$, and $m$ are constant. First term on the right-hand side of the equation describes the diffusion contribution while the second term describes non-Fickian

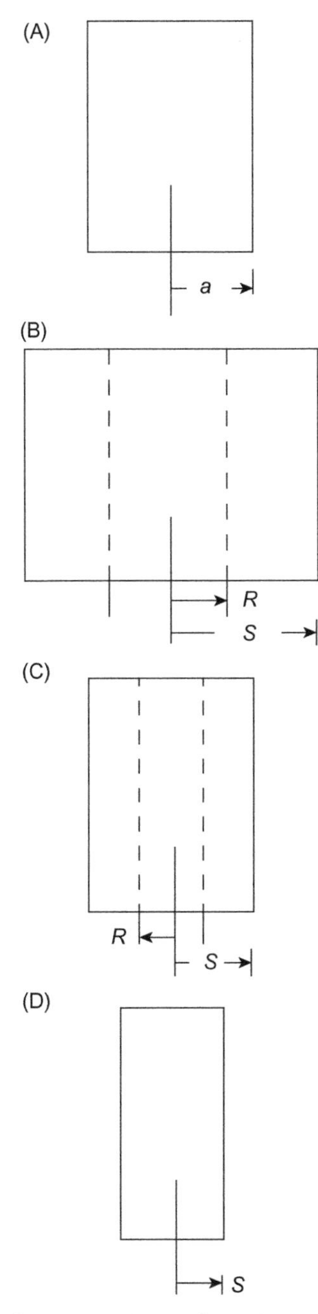


**FIGURE 15–9** Schematics of polymer swelling process in one dimension as a result of diffusion of solvent and dissolution of polymers: (A) carrier's initial thickness, (B) early time swelling with increasing rubbery/solvent interface and decreasing glassy /rubbery interface. (C) Late time swelling with decrease in both interfaces, (D) dissolution process where the entire slab consists of rubbery region, with rubbery/solvent decreasing interface ($S$: rubbery/solvent interface, $R$: rubbery/glassy interface) [102]. *[Reprinted with permission from P.I. Lee and N.A. Peppas, Prediction of polymer dissolution in swellable controlled-release systems, J. Control. Release 6 (1987) 207–215. Copyright (1987) Elsevier.]*

**Table 15–1** Interpretation of release models from polymeric matrices with different geometries [88,109].

| Model of release mechanism | Geometry of matrix | Exponent of release ($n$) |
|---|---|---|
| Fickian diffusion (Case-I diffusional) | Planar | 0.5 |
| | Cylindrical | 0.45 |
| | Spherical | 0.43 |
| Non-Fickian diffusion (anomalous transport) | Planar | $0.5 < n < 1$ |
| | Cylindrical | $0.45 < n < 0.89$ |
| | Spherical | $0.43 < n < 0.85$ |
| Case-II transport (zero-order release) | Planar | 1 |
| | Cylindrical | 0.89 |
| | Spherical | 0.85 |
| Super Case-II transport | Planar | $n > 1$ |
| | Cylindrical | $n > 0.89$ |
| | Spherical | $n > 0.85$ |



"Case-II transport." The values of the release exponents that reveal the type of diffusion taking place for a particular geometry of matrix are given in Table 15–1 [88,109].

$$\frac{M_t}{M_\infty} = k_1 t^m + k_2 t^{2m} \tag{15.26}$$

### 15.3.1.4 Models for erosion controlled release system

Polymers capable of undergoing bioerosion are highly desired for biomedical applications, specifically for the purpose of drug delivery. The kinetics of erosion can be controlled by carefully selecting the polymer and with implementation of variety of encapsulation techniques. Two ideal scenarios can be visualized to study polymer erosion (1) surface erosion and (2) bulk erosion. In the case of bulk erosion, the diameter of a nanocontainer/microsphere remains constant while fluid in surrounding medium is allowed to percolate into it. In surface erosion the diameter of nanocontainer/microsphere decreases as the polymer erodes into the surrounding medium. As theory suggests, the mechanism of erosion can be considered as a phenomenon wherein mass transfer is coupled with chemical reaction. It also involves several other phenomena, such as dissolution of active agents, degradation of polymers, creations of pores, and pH change in the microenvironment, which makes it even more complex. Mathematical models for erosion-based release process can be categorized into two subclasses:

1. Empirical models
2. Mechanistic models

The later considers phenomena of physio-chemistry involving mass transfer and processes of chemical reactions. The later can further be classified into categories such as diffusion and chemical reaction models and cellular automata models. For detailed information on the mechanistic models of erosion, there is a wide literature review done in the past [88,91].

#### 15.3.1.4.1 Empirical models

Weibull [110] developed an empirical equation, capable of describing variety of dissolution curves. It gives a relationship between cumulative fraction of release of active agents and release time, as in the following equation:

$$\frac{M_t}{M_\infty} = 1 - e^{\left(-\left(\left(t - t_{\text{lag}}\right)^b / t_{\text{scale}}\right)\right)} \tag{15.27}$$

where $t_{\text{lag}}$ is lag time, $t_{\text{scale}}$ is scale time, and $b$ is shape factor. $b = 1$ for exponential curve, $b > 1$ for sigmoid, S-shaped, with upward curvature followed by a turning point, $b < 1$ for parabolic curve, with a higher initial slope and after that consistent with the exponential [111].

Hopfenberg [112] formulated an empirical model as given in Eq. (15.28) to describe the release of active agents from surface eroding devices with various geometries, with an assumption that overall release process behaves as a zero-order release. This general mathematical equation is applicable for slabs, spheres, and infinite cylinders showing heterogeneous erosion, where $k_0$, $C_l$, and $a$ are the erosion rate constant, initial drug concentration, and initial radius of the sphere or cylinder, respectively.

$$\frac{M_t}{M_\infty} = 1 - \left[1 - \frac{k_0 t}{C_l a}\right]^n \tag{15.28}$$

Hixson and Crowell [113] noticed that there is a proportionality relationship between regular area of particles and the cube root of its volume. They derived the equation that could be described as given in Eq. (15.29), where $\kappa$ is the surface–volume relation constant.

$$W_0^{1/3} - W_t^{1/3} = \kappa\, t \tag{15.29}$$

where $W_0$ and $W_t$ represent the initial and remaining amount of active agents in the container. $\kappa$ (kappa) is a constant representing the surface–volume relation. This equation considers the change in surface area and diameter of particles during release from systems. This equation is applicable to such active agents containing nanocontainers/microspheres, where the dissolution occurs in the direction parallel to the nanocomposite. If there is a proportional reduction in the size of nanocontainers/microspheres, so that their geometry remains the same, then Eq. (15.30) can be applied [111].

$$W_0^{1/3} - W_t^{1/3} = \frac{\kappa' N^{1/3} D C_s t}{\delta} \tag{15.30}$$

where $\kappa'$ represents a constant related to the shape, surface, and density of the particle, $D$ represents the diffusion coefficient, $C_S$ represents the solubility at experimental temperature, and $\delta$ represents diffusion layer's thickness. If dissolution of particles in all sides is desired, the shape factors for spherical or cubic particles must be kept constant. This possibility does not exist for particles having different shape and as a result, the equation cannot be applied. [111].

Bhanvase et al. [10] studied the release kinetics of imidazole that is a corrosion inhibitor, loaded in cerium zinc molybdate nanocontainers and validated the experimental results using various models. They found that the Korsmeyer–Peppas and the Hixson–Crowell models can better predict the release behavior of imidazole as compared to the zero-order, first-order, and the Higuchi models. Fig. 15–10 shows how close the Korsmeyer–Peppas model predicted the corrosion inhibitor release behavior at various pH values. Also, the value of the rate constant is higher at pH 9 and 2 as compared to a neutral pH 7. This confirms better diffusion of imidazole at basic and acidic condition thus exhibiting better performance.

Similarly, release of benzotriazole from silica nanocontainer for corrosion inhibition application has been studied by Tyagi et al. [9] in which the Korsmeyer–Peppas release model is again found to be accurately predicting the data. In this study the authors also used artificial neural network (ANN) as a modeling tool and found that it can also predict the data to the same level of accuracy.

A release study, sensitive to pH, was done by Ghodke et al. [46] for the release of rosewater absolute from HNTs and it was found to increase with increase in pH from 3 to 7. The Korsmeyer–Peppas model turns out to be the best fit for the data with an $R^2$ value above 0.9 for every pH, and release exponent, $n$, to be approximately 0.4 thus confirming Fickian diffusion.

Further, release kinetics of carbidopa–levodopa extended release tablets synthesized by wet granulation technique was studied by Gouda et al. [114]. In this study, they tried to interpret the release profile of the drug using various models. They found the Higuchi square root model

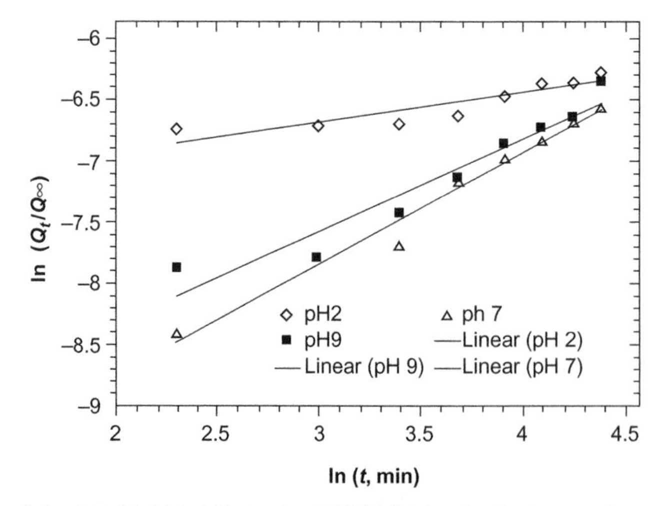


**FIGURE 15–10** Release behavior of imidazole (corrosion inhibitor) using the Korsmeyer–Peppas model at different pH conditions as done by Bhanvase et al. [10]. *[Reprinted with permission from B.A. Bhanvase, M.A. Patel and S.H. Sonawane, Kinetic properties of layer-by-layer assembled cerium zinc molybdate nanocontainers during corrosion inhibition, Corros. Sci. 88 (2014) 170–177. Copyright (2014) Elsevier.]*

to be the best fit for the experimental release data. Also, the drug release is found to follow diffusion mechanism which is of Super Case-II transport type as confirmed from the value of release exponent of the Korsmeyer–Peppas power law model. Karekar et al. [115] studied corrosion inhibitor release kinetics from zinc molybdate nanocontainers and found that the Korsmeyer–Peppas model is able to predict the release profile most accurately compared to other models. The graphs plotted for various models by the authors are as given in Fig. 15–11.

There are many other empirical models for erosion controlled release processes. Cooney [116] and El-Arini and Leuenberger [117] developed the empirical models, of which the end results are like those of the Hopfenberg model. The underlying difference between these two models and Hopfenberg model is that the later does not consider the effect of concentration gradient.

## 15.4 Simulation of release of active agents

The simulation techniques can be classified in two categories, molecular dynamics and Monte Carlo simulations, along which there exist many hybrid techniques with combined features from both the abovementioned techniques. Molecular dynamics holds advantage over Monte Carlo simulation as the system's dynamic properties can be more effectively analyzed using molecular dynamics. The trajectories of atoms and molecules are determined by numerically solving the equations of motion in molecular dynamics solutions, as evaluating such a complex system analytically is an extremely difficult task. On the other hand, Monte Carlo simulation technique is conceptually straightforward as it involves calculations of time-averaged properties and does not require calculation of forces.

A heat-driven release of ciprofloxacin drug molecules into the tissue by using carbon nanotube as a nanocarrier has been simulated by Chaban et al. [118] using molecular dynamics. In this study, the authors analyzed the effect of temperature and concentration of the drug inside the carbon nanotube on its diffusion coefficient. The main suggestion from this study was noted as the easy optical heat-assisted release of drug molecule from carbon nanotube. Panczyk et al. [119] studied the uncapping of a carbon nanotube containing drug using Monte Carlo simulations. For this, they designed a scheme of carbon nanotube capped with a magnetic nanoparticle at both its ends. Such nanocontainer can respond to a magnetic field and thus release the drug at the required location inside a living body. They studied the effect of various properties of the two components on the capping and uncapping of the nanocontainer and found that a very small magnetic nanoparticle (below 50 Å) cannot be triggered for uncapping the carbon nanotube unless a large magnetic field is applied. Also, it is found from the simulation that the process of uncapping responds rapidly to the applied field than the process of capping occurring in the absence of field. This study suggests that proper choice of sizes of carbon nanotube and the magnetic nanoparticle can help decrease the barrier for capping and uncapping the nanocontainer. In another study [120], the authors studied many other factors influencing the capping and uncapping of the same nanocontainers using Monte Carlo simulations. They analyzed the configuration of the

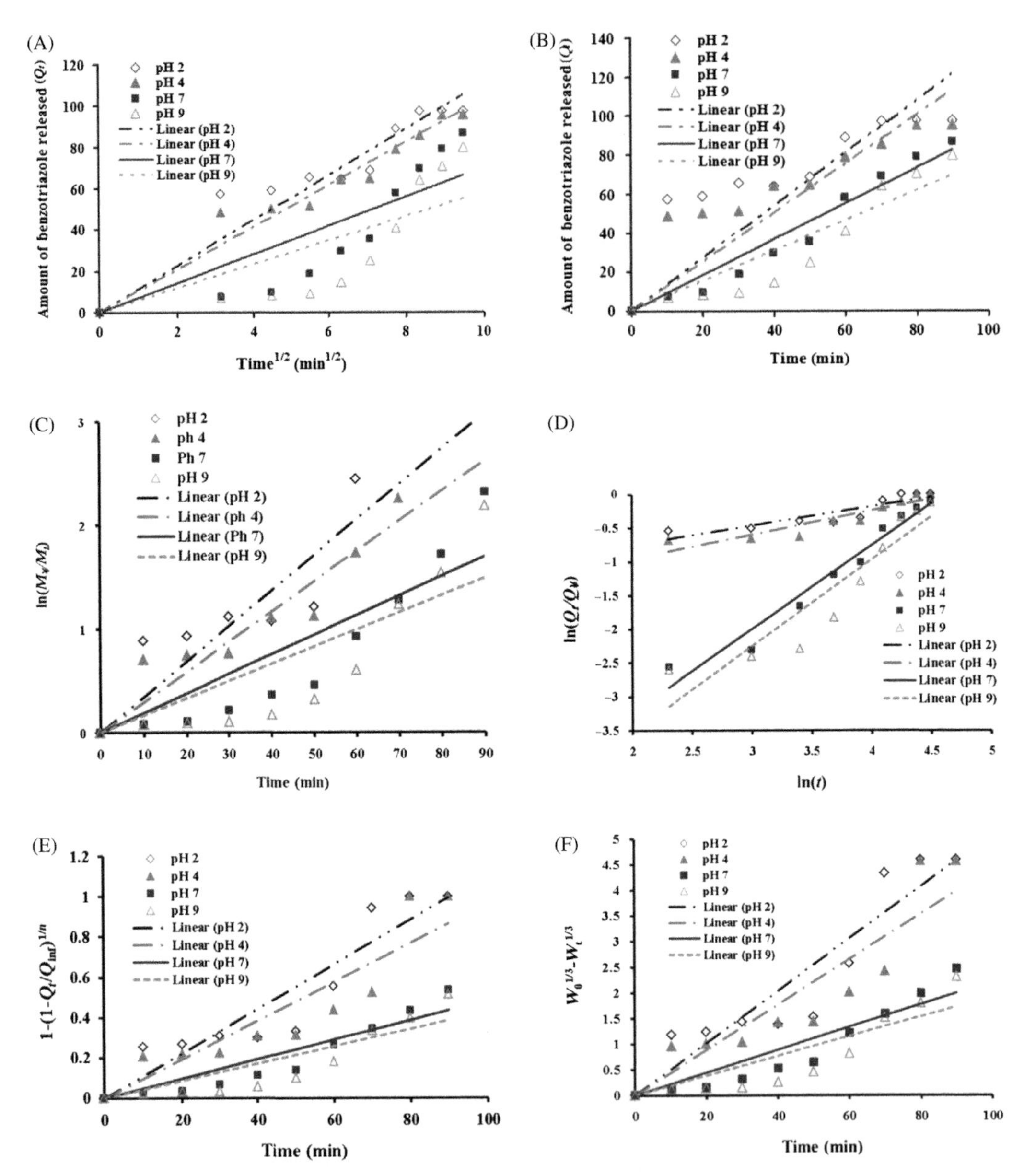


**FIGURE 15–11** Comparison between experimental and (A) Higuchi model, (B) zero-order model, (C) first-order model, (D) Korsmeyer–Peppas model, (E) Hopfenberg model, and (F) Hixson–Crowell model for corrosion inhibitor release from zinc molybdate nanocontaineras done by Karekar et al. [115]. *[Reprinted with permission from S.E. Karekar, U.D. Bagale, S.H. Sonawane, B.A. Bhanvase and D.V. Pinjari, A smart coating established with encapsulation of zinc molybdate centred nanocontainer for active corrosion protection of mild steel: release kinetics of corrosion inhibitor, Compos. Interface 25 (2018) 785–808. Copyright (2014) Taylor & Francis.]*

points of anchoring over the carbon nanotubes and found that it affects the process of capping more than uncapping. Due to the barrier for capping, the nanocontainer might get stuck in an uncapped position thus behaving in an irreversible manner. This can be avoided by increasing the length of the carbon nanotubes up to the length of sum of nanoparticle radii or by using electrostatic repulsion of the charged nanoparticles. Also, it is found that a slow increase in the intensity of magnetic field aligns the nanocontainer in the direction of the field thus reducing its effect on uncapping. This is prevented from happening by instantly increasing the strength of magnetic field.

Panczyk et al. [121] also simulated and optimized the nanocontainer and found that the state of the nanocontainer with various capped or uncapped configurations depend on the activation barriers. For this, they made a few modifications of the magnetic nanoparticles. They designed a magnetic nanoparticle protected inside a shell made of silica and alumina and found that the alumina shell can kinetically stabilize the various states of the nanocontainer, unlike the silica shell. Finally, they conclude that these computational studies are helpful in finding the best molecular parameters thus optimizing the process, without the use of expensive experimental procedures.

Further, Panczyk et al. [122] also studied the effects of forces of interaction between the same type of nanocontainers containing cisplatin molecules using molecular dynamics simulation. They found that the collision of the nanocontainers cannot result into uncapping, but it may occur at high temperatures as long as the spatial orientation if the nanocontainer during collision allows. It is also found that there are high chances of the nanocontainers to agglomerate due to large interaction energy, which can be prevented by functionalizing nanotube sidewalls. Finally, they conclude that the release of cisplatin molecules out of the nanotube follows a simple one-dimensional diffusion mechanism. Also, at high initial concentrations of cisplatin tucked inside the nanotube, there occurs a nonactivated diffusion due to the drag molecules.

Investigation of release of pyrazinamide drug from SWCNT by externally present fillers using molecular dynamics simulation has been done by Saikia et al. [123]. They found that the fullerene ($C_{60}$) present as filler drive the drug out of the SWCNT by getting encapsulated within and displacing the drug molecule and that this process becomes fast at higher temperatures. A dissipative particle dynamics simulation of release of self-healing agent from a nanocontainer by cracking was studied by Filipovic et al. [124].

Apart from the common forces considered during dissipative particle dynamics, such as the repulsive, dissipative, and random forces, an additional force of polymerization between two particles in a nanocontainer was also considered in another study done by Filipovic et al. [125]. Molecular dynamics simulation has also been a choice for simulating release of self-healing agents from nanocontainer in a coating [126]. Further, after realization of the inefficiency of the molecular dynamics simulation technique for simulating behavior of complex systems, like those we have in case of nanocontainers, a new method known as the dissipative particle dynamics was introduced [127]. There have been several studies on simulation of release of fertilizer from nanocontainers also [128,129].

Molecular dynamics simulation to study the temperature-driven desorption of stored methane from a carbon nanocapsule has been performed by Volkova et al. [78]. In this study, they demonstrated a nanocapsule made of a storage chamber and a blocking chamber connected by a junction in between them both for simulating the process. Vakhrushev et al. [130] have studied the adsorption storage and desorption of hydrogen in carbon nanocontainers using molecular dynamics simulation. They presented how the various structural forms of the nanocontainer affect these processes.

## 15.5 Summary

Nanocontainers present a lot of scope for their applications in almost every field where controlled release of active components is desired such as in drug delivery, anticorrosive self-healing coatings, wastewater treatment, fertilizers, catalysis, encapsulants in food industry, and gas storage. Further, various properties of the synthesized nanocontainer shall play an important role in determining the release behavior of the agent from inside the shell to the outside surrounding, one of which is the material of the shell. A proper selection of the encapsulating shell that acts as a protective layer carrying the required active agent inside it is very crucial part of their synthesis. This release operation is triggered with the help of various kinds of internal (diffusion, degradation, and swelling) and external stimulants (changes in temperature, pH, ionic strength, ultrasound, and electric or magnetic field). This chapter presents details of various studies done on release of agents from the nanocontainers with stress on modeling and simulation of this process. It is seen that there are various models developed that can predict the release profile of the active agents out to the surroundings by considering occurrence of different paces of kinetics. Various researchers have tried to predict their experimental release data by using different models such as the zero-order, first-order, Higuchi model, Korsmeyer–Peppas model, Hopfenberg model, and Hixson–Crowell model. Most of the studies done on release of corrosion inhibitor from a nanocontainer are found to follow the Korsmeyer–Peppas model of release of active agents. Further, simulation of release behavior of the active agents is commonly done by using Monte Carlo and molecular dynamics simulation methods. Along with this, few researchers also utilized the dissipative particle dynamics method as a nanocontainer release simulation tool. Overall, it is observed that the nanocontainers are very useful for carrying the active agents in various applications and that there are several ways to validate and simulate their release profiles.

## References


[1] S.H. Sonawane, B.A. Bhanvase, A.A. Jamali, S.K. Dubey, S.S. Kale, D.V. Pinjari, et al., Improved active anticorrosion coatings using layer-by-layer assembled ZnOnanocontainers with benzotriazole, Chem. Eng. J. 189– 190 (2012) 464–472.

[2] S.A. Kapole, B.A. Bhanvase, D.V. Pinjari, R.D. Kulkarni, U.D. Patil, P.R. Gogate, et al., Intensification of corrosion resistance of 2K epoxy coating by encapsulation of liquid inhibitor in nanocontainer core of sodium zinc molybdate and iron oxide, Compos. Interface 21 (2014) 469–486.

[3] S.A. Ghodke, S.H. Sonawane, B.A. Bhanvase, S. Mishra, K.S. Joshi, I. Potoroko, Halloysite nano containers for controlled delivery of gibberellic acid, Russ. J. Appl. Chem. 90 (2017) 120–128.

[4] S.A. Ghodke, S.H. Sonawane, B.A. Bhanvase, S. Mishra, K.S. Joshi, I. Potoroko, Functionalization, uptake and release studies of active molecules using halloysite nanocontainers, J Inst. Eng. (India): Ser. E 100 (2019) 59–70.

[5] S.G. Ghodke, B.A. Bhanvase, S.H. Sonawane, S. Mishra, K. Joshi, Nanoencapsulation and nanocontainer based delivery systems for drugs, flavours and aromas, in: A.M. Grumezescu (Ed.), Encapsulations: Nanotechnology in the Agri-Food Industry, vol. 2, Elsevier (Academic Press), 2016, pp. 673–715.

[6] S.E. Karekar, B.A. Bhanvase, S.H. Sonawane, Sustained and responsive release of corrosion inhibiting species from multilayered assembly of zinc phosphomolybdatenanocontainers with application in anticorrosive paint coating formulation (Partnered with CRC Press, a Taylor & Francis Group) in: S. Thomas, N. Kalarikkal, A.M. Stephan, B. Raneesh (Eds.), Advanced Nanomaterials Synthesis, Properties, and Applications, Apple Academic Press, 2014, pp. 329–353.

[7] R.K. Mishra, S.K. Tiwari, S. Mohapatra, S. Thomas, Efficient nanocarrier for drug-delivery systems: types and fabrication, in: S.S. Mohapatra, S. Ranjan, N. Dasgupta, R.K. Mishra, S. Thomas (Eds.), Nanocarriers for Drug Delivery. Nanoscience and Nanotechnology in Drug Delivery. A Volume in Micro and Nano Technologies, Elsevier, 2019, pp. 1–41.

[8] B.A. Bhanvase, Y. Kutbuddin, R.N. Borse, N.R. Selokar, D.V. Pinjari, P.R. Gogate, et al., Ultrasound assisted synthesis of calcium zinc phosphate pigment and its application in nanocontainer for active anticorrosion coatings, Chem. Eng. J. 231 (2013) 345–354.

[9] M. Tyagi, B.A. Bhanvase, S.L. Pandharipande, Computational studies on release of corrosion inhibitor from layer-by-layer assembled silica nanocontainer, Ind. Eng. Chem. Res. 53 (2014) 9764–9771.

[10] B.A. Bhanvase, M.A. Patel, S.H. Sonawane, Kinetic properties of layer-by-layer assembled cerium zinc molybdatenanocontainers during corrosion inhibition, Corros. Sci. 88 (2014) 170–177.

[11] J. Yang, J. Lee, J. Kang, K. Lee, J.-S. Suh, H.-G. Yoon, et al., Hollow silica nanocontainers as drug delivery vehicles, Langmuir 24 (2008) 3417–3421.

[12] J. Nicolas, P. Couvreur, Polymer nanoparticles for the delivery of anticancer drug, Med. Sci. 33 (2017) 11–17.

[13] S. Garg, A. Garg, A. Shukla, S.K. Dev, M. Kumar, A review on nano-therapeutic drug delivery carriers for effective wound treatment strategies, Asian J. Pharm. Pharmacol. 4 (2018) 90–101.

[14] V.P. Torchilin, Cell penetrating peptide-modified pharmaceutical nanocarriers for intracellular drug and gene delivery, Biopolymers 90 (2008) 604–610.

[15] U. Ikoba, H. Peng, H. Li, C. Miller, C. Yu, Q. Wang, Nanocarriers in therapy of infectious and inflammatory diseases, Nanoscale 7 (2015) 4291–4305.

[16] E. Cauchetier, M. Deniau, H. Fessi, A. Astier, M. Paul, Atovaquone-loaded nanocapsules: influence of the nature of the polymer on their in vitro characteristics, Int. J. Pharm. 250 (2003) 273–281.

[17] J. Crank, The Mathematics of Diffusion, second ed., Clarendon Press, Oxford, 1975.

[18] H.S. Yoo, T.G. Park, Biodegradable polymeric micelles composed of doxorubicin conjugated PLGA–PEG block copolymer, J. Control. Release 70 (2001) 63–70.

[19] N.A. Peppas, P. Bures, W. Leobandung, H. Ichikawa, Hydrogels in pharmaceutical formulations, Eur. J. Pharm. Biopharm. 50 (2000) 27–46.

[20] S. Kaity, J. Isaac, A. Ghosh, Interpenetrating polymer network of locust bean gum-poly(vinyl alcohol) for controlled release drug delivery, Carbohydr. Polym. 94 (2013) 456–467.

[21] S.A. Abouelmagd, H. Hyun, Y. Yeo, Extracellularly activatable nanocarriers for drug delivery to tumors, Expert Opin. Drug Deliv. 11 (2014) 1601–1618.

[22] D.G. Shchukin, D.A. Gorin, H. Mohwald, Ultrasonically induced opening of polyelectrolyte microcontainers, Langmuir 22 (2006) 7400–7404.

[23] M. Gai, J. Frueh, T. Tao, A.V. Petrov, V. Petrov, E.V. Shesterikov, et al., Polylactic acid nano- and microchamber arrays for encapsulation of small hydrophilic molecules featuring drug release via high intensity focused ultrasound, Nanoscale 9 (2017) 7063–7070.

[24] C.G. Dariva, A.F. Galio, Corrosion inhibitors-principles, mechanisms and applications, Developments in Corrosion Protection, IntechOpen Limited, London, 2014, ISBN: 978-953-51-1223-5, p. 16.

[25] K.A. Zahidah, S. Kakooei, M.C. Ismail, P.B. Raja, Halloysite nanotubes as nanocontainer for smart coating application: a review, Prog. Org. Coat. 111 (2017) 175–185.

[26] S.R. White, N.R. Sottos, P.H. Geubelle, J.S. Moore, M.R. Kessler, S.R. Sriram, et al., Autonomic healing of polymer composites, Nature 409 (2001) 794–797.

[27] E. Shchukina, D.G. Shchukin, Nanocontainer-based active systems: from self-healing coatings to thermal energy storage, Langmuir 35 (2019) 8603–8611.

[28] E.V. Skorb, D. Fix, D.V. Andreeva, H. Möhwald, D.G. Shchukin, Surface-modified mesoporous $SiO_2$ containers for corrosion protection, Adv. Funct. Mater. 19 (2009) 2373–2379.

[29] A.J. Jadhav, S.E. Karekar, D.V. Pinjari, Y.G. Datar, B.A. Bhanvase, S.H. Sonawane, et al., Development of smart nanocontainers with a zinc phosphate core and a pH-responsive shell for controlled release of immidazole, Hybrid. Mater. 2 (2015) 71–79.

[30] D.G. Shchukin, M. Zheludkevich, K. Yasakau, S. Lamaka, M.G.S. Ferreira, H. Möhwald, Layer-by-layer assembled nanocontainers for self-healing corrosion protection, Adv. Mater. 18 (2006) 1672–1678.

[31] E.V. Skorb, D.V. Sviridov, H. Möhwald, D.G. Shchukin, Light responsive protective coatings, Chem. Commun. 40 (2009) 6041–6043.

[32] E.V. Skorb, A.G. Skirtach, D.V. Sviridov, D.G. Shchukin, H. Möhwald, Laser-controllable coatings for corrosion protection, ACS Nano 3 (2009) 1753–1760.

[33] A. Vimalanandan, L. Lv, T.H. Tran, K. Landfester, D. Crespy, M. Rohwerder, Redox-responsive self-healing for corrosion protection, Adv. Mater. 25 (2013) 6980–6984.

[34] B. Bharati, A.K. Sonkar, N. Singh, D. Dash, C. Rath, Enhanced photocatalytic degradation of dyes under sunlight using biocompatible $TiO_2$ nanoparticles, Mater. Res. Express 4 (2017) 085503.

[35] S.P. Kim, M.Y. Choi, H.C. Choi, Photocatalytic activity of $SnO_2$ nanoparticles in methylene blue degradation, Mater. Res. Bull. 74 (2016) 85–89.

[36] D.M. Nasikhudin, A. Kusumaatmaja, K. Triyana, Study on photocatalytic properties of $TiO_2$ nanoparticle in various pH condition, J. Phys.: Conf. Ser. 1011 (2018) 012069.

[37] S.M. Gupta, M. Tripathi, A review of $TiO_2$ nanoparticles, Chin. Sci. Bull. 56 (2011) 1639–1657.

[38] H. Huang, G. Huang, H. Chen, Y. Lee, Immobilization of $TiO_2$ nanoparticles on Fe-filled carbon nanocapsules for photocatalytic applications, Thin Solid Films 515 (2006) 1033–1037.

[39] H. Tian, B. Li, H. Wang, Y. Li, J. Wang, S. Zhao, et al., A nanocontainer that releases a fluorescence sensor for cadmium ions in water and its biological applications, J. Mater. Chem. 21 (2011) 10298–10303.

[40] K. Cheng, Y. Zhou, Z. Sun, H. Hu, H. Zhong, X. Kong, et al., Synthesis of carbon-coated, porous and water-dispersive $Fe_3O_4$ nanocapsules and their excellent performance for heavy metal removal applications, Dalton Trans. 41 (2012) 5854–5861.

[41] S. Dib, M. Boufatit, S. Chelouaou, F. Sadi-Hassaine, J. Croissant, J. Long, et al., Versatile heavy metals removal via magnetic mesoporousnancontainers, RSC Adv. 4 (2014) 24838–24841.

[42] V. Van Tran, D. Park, L. Young-Chul, Hydrogel applications for adsorption of contaminants in water and wastewater treatment, Environ. Sci. Pollut. Res. 25 (2018) 24569–24599.

[43] A.M. Dave, M.H. Mehta, T.M. Aminabhavi, A.R. Kulkarni, K.S. Soppimath, A review on controlled release of nitrogen fertilizers through polymeric membrane devices, Polym. Technol. Eng. 38 (1999) 675–711.

[44] K. Hong, S. Park, Polyurea microcapsules with different structures: preparation and properties, J. Appl. Polym. Sci. 78 (2000) 894–898.

[45] L. Chen, Z. Xie, X. Zhuang, X. Chen, X. Jing, Controlled release of urea encapsulated by starch-*g*-poly(L-lactide), Carbohydr. Polym. 72 (2008) 342–348.

[46] S.A. Ghodke, S.H. Sonawane, B.A. Bhanvase, S. Mishra, K.S. Joshi, Studies on fragrance delivery from inorganic nanocontainers: encapsulation, release and modeling studies, J. Inst. Eng. (India): Ser. E 96 (2015) 45–53.

[47] M. Liu, Z. Jia, D. Jia, C. Zhou, Recent advance in research on halloysite nanotubes-polymer nanocomposite, Prog. Polym. Sci. 39 (2014) 1498–1525.

[48] Y. Fu, D. Zhao, P. Yao, W. Wang, L. Zhang, Y. Lvov, Highly aging-resistant elastomers doped with antioxidant-loaded clay nanotubes, ACS Appl. Mater. Interfaces 7 (2015) 8156–8165.

[49] Y.M. Lvov, D.G. Shchukin, H. Mohwald, R.R. Price, Halloysite clay nanotubes for controlled release of protective agents, ACS Nano 2 (2008) 814–820.

[50] M.C. Hermosín, R. Celis, G. Facenda, M.J. Carrizosa, J.J. Ortega-Calvo, J. Cornejo, Bioavailability of the herbicide 2,4-D formulated with organoclays, Soil Biol. Biochem. 38 (2006) 2117–2124.

[51] B. Zhong, S. Wang, H. Dong, Y. Luo, Z. Jia, X. Zhou, et al., Halloysite tubes as nanocontainers for herbicide and its controlled release in biodegradable poly(vinyl alcohol)/starch film, J. Agric. Food Chem. 65 (2017) 10445–10451.

[52] X. Zeng, B. Zhong, Z. Jia, Q. Zhang, Y. Chen, D. Jia, Halloysite nanotubes as nanocarriers for plant herbicide and its controlled release in biodegradable polymers composite film, Appl. Clay Sci. 171 (2019) 20–28.

[53] M. Patil, V. Patil, A. Sapre, T. Ambone, A.T. Torris, P. Shukla, et al., Tuning controlled release behaviour of starch granules using nanofibrillated cellulose derived from waste sugarcane bagasse, ACS Sustain. Chem. Eng. 6 (2018) 9208–9217.

[54] D. Yiamsawas, G. Baier, E. Thines, K. Landfester, F.R. Wurm, Biodegradable lignin nanocontainers, RSC Adv. 4 (2014) 11661–11663.

[55] F. Cuomo, A. Ceglie, A. De Leonardis, F. Lopez, Polymer capsules for enzymatic catalysis in confined environments, Catalysts 9 (2019) 1.

[56] M. Priebe, K.M. Fromm, Nanorattles or yolk-shell nanoparticales-what are they, how are they made, and what are they good for? Chemistry 21 (2015) 3854–3874.

[57] N. Yan, Q. Chen, F. Wang, Y. Wang, H. Zhong, L. Hu, High catalytic activity for CO oxidation of $Co_3O_4$ nanoparticles in $SiO_2$ nanocapsules, J. Mater. Chem., A 1 (2013) 637–643.

[58] Y. Yang, X. Liu, X. Li, J. Zhao, S. Bai, J. Liu, et al., A yolk-shell nanoreactor with a basic core and an acidic shell for cascade reactions, Angew. Chem. Int. Ed. 51 (2012) 9164–9168.

[59] C. Lin, X. Liu, S. Wu, K. Liu, C. Mou, Corking and uncorking a catalytic yolk-shell nanoreactor: stable gold catalyst in hollow nanosphere, J. Phys. Chem. Lett. 2 (2011) 2984–2988.

[60] M. Priebe, K.M. Fromm, One pot synthesis and catalytic properties of encapsulated silver nanoparticles in silica nanocontainers, Part. Part. Syst. Charact. 31 (2014) 645–651.

[61] J. Rubio-Cervilla, E. Gonzalez, J.A. Pomposo, Advances in single-chain nanoparticles for catalysis applications, Nanomaterials 7 (2017) 341.

[62] Q.C. Do, D.G. Kim, S.O. Ko, Controlled formation of magnetic yolk-shell structures with enhanced catalytic activity for removal of acetaminophen in a heterogeneous Fenton-like system, Environ. Res. 171 (2019) 92–100.

[63] Z. Yang, H. Wu, Z. Zheng, Y. Cheng, P. Li, Q. Zhang, et al., Tin nanoparticles encapsulated carbon nanoboxes as high-performance anode for lithium-ion batteries, Front. Chem. 6 (2018) 533.

[64] W. Zhou, Y. Yu, H. Chen, F.J. Di-Salvo, H.D. Abruna, Yolk-shell structure of polyaniline-coated sulfur for lithium-sulfur batteries, J. Am. Chem. Soc. 135 (2013) 16736–16743.

[65] M.H.A. El-Salam, S. El-Shibiny, Chapter 19: Natural biopolymers as nanocarriers for bioactive ingredients used in food industries, in: A.M. Grumezescu (Ed.), Nanotechnology in the Agri-Food Industry, Encapsulations, Elsevier, 2016, pp. 793–829.

[66] S. Gupta, S. Khan, M. Muzafar, M. Kushwaha, A.K. Yadav, A.P. Gupta, Chapter 6: Encapsulation: entrapping essential oil/flavors/aromas in food, in: A.M. Grumezescu (Ed.), Nanotechnology in the Agri-Food Industry, Encapsulations, Elsevier, 2016, pp. 229–268.

[67] A. Matalanis, O.G. Jones, D.J. McClements, Structured biopolymer-based delivery systems for encapsulation, protection, and release of lipophilic compounds, Food Hydrocoll. 25 (2011) 1865–1880.

[68] B. Shah, S. Ikeda, P.M. Davidson, Q. Zhong, Nanodispersingthymol in whey protein isolate–maltodextrin conjugate capsules produced using the emulsion–evaporation technique, J. Food Eng. 111 (2012) 79–86.

[69] M.R. Mozafari, Bioactive entrapment and targeting using nanocarrier technologies: an introduction, in: M.R. Mozafari (Ed.), Nanocarrier Technologies: Frontiers of Nanotherapy, Springer, The Netherlands, 2006, pp. 1–16.

[70] A.R. Bilia, C. Guccione, B. Isacchi, C. Righeschi, F. Firenzuoli, M.C. Bergonzi, Essential oils loaded in nanosystems: a developing strategy for a successful therapeutic approach, Evid. Based Complement. Altern. Med. 2014 (2014) 651593.

[71] J. Pandit, M. Aqil, Y. Sultana, Chapter 14: Nanoencapsulation technology to control release and enhance bioactivity of essential oils, in: A.M. Grumezescu (Ed.), Nanotechnology in the Agri-Food Industry, Encapsulations, Elsevier, 2016, pp. 597–640.

[72] A. Kumari, S.K. Yadav, S.C. Yadav, Biodegradable polymeric nanoparticles based drug delivery systems, Colloids Surf., B 75 (2010) 1–18.

[73] K.S. Soppimath, T.M. Aminabhavi, A.R. Kulkarni, W.E. Rudzinski, Biodegradable polymeric nanoparticles as drug delivery devices, J. Control. Release 70 (2001) 1–20.

[74] V.K. Bajpai, M. Kamle, S. Shukla, D.K. Mahato, P. Chandra, S.K. Hwang, et al., Prospects of using nanotechnology for food preservation, safety, and security, J. Food Drug Anal. 26 (2018) 1201–1214.

[75] Y.X. Ren, T.Y. Ng, K.M. Liew, State of hydrogen molecules confined in $C_{60}$ fullerene and carbon nanocapsule structures, Carbon 44 (2006) 397–406.

[76] A.V. Vakhrushev, M.V. Suyetin, Methane storage in bottle-like nanocapsules, Nanotechnology 20 (2009) 125602.1–125602.5.

[77] M.V. Suyetin, A.V. Vakhrushev, Nanocapsule for safe and effective methane storage, Nanoscale Res. Lett. 4 (2009) 1267–1270.

[78] E.I. Volkova, M.V. Suyetin, A.V. Vakhrushev, Temperature sensitive nanocapsule of complex structural form for methane storage, Nanoscale Res. Lett. 5 (2010) 205–210.

[79] O.V. Pupysheva, A.A. Farajian, B.I. Yakobson, Fullerene nanocage capacity for hydrogen storage, Nano Lett. 8 (2008) 767–774.

[80] J. Jin, L. Fu, H. Yang, J. Ouyang, Carbon hybridized halloysite nanotubes for high-performance hydrogen storage capacities, Sci. Rep. 5 (2015) 12429.

[81] A.P. Terzyk, S. Furmaniak, P.A. Gauden, P. Kowalczyk, Fullerene-intercalated graphenenano-containers—mechanism of argon adsorption and high-pressure $CH_4$ and $CO_2$ storage capacities, Adsorpt. Sci. Technol. 27 (2009) 281–296.

[82] T. Oku, M. Kuno, I. Narita, Hydrogen storage in boron nitride nanomaterials studied by TG/DTA and cluster calculation, J. Phys. Chem. Solids 65 (2004) 549–552.

[83] X. Ye, X. Gu, X.G. Gong, T.K.M. Shing, Z. Liu, A nanocontainer for the storage of hydrogen, Carbon 45 (2007) 315–320.

[84] C. Tang, C. Man, Y. Chen, F. Yang, L. Luo, Z. Liu, et al., Realizing the storage of pressurized hydrogen in carbon nanotubes sealed with aqueous valves, Energy Technol. 1 (2013) 309–312.

[85] R.E. Barajas-Barraza, R.A. Guirado-Lopez, Endohedral nitrogen storage in carbon fullerene structures: physisorption to chemisorption transition with increasing gas pressure, J. Chem. Phys. 130 (2009) 234706.1–234706.9.

[86] J. Shang, G. Li, R. Singh, P. Xiao, J.Z. Liu, P.A. Webley, Potassium chabazite: a potential nanocontainer for gas encapsulation, J. Phys. Chem., C 114 (2010) 22025–22031.

[87] M. Zamani, H.A. Dabbagh, H. Farrokhpour, Gas storage of simple molecules in boron oxide nanocapsules, Int. J. Quantum Chem. 113 (2013) 2319–2332.

[88] J. Siepmann, F. Siepmann, Mathematical modeling of drug delivery, Int. J. Pharm. 364 (2008) 328–343.

[89] M. Polakovič, T. Görner, R. Gref, E. Dellacherie, Lidocaine loaded biodegradable nanospheres II. Modelling of drug release, J. Control. Release 60 (1999) 169–177.

[90] J. Siepmann, N.A. Peppas, Modeling of drug release from delivery systems based on hydroxypropyl methylcellulose (HPMC), Adv. Drug Deliv. Rev. 48 (2001) 139–157.

[91] D.Y. Arifin, L.Y. Lee, C. Wang, Mathematical modeling and simulation of drug release from microspheres: implications to drug delivery systems, Adv. Drug Deliv. Rev. 58 (2006) 1274–1325.

[92] R. Baker, Controlled Release of Biologically Active Agents, John Wiley & Sons, New York, 1987.

[93] J.M. Vergnaud, Controlled Drug Release of Oral Dosage Forms, Ellis Horwood, New York, 1993.

[94] R.W. Baker, H.K. Lonsdale, Controlled release: mechanisms and rates, in: A.C. Tanquarry, R.E. Lacey (Eds.), Controlled Release of Biologically Active Agents, Plenum Press, New York, 1974, pp. 15–71.

[95] T. Higuchi, Physical chemical analysis of percutaneous absorption process from creams and ointments, J. Soc. Cosmet. Chem. 11 (1961) 85–97.

[96] T. Higuchi, Rate of release of medicaments from ointment bases containing drugs in suspensions, J. Pharm. Sci. 50 (1961) 874–875.

[97] T. Higuchi, Mechanisms of sustained action mediation. Theoretical analysis of rate of release of solid drugs dispersed in solid matrices, J. Pharm. Sci. 52 (1963) 1145–1149.

[98] S.J. Desai, A.P. Simonelli, W.I. Higuchi, Investigation of factors influencing release of solid drug dispersed in inert matrices, J. Pharm. Sci. 54 (1965) 1459–1464.

[99] S.J. Desai, P. Singh, A.P. Simonelli, W.I. Higuchi, Investigation of factors influencing release of solid drug dispersed in inert matrices II, J. Pharm. Sci. 55 (1966) 1224–1229.

[100] H. Lapidus, N.G. Lordi, Some factors affecting the release of a water-soluble drug from a compressed hydrophilic matrix, J. Pharm. Sci. 55 (1966) 840–843.

[101] H. Lapidus, N.G. Lordi, Drug release from compressed hydrophilic matrices, J. Pharm. Sci. 57 (1968) 1292–1301.

[102] P.I. Lee, N.A. Peppas, Prediction of polymer dissolution in swellable controlled-release systems, J. Control. Release 6 (1987) 207–215.

[103] O.V. Demant'eva, V.M. Rudoy, One-pot synthesis and loading of mesoporous $SiO_2$nanocontainers using micellar drug as a template, RSC Adv. 6 (2016) 36207–36210.

[104] R.W. Korsmeyer, S.R. Lustig, N.A. Peppas, Solute and pentrant diffusion in swellable polymers. I. Mathematical modeling, J. Polym. Sci., B 24 (1986) 395–408.

[105] R.W. Korsmeyer, E. von Meerwall, N.A. Peppas, Solute and penetrant diffusion in swellable polymers. II. Verification of theoretical models, J. Polym. Sci., B 24 (1986) 409–434.

[106] P.L. Ritger, N.A. Peppas, A simple equation for description of solute release. I. Fickian and non-Fickian release from nonswellable devices in the form of slabs, spheres, cylinders or discs, J. Control. Release 5 (1987) 23–26.

[107] H. Kim, R. Fassihi, Application of binary polymer system in drug release rate modulation. 2. Influence of formulation variables and hydrodynamic conditions on release kinetics, J. Pharm. Sci. 83 (1997) 323–328.

[108] N.A. Peppas, J.J. Sahlin, A simple equation for the description of solute release. III. Coupling of diffusion and relaxation, Int. J. Pharm. 57 (1989) 169–172.

[109] M. Amoli-Diva, K. Pourghazi, M.H. Mashhadizadeh, Magnetic pH-responsive poly(methacrylic acid-*co*-acrylic acid)-*co*-polyvinylpyrrolidone magnetic nano-carrier for controlled delivery of fluvastatin, Mater. Sci. Eng., C: Mater. Biol. Appl. 47 (2015) 281–289.

[110] W. Weibull, A statistical distribution of wide applicability, J. Appl. Mech. 18 (1951) 293–297.

[111] P. Costa, J.M. Sousa Lobo, Modeling and comparison of dissolution profile, Eur. J. Pharm. Sci. 13 (2001) 123–133.

[112] H.B. Hopfenberg, Controlled release from erodible slabs, cylinders, and spheres, in: Paul, D.R., Harris, F.W. (Eds.), Controlled Release Polymeric Formulations, ACS Symp. Ser. No. 33, American Chemical Society, Washington, DC, 1976, pp. 26–32.

[113] A.W. Hixson, J.H. Crowell, Dependence of reaction velocity upon surface and agitation, Ind. Eng. Chem. 23 (1931) 923–931.

[114] R. Gouda, H. Baishya, Z. Qing, Application of mathematical models in drug release kinetics of carbidopa and levodopa ER tablets, J. Dev. Drugs 6 (2017) 171.

[115] S.E. Karekar, U.D. Bagale, S.H. Sonawane, B.A. Bhanvase, D.V. Pinjari, A smart coating established with encapsulation of zinc molybdate centred nanocontainer for active corrosion protection of mild steel: release kinetics of corrosion inhibitor, Compos. Interface 25 (2018) 785–808.

[116] D.O. Cooney, Effect of geometry on the dissolution of pharmaceutical tablets and other solids: surface detachment kinetics controlling, AIChE J. 18 (1972) 446–449.

[117] S.K. El-Arini, H. Leuenberger, Dissolution properties of praziquantel–PVP systems, Pharm. Acta Helv. 73 (1998) 89–94.

[118] V.V. Chaban, T.I. Savechenko, S.M. Kovalenko, O.V. Prezhdo, Heat-driven release of a drug molecule from carbon nanotube: a molecular dynamics study, J. Phys. Chem., B 114 (2010) 13481–13486.

[119] T. Panczyk, T.P. Warzocha, P.J. Camp, A magnetically controlled molecular nanocontainer as a drug delivery system: the effects of carbon nanotube and magnetic nanoparticle parameters from montecarlo simulations, J. Phys. Chem., C 114 (2010) 21299–21308.

[120] T. Panczyk, T.P. Warzocha, P.J. Camp, Enhancing the control of a magnetically capped molecular nanocontainer: Monte Carlo studies, J. Phys. Chem., C 115 (2011) 7928–7938.

[121] T. Panczyk, P.J. Camp, G. Pastorin, T.P. Warzocha, Computational study of some aspects of chemical optimization of a functional magnetically triggered nanocontainer, J. Phys. Chem., C 115 (2011) 19074–19083.

[122] T. Panczyk, T. Da Ros, G. Pastorin, A. Jagusiak, Role of intermolecular interactions in assemblies of nanocontainerscomposed of carbon nanotubes and magnetic nanoparticles: a molecular dynamics study, J. Phys. Chem., C 118 (2014) 1353–1363.

[123] N. Saikia, A.N. Jha, R.Ch Deka, Dynamics of fullerene-mediated heat-driven release of drug molecules from carbon nanotubes, J. Phys. Chem. Lett. 4 (2013) 4126–4132.

[124] N. Filipovic, D. Petrovic, A. Jovanovic, D. Balos, M. Kojic, DPD simulation of self-healing process of nanocoating, J. Serb. Soc. Comput. Mech. 2 (2008) 42–50.

[125] N. Filipovic, D. Petrovic, V. Isailovic, A. Jovanovic, M. Kojic, Modeling of self-healing process in new nanocoating of surfaces by material with containers filled with healing agents, Contemp. Mater. 1 (2010) 129–132.

[126] S. Tyagi, J.Y. Lee, G.A. Buxton, A.C. Balazs, Using nanocomposite coatings to heal surface defects, Macromolecules 37 (2004) 9160–9168.

[127] P.J. Hoogerbrugge, J.M.V.A. Koelman, Simulating microscopic hydrodynamic phenomena with dissipative particle dynamics, Europhys. Lett. 19 (1992) 155–160.

[128] C. Du, D. Tang, J. Zhou, H. Wang, A. Shaviv, Prediction of nitrate release from polymer-coated fertilizers using an artificial neural network model, Biosyst. Eng. 99 (2008) 478–486.

[129] S.A. Irfan, R. Razali, K. KuShaari, N. Mansor, B. Azeem, A.N.F. Versypt, A review of mathematical modeling and simulation of controlled-release fertilizers, J. Control. Release 271 (2018) 45–54.

[130] A.V. Vakhrushev, M.V. Suetin, Carbon nanocontainers for gas storage, Nanotechnol. Russ. 4 (2009) 806–815.

# 16

# Flavor encapsulation and release studies in food

Shital B. Potdar[1], Vividha K. Landge[1], Shrikant S. Barkade[2], Irina Potoroko[3], Shirish H. Sonawane[4]

[1]CHEMICAL ENGINEERING DEPARTMENT, NATIONAL INSTITUTE OF TECHNOLOGY, WARANGAL, INDIA [2]CHEMICAL ENGINEERING DEPARTMENT, SINHGAD COLLEGE OF ENGINEERING, PUNE, INDIA [3]DEPARTMENT OF FOOD TECHNOLOGY AND BIOTECHNOLOGY, SCHOOL OF MEDICAL BIOLOGY, SUSU, CHELYABINSK, RUSSIA [4]CHEMICAL ENGINEERING DEPARTMENT, NATIONAL INSTITUTE OF TECHNOLOGY, WARANGAL, INDIA

## Chapter outline

**16.1 Introduction ..... 294**
16.1.1 Encapsulation of flavor ..... 294
16.1.2 Classification of encapsulation ..... 295
16.1.3 Definition of aroma and flavor: necessity of flavor encapsulation ..... 295
16.1.4 Carrier materials used for flavor encapsulation ..... 296
**16.2 Aroma extraction methods ..... 296**
16.2.1 Pretreatment methods to bioactive flavor compounds ..... 297
16.2.2 Aroma extraction using hot water bath ..... 298
16.2.3 Extraction of bioactive compounds using Soxhlet extractor ..... 298
16.2.4 Bioactive aroma extraction using microwave irradiation ..... 299
16.2.5 Maceration process for aroma extraction ..... 299
16.2.6 Aroma extraction using supercritical carbon dioxide ..... 300
16.2.7 Ultrasound-assisted extraction of bioactive aroma compounds ..... 300
**16.3 Encapsulation techniques: conventional and newer approach ..... 300**
16.3.1 Conventional techniques of encapsulation ..... 300
16.3.2 Newer intensified approach for the flavor encapsulation: nanoemulsions ..... 307
**16.4 Phenomena of encapsulated flavor release ..... 313**
16.4.1 Controlled flavor release ..... 313
16.4.2 Modes of uncontrolled flavor release ..... 315

Encapsulation of Active Molecules and their Delivery System. DOI: https://doi.org/10.1016/B978-0-12-819363-1.00016-8

**16.5 Characterization techniques for encapsulated bioactive compounds ... 315**
16.5.1 Droplet size and stability measurements of food emulsions ... 315
16.5.2 Surface morphology and topography ... 316
16.5.3 Estimation of antioxidant activity for flavor encapsulated food emulsions ... 316
16.5.4 Estimation of total phenolic content of flavor encapsulated food emulsions ... 316
16.5.5 Estimation of flavonoids concentrations in flavor encapsulated food emulsions ... 317
**16.6 Conclusion and future prospective ... 317**
**Acknowledgment ... 317**
**References ... 318**

# 16.1 Introduction

## 16.1.1 Encapsulation of flavor

Encapsulation can be defined as a process of stabilization of active compounds through the structuring of systems capable of preservation of their chemical, physical, and biological properties, as well as their release or delivery under established or desired conditions [1]. In the process of encapsulation, either one or a mixture of bioactive materials is coated with another single material or combination of materials. In encapsulation, two main terminologies are frequently used. One is the material that is being coated is termed active material or core material and another is the shell material that is termed carrier material, wall material, shell, encapsulating agent, or coating material. The basic schematic structure of core–shell encapsulation is shown in Fig. 16–1. Core material (generally lipid) is in the center, and shell material is dispersed in continuous phase.

The encapsulation is carried out in order to protect unstable, sensitive material from the environment before it's in use. It is a technique that provides better processability and protects interaction between flavor and food when stored. This technique enhances the shelf-life of the product [2]. It is observed that the bioactive compounds when encapsulated helps in controlled and sustained time release of active core material.

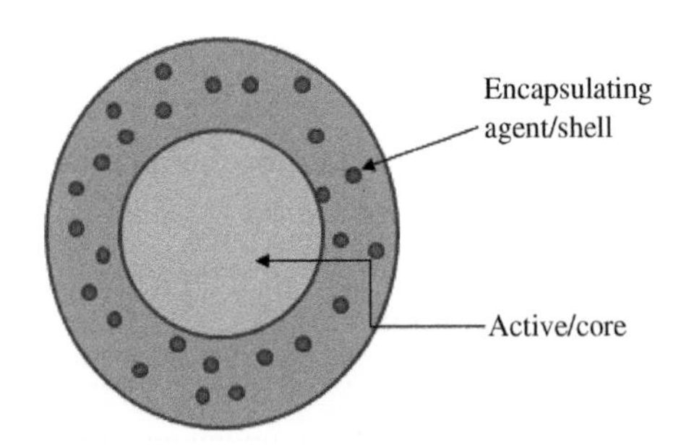


**FIGURE 16–1** Schematic of encapsulation: core–shell morphology.

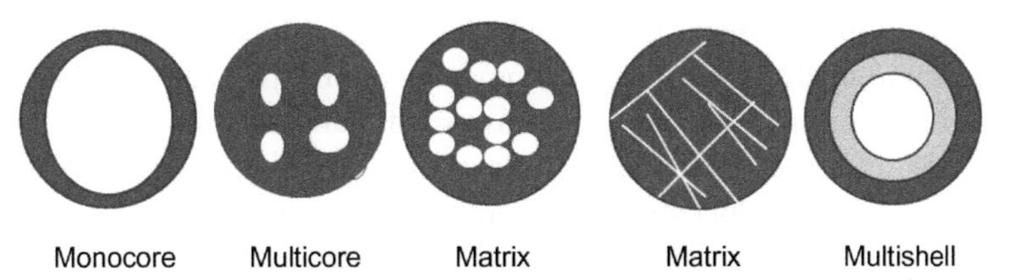


**FIGURE 16–2** Schematic of encapsulation classification.

## 16.1.2 Classification of encapsulation

The structure of encapsulation can be broadly divided into two categories: first is reservoir type and second is matrix type. In reservoir type, structure of encapsulated material is such that the active agent is having shell around it. It is also known as mono-core capsule, core shell type or single-core, whereas in matrix type, structure of encapsulated material is such that the core material is sprayed or dispersed over the outer coating layer (Fig. 16–2).

## 16.1.3 Definition of aroma and flavor: necessity of flavor encapsulation

The terms aroma and flavor are close to each other but differ in a way that flavor is a combination of the terms taste, aroma, and feel of food, whereas aroma is nothing but odor and is usually known as smell and is felt by consumer through nose.

Nowadays, consumers are aware about food. The consumers prefer natural food preservatives as compared to synthetic food preservatives. Because consumers suspect that there are side effects with synthetic preservative and it is not good for health [3]. This fact leads to a rapid or high need for the alternative search for synthetic preservative. The encapsulation technique helps to fulfill this need. The encapsulation is a process used for the preservation of food bioactive compounds. It is done with different shells and core materials, depending on the purpose to serve. The reduction in flavor is observed during the food-processing step, so it is necessary to add flavor externally. The encapsulated flavor helps to make up by bringing the flavor in its natural form [4].

Out of several reasons to encapsulate flavor, its functionality and preservation stability are of more importance [5]. The encapsulation of flavor compounds takes the attraction of consumer because of its taste and smell; moreover, they are easily digestible. In the process of encapsulation the aroma is extracted and is converted into oil, which is stored in the form of powder with different methods section in encapsulation methods. The encapsulated flavor powder is superior in a way that it is easy to handle and dust free. Fulger and Popplewell [6] observed that flavor molecules are in general low molecular weight compounds. Most of the flavor molecules are lipids in nature but some hydrophilic flavor s also exists in nature. The low molecular weight flavors are aldehydes, hydrocarbons, ketones, alcohols, esters, acids, sulfides, etc.

The encapsulated food is better over conventionally stored food in a way that encapsulated aroma is stored in the solid form and thus becomes easy to handle. The possibility of contamination of flavors in liquid forms is reduced because solids can be stored in the dust-free powder. Encapsulated food ingredients become stable decreasing the chances of bacteria growth. Improved stability in final product and during processing can be obtained because encapsulated flavor is less volatile as compared to natural flavor, and moreover, it can be released with control. Enhanced safety (e.g., decreased flammability of flavor) is also an advantage of encapsulation. Flavor is encapsulated with different technologies that allow to control its size and structure.

The abovementioned benefits are subject to minimize the following weak points of encapsulation like preserving flavor in encapsulation form will save the flavor but the cost of encapsulation synthesis is increased. The studies are going to make the process cost effective. Externally added flavor sometimes tends to change the taste of food in which it is loaded. So retaining original natural taste is also an important factor.

### 16.1.4 Carrier materials used for flavor encapsulation

The selection of carrier material is a crucial step in encapsulation. The selection depends on the purpose for which encapsulated product is being synthesized. The core material should be chosen in such a way that it is compatible with active material used as core material. It is reported that one core used with specific core material cannot form a stable product if the method of encapsulation is changed. The carrier material should be easy to handle and it should give the maximum protection for core material.

In the choice for selection of coating/carrier material, it is preferred to use the different blend of 2–3 carrier materials to increase encapsulation efficiency as well as cost effectiveness (Table 16–1).

Charve and Reineccius [7] studied the behavior of different carrier materials and came up with the conclusion that gum acacia is having highest flavor retention ability followed by modified starch having 88%, whey protein isolate 87%. They also mentioned that gum acacia is having the highest retention ability as it undergoes oxidation, whereas whey protein forms a barrier for oxidation. Vaidya et al. [8] compared the performance of various carrier materials for the encapsulation of oleoresin, they concluded that the ratio of 4:1:1 of gum arabic: maltodextrin:modified starch provides a protection, when compared with stability and protection provided by gum arabic alone. They supported the results with half lifetime period. Kanakdande et al. [9] encapsulated the cumin oleoresin with the mixture of gum arabic, maltodextrin and third compound in the blend was modified starch. Their study concluded that the Gum arabic provides better protection than maltodextrin and modified starch.

## 16.2 Aroma extraction methods

The food products that are prepared using plants form the key sources of flavonoids for humans. When food is processed for different purposes such as preservation, stability, and tastes, its flavor is reduced to some extent so it is necessary to add the flavor in the food

**Table 16–1** List of carrier materials used for encapsulation system.

| Carrier material | Examples | Properties | Limitations in use as carrier material |
|---|---|---|---|
| Carbohydrate | Starch, maltodextrin, cyclodextrin gums | Act as flavor binder<br>Low cost<br>Low viscosity<br>Good solubility | – |
| Proteins | Sodium caseinate, whey protein<br>Soy protein isolates gelatin | Good solubility in aqueous phase<br>Less viscous<br>Acts as an emulsifier and ability to form film on core material | High cost of proteins leads to limited application |
| Starch | Modified starches maltodextrins, b-cyclodextrins | Acts as aroma carriers fat replacers performs as a stabilizer in emulsion | |
| Maltodextrin | Corn flour with acids or enzymes | Capability to form matrices/less costly with good carrier properties<br>When used with other carrier material reduces cost of encapsulation less viscous at a higher solid ratio maltodextrin with different molecular weight are available | Does not act as an emulsifier<br>Flavor retention capacity is low |
| Gums | Gum arabic | Good effect on the taste and flavor of foods<br>Solubility, viscosity, and flavor retention properties | High cost in comparison with maltodextrin and its availability is also an issue |

externally. This fact makes the necessity to extract flavors from natural bioactive compound and use in food as an additive. The methods in this section describe the flavor extraction technique from natural herbs. Finally, the selection of extraction method depends on the flavor to be encapsulated. For example, heat-sensitive material cannot be used using hot water bath extraction. Product characteristic such as particle size, flavonoids content decides the extraction method.

## 16.2.1 Pretreatment methods to bioactive flavor compounds

When the purpose is to encapsulate the flavor, it is extracted from natural sources such as ginger, spices, and fruit peels. The very first step in this process is, the bioactive aroma compounds are collected and given pretreatment such as cleaning to remove foreign material, if present. In some cases, it is desired to use raw material in fresh form for extraction because its extraction efficiency is higher, but in most of pretreatment methods, bioactive aroma compound is dried using various methods, such as sun drying and freeze drying.

Drying is done in order to avail some of the characteristics of dried material that the dried material gives increased yield per unit weight of raw material. Drying of materials not only increases the shelf-life of aroma but also facilitates ease in handling and storage. In the next step, the dried aroma is crushed using domestic mixer or using ball mill. In grinding process the size of dried aroma is decreased to an approximate size of 1 mm and sent for extraction with different extraction methods, depending upon the purpose of extraction, nature of the material to be extracted. These extraction methods are well explained in the next section. Throughout, the process care is taken to assure that bioactive components are not lost, altered, or destroyed in the extraction from bioactive source.

### 16.2.2 Aroma extraction using hot water bath

In the extraction using hot water bath, extracting solvent is chosen, depending upon the nature of bioactive compound (hydrophilic/lipophilic). Deionized or tap water is the best choice for hot water bath extraction. The process of hot water bath extraction is necessary to remove oxygen present in the water. To do so, ultrasound or helium gas purging is used. Ultrasonication is a cost-effective method for degassing when the choice is available. For complete degassing a 60-min time interval is required. In hot water bath, pressure is maintained either by adding water with pump or adding water manually to increase the water level. To provide heat necessary for extraction of bioactive compound from water, convection of heat from water to bioactive material is necessary, which is accomplished using stirrer in the vessel. For heating the water heating jacket or oven is used.

### 16.2.3 Extraction of bioactive compounds using Soxhlet extractor

The Soxhlet extraction is an old-age technique for lipid extraction. It is a laboratory technique to extract lipid from a solid material. It is used when the desired flavor to be extracted from bioactive material is having less solubility in a solvent. The equipment has three main parts—distillation flask, extraction chamber, and condenser. The flavor base material to be extracted is filled in a distillation flask which is stirred and heated. The vapor passes through distillation path and condenses in the condenser chamber. In condenser the solvent is cooled and sent back to the middle portion, that is, extraction chamber which houses the solid material. When this chamber is full, it is emptied in distillation flask with the help of siphon. The thimble is used to move the solvent rapidly. The characteristic feature of thimble is that it passes the solvent with ease but restricts the transformation of solids inside still pot. In order to get concentrated flavor the extraction cycle is repeated many times. Once the solvent is used for one cycle, it is recycled in the next cycle. In the Soxhlet extractor, after extraction process, the desired sample gets accumulated and concentrated in distillation flask. The solvent present in the distillation flask is removed with the help of rotary evaporator. The advantages of Soxhlet extraction include one magnitude higher recovery of flavor from its base herb, the process is stable, extraction can be carried out in a continuous manner, and it requires lower energy in comparison with other extraction techniques. All these advantages make its use in extraction, but this process also carries disadvantages, such as the solvents used in extraction are hazardous and easily flammable. The solvent for extraction must be of pure grade, and as the process is operated in many cycles, the total duration of extraction is increased. Thermally sensitive food material cannot be extracted with Soxhlet extraction.

The Soxhlet extraction method has many advantages over other extraction methods. The extraction process requires organic solvents that are hazardous. The general time of extraction in Soxhlet extractor is 24 h and the quantity of solvent depends on the weight of plant material. Root and stem parts of the tree are heavier than leaf part. So, the solvent required to settle well in the extractor is more for the heavier part in comparison to latter part. Depending upon the material to be extracted, the solvent is chosen, that is, hexane is the best solvent used to remove oily and fatty materials (nonpolar), whereas chloroform or methanol are useful solvents for the extraction of chlorophyll (slightly polar), ethyl acetate is most common solvent for extraction of phenolic compounds. Acetone or methanol is useful to remove all components (polar). After solvent extraction, water extraction of the plant material also useful. These solvents are costly and need to be used in a pure state. This is a lab-scale process and difficult to transfer at high scale. The extraction is generally done in continuous mode, and thus, the time required for complete extraction is more. With the above limitations the Soxhlet extraction process is still in use because the recovery is high, process is simple and thermally stable. Also, the energy required for extraction is less. The limitations of the method are, the solvents used for extraction are hazardous and flammable. Also, potential toxic emissions during extraction take place. Solvents used in Soxhlet extraction are costly, and it is a laborious procedure. A complete extraction process requires time in days as compared to methods such as ultrasound extraction that requires time in hours.

### 16.2.4 Bioactive aroma extraction using microwave irradiation

Microwave-assisted extraction is a newer approach of extraction. In this approach, extraction rate is high and the extraction process is safe and cheap. This method was first adopted by Ganzler et al. [10] to extract different compounds from different types of soils, seeds, and foods. In microwave-assisted extraction the food material to be extracted is crushed/grounded. The solvent such as methanol is added to crushed powder. The samples are kept in microwave for irradiation for 30–50 s. Care must be taken that sample is not boiling inside the microwave. After irradiation, sample is taken out and cooled. The cycle needs to be repeated several times in order to assure complete extraction. In the next step, extracted sample is sent for centrifuge to remove the biomass. Microwave-assisted extraction is rapid, safe, and cheap. In this method, it is not essential for the sample to be free from water. It is a good method for separating the isolating lipids and pesticides from seeds, foods, feeds, and soil.

### 16.2.5 Maceration process for aroma extraction

In maceration extraction process the bioactive aromatic plant material is crushed or cut into small pieces to make it's powder (particle size = 1 mm) and is kept in a closed container to which extracting solvent is added and allowed to stand for a long duration (7–8 days), and the sample is occasionally shook. The completion of extraction is observed visually from color change to desired level. After extraction the liquid is strained off and solid biomass is separated. Extracted supernant is separated and stored for further use in encapsulation as core material.

### 16.2.6 Aroma extraction using supercritical carbon dioxide

In the supercritical fluid (SCF) extraction technique the critical temperature and pressure conditions are applied to fluid at ambient condition to achieve SCF properties. In this process of extraction the bioactive compound (extraction source) is brought in contact with SCF, the parameters such as temperature, pressure, and time are maintained and extraction is done. After extraction is complete the oil (from bioactive) and SCF are separated by drop in solution pressure. The advantage of SCF-extraction technique over technique of extraction with solvent and mechanical-pressing technique is that the aroma in the bioactive aromatic compound is retained to better extent.

### 16.2.7 Ultrasound-assisted extraction of bioactive aroma compounds

In the extraction using ultrasound technique the waves in the range of 20 kHz are passed through the solution of bioactive compounds to be extracted and extracting solvent. Once the extraction is complete, biomass and extracted component are separated by centrifuging at around 8000–9000 rpm. Supernant is separated and solvent from it is removed by evaporation.

The extraction efficiency of ultrasound-assisted extraction is higher amongst all other extraction techniques. Extraction time is less (nearly 1 h) as compared to many hours by conventional processes; it is suitable for thermally sensitive materials. The schematic of flow diagram for the extraction of ginger by wet and dry method is shown in Fig. 16–3.

## 16.3 Encapsulation techniques: conventional and newer approach

### 16.3.1 Conventional techniques of encapsulation

The conventional encapsulation techniques are of two types: chemical and physical. There are three methods of chemical encapsulation—coacervation, molecular inclusion, and cocrystallization. Physical encapsulation is also known as mechanical encapsulation. Spray drying, extrusion, freeze drying and vacuum drying, spray cooling/chilling, fluidized bed coating are the methods of physical encapsulation. The abovementioned methods are described in this section. The choice of method depends on the properties of core and shell material and another important parameter in selection is that method should give higher encapsulation efficiency and loading capacity. Knowing the details of the process is a way to select the process for encapsulation.

#### *16.3.1.1 Chemical encapsulation techniques*

**16.3.1.1.1 Encapsulation by coacervation**

The coacervation is the very first and original technique used for microencapsulation. It is also known as phase-separation technique. Coacervation is one of the chemical methods for incorporating the polymer droplets in suspension. The main principle is the separation of

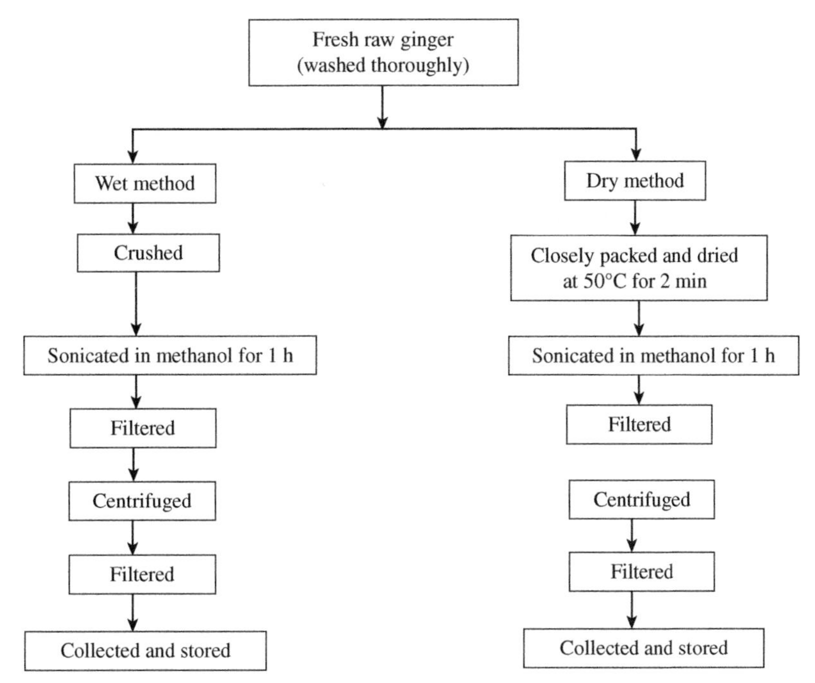


**FIGURE 16–3** Schematic of ultrasound-assisted extraction.

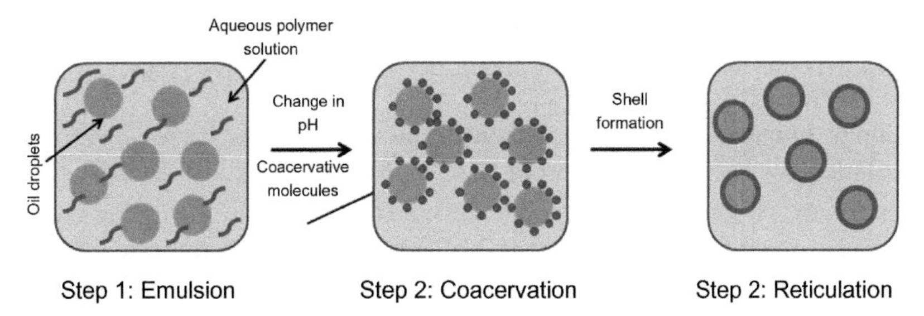


**FIGURE 16–4** Schematic of coacervation in three steps.

two liquid phases into a single concentrated colloidal phase, being the coacervate, and another highly dilute colloidal phase [11]. Here, liquid phase is used as a coating and used to coat lipid phase that is suspended in a solution. The polymer-containing phase is known as coacervate, and the other is referred to as equilibrium solution [12]. Aqueous phase used is generally consisting of gelatine alone or combination of gelatine and gum acacia. The necessary condition in choice of coating material is that it should not have any solubility in core (lipid) particles. Fig. 16–4 depicts the main three steps in coacervation—emulsion preparation, formation of coacervation, and reticulation. Once the microcapsules are formed, they are not stable. So, further processing is required wherein aqueous/continuous phase is separated by filtration or centrifugation. In the next step, these microcapsules are washed with some solvent and dried.

The coacervation process of encapsulation is basically categorized into two types:

1. *Simple coacervation:*

   In the simple coacervation only addition of poor solvent takes place, a poor solvent is added to a hydrophilic colloidal solution resulting in the creation of two phases: in one phase, polymer is used which is called coacervative and the other is free from polymer.
2. *Complex coacervation:*

   In the complex coacervation, two or more polymers are used. In complex coacervation, oppositely charged polymers such as gelatine (cationic) and gum (anionic) are used. In complex coacervation emulsion of lipid and aqueous phase is prepared. The aqueous phase contains two different polymers. The emulsion gets divided into two polymer phases—one is soluble in water and the other is insoluble. Aqueous phase adjusts itself in two phases. In the next step, coating layer is formed around the hydrophobic core material which is nothing but an encapsulation process. With coacervation process, capsules are formed when wall material around the core becomes hard. Meyer et al. reported that complex coacervation is a suitable technique for encapsulating materials with small solubility in water. Flavor encapsulation by coacervation has limitations such as evaporation of volatile compounds in the heat-treatment process, dissolution of active compound into the processing solvent, oxidation of product. The complex coacervate is a very unstable process, so chemical agents, such as glutaraldehyde, are crucial to stabilize it. Even though complex coacervation has many benefits in food industries for flavor encapsulation, it is not yet industrially accepted because it is complex process and expensive [13].

#### 16.3.1.1.2 Flavor encapsulation by molecular inclusion

The encapsulation by inclusion consists of host–guest-like structure. In this method internal hydrophobic cavity is the key feature for complex formation. The host (shell) generally hydrophilic has a cavity into which guest (core) (hydrophobic) compound can be accommodated. Cho and Park [14] explained the interaction between the shell and the core is purely due to van der Waals forces. Cyclodextrins, which are enzymatically modified starch molecules, used to encapsulate fragrance material inside as complex and entrap molecules. Cyclodextrins are cyclic oligosaccharides consisting of glucopyranose. In the process of encapsulation by molecular inclusion, retention of flavor is based on the molecular weight, shape, chemical functionality, polarity, and volatility of the core material.

#### 16.3.1.1.3 Cocrystallization process for extraction of flavor

The encapsulation of flavor compounds by cocrystallization takes place when spontaneous crystallization takes place. This encapsulation process takes either by inclusion or by entrapment. In spontaneous crystallization the core materials such as aroma compounds retain much of volatile oil during the process. The product obtained is in the form of dry powder. Only limitation of this method is that it restricts the addition of external antioxidant to minimize the oxidation during storage. Beristain et al. [15] used cocrystallization process for encapsulation of orange peel oil using sucrose as an encapsulating agent. The process begins with concentrating the sucrose syrup from 70°Bx to 95°Bx using hot plate and magnetic stirrer. Once the desired level of concentration of sucrose is reached, they made an addition of orange peel oil as an active material in various weight ratios of oil/sugar. As soon as crystallization started, the heating was then stopped. The heat required for removal of water was utilized from heat from the crystallization and thus dry product in the form of flowing powder was obtained. Butylated hydroxyanisole addition was before adding core material for avoiding the oxidation of formed encapsulation.

### *16.3.1.2 Physical methods of encapsulation*

#### 16.3.1.2.1 Spray-drying method of flavor encapsulation

The spray-drying technique for encapsulation is well practiced in food industries [5,16]. The basic principle of spray drying is that liquid sample is converted into solid powder by atomization in hot gas. For most of the cases, air and nitrogen gas are used to dry the liquid as they are inert toward the material to be dried. In the process of encapsulation with spray-drying technology core (bioactive-aroma extract compound) is dissolved or dispersed in outer shell/carrier material. Fig. 16–5 describes the microencapsulation process by spray drying. The stable emulsion is prepared using high-pressure homogenization (HPH), microfluidization or using ultrasonic probe reactor. The formed emulsion is kept for observation in order to check its stability. Upon the stability confirmation the emulsion is sent for spray-drying process. The key benefit of this technique is that it provides cost effectiveness of the process.

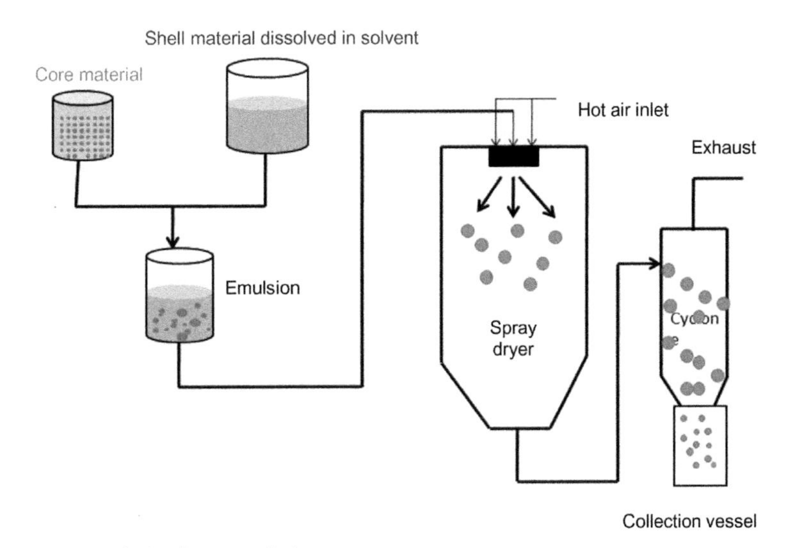


**FIGURE 16–5** Microencapsulation by spray drying.

In the spray-drying process, active flavor material is dispersed in coating material. The coating materials such as gum arabic, maltodextrin, and whey protein are dissolved in water. Sometimes, the blend of coating material is preferred as compared to single-coating material. Baranauskiene et al. [17] studied the flavor encapsulation of peppermint essential oil with modified starch as encapsulating agent using spray-drying technology. They studied the flavor release mechanism. They observed that the release of aromas from encapsulated powder products during storage increased with increasing water activity. Bruckner et al. [18] investigated the stabilization of a volatile aroma compound using a combined emulsification and spray-drying process.

#### 16.3.1.2.2 Extrusion method of flavor encapsulation

The flavor encapsulation by extrusion is used widely after spray-drying technique [19]. Encapsulation of flavors via extrusion has been used for volatile and unstable flavors in glassy carbohydrate matrices. The basic principle in all techniques is of extrusion: the bioactive compound is incorporated into shell material and it is passed through an aperture [lab-scale: needle of syringe-beads and capsules structure] and droplet creation at the nozzle discharge point. The methods of extrusion are classified into electrostatic simple dripping extrusion, jet cutting, coaxial airflow and spinning disk atomization, vibrating jet/nozzle. Coaxial double capillary and centrifugal extrusion are the two types of devices used to pass core and carrier material. The droplets formed in this step get solidified immediately to form encapsulation structure. Chemical (gelation) and physical (cooling or heating) are the two approaches to solidify the droplet formed [13]. The particles formed by extrusion are larger in size as compared to particles formed by spray drying. Wolf [20] encapsulated orange peel oil using the extrusion process. The blend of carrier materials such as glucose, sucrose, maltodextrin, and starch is used. Most commonly starch-based materials are encapsulated with extrusion [21].

Advantages of the encapsulation using extrusion encapsulation process are discussed next.

The pressure and temperature conditions required for extrusion process are below 100 psi and 118°C, respectively. As these conditions are not severe, heat-sensitive food materials can be encapsulated using this technique [22]. Encapsulated materials formed by extrusion have less porous structure as compared to the one formed by spray drying. As in the process, oxygen is not allowed to enter, the oxidation of bioactive compound is avoided. Thus shelf-life of product increases to 4–5 years against shelf-life of few months for unencapsulated food [23]. High viscous polymers are easily handled without the use of solvents. It offers the ability to control residence times and degrees of mixing through multiple injections. The disadvantages of extrusion process can be listed: the process of encapsulation by extrusion is difficult to scale up, microgel particle formed in passing the core-carrier material through syringe needle must be separated.

#### 16.3.1.2.3 Freeze-drying and vacuum-drying method for flavor encapsulation

Freeze drying is the encapsulation technique used to encapsulate heat-sensitive flavors. The freeze-drying technique is useful in a way that it is an effective method to preserve the integrity of the food material while retaining its color, flavor, and shape. The very basic principle of freeze drying is that the water is removed from a flavor bioactive compound in three steps: the first step is freezing in which the aim is to freeze the mobile water of the aroma compound. The second step is primary drying; in this phase, the chamber pressure is reduced and heat is applied to cause the frozen mobile water to sublime. The third step is secondary drying; depending upon the characteristic of material the time of secondary drying is decided to ensure complete removal of water. Once the freeze drying of bioactive material is complete plugging system in the chamber is used to plug the vials to prevent the freeze-dried bioactive aroma compound from oxidation and water absorption. The encapsulation is completed after filling nitrogen gas into the chamber. The main reason that restricts its wide use is that it requires higher energy for freezing and drying time is several times (almost 50 times) higher than spray drying.

#### 16.3.1.2.4 Spray chilling or spray cooling techniques of flavor encapsulation

Spray cooling also known as spray chilling is an encapsulation technique suitable to treat the aromas that are in liquid forms. The key advantage of this method is that it is least expensive and useful to treat liquid flavor and transform into free-flowing powder. Encapsulation by spray cooling/spray chilling differs from encapsulation by spray drying in the fact that in former method, carrier material is cooled down, and instead, it is evaporated in the next method [24]. Furthermore, spray cooling differs from spray chilling where spray chilling is carried out at 34°C–42°C temperature and spray cooling is carried out at elevated temperature of 45°C–122°C. The spray cooling is operated in both batch and continuous mode. The key feature of this technique is that the yield obtained is high, whereas in spray chilling, the particles are set to low temperature. The cost effectiveness of this method makes its increased use. When the melting point is below 40°C the spray cooling is used and when it is in the range of 45°C–120°C spray chilling is used. In the process the active (flavor core) material and carrier material are emulsified and subjected to solidification process. The particle size in solidification process depends on various factors. For example, atomization fountain nozzle is suitable to produce particle size in the range of 50–150 μm. By using low-speed rotary and prilling wheel we can get the particles of size 500–2000 μm.

#### 16.3.1.2.5 Fluidized bed coating: a technique of flavor encapsulation

The fluidized bed coating method differs from other methods that this is the only method that uses core and shell material in the form of powder (Fig. 16–6). Key feature of this encapsulation technique is that uniform layer of encapsulating material forms on core material. In the process of encapsulation by fluidized bed coating the active flavor ingredient is suspended (fluidized) in the chamber and coating material is sprayed onto it. The conditions in the selection of carrier material are that it should be enough viscous to get sprayed in the reactor. Because highly viscous fluid does not move freely in the fluidized bed, very less viscous fluid (carrier material) may get escaped through the chamber without coating to bioactive aroma core material. The factors affecting the encapsulation efficiency in the fluidized bed reactor depend on some of the factors which can be listed as follows.

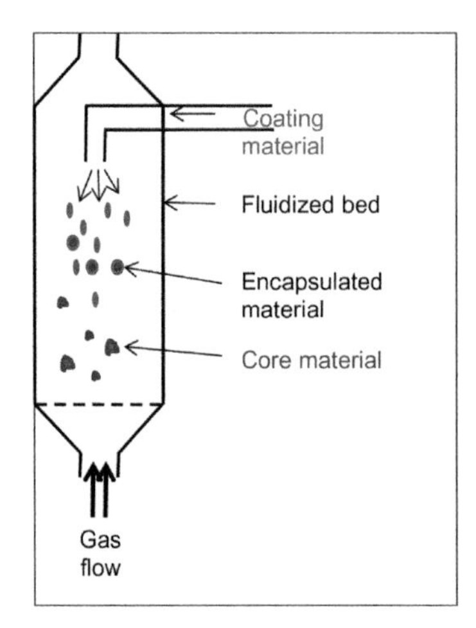


**FIGURE 16–6** Encapsulation by fluidized bed coating.

The flow rate of encapsulating material will decide its residence time in the reactor and affects the encapsulation efficiency. In encapsulation with fluidized bed the better temperature control can be achieved which is beneficial for treating food products and their encapsulation. The only drawback of encapsulation in the fluidized bed is that at higher velocity, core and shell material get entrained out of the reactor, pressure drop in the bed is also a considerable issue of fluidized bed reactor for encapsulation. The technique of fluidized bed coating for encapsulation is used to treat emulsifiers, hot-melt coatings, hydrogenated vegetable oil, fatty acids, and for the coatings in which bifunctional material like maltodextrin gum acacia acting both as solvent and carrier.

#### 16.3.1.2.6 Agglomeration or granulation process of encapsulation

All the encapsulation techniques mentioned above are useful when the encapsulated aroma is required in smaller size (Fig. 16–7). But sometimes the larger size (0.3–0.9 mm) of aroma particle size is required. In such a condition the method called agglomeration is useful. In the process of encapsulation by granulation the spray-dried powder in the first step is not completely dried and thus allows the agglomeration of the second phase causing an increase in aroma core material which is spray dried onto another carrier material. In the last step, water is used to bind and sprayed onto dried product. The mentioned process is called wet agglomeration and it has a drawback that it contains moisture/water in the last step, and as this moisture gets held in aroma compound for a longer time, it leads to volatile losses of aroma material which is not desired. To overcome this limitation, pressure agglomeration is used in which high pressure is used along with carrier material (0.7–3.0 mm). These agglomerated encapsulated aromas are useful in certain applications where it is necessary that aroma should not get separated from food material.

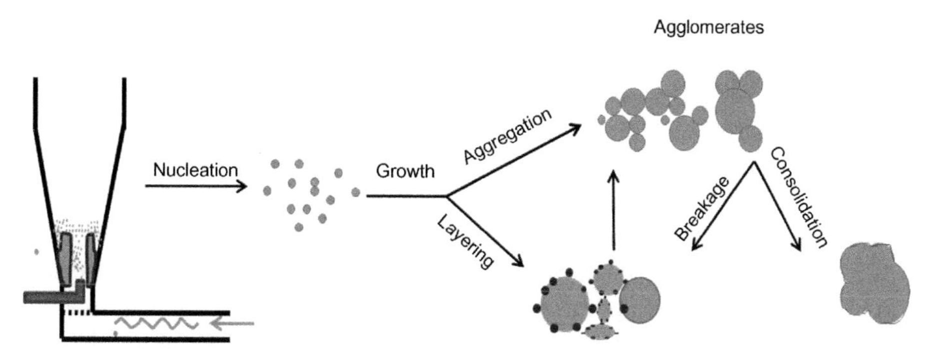


**FIGURE 16–7** Representation of encapsulation by agglomeration.

## 16.3.2 Newer intensified approach for the flavor encapsulation: nanoemulsions

Nowadays, the application of nanoemulsions is growing day by day. Nanoemulsion-based drug-delivery system is one of the best examples for the application of nanoemulsions, where we can provide the availability of bioactive-encapsulated materials exactly at the location of need and as the nano-sized encapsulation requires low amount of bioactive compound for cure which will ultimately optimize the use of bioactive compounds for cure. These nanoemulsions encapsulated with bioactive compounds are also applicable in food matrix to ensure the high shelf-life and safety of food material [1]. Synthesis of these emulsions demand high-energy exchange between system and surroundings, and at the same time, it is a highly constrained process. Energy is required for converting one phase into a dispersed form such as liquid droplets of size micro or nano. For example, in the homogenization process at high-pressurized condition, immiscible liquid phase is dispersed into another by forming small droplets [25]. The droplet size generated in emulsion state will be dependent on the amount of energy utilized. Emulsions with the droplet size ranging from 5 to 300 nm are called nanoemulsions [26,27].

The synthesis of nanoemulsions is classified based on the energy requirement for disruptive force delivery, by the virtue of which dispersion of one phase into another continuous phase takes place. Hence high- and low-energy methods are two ways by which we can synthesize the nanoemulsions. Disruptive forces are delivered through mechanical devices in high-energy methods [28], whereas no external force is needed in low-energy methods.

Initially, nanoemulsion studies were restricted to the high-energy methods and thus the most widely used methods were high-energy stirring and ultrasonic emulsification [29]. Physical separation with high mechanical forces can cause damage to the encapsulated materials; thus nowadays considerable attention has been drawn by low-energy methods [30]. The emulsion stability depends upon the droplet interactions, their size, composition of the system, and the method by which emulsion is synthesized. In the present section, an overview is given for the most commonly used methodologies for the synthesis of stable emulsion.

### *16.3.2.1 Encapsulation of flavor by emulsification with higher energy*

The physical destruction of immiscible liquid phase in small size droplets is a very ancient technology for the preparation of emulsion. Processes such as high-pressure homogenizer, ultrasound sonicator, or microfluidizers are very commonly used for the synthesis of nanoemulsion. Mechanical (impact) energy can be used for the physical damage and colloidal phase is achieved; also, high-energy utilization ensures the synthesis of droplet size in 50–300 nm range that is nanoemulsion [31]. Two immiscible phases such as oil in water require strong forces for the synthesis of droplets of low concentration phase for getting stable form of emulsion. When low-quantity oil phase is blended in high-quantity water phase, mechanical devices need high energy to distribute nano-sized oil droplets uniformly in the latter phase. In order to get emulsion droplet size (EDS) in the range of 50–300 nm, energy density of about 100–1500 W/kg has to be delivered to the oil–water system for a short period of time [28].

More recently used mechanical devices for the synthesis of emulsifications like high--pressure homogenizers are able to provide high density of energy and their smooth operations make them more admirable for the process [26]. Other parameters such as use of surfactants, concentration of system, and temperature affect the energy demand of emulsification. High-pressure homogenizers require high amount of energy, which is also associated with inefficient utilization of energy. It has been reported that with suitable surfactant at low concentrations, ultrasound is a cost-efficient process. Ultrasound optimizes cost in terms of energy savings and requirement of surfactant, the only constraint associated with it is quantity, whereas methods such as microfluidization are industrially used techniques as they provide flexible control over distribution of EDS, and their capability to synthesize fine emulsions [32].

#### **16.3.2.1.1 Encapsulation by emulsification with high-pressure homogenization**

The HPH is a potential technique among the nonthermal technologies, to develop new functional applications, as it can be easily implemented at industrial level. It is a purely mechanical process in which a process fluid is forced to flow through low cross-sectional area such as orifice or tapered section of venturi-meter at a high pressure (300–400 MPa). High shear stress caused by narrower flow area and high pressure applied is responsible for the generation of droplets with small sizes of oil-in-water emulsion.

The principle of HPH is simple: a coarse emulsion (produced via conventional methods) is passed with high-pressure systems through thin valve for a very short period of time where the pressure difference, which is in the range of 100–250 MPa, plays an important role [33]. HPH represented by Fig. 16–8 shows emulsification process that is completed in two steps: (1) In the first step, due to high-pressure drop the droplets in coarse emulsion get disrupted and the formation of smaller size droplets increases. (2) Stabilization of smaller sized droplets by virtue of emulsifiers. The smaller size droplets generated have a tendency to agglomerate with great speed than that of the coarse droplets. Emulsifiers are used to stabilize the smaller size droplets, and adsorption of emulsifiers on the droplet surface generates the films surrounding the droplet that avoid the agglomeration. Due to smaller size droplet, the

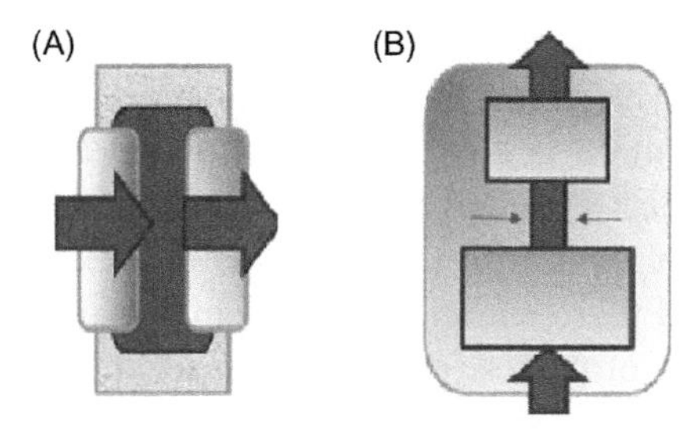


**FIGURE 16–8** Homogenizer classification nozzles flow systems: (A) inertial shear force–induced shear stress and (B) shear force–induced shear stress.

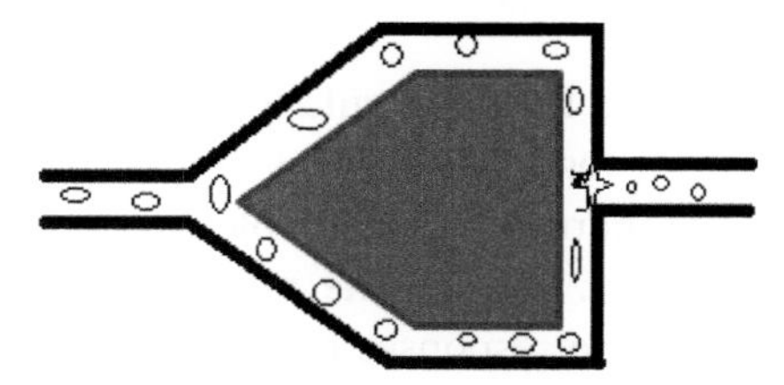

**FIGURE 16–9** Schematic representation of microfluidization technique.

average surface area of the emulsion synthesized is high as compared to the coarser emulsion, prepared by the conventional agitation method [34]. Deficiencies associated with conventional homogenization process are overcome by HPH for the synthesis of food emulsion [35]. Morphological and rheological properties of the food emulsions are improved to a great extent due to HPH. For example, the texture of cheese and yogurt prepared by HPH was improved [36]. The protein and triglyceride emulsifiers reduce the ripening time of cheese [37]. It was also reported that due to HPH exposures, alteration in activity of enzymes was found [38,39] which reduces biogenic amine content of cheese.

#### 16.3.2.1.2 Microfluidization for nanoemulsions synthesis

High-pressure displacement pump is used to generate fine nanoemulsions through microfluidization [40]. Device used for microfluidization is called microfluidizer that provides high pressures. The design of microfluidizer is similar to HPH, a small annular path through which coarse emulsion is passed with high pressure that makes easy disruption of droplets through the continuous phase. It differs only in the channels in which the emulsion flows. Two different annular cross-sectional limbs are utilized through which dispersed droplet flows with a continuous phase at high pressure. Collision of these droplets with high-pressure zone flow produces smaller size droplets that are stabilized with the emulsifiers. Schematic representation of microfluidizer is shown in Fig. 16–9 [41]. Collision takes place in the chest box which is the heart of this device, two limbs are connected at chest box of the fluidizer, two limbs with high-pressure flow collides with each other at chest box, and destruction of droplets takes place [40,42]. The coarser emulsion is divided with high-pressure pumps and divider at a pressure range of 200–700 kPa. This pressure causes huge amount of shearing action that causes for the microdroplets distribution in continuous phase with high-pressure flow. High impact and shearing action in fluidizer channels produce fine emulsion. High shearing rate induces inertial force at the chest box along with the cavitation effect where temperature and pressure gradients are much high and these effects are the main cause for the fine droplet distribution in channels [43,44]. Number of passes, concentration of emulsifiers, and pressure by which fluid flows greatly affect the droplet size of emulsion.

The pharmaceutical industries generally use microfluidization technology for the synthesis of colloidal solutions, for example, edible nutritional emulsions, food emulsions, and colloidal phase flavored milk [26]. The encapsulation of enzymes or bioactive compounds can also be achieved by using microfluidizers [45].

#### 16.3.2.1.3 Ultrasonic emulsification for encapsulation of flavor

The ultrasonic emulsification can result in efficient droplet size reduction. In this method, vibrating probe called sonication horn plays a key role as it provides the required energy to the liquid mixture as shown in Fig. 16–10. The mechanical vibrations synthesized in the probes are because of the alternative responses of quartz crystal for contraction and expansion. The mechanical vibrations are generated due to high amplitude sound generation at quartz crystal. The high-frequency sound waves transferred into liquid surface produce high mechanical vibration and cavitation effect. The unstable cavities, generated initially, grow up to certain limit and then collapse, and all this process takes minor time (in milliseconds). The collision/collapse of droplets causes high-pressure and high-temperature zone, which is responsible for the size reduction of the droplets. The only drawback for the method is the inefficiency to handle a large amount of liquid for the synthesis of nanoemulsion [46]. Ultrasound technology is effective with certain constraints, for example, one must avoid shear induced in droplets in emulsion, which causes coagulation of droplets and large diameter droplets can be formed [47,48].

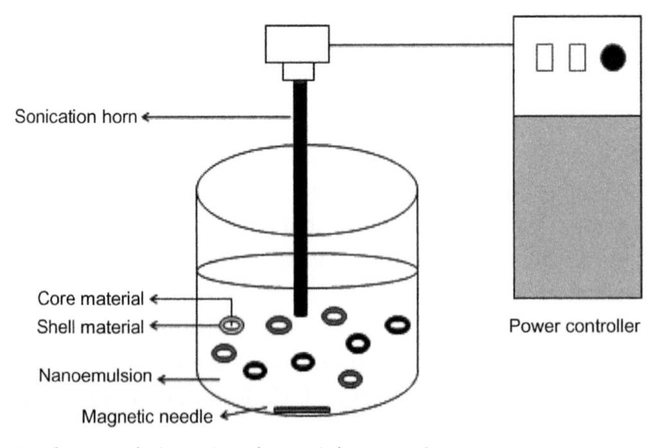


**FIGURE 16–10** Schematic of encapsulation using ultrasonic homogenizer.

For the nanoemulsions produced via sonication, EDS decreases with increasing power, extent of time duration, and surfactant concentrations [26]. Nanoemulsions with large volumes cannot be produced by sonication, which is its major drawback [40].

#### 16.3.2.1.4 Emulsification by high-shear stirring for flavor encapsulation

In the emulsification by high-shear stirring method, rotor-stator systems, and high-energy mixers are used for the production of nanoemulsions. Considerable decrease in the droplet size of the internal phase is observed by raising the intensity of mixing. However, it is reported that with this method, it is difficult to obtain EDS less than 200–300 nm [29].

### *16.3.2.2 Encapsulation of flavor by emulsification with low-energy*

Some systems possess high inertial energy such as chemical energy due to the presence of unreacted/unstable reactants, high potential energy by the virtue of its position and all. This high inertial energy is utilized for the synthesis of nanoemulsion. Low-energy emulsification methods are based on this principle [49]. Surrounding conditions and composition of the reaction as well as phase inversion play an important role in the low-energy synthesis processes, as the reaction rate, energy associated with reaction, depends upon these two parameters greatly. Temperature of surroundings affects the droplet size in emulsion [50]. Two immiscible phase interactions are greatly affected by the composition of two phases, surfactant utilized, surrounding conditions such as temperatures, pressures, and method applied for synthesis [30,51].

Small and uniform droplets can be obtained via low-energy methods such as phase-inversion temperature (PIT) and phase-inversion composition (PIC) with the help of physicochemical properties of the system as compared with high-energy methods [27]. Whereas low-energy methods are applicable to restricted types of emulsifiers and oils, polysaccharides or proteins, for example, cannot be utilized as emulsifiers. Also, the requirement of synthetic surfactants with high concentrations to form nanoemulsions with low-energy approaches limits their usage in food applications [12].

#### 16.3.2.2.1 Spontaneous emulsification for flavor encapsulation

In the spontaneous emulsification technique, spontaneous production of nanoemulsion at room temperature without any special devices takes place [27,30]. It proceeds through three steps: (1) synthesis of uniform organic solution made of lipophilic surfactant and oil-in-water miscible solvent and hydrophilic surfactant; (2) under continuous magnetic stirring, the organic phase is introduced in aqueous phase to form o/w emulsion; and (3) removal of the aqueous phase under reduced pressure by evaporation [26,52]. These mechanisms are greatly affected by the composition of system and their physicochemical properties [53]. The control over droplet size can be obtained by changing the compositions of initial phases and mixing conditions [48]. Systems produced by this technique are usually termed self-nanoemulsifying drug-delivery systems in the pharmaceutical industries [28]. McClements and Rao compared spontaneous emulsification with high-energy microfluidization for preparing nanoemulsions [48]. They obtained EDS of 110 nm via microfluidization, whereas spontaneous emulsification resulted in the production of droplets with around 140 nm size.

This indicates that nanoemulsions can be efficiently prepared by spontaneous emulsification method, if the system is optimized in terms of composition, that is, water, oil, and surfactant contents [41].

#### 16.3.2.2.2 Flavor encapsulation by phase inversion emulsification

The phase inversion is also termed phase condensation method and is based upon the phase transitions that occur during emulsification process [40]. Here, the chemical energy released because of the phase transitions during emulsification process is responsible for the formation of emulsion [54]. The phase transitions occur due to changes in the spontaneous curvature of the surfactant and can be obtained by (1) the PIT method or (2) PIC method. These methods have drawbacks such as complexity, high precision requirements, and use of synthetic surfactants [55,56].

1. Emulsification with PIT

   In this method, constant composition is maintained and temperature is varied. Nonionic surfactants play a key role in this method as their solubility is greatly affected by temperature. For preparing nanoemulsions by PIT method, it is essential to bring sample temperature to its PIT level [54]. It is reported that fine and stable emulsion droplets can be prepared by rapid cooling of the emulsion near the temperature of PIT [52,57].
2. Emulsification with PIC

   In this method, composition varies at constant temperature. It varies by the consistent addition of oil or water in the mixture of water- or oil-surfactant. The large-scale production can be obtained by PIC method as compared to the PIT method as an addition of one component is easier than the generation of abrupt change in temperature [58]. A transition composition is achieved by the addition of water to the system. When the composition exceeds transition composition, small, metastable oil-in-water droplets are formed because of the separation of the structures that have zero curvature [54].

#### 16.3.2.2.3 Membrane technology for flavor encapsulation

In the encapsulation by membrane technology, a dispersed phase is formed into a continuous phase through a membrane [41]. This method can produce emulsions with narrow-size distribution with less consumption of surfactant as compared to high-energy methods. However, scale-up of this technique is difficult as the flux of the dispersed phase through the membrane is low [59].

#### 16.3.2.2.4 Emulsification through solvent displacement method for flavor encapsulation

The emulsification through solvent displacement method is based on the quick diffusion of water-miscible organic solvents having lipophilic functional compounds in the aqueous phase supporting the generation of nanoemulsions. At low-energy input the rapid diffusion facilitates the one-step synthesis of nanoemulsion with high yield of encapsulation. Finally, the evaporation of organic solvent from the nanodispersion under vacuum has to be done [60,61]. But, this technique is restricted to water-miscible solvents. A high solvent-to-oil ratio needed to produce small-sized droplets limits the application of this technique [62,63].

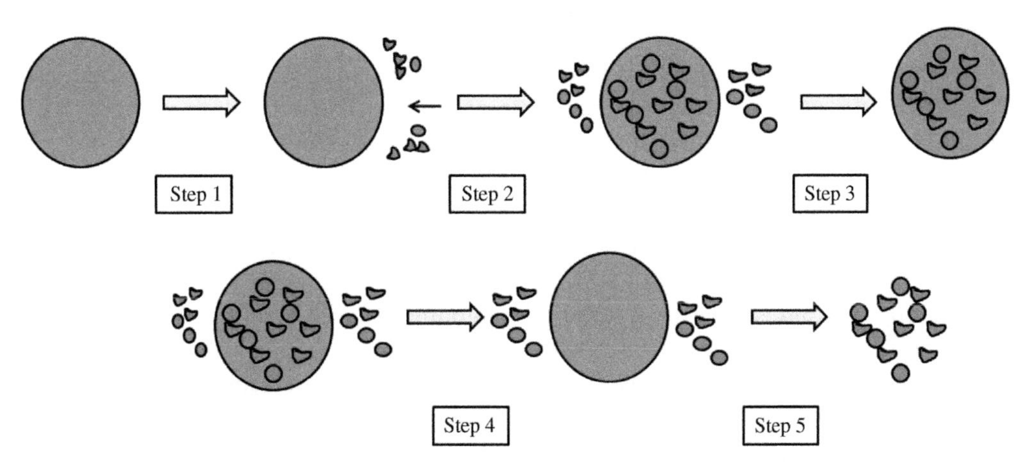


**FIGURE 16–11** Schematic representation of flavor release.

# 16.4 Phenomena of encapsulated flavor release

Flavor release is the process in which flavor migrates from one environment to another particular molecular environment within a food and into the surrounding saliva or vapor phase. Rate of flavor release can be described as the flavor migrated from one environment to another per unit time. Flavor release can be a controlled release or an uncontrolled release. This is well explained in the following section. There are five steps in the process of flavor encapsulation and its release is as shown in Fig. 16–11. In the first step encapsulation begins, flavor compound continues to enter in shell in step two. Encapsulation is completed in the third step. Flavor release from shell material is the fourth and flavor is completely released in the fifth step.

The study shows that the flavor release rate is higher in lipids as compared to its product environment, and resistance to mass transfer also increases resulting in improving the flavor retention. Goubet et al. [64] with experiment found that molecular weight is the factor to control retention of the flavor. The relation between these two parameters is that retention increases with the increase in molecular weight of the coating material up to certain value beyond which retention decreases with the increase in the molecular weight of coating material. Kant et al. [65] pointed out with the practical example of encapsulation of lemon with cyclodextrins and its loading in yogurt (fat free) resulted in the improvement of flavor and compared its release with regular yogurt. They found that both yogurts have similar release rate.

## 16.4.1 Controlled flavor release

Controlled release is the one in which bioactive aroma agent is released at desired location or section of body and at required time with particular rate in the body. The encapsulation helps in controlling the flavor release rate at a specific rate under certain conditions [66]. Controlled release takes place by different mechanisms.

#### *16.4.1.1 Factors affecting the release of flavor*

Release rate of flavor depends on several factors. Some of them can be mentioned as properties of the wall material, Nature of the encapsulated flavor, Method employed for encapsulation, physical factors of the capsule, Chemical structure (size and molecular weight) of formed encapsulation, Degradation/dissolution of the food matrix material, The thermodynamic factor-the volatility of the flavor compounds, kinetic factor-resistance to flavor mass transport from core to emulsion and from emulsion to environment, Properties of solvent forming encapsulation emulsion such as melting, diffusion, degradation, The percentage quantity of oil compound (flavor essential oil) used to form encapsulation emulsion affects the release. The relation is that the increase in oil percentage decreases the flavor release rate with the exception of hydrophilic compounds, The salts present in food increases the volatility of flavor. Liu et al. used a droplet drying technique to study that how oil flavor escapes from core and release in core. They reported that the flavor retention is mainly dependent on nature of flavor and the emulsion stability. Malone et al. [67] in their work explained the importance of change in fat content level as it directly affects the flavor release of food. Viscosity has very less impact on flavor release.

#### *16.4.1.2 Flavor release mechanism by diffusion*

The flavor release by diffusion gets influenced by percentage solubility of aroma in food matrix and its mobility through the matrix. The release mechanism involves two main steps. The first step is active agent's diffusion to the outer layer of the food matrix; and in the second step is separation of the volatile component between the matrix and the surrounding food and transport away from the matrix surface. The aroma release by diffusion is limited because some food have very less or almost no solubility in matrix diffusion. There are basically two subtypes of flavor diffusion—one is static diffusion and the other is eddy or convective diffusion. The static diffusion is due to the random movement of the molecules in the stagnant fluid. In convective diffusion the transport of an element of the flavor takes place from one location to another, carrying with them the dissolved solute.

#### *16.4.1.3 Flavor release mechanism by erosion and degradation*

The transport of an active ingredient from a core can also be governed by diffusion or erosion or a combination of both. Homogeneous and heterogeneous erosion are both detectable. Heterogeneous erosion can be defined as the degradation limited to thin layer present at the delivery surface, and homogenous erosion can be defined as the one in which degradation of active compound occurs equally throughout food matrix.

#### *16.4.1.4 Flavor release mechanism by swelling*

The flavor release by swelling mechanism takes place when flavor release by dissolution or by dispersion is not possible. When the encapsulated emulsion is kept in thermodynamically compatible medium, the swelling of polymer is due to fluid absorption from the medium.

In the next step, the flavor in the swollen part diffuses out. Water absorption and presence of solvents are the two factors, which control the degree of swelling. Bouquer et al. [68] studied the flavor release by swelling. In their study, glassy particles were subjected to water in which swelling and erosion mechanism takes place.

#### *16.4.1.5 Flavor release mechanism by melting*

The encapsulated matrix is melted to release the aroma compound. In food industries most of the flavor compound release takes place by melting. The basic principle is that unless desired the encapsulated food/flavor is stored at lower temperature and melted when desired in cooking or in food preparation.

### 16.4.2 Modes of uncontrolled flavor release

Uncontrolled flavor release may cause due to rapid leakage of bioactive flavor material from shell material in surrounding medium. The uncontrolled flavor release is also attributed to continue and slow diffusion of flavor from the core of encapsulation to shell and surrounding medium. This phenomenon occurs by any of the triggering mechanisms such as diffusion release, barrier release, pressure-activated release, solvent-activated release, osmotically release, temperature-sensitive release, melting-activated release, and combined system.

Uncontrolled flavor release has some limitations such as flavor can be released early, even before consumption because of some of the environmental conditions which is not desirable. In order to overcome these problems, attempts are being taken to make the controlled flavor release.

## 16.5 Characterization techniques for encapsulated bioactive compounds

### 16.5.1 Droplet size and stability measurements of food emulsions

The EDS and distribution of droplet size in dispersion are often determined by *dynamic light scattering* technique. It is based on the measurement of Doppler shift which is then related to the particle size. The Doppler shift is caused as soon as the light hits the moving particle, which results in the change of the incoming light wavelength [69].

Food emulsion stability can be determined using *zeta potential*. The zeta potential is based on electrophoresis and is estimated using Henry's equation by electrophoretic mobility. If all the droplets in emulsions have a large positive or negative zeta potential, they will have a tendency to repel each other and thus indicating no affinity to flocculate or aggregate. On the other hand, if droplets in emulsions have low values of zeta potential, they tend to flocculate or aggregate resulting in unstable emulsion [70].

## 16.5.2 Surface morphology and topography

### *16.5.2.1 Optical microscopy of flavor encapsulated emulsion*

The optical microscope is one of the most important tools for observing the morphology of emulsions. It provides valuable information about the droplet size distribution of emulsions and the extent of encapsulation done. It is also used to differentiate between coalescence and flocculation, which is often difficult using other techniques that are based on particle counting or light scattering.

### *16.5.2.2 Morphology characterization of food emulsions with fluorescence microscopy*

The fluorescence microscopy is used to study the morphology of emulsions which cannot be detected by conventional optical microscopy such as emulsions with very small droplet concentrations. Moreover, it can also highlight particular components of emulsion by selecting fluorescent dyes that bind to them [35].

### *16.5.2.3 Scanning and transmission electron microscopy analysis*

The scanning electron microscope (SEM) and transmission electron microscope (TEM) are common types of electron microscopes. Using SEM, study of surface morphology and topography, and composition of very fine emulsions having EDS in the range of nanometers (nm) are also possible. It is widely used as it is capable of producing three-dimensional-like surface images. In SEM a highly energized beam of electrons scans the surface of the object to produce its image on the other hand in TEM beam of electrons transmits the object and projects the image of an entire object including the internal structures and surface. Thus TEM is able to characterize objects with diverse internal structures as they result in different projections and are advantageous over SEM.

## 16.5.3 Estimation of antioxidant activity for flavor encapsulated food emulsions

By estimating the antioxidant property of essentials, oils/aroma-encapsulated food emulsions can be obtained with free-radical capacity scavenging assay. The 2,2-diphenyl-1-picrylhydrazyl (DPPH*) radical scavenging capacity assay method is used for obtaining the antioxidant capacity, and it is based on the scavenging of the stable DPPH radical by the antioxidant. It can be used in aqueous and nonpolar organic solvents and can also be exploited to examine both hydrophilic and lipophilic antioxidants.

## 16.5.4 Estimation of total phenolic content of flavor encapsulated food emulsions

The spectrophotometric method reported by Singleton et al. [71] is used to determine the concentration of phenolic content in encapsulated nanoemulsions. Based on the measured absorbance, the concentration of phenolic content has to be read (mg/mL) from the calibration curve [72].

### 16.5.5 Estimation of flavonoids concentrations in flavor encapsulated food emulsions

The spectrophotometric method reported by Quettier et al. [73] is used for the estimation of flavonoids concentrations in encapsulated nanoemulsions. Based on the measured absorbance, the concentration of flavonoids can be determined (mg/mL) on the calibration line [72].

## 16.6 Conclusion and future prospective

The starting point of every food industries is flavor of food. The flavor of food attracts the attention of flavor. So, various attempts are made to externally add the flavor. Encapsulation gives a stable form of flavor, which is suitable for application in many food materials. The various methods of flavor encapsulation possess its own benefits and limitations. Like spray drying, encapsulation is used to obtain final product in powder form. When the aim is to load liquid flavor in food, technique such as ultrasonic encapsulation is useful. Sometimes, it is beneficial to use combination of two methods—homogenization and spray drying. The combination of these two techniques will increase the product yield and encapsulation efficiency. Most of the essential oil performs two activities—first is they act as a preservative and second they add flavor to food. So it is desired to incorporate these oils using shell material. In core–shell structure of encapsulation, combination of core material as well as single or combination of shell material can be used depending upon the purpose to serve, cost effectiveness, and method of application. Some of the shell materials have bifunctional activity such as gum arabic, whey protein isolate acts as a shell as well as it possesses emulsifier properties. Two major emulsion forms are synthesized—micro and nano. Nanoemulsion possesses high loading capacity as compared to the latter. Some flavor materials such as ginger have high antioxidant and anticancer properties. So it is very much essential to encapsulate these flavor materials and release them in a controlled manner as described. The growing use of bioactive flavors will give new insight into food industries.

The encapsulation of various natural flavors with the combination of different methods, core material, and shell material and its different loading percentage in food products is essential. Industrial-scale encapsulation process needs to be developed to meet the consumer demands.

## Acknowledgment

An author acknowledges the Department of Science and Technology for DST RFBR grants no INT/RUS/RFBR/324.

# References


[1] H.D. Silva, M.A. Cerqueira, A.A. Vicente, Nanoemulsions for food applications: development and characterization, Food Bioprocess Technol. 5 (2012) 854–867.

[2] T.A. Reineccius, G.A. Reineccius, T.L. Peppard, Encapsulation of flavours using cyclodextrins: comparison of flavour retention in alpha, beta, and gamma types, J. Food Sci. 67 (9) (2002) 3271–3279.

[3] M.I. Teixeira, L.R. Andrade, M. Farina, M.H.M. Rocha-Leao, Characterization of short chain fatty acid microcapsules produced by spray drying, Mater. Sci. Eng., C 24 (5) (2004) 653–658.

[4] R.G. Berger, Aroma Biotechnology, Springer Science & Business Media, 2012.

[5] X. Wang, Y. Yuan, T. Yue, The application of starch based ingredients in flavour encapsulation, Starch Starke 67 (3–4) (2015) 225–236.

[6] C.V. Fulger, L.M. Popplewell, US Patent No. 5,601,865, US Patent and Trademark Office, Washington, DC, 1997.

[7] J. Charve, G.A. Reineccius, Encapsulation performance of proteins and traditional materials for spray dried flavours, J. Agric. Food Chem. 57 (6) (2009) 2486–2492.

[8] S. Vaidya, R. Bhosale, R.S. Singhal, Microencapsulation of cinnamon oleoresin by spray drying using different wall materials, Drying Technol. 24 (8) (2006) 983–992.

[9] D. Kanakdande, R. Bhosale, R.S. Singhal, Stability of cumin oleoresin microencapsulated in different combination of gum arabic, maltodextrin and modified starch, Carbohydr. Polym. 67 (4) (2007) 536–541.

[10] K. Ganzler, A. Salgo, K. Valkó, Microwave extraction: a novel sample preparation method for chromatography, J. Chromatogr. A 371 (1986) 299–306.

[11] K.S. Mayya, A. Bhattacharyya, J.F. Argillier, Micro-encapsulation by complex coacervation: influence of surfactant, Polym. Int. 52 (4) (2003) 644–647.

[12] P. Kaushik, K. Dowling, C.J. Barrow, B. Adhikari, Microencapsulation of omega-3 fatty acids: a review of microencapsulation and characterization methods, J. Funct. Foods 19 (2015) 868–881.

[13] V. Dordevic, B. Balanc, A. Belscak-Cvitanovic, S. Levic, K. Trifkovic, A. Kalusevic, et al., Trends in encapsulation technologies for delivery of food bioactive compounds, Food Eng. Rev. 7 (2014) 452–490.

[14] Y.H. Cho, J. Park, Encapsulation of flavour by molecular inclusion using β-cyclodextrin: comparison with spray-drying process using carbohydrate-based wall materials, Food Sci. Biotechnol. 18 (1) (2009) 185–189.

[15] C.I. Beristain, A. Vazquez, H.S. Garcia, E.J. Vernon-Carter, Encapsulation of orange peel oil by co-crystallization, LWT—Food Sci. Technol. 29 (7) (1996) 645–647.

[16] C. Belingheri, A. Ferrillo, E. Vittadini, Porous starch for flavour delivery in a tomato-based food application, LWT—Food Sci. Technol. 60 (1) (2015) 593–597.

[17] R. Baranauskiene, E. Bylaite, J. Zukauskaite, R.P. Venskutonis, Flavour retention of peppermint (*Mentha piperita* L.) essential oil spray-dried in modified starches during encapsulation and storage, J. Agric. Food Chem. 55 (8) (2007) 3027–3036.

[18] M. Bruckner, M. Bade, B. Kunz, Investigations into the stabilization of a volatile aroma compound using a combined emulsification and spray drying process, Eur. Food Res. Technol. 226 (1–2) (2007) 137–146.

[19] B. Bhandari, N. Bansal, M. Zhang, P. Schuck, Handbook of Food Powders: Processes and Powders, Woodhead Publishing, Philadelphia, PA, 2013.

[20] B. Wolf, Polysaccharide functionality through extrusion processing, Curr. Opin. Colloid Interface Sci. 15 (2010) 50–54.

[21] R. Ashady, Microcapsules for food, J. Microencapsul. 10 (1993) 413–435.

[22] M.W. Tackenberg, R. Krauss, H.P. Schuchmann, P. Kleinebudde, Encapsulation of orange terpenes investigating a plasticisation extrusion process, J. Microencapsul. 32 (4) (2015) 408–417.

[23] G.A. Reineccius, Flavour Chemistry and Technology, second ed., CRC Press, New York, 2005.

[24] A. Madene, M. Jacquot, J. Scher, S. Desobry, Flavour encapsulation and controlled release—a review, Int. J. Food Sci. Technol. 41 (1) (2006) 1–21.

[25] I.O. Ocampo-Salinas, D.I. Tellez-Medina, C.D.-O.G. Jimenez-Martinez, Application of high pressure homogenization to improve stability and decrease droplet size in emulsion-flavor systems, Int. J. Environ. Agric. Biotechnol. 1 (2016) 646–662.

[26] C. Solans, P. Izquierdo, J. Nolla, N. Azemar, M.J. Garcia-Celma, Nano-emulsions, Curr. Opin. Colloid Interface Sci. 10 (2005) 102–110.

[27] G. Caldero, M.J. Garcia-Celma, C. Solans, Formation of polymeric nano-emulsions by a low-energy method and their use for nanoparticle preparation, J. Colloid Interface 353 (2011) 406–411.

[28] K. Cinar, A review on nanoemulsions: preparation methods and stability, Trak. Univ. J. Eng. Sci. 18 (2017) 73–83.

[29] M.Y. Koroleva, E.V. Yurtov, Nanoemulsions: the properties, methods of preparation and promising applications, Russ. Chem. Rev. 81 (2012) 21–43.

[30] N. Anton, J. Benoit, P. Saulnier, Design and production of nanoparticles formulated from nano-emulsion templates—a review, J. Control. Release 128 (2008) 185–199.

[31] I. Sole, C. Solans, A. Maestro, et al., Study of nano-emulsion formation by dilution of microemulsions, J. Colloid Interface Sci. 376 (2012) 133–139.

[32] O. Kaltsa, C. Michon, S. Yanniotis, I. Mandala, Ultrasonic energy input influence on the production of sub-micron o/w emulsions containing whey protein and common stabilizers, Ultrason. Sonochem. 20 (2013) 881–891.

[33] S. Mahdi, J. Yinghe, H. Bhesh, Optimization of nano-emulsions production by microfluidization, Eur. Food Res. Technol. 255 (2007) 733–741.

[34] L. Siroli, F. Patrignani, D.I. Serrazanetti, R. Lanciotti, Potential of high pressure homogenization and functional strains for the development of novel functional dairy foods, Technological Approaches for Novel Applications in Dairy Processing, Intech Open, 2018, pp. 36–50.

[35] D.J. Mc-Clements, Food Emulsions, third edition, CRC Press, 2016.

[36] J. Flourya, J. Bellettre, J. Legrand, A. Desrumaux, Analysis of a new type of high pressure homogeniser. A study of the flow pattern, Chem. Eng. Sci. 59 (2004) 843–853.

[37] R. Lanciotti, L. Vannini, P. Pittia, M. Elisabetta, Suitability of high-dynamic-pressure-treated milk for the production of yoghurt, Food Microbiol. 21 (2004) 753–760.

[38] F. Patrignani, D.I. Serrazanetti, J.M. Mathara, L. Siroli, F. Gardini, W.H. Holzapfel, et al., Use of homogenisation pressure to improve quality and functionality of probiotic fermented milks containing *Lactobacillus rhamnosus* BFE 5264, Int. J. Dairy Technol. 69 (2016) 262–271.

[39] P. Burns, F. Patrignani, D. Serrazanetti, G.C. Vinderola, J.A. Reinheimer, R. Lanciotti, et al., Probiotic crescenza cheese containing *Lactobacillus casei* and *Lactobacillus acidophilus* manufactured with high-pressure homogenized milk, J. Dairy Sci. 91 (2008) 500–512.

[40] H. Jasmina, O. Dzana, E. Alisa, V. Edina, R. Ognjenka, Preparation of nanoemulsions by high-energy and low-energy emulsification methods, IFMBE Proceedings 62, Springer Nature Singapore Pte Ltd, 2017, pp. 317–322.

[41] S.M. Ezzat, M.A. Salem, Nanoemulsions in food industry, Some New Aspects of Colloidal Systems in Foods, IntechOpen, 2018, pp. 31–51.

[42] L. Vannini, R. Lanciotti, D. Baldi, M.E. Guerzoni, Interactions between high pressure homogenization and antimicrobial activity of lysozyme and lactoperoxidase, Int. J. Food Microbiol. 94 (2004) 123–135.

[43] D.W. Olson, C.H. White, R.L. Richter, Effect of pressure and fat content on particle sizes in microfluidized milk, J. Dairy Sci. 87 (2004) 3217–3223.

[44] B.S. Schultz, G. Wagner, K. Urban, J. Ulrich, High-pressure homogenization as a process for emulsion formation, Chem. Eng. Technol. 27 (2004) 361–368.

[45] O. Robin, V. Blanchot, J. Vuillemard, P. Paquin, Micro fluidization of dairy model emulsions. I. Preparation of emulsions and influence of processing and formulation on the size distribution of milk fat globules, Lait 72 (1992) 511–531.

[46] Y. Maa, C.C. Hsu, Performance of sonication and microfluidization for liquid – liquid emulsification, Pharm. Dev. Technol. 4 (1999) 233–240.

[47] U. Buranasuksombat, Y.J. Kwon, M. Turner, B. Bhandari, Influence of emulsion droplet size on antimicrobial properties, Food Sci. Biotechnol. 20 (2011) 793–800.

[48] D.J. McClements, J. Rao, Food-grade nanoemulsions: formulation, fabrication, properties performance, biological fate, and potential toxicity, Crit. Rev. Food Sci. Nutr. 51 (2011) 285–330.

[49] L. Salvia-Trujillo, M.A. Rojas-Grau, R. Soliva-Fortuny, O. Martín-Belloso, Effect of processing parameters on physicochemical characteristics of micro fluidized lemongrass essential oil-alginate nanoemulsions, Food Hydrocolloids 30 (2013) 401–407.

[50] M. Jaiswal, R. Dudhe, P.K. Sharma, Nanoemulsion: an advanced mode of drug delivery system, 3 Biotech. 5 (2015) 123–127.

[51] T. Delmas, H. Piraux, A.-C. Couffin, I. Texier, F. Vinet, P. Poulin, et al., How to prepare and stabilize very small nanoemulsions, Langmuir 27 (2011) 1683–1692.

[52] T. Tadros, P. Izquierdo, J. Esquena, C. Solans, Formation and stability of nano-emulsions, Adv. Colloid Interface Sci. 108 (2004) 303–318.

[53] K. Bouchemal, S. Briancon, E. Perrier, H. Fessi, Nano-emulsion formulation using spontaneous emulsification: solvent, oil and surfactant optimisation, Int. J. Pharm. 280 (2004) 241–251.

[54] C. Anandharamakrishnan, Techniques for Nanoencapsulation of Food Ingredients, Springer, New York; Heidelberg; Dordrecht; London, 2014.

[55] A. Forgiarini, J. Esquena, C. Gonza, C. Solans, Formation of nano-emulsions by low-energy emulsification methods at constant temperature, Langmuir 17 (2001) 2076–2083.

[56] P. Izquierdo, J. Feng, J. Esquena, et al., The influence of surfactant mixing ratio on nano-emulsion formation by the pit method, J. Colloid Interface Sci. 285 (2005) 388–394.

[57] R. Rajalakshmi, K. Mahesh Ckak, A critical review on nano emulsions, Int. J. Innov. Drug Discov. 1 (2011) 1–8.

[58] C. Solans, I. Sole, Nano-emulsions: formation by low-energy methods, Curr. Opin. Colloid Interface Sci. 17 (2012) 246–254.

[59] P. Sanguansri, M.A. Augustin, Nanoscale materials development—a food industry perspective, Trends Food Sci. Technol. 17 (2006) 547–556.

[60] B. Hu, C. Soi, A. Suk, A. Min, Preparation and characterization of carotene nanodispersions prepared by solvent displacement technique, J. Agric. Food Chem. 55 (2007) 6754–6760.

[61] L. Yin, B. Chu, I. Kobayashi, M. Nakajima, Performance of selected emulsifiers and their combinations in the preparation of b-carotene nanodispersions, Food Hydrocolloids 23 (2009) 1617–1622.

[62] M.M. Fryd, T.G. Mason, Advanced nanoemulsions, Annu. Rev. Phys. Chem. 63 (2012) 493–518.

[63] S. Setya, S. Talegaonkar, B.K. Razdan, Nanoemulsions: formulation methods and stability aspects, World J. Pharm. Pharm. Sci. 3 (2014) 2214–2228.

[64] I. Goubet, J.L. Le Quere, A.J. Voilley, Retention of aroma compounds by carbohydrates: influence of their physicochemical characteristics and of their physical state. A review, J. Agric. Food Chem. 46 (5) (1998) 1981–1990.

[65] A. Kant, R.S. Linforth, J. Hort, A.J. Taylor, Effect of β-cyclodextrin on aroma release and flavour perception, J. Agric. Food Chem. 52 (7) (2004) 2028–2035.

[66] M.R.I. Shishir, L. Xie, C. Sun, X. Zheng, W. Chen, Advances in micro and nano-encapsulation of bioactive compounds using biopolymer and lipid-based transporters, Trends Food Sci. Technol. 78 (2018) 34–60.

[67] M.E. Malone, I.A.M. Appelqvist, I.T. Norton, Oral behaviour of food hydrocolloids and emulsions. Part 2. Taste and aroma release, Food Hydrocolloids 17 (6) (2003) 775–784.

[68] P.E. Bouquer, S. Maio, V. Normand, S. Singleton, D. Atkins, Swelling and erosion affecting flavour release from glassy particles in water, AIChE J. 50 (12) (2004) 3257–3270.

[69] M. Schneider, T.F. Mckenna, Comparative study of methods for the measurement of particle size and size distribution of polymeric emulsions, Part. Part. Syst. Charact. 19 (2002) 28–37.

[70] X. Zhang, J. Liu, Effect of Arabic gum and xanthan gum on the stability of pesticide in water emulsion, J. Agric. Food Chem. 59 (2011) 1308–1315.

[71] V.L. Singleton, R. Orthofer, R.M. Lamuela-Raventós, [14] Analysis of total phenols and other oxidation substrates and antioxidants by means of folin-ciocalteu reagent, Methods Enzymol. 299 (1999) 152–178.

[72] M.S. Stankovi, Total phenolic content, flavonoid concentration and antioxidant activity of *Marrubium peregrinum* L. extracts, Kragujevac J. Sci. 33 (2011) 63–72.

[73] D.C. Quettier, B. Gresseier, J. Vasseur, T. Dine, C. Brunet, M.C. Luyckx, Phenolic compounds and antioxidant activities of buckwheat (*Fagopyrum esculentum* Moench) hulls and flour, J. Ethnopharmacol. 72 (2000) 35–42.

17

# Encapsulation and delivery of antiparasitic drugs: a review

Santanu Sasidharan[1], Prakash Saudagar[2]

[1]DEPARTMENT OF BIOTECHNOLOGY, NATIONAL INSTITUTE OF TECHNOLOGY, WARANGAL, INDIA [2]DEPARTMENT OF BIOTECHNOLOGY, NATIONAL INSTITUTE OF TECHNOLOGY WARANGAL, HANAMKONDA, INDIA

## Chapter outline

**17.1 Introduction: encapsulation and techniques** ..... **324**
17.1.1 Spray drying ..... 325
17.1.2 Spray cooling or spray chilling ..... 325
17.1.3 Simple extrusion ..... 325
17.1.4 Centrifugal extrusion ..... 325
17.1.5 Ionic gelation ..... 325
17.1.6 Thermal gelation ..... 325
17.1.7 Fluidized bed coating ..... 326
17.1.8 Lyophilization/freeze-drying ..... 326
17.1.9 Inclusion complexation ..... 326
17.1.10 Emulsion polymerization ..... 326
17.1.11 Liposome entrapment ..... 326
17.1.12 Coacervation ..... 326
**17.2 Need for encapsulated drugs against parasite** ..... **327**
**17.3 Encapsulation of drugs in various parasites** ..... **328**
17.3.1 Leishmaniasis ..... 328
17.3.2 Malaria ..... 330
17.3.3 Trypanosomiasis ..... 332
17.3.4 Other parasites ..... 334
**17.4 Encapsulated drugs in clinical trials and commercial usage** ..... **335**
**17.5 Summary and future outlook** ..... **336**
**References** ..... **336**

Encapsulation of Active Molecules and their Delivery System. DOI: https://doi.org/10.1016/B978-0-12-819363-1.00017-X

## 17.1 Introduction: encapsulation and techniques

In the past few years, many drugs have been discovered to treat various diseases and in the worst-case, its symptoms accordingly. The major problem that persists in the treatment is the failure to administer the drugs directly. A large proportion of the synthesized drugs of today are hydrophobic that causes hindrance in delivery in our body. Drug molecules face difficulty in maintaining stability through weak noncovalent interactions and electrostatic interactions. There are various other factors such as temperature, metal ions, pH, and adsorption that play an important role in the stability of the drugs. Some drugs are degraded by enzymes at the administration site or in the path to the site of action. Due to these reasons, drugs fail to treat diseases along with the symptoms of the disease. Therefore there is a need of the hour to develop methods to preserve and deliver drugs in the body, thereby maximizing its potential as drugs. Encapsulation methods offer a solution to both the preserving and delivering of drug molecules.

Encapsulation is a method of caging the molecules of interest using synthetic polymers. The polymers used in the technique protect the molecules from the environment in which the drug is exposed to in general. In this method, natural polymers are rarely employed because of their purity problems as well as the need for cross-linking. Therefore synthetic polymers such as poly glycolic acid, poly lactic acid (PLA), poly lactide-*co*-glycolide are mostly used in the technique owing to its biocompatibility, biodegradability, and its resorption into various pathways. Another added advantage in encapsulation is the customization of degradability rate and release rate of a drug by controlling the ration of polymers utilized. This, however, does not follow a standard rule and therefore has to be optimized in detail before use. The drugs that have been encapsulated have shown increased bioavailability, controlled delivery, and bioactivity, which gives them an upper hand in the treatment of diseases. The drugs that are encapsulated are capable of transporting the drugs to the target region and deliver them in optimum dosage for an extended period of time. This profoundly increases the efficiency of the drugs by decreasing the toxicity and increasing the stability and solubility. The size and its distribution of encapsulated drugs affect the membrane interactions and drug-barrier penetration.

Encapsulation usually produces nano- or microparticles and the reduced size promotes its delivery to body parts for a long time. Microencapsulation is a technique in which tiny droplets of solid or liquid are coated by a continuous layer of polymeric material. This method, introduced by Bungen burg de Jon and Kan in 1931, produced products of diameter between 1 and 1000 μm. Nano-encapsulation method differs in the encapsulation of bioactive substances in the nanoscale range typically from 10 to 1000 nm. Nano-encapsulation has greater potential when compared to microencapsulation due to its enhanced bioavailability, improved control in the release of drugs, and targeting precisely. There are several methods of encapsulating drugs in polymeric materials and few of these methods are listed in brief.

### 17.1.1 Spray drying

The method involves the dispersion of core material in an entrapment material. This is followed by the atomization and sequentially by spraying of the encapsulated mixture along with hot air desiccant into a chamber. The method is advantageous because of low processing cost, choice of coating material being wide, improved encapsulation efficiency and stability. But, the method degrades thermos-labile drugs and controlling particle size also becomes difficult.

### 17.1.2 Spray cooling or spray chilling

The spray cooling technique is similar to the spray drying method and differs in the use of cold air as a desiccant in the chamber. The disadvantage of degradation of thermos-labile drugs in spray drying method is overcome in this method. But the particle size cannot be controlled in this method too and the technique involved skilled handling and storage conditions.

### 17.1.3 Simple extrusion

In this method the material is forced through a die with coating mass into a desiccant liquid. The coating material hardens when it comes in contact with the desiccant liquid, thereby entrapping the core material of interest. The encapsulation coat is complete and residual material of interest is removed in this method. But, the method has disadvantages such as the need for separation of capsules from the liquid bath as well as difficulty when the capsules are in the viscous carrier material.

### 17.1.4 Centrifugal extrusion

The method is similar to the simple extrusion method wherein there is a unified jet flow through a nozzle at the end and the force employed is a centrifugal force.

### 17.1.5 Ionic gelation

The gelation in this method is made by ionic interactions where the coating material and the core material (drug) are dissolved together and extruded in drops into ionic solution. The method is relatively simple as physical factors such as temperature and pH are avoided along with the usage of organic solvents. Scaling up of ionic gelation is difficult and the capsules have high porosity. The highly porous nature of the capsule causes bursting when handled improperly.

### 17.1.6 Thermal gelation

The gelation is similar to ionic gelation but there is no necessity of ionic solution. The gelation is completely due to thermal parameters. Thermally labile drugs cannot be encapsulated by this method.

### 17.1.7 Fluidized bed coating

In fluidized bed coating the coating material is sprayed into a fluidized bed of drug of interest in a hot environment. The process is economical and allows control of size distribution and porosity. Thermally labile compounds cannot be encapsulated by this method.

### 17.1.8 Lyophilization/freeze-drying

Freeze-drying of an emulsion solution composed of the coating material and the drug of interest is carried out. The method is suitable for thermos-sensitive drugs and drugs that are unstable in an aqueous environment. Lyophilization process is expensive and requires long processing time. The storage and transport of the lyophilized and encapsulated drugs should be done carefully and involved high costs.

### 17.1.9 Inclusion complexation

Apolar drug molecules are entrapped in the β-cyclodextrin cavity through hydrophobic interactions. The technique is efficient in encapsulating unstable and apolar drugs of interest, but the β-cyclodextrin is very expensive and the controlled release of drugs cannot be achieved.

### 17.1.10 Emulsion polymerization

Drug compounds are dissolved into polymer solution and the monomers polymerize to form capsules in aqueous solution. Micro- and nano-encapsulation can be achieved but size distribution is difficult to achieve in emulsion polymerization.

#### *17.1.10.1 Emulsion phase separation*

The drug of interest is added to a polar/apolar layer of either oil in water or water in oil emulsion. The emulsions used in this method are usually made using a surfactant. Polar, apolar, and amphiphilic drugs can be manufactured by this method. These drugs are prone to environmental stresses such as temperature and drying and only a few emulsifiers can be used for this purpose.

### 17.1.11 Liposome entrapment

Liposomes are formed spontaneously when phospholipids are dispersed evenly onto an aqueous phase. The drug is thus entrapped into the core. The drug material can be dissolved either in the aqueous or lipid material and the liposomes are highly suitable for high water activity. Liposomes are also efficient in controlled release and delivery of drugs inside the body. The liposomes are usually prepared on a laboratory scale.

### 17.1.12 Coacervation

The encapsulation of the drug compound is due to the deposition of a coating liquid over the drug by the formation of electrostatic interactions. Thermolabile drugs can be encapsulated by this method at room temperature. The technique employs toxic chemicals, residual solvents, and coacervating agents on the capsule surface and therefore can be used widely. Coacervation method is also complex and expensive at the production level.

## 17.2 Need for encapsulated drugs against parasite

The primary reason for encapsulation is the prolonged and sustained release of the drug compounds into the human body. The targets of encapsulated drugs are illustrated in Fig. 17–1. Conventional drugs are cleared from the body by mononuclear phagocytic systems. They are recognized as foreign particles and are immediately removed from the system. More than 40% of oral drugs are sparingly soluble in water and 90% of the new molecules that are discovered have poor oral bioavailability. The parasite usually attacks the host and resides in the organelles and organs of the host system. Leishmania parasite has two forms, promastigotes and amastigotes. Amastigote form dwells in the macrophage of the human host, thereby avoiding expulsion from the host. Malaria parasite infects the human host by the blood meal of an Anopheles mosquito and then resides in the liver cells.

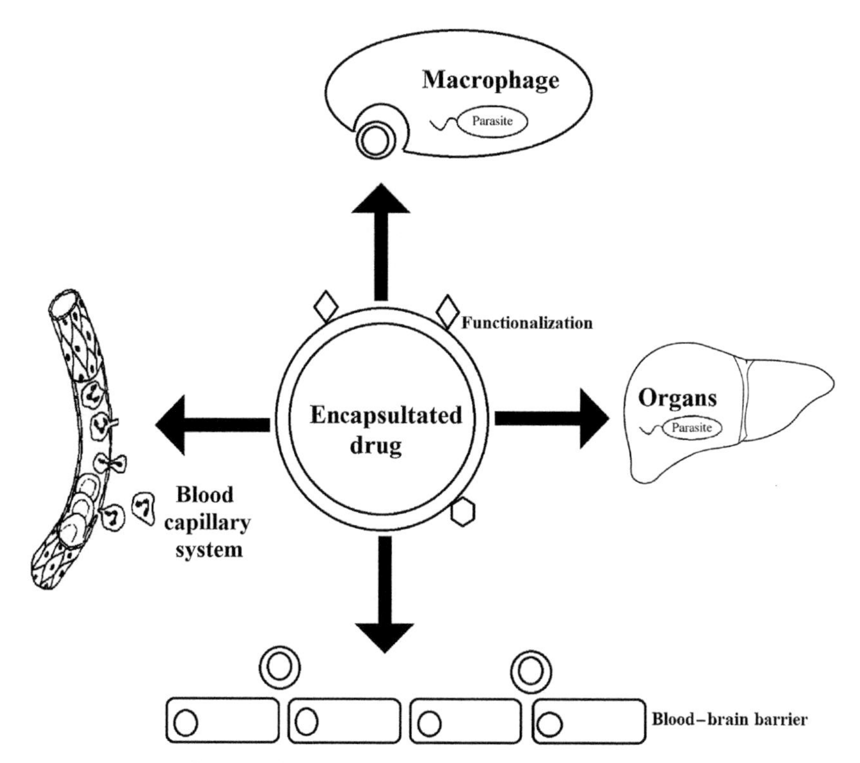


**FIGURE 17–1** Various targets of encapsulated drugs are depicted in the figure. The encapsulated drugs take various routes to macrophages, organs, and blood capillary system to fight the parasite. The encapsulated drugs also try to cross the blood–brain barrier for treatment of the diseases.

The sporozoites infect liver cells and over time they mature into schizonts that rupture the liver cells and release merozoites. Toxoplasma is yet another parasitic group that has two forms, tachyzoites and bradyzoites. The tachyzoites after infection of human host localize in the neural and muscular tissues. The tachyzoites proliferate rapidly and convert themselves into bradyzoites. Bradyzoites are formed inside latent intracellular tissue cysts in muscles and brain. They are nonsusceptible to antibiotics too. Trypanosome parasite enters the human host by the bite of a tsetse fly and resides in the mucous membranes such as nose, mouth, or eye. Trypomastigotes then reproduce by binary fission in various fluids in the body such as blood, lymph, and spinal fluid. They are commonly found in the blood, intestines, and intracellular environment. The drugs that are discovered are successful in preventing these diseases in vitro. But, when it comes to the application of the drug in the host system, most of the drugs fail. The reason for failure is mainly due to the two following reasons:

1. The drug does not target the specific organelle or organ of parasite attack causing side effects.
2. Uncontrolled release of drugs causing toxicity.

To combat these issues, encapsulation is the most favorable and viable option. By encapsulation, phagocytosis can be prevented, the surface can be modified along with the charge carried on the surface. This allows specific targeting depending upon charged interactions. Along with the abovementioned advantages, encapsulation also has various others, which are as follows:

- Masking the taste and odor of drugs making it easier for patients to consume.
- Liquid drugs that require more care can be converted into a powdered or solid form.
- Light-, oxygen-, and moisture-sensitive drugs can be preserved for a long time.
- Volatile and unstable drugs at room temperature can be encapsulated into solid forms.
- Prevention of GI irritation.
- Extended circulation time can be achieved.
- Fluorescent labeling for cell imaging and drug labeling.

## 17.3 Encapsulation of drugs in various parasites

Many drugs have been encapsulated with a wide array of coating materials to enhance the solubility, delivery, and their respective therapeutic activity. In more specific encapsulation requirements, the surface has been modified to increase the circulation time of the drug in the blood fluid too. The various drugs that have been used to date for the treatment of parasites and the advantage of encapsulating the drugs are discussed broadly in the following sections.

### 17.3.1 Leishmaniasis

Leishmaniasis is a vector-borne disease that is found in several countries worldwide. The disease is caused by the various species of the intracellular parasite of the genus *Leishmania.*

The parasite resides in the phagocytic system of hosts and is therefore difficult to target. The mortality caused by the parasite is second to malaria and a third common cause of morbidity. Various drugs such as pentavalent antimonials ($Sb^V$) Glucantime and Pentosam along with diamidines such as pentamidines are being used. Antifungals such as amphotericin-B, ketoconazole, paromomycin, and dapsone are also prescribed to treat the parasite. In the 21st century, miltefosine was approved as the first oral drug against leishmaniasis [1]. But, these drugs come with their own side effects. Liver and kidney toxicity was found to be a common side effect when treated with these drugs. Intravenous treatment with $Sb^V$ produces adverse effects in the body such as pain at the injection site, cardiac arrhythmias, and rashes [2–4].

Encapsulation of drugs has been done to treat the parasitic infection along with reducing the adverse effects exerted by the drugs on the human body. Meglumine containing liposome formulations were found to be10-fold more effective than the free drug. A fivefold increase in the selectivity index, activity, and also reduced macrophage toxicity was also recorded with the liposomal formulation. The study also found that the concentration of meglumine encapsulated liposome with phosphatidylserine required to kill 100% of amastigotes of *Leishmania* was 40 times less when compares to free meglumine drug [5]. Babassu mesocarp was loaded into poly(lactic-*co*-glycolic acid) to form microparticles of size 3–6.4 μm. These microencapsulated particles with a zeta potential of −25 mV were found to be 10-fold higher in efficacy when compared to free babassu mesocarp extract. The microencapsulated extract also reduced the parasite infectivity by a possible mechanism of increased nitric oxide, hydrogen peroxide, and TNF-α production [6].

The effects of Amphotericin-B have been reduced in large by the employment of lipid formulations. The potential of liposomes for the delivery of AmB was first analyzed by New et al. [7]. Deoxycholate has been substituted by other lipid molecules that mask the amphotericin molecules from sensitive tissues, thus reducing the side effects. Three lipid formulations are currently used: liposomal AmB, lipid complex AmB, and colloidal dispersion AmB [8]. In liposomal AmB the AmB is added at 10 mol.% into the bilayer and has a very stable composition and low toxicity. These liposomes were capable of penetrating into the tissue as they were 80 nm in size [9]. Larabi et al. produced lipid-based amphotericin formulations using phospholipids, which reduced toxicity both in vitro and in vivo, but they were not as efficient as the former AmBisome performance [10]. A lipid system that resembled high-density lipoprotein particles were fabricated by Oda et al., which reduced the toxicity of AmB and also exhibited activity in mice infected with *Leishmania major* [11,12]. "Emulsomes" were another solid lipid nanoparticles based on triglycerides that delivered AmB. These systems had increased specificity toward liver, spleen, and lungs and amastigotes [13]. The system was also studied using other ligands too. Another study involved the loading of AmB into gelatin nanoparticles that were functionalized using mannose molecules [14]. AmB delivery based on polysaccharides such as dextran sulfate or chitosan has also been proposed before for the treatment of leishmaniasis [15].

Microspheres of AmB based on albumin and polymer were demonstrated to have activity in hamster infected with *L. infantum*. The albumin microspheres exhibited higher expression of proinflammatory cytokines such as IFN-γ and TNF-α and reduced expression of TGF-β [16–18]. Layer-by-layer approach for the encapsulation of chalcones have been performed and they were found to have controlled and sustained release up to 824 min [19]. Liposomal encapsulation of 17-allylamino-17-demthoxygeldanamycin that is a potent inhibitor of HSP90 was complexed with hydroxypropyl-β-cyclodextrin. These liposomes were less than 200 nm in size and completely cleared the amastigotes from macrophages at the end of 48 h at a concentration of 0.006 nM [20]. A liposomal formulation of MPEG-PLA was also made for doxorubicin and mitomycin C, which had a special shape capable of releasing the drug at a controlled pace. The encapsulation also led to a reduction in the toxicity of mouse macrophages [21].

## 17.3.2 Malaria

Malaria is a disease caused by the protozoan that belongs to the genus *Plasmodium*. There are five reported species of *Plasmodium* that causes disease in humans, among which *Plasmodium falciparum* causes over 90% mortality [22]. The life cycle of *Plasmodium* occurs in two hosts, one being mosquito and the other being human. The female *Anopheles* mosquito transmits the parasite and requires a blood meal for egg development. The human hosts incubate three stages of the parasite. Sporozoites infect the liver upon entry into the human host and morphologically changes into schizonts. Sometimes, the schizonts remain dormant inside the liver and are called hypnozoites. Inside RBCs, the parasite multiplies rapidly into merozoites and emerges when the host cell lysis takes place. Trophozoites are a divergent stage that occurs when the parasite matures in the erythrocytes and further develops into gametocytes. The major drugs used to treat the various stages of malaria are grouped into four chemical classes, 4-aminoquinolines, 8-aminoquinolines, aryl amino alcohols, and artemisinins. These groups share a common pharmacophore and are modified accordingly. But today, malaria is combatted by the use of combinatorial drugs where two or more drugs are prescribed simultaneously that has different mode of actions and different targets inside the parasite [23,24]. These drugs, even though effective, come with various side effects such as hearing impairment, rashes, abdominal pain, nausea, vomiting, hypotension, hypoglycemia, renal toxicity, cardiotoxicity, and problems during pregnancy. The drugs sometimes have short plasma half-life, high recrudescent rate, and poor solubility in water.

Liposomal drug encapsulation has been a breakthrough technology in improving the pharmacological potentials of antimalarial drugs that are hydrophobic [25,26]. Marques et al. had studied the effective nature of liposomes to carry antimalarial drugs to the target RBC cells infected with *Plasmodium* [27]. Primaquine was loaded into a liposomal formulation containing antimalarial lipid 1,2-dipalmitoyl-sn-glycero-3-phosphoethanolamine-*N*-(4-(*p*-maleimidophenyl)butyramide). The liposomes were found to be capable of transporting the drug with minimal loss. Another study was carried out to encapsulate chloroquine and primaquine using amphiphilic dendritic derivatives [28]. The dendrimer encapsulated drugs were found to be able to discern between infected RBCs and noninfected RBCs. These directly reflected in the reduction of drug concentration required to cause parasite mortality.

Lipid nanoparticle encapsulation systems have been tried on both hydrophilic and hydrophobic drug compounds [29–31]. Solid lipid nanoparticle (SLN) encapsulation of artemether was produced by high-pressure homogenization [32]. The encapsulated particles were 100 nm in size and the efficiency of encapsulation was around 68%. Bioavailability of the SLN encapsulated drug was 170% higher than the free drug administered. The encapsulation provided an advantage as artemisinin and their derived drugs are toxic in nature and need to be administered at a slow rate. SLN-based drug release was slow and survived comfortably through the gastric environment in the host digestive system. Lipid-based delivery vehicles called pheroids were also manufactured and this technique improved the pharmacological properties of several antimalarial drugs [33,34]. Mefloquine encapsulated by pheroids exhibited increased antimalarial activity while largely decreasing the toxicity of the drug molecule [35]. Lumefantrine that was encapsulated in a pheroid lipid system had enhanced solubility, absorption, efficacy (46.8% higher), and bioavailability of the drug (3.5 times higher) [34]. Nanostructured lipid carriers (NLC) are another type of lipid-based delivery system, which, unlike SLNs, are manufactured from various lipids in different phases, which often include saturated as well as unsaturated fatty acids [36,37]. Lumefantrine encapsulation by NLC by the method of ultrasonication was developed and the encapsulated drugs were 213 nm in size. The encapsulated drug was easily absorbed into the blood rapidly when compared to the free drug thus increasing the therapeutic efficacy [38]. Artemether–lumefantrine combinatorial drug was formulated into NLC and the formulation was found to be available for both oral and intravenous delivery [39,40]. The formulation showed 100% parasite clearance at just 10% the normal free drug dose in both fasted as well as fed animals. A dose of 600 mg/kg daily dose of artemether–lumefantrine with NLC displayed no toxic effects when compared to the control animals. When the same formulation was tested for cerebral malaria in an intravenously delivered system, the pharmacological properties of the formulation were found to be similar to that of the orally administered formulation [40]. The formulation was stable when autoclaved thus making the system the only existing intravenous formulation for artemisinin-based combination therapy. An experimental drug tafenoquine was encapsulated by microemulsion with a particle size of 20 nm [41]. The encapsulation resulted in an increase of bioavailability of the drug from 55% to 99% in mice. Coencapsulation of quinine and curcumin was carried out by loading onto polysorbate-coated polymeric nanocapsules. The nanocapsules were 200 nm in size and displayed a significant reduction in the parasite when compared to individual free drugs. It was also found that the loaded capsules do not affect *Caenorhabditis elegans* worms and showed higher survival and longevity [42]. Immunoliposomized aminoquinoline was found to be specific and decreased the blood parasitemia, whereas the free drug was not able to produce the same [43]. Two new nano-biomaterial MCM-41 and 3-phenylpropyl silane functionalized MCM-41 was used to encapsulate quinine. This encapsulated quinine was found to have high antimalarial activity $ED_{50} < 0.065$ mg/kg of body weight and also high mean survival time in the body [44]. Artefenomel, a synthetic drug was encapsulated by flash nanoprecipitation by polymer directed self-assembly. The nano-encapsulated

Artefenomel produced sustained release of the drug compared to uncapsulated drug. The encapsulated drug was found to be in an amorphous form thus corroborating the dissolution kinetics [45].

Encapsulation of antimalarial drugs definitely enhanced the absorption, bioavailability, stability, and solubility while reducing the toxicity and dosage concentration. But, the kinetics involved in the release of these drugs have not yet been clearly studied. Encapsulated artemether has been found to be released less than 10% in 2 h in SLNs, whereas approximately 20% release was recorded in the oral and intravenous formulation [32,39,40,46]. The release might be useful for hydrophobic drugs, but the diffusion of hydrophobic drugs into aqueous solution may not be favored [34].

### 17.3.3 Trypanosomiasis

Trypanosomiasis is a parasitic disease that causes devastation in both human and animals. The parasite is classified as a neglected tropical disease and affects many countries worldwide. The transmission of the parasite happens mainly between the mammalian host and a bloodsucking fly belonging to the genus *Glossina*. There exists a polymorphism in the life cycle of the trypanosomes like other parasites. The metacyclic trypomastigotes are transmitted from the salivary glands of the vector to the mucocutaneous tissues of the mammalian host. The early stage in which the parasite becomes slender and resides extracellular is called the hemolymphatic stage. The parasite resides in the bloodstream, lymph nodes, organs, and central nervous system. The parasite crosses the blood–brain barrier and takes over the cerebrospinal fluid, thereby protecting the pathogen from drugs. Animals have been treated with homidium chloride, isometamidium chloride, and diminazene aceturate [47,48]. Humans are treated by a wide range of drugs such as pentamidine isethionate, suramin sodium, melarsoprol and eflornithine [49–51]. Nifurtimox is another approved drug being used to combat this disease. Today, the parasite is controlled by the administration of combined drug therapy of nifurtimox and eflornithine. However, the drugs carry a drawback of faster diffusion, poor half-life, poor solubility, and low therapeutic index. The necessity of the drugs to enter the blood–brain barrier also makes it difficult to control the parasite.

Liposomes have been found to cause toxicity to various different cells depending upon nature and charge [52,53]. The liposomes that contained stearylamine: egg phosphatidylcholine: cholesterol or dipalmitoyl phosphatidylcholine: cholesterol were effective in causing cytotoxicity against *Trypanosoma* sp. [54–57]. The effect of inhibition included a deformation in the morphological characteristics of the parasite along with decreased motility in the bloodstream. The mechanism behind the activity was deciphered to be the interaction between the positively charged liposomes and the negatively charged cell wall of the parasite, which resulted in the cell lysis by osmosis [58]. The individual usage of stearylamine and phosphatidyl choline (PC) did not show any trypanocidal activity and moreover was found to be toxic to human cells. Therefore it is important to note that the vesicular structure of the liposomes is very important for the activity [59]. Dipalmitoyl phosphatidylcholine liposomes showed high in vitro toxicity and also low efficacy improvement with 20% survival.

Diminazene aceturate was encapsulated in stearylamine-bearing liposomes that showed higher activity of the encapsulated liposome than the free drug. Moreover, the encapsulated diminazene displayed slow absorption, slow elimination, less toxicity, and longer efficacy duration. The mechanism of activity was found to be the fusion of liposomal membrane onto the parasite plasma membrane thereby causing cell lysis [57]. Arsonoliposomes or arsonolipids (1,2-diacyloxypropyl-3-arsonic acids) are analogs of phospholipids that contain an arsenic group instead of a phosphate group at the polar head. The mechanism of action of the arsonolipids is the ability of pentavalent arsenic at the polar head to reduce to trivalent arsonolipids by thiols [60,61]. The activity of arsonolipids against trypanosomes has been evaluated before containing a palmitic acid acyl chain. Three different arsonolipids were synthesized: C16-arsonolipid/egg PC/Chl; C16-arsonolipid/DiStearyl Phosphatidyl Choline (DSPC)/Chl and C16-arsonolipid /DSPC/Chl + 1,2-DiStearoyl-sn-glycero-3-Phospho Ethanolamine-PEG2000. These liposomes were found to be in the range of 80–100 nm. The in vivo tests on mice using an intraperitoneal injection of the PC drug showed cure without any toxic effects. But, the dosage of free drug, that is, potassium melarsonyl was required much less than the dosage required for the encapsulated drug. The DSPC encapsulated drug showed no effect at dosages five times higher than the usual concentration. DSPC and PEG encapsulated drugs showed a loss after 5 min of injection into the body and that no arsenic was observed in the brain. This might be due to the composition of the liposome that does not allow it to cross the blood–brain barrier [62–65]. Another strategy to encapsulate miltefosine using arsonoliposome was carried out where miltefosine:PC:SA was used in a 10:10:0.1 ratio. The liposomes were 104 nm in size and the activity was two times more in Trypanosoma *brucei brucei* and the activity decreased slightly in the in vivo studies. The decrease in efficacy was suggested to be due to the large size of the liposomes [66].

Diminazene was encapsulated in poly(amino acid)-based micelles. The drug formed electrostatic interaction with poly(aspartate) and this increased the biodegradability, biocompatibility, and versatility in the structure [67–70]. The micelles were formed with the help of diminazene formed the hydrophobic core and PEG formed the hydrophilic shell. The PEG-PAsp$^{-}$diminazene formed uniform size distributed micelles and uniform shape. The combination however caused rapid drug release thus pushing it to strengthen using PPhen. A new micellar structure using PEG-P(Asp-stat-Phen)-diminazene was formulated where the size of the micelle decrease to 50 nm and the drug amount was increased [71]. This strengthening caused the reduced release of drugs while preserving other advantageous features. Other block encapsulation of diminazene included the use of carboxymethyldextran-PEG and poly (ethylene oxide)-PGlu and hydroxyl ethyl cellulose-PAsp [69,72]. These micelles were 36–50 nm in size, but they were sensitive to pH, temperature, and concentration.

Nanoparticle encapsulation of antitrypanosomal drugs has also been studied. Melarsoprol nanosuspensions with poloxamer 188 or 407 and mannitol were formed with a size distribution of 324 and 407 nm [73]. These formulations could be stored after freeze-drying and no aggregation was recorded. The administration of these drugs showed 5–9-fold higher targeting of the liver when compared to the free drug injection. Lipid drug conjugation of diminazene was carried out using high-pressure homogenization of stearic acid and fatty acid drug [74–76]. Polysorbate 80 was used as the surfactant that helped improve the brain targeting ability. Pentamidine methanesulfonate has been loaded into methacrylate polymers by ionic process wherein carboxylic acids are involved [77–81]. The efficacy was found to be six times better than the free drug with higher tolerability. The nanopolymer drug had low biodegradability and therefore was formulated with poly(D,L-lactide), which was biocompatible and biodegradable too [82].

### 17.3.4 Other parasites

Toxoplasmosis is caused by an obligate intracellular parasite that is caused by *Toxoplasma gondii*. The parasite affects over 1 billion people worldwide and has three infectious stages sporozoites, tachyzoites, and bradyzoites. The parasite affects the CNS mainly and the resistant cyst form of the parasite is resistant to both the immune system and the drugs. The most common treatment for *Toxoplasma* is by using pyrimethamine and sulfadiazine. But these drugs have side effects such as neutropenia, thrombocytopenia, leucopenia, creatine elevation in serum, and hypersensitivity reactions [83–85]. Liposomal formulation of triclosan exhibited requirement of lower dosage and reduced the adverse effects too [86]. Encapsulation of drug pyrimethamine was found to enhance the efficacy and tolerability of the compound against the parasite [82]. Rhodamine B–labeled polystyrene latex particles were coated by polybutyl cyanoacrylate along with the addition of two new drugs against toxoplasmosis. The carrier was characterized to be in nanolevel and was able to achieve 85% therapy efficiency when compared to the free drug [87]. Nanocapsules formed by alginate chitosan calcium phosphate (AEC-CCo-CP-NCs) along with lactoferrin and the formulation were tested against *T. gondii*. The authors noted that the treatment with NCs caused an increase in nitric oxide and decreased parasitemia. The nanocapsules were found to increase the stability of the protein while maintaining the therapeutic potential of lactoferrin protein [88].

Schistosomiasis is another neglected tropical disease that is deemed the most important helminthic parasite when morbidity and mortality rates are taken into consideration. Both the male and female schistosomes reside in the mesenteric veins or venous plexus and lay numerous eggs. These eggs travel through the host intestinal walls or else bladder walls and finally exit through the feces or urine. The drug of interest for the treatment of the parasite is praziquantel. The drug has a broad spectrum of activity but has side effects such as poor bioavailability when administered orally [89]. The drug also subjects itself to metabolism in the liver, therefore, leading to the limitation in circulation [90]. SLN has been employed to increase bioavailability and biodegradability [91,92]. The liquid component of the encapsulation was increased to improve drug efficiency. Ninosomes using span 60, cholesterol, and peceol were also formed using praziquantel in the core. These ninosomes were capable of producing 30%–50% death of adult parasites whereas free drug was capable of only 10% death. In vivo studies also reflected similar results [93]. Liposomal encapsulation of praziquantel was also found to be several folds efficient that the free praziquantel in in vitro and in vivo studies and higher bioavailability [94,95]. Lipid nanocapsules of praziquantel had a size of 46–62 nm and significant improvement in the worm burden reduction, amelioration of liver pathology, and increased damage to the fluke suckers and tegument was observed [96].

Echinococcosis is another helminthic parasite that is caused by *Echinococcus* sp. and causes serious damage to the liver, lungs, and other organs of the human host. The parasite gets lethal if it left untreated. The disease is transmitted through foxes predominantly and is found commonly throughout the world. The treatment options include surgery followed by pre- and postoperative treatments. Alternatively, the most commonly prescribed drug is albendazole. The drug has limited solubility in water and therefore is poorly absorbed in the intestine. Along with the poor availability, the drug is poorly available in the plasma level and in liver distribution [97]. Chitosan microspheres of albendazole along with liposomal albendazole were experimentally studied and the parasite was suppressed in the treatment with chitosan microspheres. Higher plasma levels of albendazole metabolites were also observed when administered with chitosan or liposomal albendazole. The treatment also displayed a shift from Th2 dominant response to Th1 immune response [98]. Lipid encapsulation of benzimidazole albendazole in multilamellar liposomes resulted in 75%–87% entrapment efficiency and the stability of the formulation. The increased oral bioavailability of the drug was attributed to the interaction with phospholipids. A reduction in the 75%–94% of biomass was observed when treated with the liposomal drug. The study also found a higher therapeutic value when administered with cimetidine [99].

## 17.4 Encapsulated drugs in clinical trials and commercial usage

The encapsulated drug market will touch the 20 billion dollar mark by 2025 and the rising demand for various encapsulated antiparasitic drugs is expected to push the industrial growth forward. Controlled and sustained release of antiparasitic drugs is the main reason for the increased demand in the pharmaceutical industry. The encapsulation of the drugs for masking the unfavorable odor and taste is another reason that will cause the expansion of this market.

Few of the liposomal formulation of antileishmanial drugs available are Ambisome, Amphocil, and Albecet. These drugs are licensed for clinical trials, but only Ambisome is used in the case of unresponsiveness to antimonial. Ambisome is effective in treatment when given through intravenous route although the dosage for cutaneous leishmaniasis is higher than visceral leishmaniasis. Ambisome and Amphocil were found to be effective due to their smaller size (100 nm) but not Albecet [100]. A clinical trial of liposomal albendazole compared with tablet albendazole showed superior efficacy of liposomal albendazole [101]. Apart from the abovementioned facts, various other encapsulated drugs are still in the clinical trials or have failed the clinical trials. Another reason for the nonavailability of the encapsulated drug is the nontranslation of the encapsulated drugs from the laboratory scale to the industrial scale.

## 17.5 Summary and future outlook

The application of encapsulated drugs has been widened from its discovery in the 20th century. Encapsulated drugs are composed of various materials such as different lipids, nanoparticles, and synthetic or naturally occurring polymers with each bearing its own uses, advantages, and disadvantages. Encapsulated antiparasitic drugs have different composition, shape, and size. They have their own benefits such as improved pharmacodynamics, pharmacokinetics, therapeutic efficacy, drug-target selectivity, and decreased toxicity against the parasite. But the encapsulated drugs have disadvantages such as narrowness in constituent selection, charge, shelf-life, stability, sterilization, and encapsulation efficacy. Another major obstacle of the encapsulated drug is the interaction of the encapsulated drug with the parasite.

Considering the abovementioned disadvantages in mind, the future success of encapsulation of antiparasitic drugs is by the optimization of the methods and features. Specific characteristics include permeability of the encapsulation, the stability of the encapsulation both chemically and physically, cell viability, the controlled release rate of the encapsulated drug, targeted delivery of the drug, drug stability, shelf-life, and most importantly the possibility of the scale-up of the encapsulation process to the industrial scale. These optimizations should be considered to explore the encapsulation of various antiparasitic drugs for their use in the pharmaceutical industry.

## References

[1] S.L. Croft, S. Sundar, A.H. Fairlamb, Drug resistance in leishmaniasis, Clin. Microbiol. Rev. 19 (1) (2006) 111–126.

[2] H.W. Murray, Clinical and experimental advances in treatment of visceral leishmaniasis, Antimicrob. Agents Chemother. 45 (8) (2001) 2185–2197.

[3] P.C. Melby, Recent developments in leishmaniasis, Curr. Opin. Infect. Dis. 15 (5) (2002) 485–490.

[4] E. Palumbo, Current treatment for cutaneous leishmaniasis: a review, Am. J. Ther. 16 (2) (2009) 178–182.

[5] S.E.T. Borborema, R.A. Schwendener, J.A.O. Junior, H.F. de Andrade Junior, N. do Nascimento, Uptake and antileishmanial activity of meglumine antimoniate-containing liposomes in Leishmania (Leishmania) major-infected macrophages, Int. J. Antimicrob. Agents 38 (4) (2011) 341–347.

[6] M.C.P. da Silva, J.M. Brito, A.D.S. Ferreira, A.A.M. Vale, A.P.A. dos Santos, L.A. Silva, et al., Antileishmanial and immunomodulatory effect of babassu-loaded PLGA microparticles: a useful drug target to *Leishmania amazonensis* infection, Evid.-Based Complement. Altern. Med. 2018 (2018) 1–14.

[7] R. New, M. Chance, S. Heath, Antileishmanial activity of amphotericin and other antifungal agents entrapped in liposomes, J. Antimicrob. Chemother. 8 (5) (1981) 371–381.

[8] S. Sundar, M. Chatterjee, Visceral leishmaniasis-current therapeutic modalities, Indian J. Med. Res. 123 (3) (2006) 345.

[9] J. Adler-Moore, R.T. Proffitt, Ambisome: lipsomal formulation, structure, mechanism of action and pre-clinical experience, J. Antimicrob. Chemother. 49 (1) (2002) 21–30.

[10] M. Larabi, V. Yardley, P.M. Loiseau, M. Appel, P. Legrand, A. Gulik, et al., Toxicity and antileishmanial activity of a new stable lipid suspension of amphotericin B, Antimicrob. Agents Chemother. 47 (12) (2003) 3774–3779.

[11] M.N. Oda, P.L. Hargreaves, J.A. Beckstead, K.A. Redmond, R. van Antwerpen, R.O. Ryan, Reconstituted high density lipoprotein enriched with the polyene antibiotic amphotericin B, J. lipid Res. 47 (2) (2006) 260–267.

[12] K.G. Nelson, J.V. Bishop, R.O. Ryan, R. Titus, Nanodisk-associated amphotericin B clears Leishmania major cutaneous infection in susceptible BALB/c mice, Antimicrob. Agents Chemother. 50 (4) (2006) 1238–1244.

[13] S. Gupta, A. Dube, S.P. Vyas, Antileishmanial efficacy of amphotericin B bearing emulsomes against experimental visceral leishmaniasis, J. Drug Target. 15 (6) (2007) 437–444.

[14] M. Nahar, V. Dubey, D. Mishra, P.K. Mishra, A. Dube, N.K. Jain, In vitro evaluation of surface functionalized gelatin nanoparticles for macrophage targeting in the therapy of visceral leishmaniasis, J. Drug Target. 18 (2) (2010) 93–105.

[15] W. Tiyaboonchai, N. Limpeanchob, Formulation and characterization of amphotericin B–chitosan–dextran sulfate nanoparticles, Int. J. Pharm. 329 (1–2) (2007) 142–149.

[16] J. Sanchez-Brunete, M. Dea, S. Rama, F. Bolás, J. Alunda, R. Raposo, et al., Treatment of experimental visceral leishmaniasis with amphotericin B in stable albumin microspheres, Antimicrob. Agents Chemother. 48 (9) (2004) 3246–3252.

[17] J. Sánchez-Brunete, M. Dea, S. Rama, F. Bolas, J. Alunda, S. Torrado-Santiago, et al., Influence of the vehicle on the properties and efficacy of microparticles containing amphotericin B, J. Drug Target. 13 (4) (2005) 225–233.

[18] S.R. Iniguez, M. Dea-Ayuela, J. Sanchez-Brunete, J. Torrado, J. Alunda, F. Bolas-Fernandez, Real-time reverse transcription-PCR quantification of cytokine mRNA expression in golden Syrian hamster infected with *Leishmania infantum* and treated with a new amphotericin B formulation, Antimicrob. Agents Chemother. 50 (4) (2006) 1195–1201.

[19] U.M. Bhalerao, J. Acharya, A.K. Halve, M.P. Kaushik, Controlled drug delivery of antileishmanial chalcones from layer-by-layer (LbL) self assembled PSS/PDADMAC thin films, RSC Adv. 4 (10) (2014) 4970–4977.

[20] A.L.O.A. Petersen, T.A. Campos, D.A. dos Santos Dantas, J. de Souza Rebouças, J.C. da Silva, J.P. de Menezes, et al., Encapsulation of the HSP-90 chaperone inhibitor 17-AAG in stable liposome allow increasing the therapeutic index as assessed, in vitro, on *Leishmania* (L) *amazonensis* amastigotes-hosted in mouse CBA macrophages, Front. Cell. Infect. Microbiol. 8 (2018).

[21] A.K. Shukla, S. Patra, V.K. Dubey, Nanospheres encapsulating anti-leishmanial drugs for their specific macrophage targeting, reduced toxicity, and deliberate intracellular release, Vector-Borne Zoonotic Dis. 12 (11) (2012) 953–960.

[22] A.M. Thu, A.P. Phyo, J. Landier, D.M. Parker, F.H. Nosten, Combating multidrug-resistant *Plasmodium falciparum* malaria, FEBS J. 284 (16) (2017) 2569–2578.

[23] W.H. Organization, Global Technical Strategy for Malaria 2016-2030, World Health Organization, 2015.

[24] A. Bosman, K.N. Mendis, A major transition in malaria treatment: the adoption and deployment of artemisinin-based combination therapies, Am. J. Trop. Med. Hyg. 77 (6_Suppl) (2007) 193–197.

[25] M. Kuentz, Lipid-based formulations for oral delivery of lipophilic drugs, Drug Discov. Today: Technol. 9 (2) (2012) e97–e104.

[26] C.W. Pouton, Formulation of poorly water-soluble drugs for oral administration: physicochemical and physiological issues and the lipid formulation classification system, Eur. J. Pharm. Sci. 29 (3–4) (2006) 278–287.

[27] J. Marques, J.J. Valle-Delgado, P. Urbán, E. Baró, R. Prohens, A. Mayor, et al., Adaptation of targeted nanocarriers to changing requirements in antimalarial drug delivery, Nanomed.: Nanotechnol., Biol. Med. 13 (2) (2017) 515–525.

[28] J. Movellan, P. Urbán, E. Moles, M. Jesús, T. Sierra, J.L. Serrano, et al., Amphiphilic dendritic derivatives as nanocarriers for the targeted delivery of antimalarial drugs, Biomaterials 35 (27) (2014) 7940–7950.

[29] W.N. Omwoyo, B. Ogutu, F. Oloo, H. Swai, L. Kalombo, P. Melariri, et al., Preparation, characterization, and optimization of primaquine-loaded solid lipid nanoparticles, Int. J. Nanomed. 9 (2014) 3865.

[30] J.O. Muga, J.W. Gathirwa, M. Tukulula, W.G. Jura, In vitro evaluation of chloroquine-loaded and heparin surface-functionalized solid lipid nanoparticles, Malar. J. 17 (1) (2018) 133.

[31] N. Naseri, H. Valizadeh, P. Zakeri-Milani, Solid lipid nanoparticles and nanostructured lipid carriers: structure, preparation and application, Adv. Pharm. Bull. 5 (3) (2015) 305.

[32] P. Dwivedi, R. Khatik, K. Khandelwal, I. Taneja, K.S.R. Raju, S.K. Paliwal, et al., Pharmacokinetics study of arteether loaded solid lipid nanoparticles: an improved oral bioavailability in rats, Int. J. Pharm. 466 (1–2) (2014) 321–327.

[33] J.D. Steyn, L. Wiesner, L.H. du Plessis, A.F. Grobler, P.J. Smith, W.-C. Chan, et al., Absorption of the novel artemisinin derivatives artemisone and artemiside: potential application of Pheroid™ technology, Int. J. Pharm. 414 (1–2) (2011) 260–266.

[34] L.H. Du Plessis, K. Govender, P. Denti, L. Wiesner, In vivo efficacy and bioavailability of lumefantrine: evaluating the application of pheroid technology, Eur. J. Pharm. Biopharm. 97 (2015) 68–77.

[35] L.H. Du Plessis, C. Helena, E. van Huysteen, L. Wiesner, A.F. Kotzé, Formulation and evaluation of pheroid vesicles containing mefloquine for the treatment of malaria, J. Pharm. Pharmacol. 66 (1) (2014) 14–22.

[36] C. Carbone, A. Leonardi, S. Cupri, G. Puglisi, R. Pignatello, Pharmaceutical and biomedical applications of lipid-based nanocarriers, Pharm. Pat. Anal. 3 (2) (2014) 199–215.

[37] C.-L. Fang, S.A. Al-Suwayeh, J.-Y. Fang, Nanostructured lipid carriers (NLCs) for drug delivery and targeting, Recent Pat. Nanotechnol. 7 (1) (2013) 41–55.

[38] J. Liu, L. Tian, S. Chen, G. Huang, Novel Nanostructured Lipid Carrier for Oral Delivery of a Poorly Soluble Antimalarial Agent Lumefantrine: Characterization and Pharmacokinetics Evaluation, 2018.

[39] P. Prabhu, S. Suryavanshi, S. Pathak, S. Sharma, V. Patravale, Artemether lumefantrine nanostructured lipid carriers for oral malaria therapy: enhanced efficacy at reduced dose and dosing frequency, Int. J. Pharm. 511 (1) (2016) 473–487.

[40] P. Prabhu, S. Suryavanshi, S. Pathak, A. Patra, S. Sharma, V. Patravale, Nanostructured lipid carriers of artemether–lumefantrine combination for intravenous therapy of cerebral malaria, Int. J. Pharm. 513 (1–2) (2016) 504–517.

[41] P. Melariri, L. Kalombo, P. Nkuna, A. Dube, R. Hayeshi, B. Ogutu, et al., Oral lipid-based nanoformulation of tafenoquine enhanced bioavailability and blood stage antimalarial efficacy and led to a reduction in human red blood cell loss in mice, Int. J. Nanomed. 10 (2015) 1493.

[42] K. Velasques, T.R. Maciel, A.H. de Castro Dal Forno, F.E.G. Teixeira, A.L. da Fonseca, F. de Pilla Varotti, et al., Co-nanoencapsulation of antimalarial drugs increases their in vitro efficacy against *Plasmodium falciparum* and decreases their toxicity to *Caenorhabditis elegans*, Eur. J. Pharm. Sci. 118 (2018) 1–12.

[43] E. Moles, S. Galiano, A. Gomes, M. Quiliano, C. Teixeira, I. Aldana, et al., ImmunoPEGliposomes for the targeted delivery of novel lipophilic drugs to red blood cells in a falciparum malaria murine model, Biomaterials 145 (2017) 178–191.

[44] S.A. Amolegbe, Y. Hirano, J.O. Adebayo, O.G. Ademowo, E.A. Balogun, J.A. Obaleye, et al., Mesoporous silica nanocarriers encapsulated antimalarials with high therapeutic performance, Sci. Rep. 8 (1) (2018) 3078.

[45] H.D. Lu, K.D. Ristroph, E.L. Dobrijevic, J. Feng, S.A. McManus, Y. Zhang, et al., Encapsulation of OZ439 into nanoparticles for supersaturated drug release in oral malaria therapy, ACS Infect. Dis. 4 (6) (2018) 970–979.

[46] D. Parashar, M.R.S. Rayasa, Development of artemether and lumefantrine co-loaded nanostructured lipid carriers: physicochemical characterization and in vivo antimalarial activity, Drug Deliv. 23 (1) (2016) 123–129.

[47] L. Kinabo, Pharmacology of existing drugs for animal trypanosomiasis, Acta Trop. 54 (3–4) (1993) 169–183.

[48] J. Williamson, Chemotherapy and chemoprophylaxis of African trypanosomiasis, Exp. Parasitol. 12 (5) (1962) 323–367.

[49] R. Brun, J. Blum, F. Chappuis, C. Burri, Human African trypanosomiasis, Lancet 375 (9709) (2010) 148–159.

[50] M.P. Barrett, D.W. Boykin, R. Brun, R.R. Tidwell, Human African trypanosomiasis: pharmacological re-engagement with a neglected disease, Br. J. Pharmacol. 152 (8) (2007) 1155–1171.

[51] J. Pepin, F. Milord, A. Khonde, T. Niyonsenga, L. Loko, B. Mpia, Gambiense trypanosomiasis: frequency of, and risk factors for, failure of melarsoprol therapy, Trans. R. Soc. Tropical Med. Hyg. 88 (4) (1994) 447–452.

[52] P.I. Campbell, Toxicity of some charged lipids used in liposome preparations, Cytobios 37 (145) (1983) 21–26.

[53] F. Szoka, D. Milholland, M. Barza, Effect of lipid composition and liposome size on toxicity and in vitro fungicidal activity of liposome-intercalated amphotericin B, Antimicrob. Agents Chemother. 31 (3) (1987) 421–429.

[54] C. Alving, Liposomes as drug carriers in leishmaniasis and malaria, Parasitol. Today 2 (4) (1986) 101–107.

[55] N. Kuboki, N. Yokoyama, N. Kojima, T. Sakurai, N. Inoue, C. Sugimoto, Efficacy of dipalmitoylphosphatidylcholine liposome against African trypanosomes, J. Parasitol. (2006) 389–393.

[56] T. Souto-Padron, E. Chiari, Further studies on the cell surface charge of *Trypanosoma cruzi*, Acta Trop. 41 (3) (1984) 215–225.

[57] H. Tachibana, E. Yoshihara, Y. Kaneda, T. Nakae, In vitro lysis of the bloodstream forms of *Trypanosoma brucei* gambiense by stearylamine-bearing liposomes, Antimicrob. Agents Chemother. 32 (7) (1988) 966–970.

[58] E. Yoshihara, H. Tachibana, T. Nakae, Trypanocidal activity of the stearylamine-bearing liposome invitro, Life Sci. 40 (22) (1987) 2153–2159.

[59] J. Senior, Fate and behavior of liposomes in vivo: a review of controlling factors, Crit. Rev. Ther. Drug Carr. Syst. 3 (2) (1987) 123–193.

[60] G.M. Tsivgoulis, D.N. Sotiropoulos, P.V. Ioannou, 1,2-Dihydroxypropyl-3-arsonic acid: a key intermediate for arsonolipids, Phosphorus Sulfur Silicon Relat. Elem. 57 (3–4) (1991) 189–193.

[61] D. Timotheatou, P.V. Ioannou, A. Scozzafava, F. Briganti, C.T. Supuran, Carbonic anhydrase interaction with lipothioars enites: a novel class of isozymes I and II inhibitors, Metal-Based Drugs 3 (6) (1996) 263–268.

[62] S.G. Antimisiaris, P.V. Ioannou, P.M. Loiseau, In-vitro antileishmanial and trypanocidal activities of arsonoliposomes and preliminary in-vivo distribution in BALB/c mice, J. Pharm. Pharmacol. 55 (5) (2003) 647–652.

[63] P. Zagana, P. Klepetsanis, P.V. Ioannou, P.M. Loiseau, S. Antimisiaris, Trypanocidal activity of arsonoliposomes: effect of vesicle lipid composition, Biomed. Pharmacother. 61 (8) (2007) 499–504.

[64] P. Zagana, M. Haikou, P. Klepetsanis, E. Giannopoulou, P.V. Ioannou, S.G. Antimisiaris, In vivo distribution of arsonoliposomes: effect of vesicle lipid composition, Int. J. Pharm. 347 (1–2) (2008) 86–92.

[65] S. Piperoudi, P.V. Ioannou, P. Frederik, S.G. Antimisiaris, Arsonoliposomes: effect of lipid composition on their stability and morphology, J. Liposome Res. 15 (3–4) (2005) 187–197.

[66] A. Papagiannaros, C. Bories, C. Demetzos, P. Loiseau, Antileishmanial and trypanocidal activities of new miltefosine liposomal formulations, Biomed. Pharmacother. 59 (10) (2005) 545–550.

[67] C. Atsriku, D. Watson, J. Tettey, M. Grant, G. Skellern, Determination of diminazene aceturate in pharmaceutical formulations by HPLC and identification of related substances by LC/MS, J. Pharm. Biomed. Anal. 30 (4) (2002) 979–986.

[68] A. Thunemann, D. Schutt, R. Sachse, H. Schlaad, H. Mohwald, Complexes of poly(ethylene oxide)-block-poly(L-glutamate) and diminazene, Langmuir 22 (5) (2006) 2323–2328.

[69] G.M. Soliman, F.M. Winnik, Enhancement of hydrophilic drug loading and release characteristics through micellization with new carboxymethyldextran-PEG block copolymers of tunable charge density, Int. J. Pharm. 356 (1–2) (2008) 248–258.

[70] T. Govender, S. Stolnik, C. Xiong, S. Zhang, L. Illum, S.S. Davis, Drug–polyionic block copolymer interactions for micelle formation: physicochemical characterisation, J. Control. Release 75 (3) (2001) 249–258.

[71] K. Prompruk, T. Govender, S. Zhang, C. Xiong, S. Stolnik, Synthesis of a novel PEG-block-poly(aspartic acid-stat-phenylalanine) copolymer shows potential for formation of a micellar drug carrier, Int. J. Pharm. 297 (1–2) (2005) 242–253.

[72] H. Dou, M. Jiang, Fabrication, characterization and drug loading of pH-dependent multi-morpho-logical nanoparticles based on cellulose, Polym. Int. 56 (10) (2007) 1206–1212.

[73] S.B. Zirar, A. Astier, M. Muchow, S. Gibaud, Comparison of nanosuspensions and hydroxypropyl-β-cyclodextrin complex of melarsoprol: Pharmacokinetics and tissue distribution in mice, Eur. J. Pharm. Biopharm. 70 (2) (2008) 649–656.

[74] C. Olbrich, A. Gessner, W. Schröder, O. Kayser, R.H. Müller, Lipid–drug conjugate nanoparticles of the hydrophilic drug diminazene—cytotoxicity testing and mouse serum adsorption, J. Control. Release 96 (3) (2004) 425–435.

[75] C. Olbrich, A. Gessner, O. Kayser, R.H. Müller, Lipid-drug-conjugate (LDC) nanoparticles as novel carrier system for the hydrophilic antitrypanosomal drug diminazenediaceturate, J. Drug Target. 10 (5) (2002) 387–396.

[76] A. Gessner, C. Olbrich, W. Schröder, O. Kayser, R. Müller, The role of plasma proteins in brain targeting: species dependent protein adsorption patterns on brain-specific lipid drug conjugate (LDC) nanoparticles, Int. J. Pharm. 214 (1–2) (2001) 87–91.

[77] R. Durand, M. Paul, D. Rivollet, H. Fessi, R. Houin, A. Astier, et al., Activity of pentamidine-loaded poly (D,L-lactide) nanoparticles against *Leishmania infantum* in a murine model, Parasite 4 (4) (1997) 331–336.

[78] R. Durand, M. Paul, D. Rivollet, R. Houin, A. Astier, M. Deniau, Activity of pentamidine-loaded methacrylate nanoparticles against *Leishmania infantum* in a mouse model, Int. J. Parasitol. 27 (11) (1997) 1361–1367.

[79] T. Fusai, Y. Boulard, R. Durand, M. Paul, C. Bories, D. Rivollet, et al., Ultrastructural changes in parasites induced by nanoparticle-bound pentamidine in a Leishmania major/mouse model, Parasite 4 (2) (1997) 133–139.

[80] T. Fusai, M. Deniau, R. Durand, C. Bories, M. Paul, D. Rivollet, et al., Action of pentamidine-bound nanoparticles against Leishmania on an in vivo model, Parasite 1 (4) (1994) 319–324.

[81] M. Paul, R. Durand, Y. Boulard, T. Fusai, C. Fernandez, D. Rivollet, et al., Physicochemical characteristics of pentamidine-loaded polymethacrylate nanoparticles: implication in the intracellular drug release in Leishmania major infected mice, J. Drug Target. 5 (6) (1998) 481–490.

[82] R.K. Kulkarni, E. Moore, A. Hegyeli, F. Leonard, Biodegradable poly(lactic acid) polymers, J. Biomed. Mater. Res. 5 (3) (1971) 169–181.

[83] L.H. Bosch-Driessen, F.D. Verbraak, M.S. Suttorp-Schulten, R.L. van Ruyven, A.M. Klok, C.B. Hoyng, et al., A prospective, randomized trial of pyrimethamine and azithromycin vs pyrimethamine and sulfadiazine for the treatment of ocular toxoplasmosis, Am. J. Ophthalmol. 134 (1) (2002) 34–40.

[84] D.R. Schmidt, B. Hogh, O. Andersen, S.H. Hansen, K. Dalhoff, E. Petersen, Treatment of infants with congenital toxoplasmosis: tolerability and plasma concentrations of sulfadiazine and pyrimethamine, Eur. J. Pediatr. 165 (1) (2006) 19–25.

[85] C. Silveira, R. Belfort Jr, C. Muccioli, G.N. Holland, C.G. Victora, B.L. Horta, et al., The effect of long-term intermittent trimethoprim/sulfamethoxazole treatment on recurrences of toxoplasmic retinochoroiditis, Am. J. Ophthalmol. 134 (1) (2002) 41–46.

[86] L.A. El-Zawawy, D. El-Said, S.F. Mossallam, H.S. Ramadan, S.S. Younis, Triclosan and triclosan-loaded liposomal nanoparticles in the treatment of acute experimental toxoplasmosis, Exp. Parasitol. 149 (2015) 54–64.

[87] S. Leyke, W. Köhler-Sokolowska, B.-R. Paulke, W. Presber, Effects of nanoparticles in cells infected by *Toxoplasma gondii*, e-Polymers 12 (1) (2012) 121–139.

[88] N. Anand, R. Sehgal, R.K. Kanwar, M.L. Dubey, R.K. Vasishta, J.R. Kanwar, Oral administration of encapsulated bovine lactoferrin protein nanocapsules against intracellular parasite *Toxoplasma gondii*, Int. J. Nanomed. 10 (2015) 6355.

[89] R. Bergquist, J. Utzinger, J. Keiser, Controlling schistosomiasis with praziquantel: how much longer without a viable alternative? Infect. Dis. Poverty 6 (1) (2017) 74.

[90] P. Olliaro, P. Delgado-Romero, J. Keiser, The little we know about the pharmacokinetics and pharmacodynamics of praziquantel (racemate and R-enantiomer), J. Antimicrob. Chemother. 69 (4) (2014) 863–870.

[91] F. Kolenyak-Santos, C. Garnero, R.N. De Oliveira, A.L. De Souza, M. Chorilli, S.M. Allegretti, et al., Nanostructured lipid carriers as a strategy to improve the in vitro schistosomiasis activity of praziquantel, J. Nanosci. Nanotechnol. 15 (1) (2015) 761–772.

[92] A.M. De Campos, A. Sánchez, M.A.J. Alonso, Chitosan nanoparticles: a new vehicle for the improvement of the delivery of drugs to the ocular surface. Application to cyclosporin A, Int. J. Pharm. 224 (1-2) (2001) 159–168.

[93] H.S. Zoghroban, S.I. El-Kowrany, I.A.A. Asaad, G.M. El Maghraby, K.A. El-Nouby, M.A.A. Elazeem, Niosomes for enhanced activity of praziquantel against *Schistosoma mansoni*: in vivo and in vitro evaluation, Parasitol. Res. 118 (1) (2019) 219–234.

[94] S.C. Mourão, P.I. Costa, H.R. Salgado, M.P.D. Gremião, Improvement of antischistosomal activity of praziquantel by incorporation into phosphatidylcholine-containing liposomes, Int. J. Pharm. 295 (1–2) (2005) 157–162.

[95] T.F. Frezza, M.P.D. Gremião, E.M. Zanotti-Magalhães, L.A. Magalhães, A.L.R. de Souza, S.M. Allegretti, Liposomal-praziquantel: efficacy against *Schistosoma mansoni* in a preclinical assay, Acta Trop. 128 (1) (2013) 70–75.

[96] R.O. Amara, A.A. Ramadan, R.M. El-Moslemany, M.M. Eissa, M.Z. El-Azzouni, L.K. El-Khordagui, Praziquantel–lipid nanocapsules: an oral nanotherapeutic with potential *Schistosoma mansoni* tegumental targeting, Int. J. Nanomed. 13 (2018) 4493.

[97] K. Daniel-Mwambete, S. Torrado, C. Cuesta-Bandera, F. Ponce-Gordo, J. Torrado, The effect of solubilization on the oral bioavailability of three benzimidazole carbamate drugs, Int. J. Pharm. 272 (1-2) (2004) 29–36.

[98] M. Abulaihaiti, X.-W. Wu, L. Qiao, H.-L. Lv, H.-W. Zhang, N. Aduwayi, et al., Efficacy of albendazole-chitosan microsphere-based treatment for alveolar echinococcosis in mice, PLoS Negl. Trop. Dis. 9 (9) (2015) e0003950.

[99] H. Wen, R. New, M. Muhmut, J. Wang, Y. Wang, J. Zhang, et al., Pharmacology and efficacy of liposome-entrapped albendazole in experimental secondary alveolar echinococcosis and effect of co-administration with cimetidine, Parasitology 113 (2) (1996) 111–121.

[100] S. Espuelas, Delivery systems for the treatment and prevention of leishmaniasis, Gaz. Médica da Bahia 79 (143) (2009) 134–146.

[101] H. Li, T. Song, Y. Shao, T. Aili, A. Ahan, H. Wen, Comparative evaluation of liposomal albendazole and tablet-albendazole against hepatic cystic echinococcosis: a non-randomized clinical trial, Medicine 95 (4) (2016) 1–6.

# Index

*Note*: Page numbers followed by "*f*" and "*t*" refer to figures and tables, respectively.


**A**

A375 cells, 237–238
Acidic conditions, 64, 263–264
Actinomycetes, 137
Active agents, 258
  modeling of release, 268–281
    diffusion model, 269–281
  simulation of release, 281–284
Active material. *See* Core material
Active molecules
  corrosion inhibitor, 187–188
  drug, 188
  perfume, 189–190
  release of, 186–187
Active targeting, 148
Adriamycin, 137
Aerosil 200, 218–220, 219*f*, 220*f*, 223*f*, 225–227
Affinity-based modifications, 203
*Ageratum conyzoides*, 136
Agglomeration process of encapsulation, 306, 307*f*
Agricultural processes, 132–133
  challenges and future prospects, 138
  encapsulation material, 133–134
  encapsulation of active ingredients, 135–137
  encapsulation techniques, 134–135
Agrochemicals, 6, 131–132
Albecet, 335
Alexa Fluor, 201–202
Alginate, 118–120
Alginate–whey protein isolate layers interaction, 41–43
17-Allylamino-17-demthoxygeldanamycin, 329–330
α-cyclodextrines (αCD), 90
Amastigote, 327–328
AmB. *See* Amphotericin-B (AmB)
Ambisome, 335
Amine group (–$NH_2$), 66–67
Amino acids, 134
Ammonium citrate, 148
Amphocil, 335
Amphotericin-B (AmB), 329–330
Animal cells, 198
Animal tests, 86
Anodic protections, 154
Anthocyanins, 106
Antibacterial activity of nanofluids, 148
Anticancer formulations, 234
Anticorrosive self-healing coatings, 261–262
Antifungals, 328–329
Antigen-presenting cells (APCs), 203–204
Antileishmanial drugs, 335
Antioxidants, 37, 45–50, 104
  activity, 46
    estimation, 316
Antiparasitic drugs
  controlled and sustained release of, 335
  encapsulated drugs
    in clinical trials and commercial usage, 335
    need against parasite, 327–328
  encapsulation of drugs in various parasites, 328–335
APCs. *See* Antigen-presenting cells (APCs)
Apparent partition coefficient studies, 213–214, 221
Arginine, glycine, and aspartate (RGD sequence), 118
Aroma, 103
  extraction methods, 296–300
    aroma extraction using supercritical carbon dioxide, 300
    bioactive aroma extraction using microwave irradiation, 299
    bioactive compounds extraction using Soxhlet extractor, 298–299
    using hot water bath, 298

Aroma (*Continued*)
maceration process for, 299
pretreatment methods to bioactive flavor compounds, 297–298
ultrasound-assisted extraction of bioactive aroma compounds, 300
and flavor, 295–296
Aromatherapy products, 189
Arsonolipids, 333
Arsonoliposomes, 333
Artefenomel, 331–332
Aspartate-based modifications, 202
Azadirachtin, 137
*Azospirillum brasilense*, 137

**B**

B-cell receptor (BCR), 204
*Bacillus pumilus*, 137
*Bacillus sphaericus*, 7
BASs. *See* Biologically active substances (BASs)
BBB. *See* Blood–brain barrier (BBB)
BCR. *See* B-cell receptor (BCR)
2-(Benzothiazol-2-ylsulfanyl)-succinic acid, 164–165
Benzotriazole (BTA), 157, 179, 183–184, 280
low molecular weight, 187
1-H-Benzotriazole-4-sulfonic acid (1-BSA), 154–155
β-cyclodextrines (βCD), 90, 179
simulation of taxifolin complexes with, 91–92
Bioactive aroma extraction using microwave irradiation, 299
Bioactive compounds, 103–107
extraction using Soxhlet extractor, 298–299
flavonoids, 105–107
phenolic compounds, 104
polyamines, 105
tannins, 107
Bioactive factors, 113–114
Bioactive flavor compounds, pretreatment methods to, 297–298
Bioactive molecules, 134
and/or cell encapsulation, 115–122
controlled delivery, 112–115
design aspect of hydrogels for encapsulation, 118–120
essential requirements for encapsulation, 115–118, 117*f*
multifunctional cell/bioactive molecule encapsulation system, 120–122
bioactive molecule–loading mechanism, 120
for bone and cartilage, 122–126, 122*t*
bioactive molecules, 122–124
encapsulation of cells, 124–126
chemical immobilization of, 120
Bioavailability and bioactivity of taxifolin–β-cyclodextrine conjugates, 97–100
Biocompatibility
of material, 116–118
and toxicity of ceramic nanoparticles, 77–79
Biodegradable polymer, 264
Biologically active substances (BASs), 85
Biomacromolecules denaturation, 55
Biomaterial–molecular interactions, 113–114
Biomaterials, 111–112
Biomimetic synthesis, 63
Biopolymers, 27–28
Biotic and abiotic factors, 132
1,8-Bis(cetyltrimethylammonium) octane dibromide, 146
1,3-Bis(cetyltrimethylammonium) propane dibromide, 146
Bleomycin (BLM), 205
Blood–brain barrier (BBB), 197–198
BMP-2. *See* Bone morphogenetic protein-2 (BMP-2)
Bone
bioactive molecule and/or cell encapsulation for, 122–126, 122*t*
homeostasis, 122–124
marrow stromal cells, 122–124
Bone morphogenetic protein-2 (BMP-2), 113–114, 122–124
Bone TE (BTE), 111–112
Bottom-up synthesis methods, 62, 62*f*, 62*t*
Bradyzoites, 327–328
Bridging flocculation, 35–36, 36*f*
Brownian motion, 143
1-BSA. *See* 1-H-Benzotriazole-4-sulfonic acid (1-BSA)
BTA. *See* Benzotriazole (BTA)
BTA–PANI–iron oxide nanoparticles, 167

BTE. *See* Bone TE (BTE)
Bulk erosion, 278

## C

CA. *See* Contrast agent (CA)
Caco-2 colorectal cancer cells, 104
*Caenorhabditis elegans*, 12–15, 331–332
  movement index in PEG hydrogel, 14*f*
Caffeine, 104
Calcium phosphate (CaP), 122–124
Cancer, 154–155. *See also* Colon cancer
  treatment, 149
CaP. *See* Calcium phosphate (CaP)
Carbon nanostructures, 267–268
Cargo loading and release ability, 201–202
Carrier, 9–10
  material. *See* Shell material
  materials used for flavor encapsulation, 296, 297*t*
Cartilage tissue engineering, 111–113, 124–126
  cell encapsulated systems for, 125*t*
  encapsulation of bioactive molecules for bone and, 122*t*
Castor oil plant, 136
Catalase (CAT), 107
Catalysis, 264–266
Catalyst, 156–157
Catalytic functionalities, 154
Catechins, 104, 106
  encapsulation, 16, 16*f*
Cathodic protections, 154
Cationic
  amphiphilic starch nanoparticles, 239–240
  gold nanoparticles, 144
  polymers, 118
Cavitation, 29–30, 29*f*
  ultrasound, 30
CCMV. *See* Cowpea chlorotic mottle virus (CCMV)
CDs. *See* Cyclodextrines (CDs)
Cell adhesion motifs, 118
Cell encapsulation, 116–118, 117*f*
Cell-binding motifs, 118
Cell-laden scaffolds, 124–126
Cellular endocytosis, 199–200
Cellulose nanofibers (CNF), 264
Centrifugal extrusion, 325
Centrifugation method, 147–148
Ceramic nanocontainers, 155–156
Ceramic NPs, 54–69. *See also* Silica nanoparticles (Silica NPs)
  biocompatibility and toxicity of ceramic nanoparticles, 77–79
  MSNs, 68
  as nanocarriers, 55–56
  size, 67–68
Ceria nanocontainers, 155–156
Cerium molybdate nanocontainers ($Ce_2(MoO_4)_3$), 155–156
Cetyltrimethylammonium bromide (CTAB), 157–158
Chemical encapsulation techniques, 300
  cocrystallization process for flavor extraction, 302–303
  encapsulation by coacervation, 300–302
  flavor encapsulation by molecular inclusion, 302
Chemical immobilization of bioactive molecules, 120
Chemical reaction nanocontainer, 161–164
  characterization of nanoparticle and nanocontainer, 167
  electrochemical analytical studies, 170, 170*f*
  interfacial polyaddition/interfacial polycondensation, 162
  release and release rate of benzotriazole, 169, 169*f*
  results and discussions, 167–169
  in situ emulsion polymerization, 163–164
Chimeric VLPs, 205
Chitosan microspheres, 335
Chondrogenic gene expression, 122–124
Chondrogenic growth factor, 122–124
CHP. *See* Cumene hydroperoxide (CHP)
Chromium(VI), 154–155
CI. *See* Creaming index (CI)
Ciprofloxacin, 281–283
Cisplatin, 283
Classical Stöber method, 75–76
CNF. *See* Cellulose nanofibers (CNF)

Coacervation, 144–145, 300–302, 301*f*, 326–327
encapsulation by, 300–302
Coalescence, 25
Coated pills, 112–113
Coating, 9–10
ionic strength, 262
matrix, 164–165
of SPIONs, 74–77
Cocrystallization process for flavor extraction, 302–303
Coffee, 104
Collagen-II/hyaluronic acid in situ hydrogels, 124–126
Colloidal route, 63
Colon cancer, 234
cell selectivity of folic acid–functionalized systems in vitro, 241
colon tumor selectivity of folic acid–functionalized systems in vivo, 241–245
folic acid functionalization on oral delivery systems, 238–240
folic acid–functionalized systems for, 240–245
folic acid–functionalized uptake by FR-mediated endocytosis, 237–238
FR expression in normal and malignant tissues, 236–237
oral delivery systems and gastrointestinal digestion, 245–249
structure and function of FRs, 235–236
Colorectal cancer. *See* Colon cancer
Complex coacervation process, 302
Condensation reactions, 64, 64*f*
Condensed tannins, 107
Conductive functionalities, 154
Container formation, 187
Container-based internal physical phenomena, 160–161
based on chemical reaction nanocontainer, 161–164
interfacial solvent, 160
LBL method container, 160–161
pickering emulsion container, 161
Contrast agent (CA), 69–70
Controlled delivery, 112–115
Controlled flavor release, 313–315
factors affecting release of flavor, 314
mechanism
by diffusion, 314
by erosion and degradation, 314
by melting, 315
by swelling, 314–315
Controlled release mechanism, 4
Conventional drugs, 327–328
Conventional techniques of encapsulation, 300–306
chemical encapsulation techniques, 300–303
physical methods of encapsulation, 303–306
Copper–methanol nanofluid, 145–146
Coprecipitation method, 72–73
Core material, 1–2, 294, 296
Core phase, 9–10
Core shell type, 295
Core–shell structures and properties of SPIONs, 77
Corrosion
inhibitor, 156–157, 187–188
problem, 154
Corrosion and nanocontainer-based delivery system
applications and future, 170–171
case studies, 165–170
container and method preparation/fabrications, 156–164
container-based internal physical phenomena, 160–161
direct or inverse emulsion based container, 159
layer double hydroxide base micro-and nanocontainer, 156–157
polymer shell and polyelectrolyte with ceramic core container, 157–158
stimuli response with ceramic core, 158–159
container approach for corrosion prevention, 154–156
micro-or nanocontainer approach, 154–156
distribution and performance of container for protective coating, 164–165
preparation of iron oxide nanocontainer by LBL, 165–170

release of active compounds from container, 165
Cowpea chlorotic mottle virus (CCMV), 144
Creaming, 25
Creaming index (CI), 39
Creatine elevation, 334
Cryogenic techniques, 12–15
Crystallinity of encapsulation systems, 15–17
CTAB. *See* Cetyltrimethylammonium bromide (CTAB)
CU-Oil. *See* Curcumin oil (CU-Oil)
Cumene hydroperoxide (CHP), 39–40
Curcumin, 45
encapsulation in multilayer emulsions, 45–50
loaded emulsions, 45–46
Curcumin oil (CU-Oil), 48–49
Current density ($I_{corr}$), 170
Cyclodextrines (CDs), 89–90
αCD, βCD, and γCD, 90
characteristics, 90*t*
possibility of encapsulating taxifolin into, 89–90
simulation of taxifolin complexes with β-cyclodextrine, 91–92
Cyclodextrins, 302
Cyclooxygenase, 106
Cysteine-based modifications, 202
Cytarabines, 137

**D**

DBTL. *See* Dibutyltin dilaurate (DBTL)
DCC/DMAP conjugation chemistries, 238–239
DCC/NHS. *See* 1,3-Dicyclohexylcarbodiimide/1-hydroxysuccinimide (DCC/NHS)
DCFH2-DA. *See* 2,7-Dichlorodihydrofluorescein-diacetate (DCFH2-DA)
DCM. *See* 2,7-Dichlorofluorescein (DCM)
Degradation
flavor release mechanism by, 314
kinetics, 120
Delivery system. *See also* Corrosion and nanocontainer-based delivery system; Nanofluids-based delivery system; Self-microemulsifying drug delivery systems (SMEDDSs)
double, 264
emulsions, 23–25
emulsion-based, 23–24
silica-based, 182–183
Density function theory, 268
Depletion flocculation, 35–36, 36*f*
Depression, 104
DETA. *See* Diethylenetriamine (DETA)
Dextran, 76, 241–245
Dextran-coated FSLN (DFSLN), 241–245
DFSLN. *See* Dextran-coated FSLN (DFSLN)
1,2-Diacyloxypropyl-3-arsonic acids. *See* Arsonolipids
Diazotrophic bacteria, 137
Dibutyltin dilaurate (DBTL), 162
2,7-Dichlorodihydrofluorescein-diacetate (DCFH2-DA), 97
2,7-Dichlorofluorescein (DCM), 98
1,3-Dicyclohexylcarbodiimide/1-hydroxysuccinimide (DCC/NHS), 237–238
Diethylenetriamine (DETA), 163–164
Differential scanning calorimetry (DSC), 15–16, 17*f*, 213, 215, 218–220, 220*f*, 225–227
Diffusion, 258
diffusion-based delivery, 113–114
flavor release mechanism by, 314
mechanism, 120
model, 269–281
for erosion controlled release system, 278–281
for matrix (monolithic) systems, 273–275
primarily diffusion-controlled drug delivery systems, 271*f*
for reservoir system, 271–273
for swelling controlled release systems, 276–278
Digestive enzymes, 245–247
Dihydroquercetin, 86
2,5-Dimercapto-1,3,4-thiadiazolate, 157
Dimethyl phthalate, 264
4-Dimethylaminopyridine (DMAP), 238–239
Dimethyldineodecanoate (DMDNT), 162
Diminazene, 333
Diminazene aceturate, 333

Dipalmitoyl phosphatidylcholine liposomes, 332–333
1,1-Diphenyl-2-picrylhydrazyl (DPPH), 46, 94
  free radical, 46
  scavenging activity of curcumin, 48–49
Dispersion model, 275
Dissipative particle dynamics simulation, 283
Dissolution
  media, 216
  model, 273–275
DiStearyl Phosphatidyl Choline (DSPC), 333
DLS. *See* Dynamic light scattering (DLS)
DMAP. *See* 4-Dimethylaminopyridine (DMAP)
DMDNT. *See* Dimethyldineodecanoate (DMDNT)
Dodecyl benzene, 154–155
Dopamine, 73, 104
Doppler shift, 315
Double delivery system, 264
Doxorubicin (DOX), 183, 201–202, 205
DPPH. *See* 1,1-Diphenyl-2-picrylhydrazyl (DPPH)
Droplet size measurement of food emulsions, 315
Droplet-gelation, 135
Drug, 188
  content determination, 216, 228
  dissolution, 270–271
  drug-delivery
    application, 182
    system, 155–156, 170–171, 259–261
  effect on phase diagram, 222, 223*t*
  molecules, 324
DSC. *See* Differential scanning calorimetry (DSC)
DSPC. *See* DiStearyl Phosphatidyl Choline (DSPC)
Dual-gelling macromeres, 121
Dynamic light scattering (DLS), 10–12, 146, 315
Dynamin-like GTPase, 199–200

**E**

EC. *See* Ethyl cellulose (EC)
Echinococcosis, 335
ECM. *See* Extracellular matrix (ECM)
Ectopic bone regeneration, 122–124
EDC. *See* 1-Ethyl-3-(3-dimethylaminopropyl) carbodiimide (EDC)
EDS. *See* Emulsion droplet size (EDS)
%EE. *See* Encapsulation efficiency (%EE)
Eflornithine, 332
Electrochemical workstation, 167
Electron microscopy (EM), 12–15
Electrostatic interaction, 118
Electrostatic stabilization method (ESM), 145–146
Ellipsoidal shaped SPIONs, 76
EM. *See* Electron microscopy (EM)
24-EMI. *See* 2-Ethyl-4-methylimidazole (24-EMI)
Empirical models, 279–281
Emulsification, 30–31. *See also* Flavor encapsulation by emulsification
  efficiency of surfactants, 220*t*
  through solvent displacement method, 312
  study of surfactant and cosurfactant combinations, 221*t*
  technology, 24
  time, 220–221
  using high-pressure homogenizer and microfluidizer, 31*f*
Emulsifiers, 26–28
Emulsion, 12–15
  diffusion method, 145
  emulsion-based delivery systems, 23–24
  as encapsulating and delivery system, 23–25
  phase separation, 326
  polymerization, 179, 326
  stability, 25
  stabilization of, 25–29
    homogenization, 28–29, 28*f*
    role of emulsifiers, 26–28
  systems, 24
Emulsion droplet size (EDS), 308
Emulsomes, 329
Encapsulants in food industry, 266–267
Encapsulated bioactive compounds
  droplet size and stability measurements of food emulsions, 315
  flavor encapsulated food emulsions
    antioxidant activity estimation for, 316
    flavonoids concentrations estimation in, 317
    total phenolic content estimation of, 316
  surface morphology and topography
    morphology characterization of food emulsions, 316

optical microscopy of flavor encapsulated emulsion, 316
SEM and TEM analysis, 316
Encapsulated drugs
in clinical trials and commercial usage, 335
need against parasite, 327–328
targets of, 327*f*
Encapsulated materials
imaging of, 12–15
rheology of, 17–20
Encapsulation, 1–4, 9–10, 136
of active ingredients, 135–137
of active molecules and release mechanism, 4*f*
of active substrate and target applications, 6–7
applications, 5*f*
areas, 5*f*
challenges, 145
nanocapsules, 143–144
nanoparticles encapsulation techniques, 144–145
classification, 295
conventional techniques of, 300–306
crystallinity of Encapsulation systems, 15–17
of curcumin in multilayer emulsions, 45–50
current trends, 4–6
design aspect of hydrogels for, 118–120
of drugs in various parasites, 328–335
emulsion, 314
of flavor, 294, 294*f*
material, 133–134
of nanomaterials, 143–145
newer intensified approach for flavor encapsulation, 307–312
by emulsification with higher energy, 308–311
by emulsification with low-energy, 311–312
protection from biotic and abiotic factors by, 132*f*
stability, 45–50
techniques, 134–135, 324–327
and release mechanisms, 3*f*
Encapsulation efficiency (%EE), 12, 13*f*, 46–47, 186
of PE and SE, 47*f*
Endocytosis, FR-mediated, 237–238
Engineered NPs, 55
Enveloped VLPs (eVLPs), 199, 199*f*
Epicatechin, 104
Epicatechin gallate, 104
Epigallocatechin, 104
Epigallocatechin gallate, 104
Equilibrium solubility studies, 212–213
Equilibrium solution, 300–302
Erosion
controlled release system models, 278–281
empirical models, 279–281
flavor release mechanism by, 314
ESM. *See* Electrostatic stabilization method (ESM)
Essential oils, 36, 189–190, 266–267
Ethyl cellulose (EC), 135
Ethyl xanthate, 157
1-Ethyl-3-(3-dimethylaminopropyl)carbodiimide (EDC), 238–239
2-Ethyl-4-methylimidazole (24-EMI), 163–164
eVLPs. *See* Enveloped VLPs (eVLPs)
Extracellular matrix (ECM), 124–126
Extrusion method of flavor encapsulation, 304

**F**

*Fagaceae*, 107
Fatty oils, 24
$Fe_3O_4$ nanoparticles, 263–264
Fertilizers, 264
FGF-2, 113–114
Fibrin-based hydrogels, 118–120
Fibrin-based scaffolds, 124–126
Fick's law, 270–271
second law, 269
Fickian model, 276
"Fingerprinting" particle-size analysis, 10–12, 11*f*
Flavanols, 106
Flavonoids, 105–107
concentrations estimation in flavor encapsulated food emulsions, 317
subclasses and compounds with natural sources, 106*t*
Flavor encapsulation by emulsification. *See also* Emulsification
higher energy, 308–311
emulsification by high-shear stirring, 311

Flavor encapsulation by emulsification (*Continued*)
with HPH, 308–309
microfluidization for nanoemulsions synthesis, 309–310
ultrasonic emulsification, 310–311
with low-energy, 311–312
emulsification through solvent displacement method, 312
flavor encapsulation by phase inversion emulsification, 312
membrane technology, 312
spontaneous emulsification, 311–312
Flavors, 24, 103
encapsulation, 294
aroma and flavor, 295–296
aroma extraction methods, 296–300
carrier materials used for, 296, 297*t*
characterization techniques for encapsulated bioactive compounds, 315–317
conventional techniques of encapsulation, 300–312
newer intensified approach for, 307–312
phenomena of encapsulated flavor release, 313–315
release, 313
controlled, 313–315
phenomena of encapsulated, 313–315
uncontrolled flavor release modes, 315
Flocculation, 25
Fluidized bed, 135
coating, 305–306, 326
Fluorescein, 201–202, 262
Fluorescent/fluorescence
markers, 15
microscopy, 12–15, 316
sensor-releasing nanocontainer, 263–264
Fluorouracil, 137
Folate conjugate systems in vitro, 241
Folate receptor (FRs), 234–235
expression in normal and malignant tissues, 236–237
structure and function of folate receptor, 235–236
FR-α, FR-β and FR-γ/γ′, 236*f*
Folic acid, 234–237
colon tumor selectivity, 241–245
experimental folic acid–functionalized oral delivery systems, 242*t*
functionalization on oral delivery systems, 238–240
for colon cancer, 240–245
effect on hydrodynamic size of systems, 239–240
effect on surface charge of systems, 240
functionalized oral delivery systems, 235
functionalized uptake via FR, 237–238
Food, 296–297
emulsions, 315
encapsulants in food industry, 266–267
functional, 18–19
matrix, 87–88
system components, 87
2-(2-Formylquinolin-8-yloxy)-*N,N*-diisopropylacetamide, 263–264
Fourier transform-infrared spectroscopy (FTIR spectroscopy), 213, 218, 219*f*
Free radicals, 104
Freeze-drying, 47–48, 326
for flavor encapsulation, 304–305
Freeze-thawing, 216, 228
FRs. *See* Folate receptor (FRs)
FSLN. *See* Functionalized solid lipid nanoparticles (FSLN)
FTIR spectroscopy. *See* Fourier transform-infrared spectroscopy (FTIR spectroscopy)
Fullerene ($C_{60}$), 283
Functional foods, 18–19
Functionalized mono-acryloxyethyl phosphate hydrogels, 121
Functionalized silica nanocontainers, 183
Functionalized solid lipid nanoparticles (FSLN), 241–245

## G

γ-cyclodextrines (γCD), 90
Gas storage, 170–171, 267–268
Gases, 268
Gastrointestinal digestion, 245–249

Gastrointestinal tract, 248–249
Gelation mechanism, 121
Gene transfection efficiency, 122–124
Gene-activated scaffolds, 122–124
Genetic modifications, 203
Globule size, 215, 225
*Glossina*, 332
Glucantime, 328–329
Glutathione peroxidase (GPx), 107
Glycophospholipid anchor, 235
Gmelin larch-tree (*Larix gmelinii* (Rupr.), 86
GPx. *See* Glutathione peroxidase (GPx)
Graft polymerization, 134
Granulation process of encapsulation, 306
Graphene-based nanosensors, 149
Grashof number, 143
Growth factors, 112–114, 122–124
Gut microbiota, 105

**H**

Halloysite nanocontainers, 183–185
Halloysite nanotubes (HNTs), 155–156, 264
HBsAg-based HBV-VLPs, 205
HBV. *See* Hepatitis B virus (HBV)
HCC. *See* Hepatocellular carcinoma (HCC)
Hepatitis B virus (HBV), 198
Hepatocellular carcinoma (HCC), 199–200
Herbicides, 135, 264
Herbology ester, 97–98
Herbs, 103
Hermospermine, 105
Herschel–Bulkley equation, 18–19
Heterogeneous catalysis, 155–156
Heterogeneous erosion, 314
High surface area–to-volume ratio, 116
High-pressure homogenizers/homogenization (HPH), 28–29, 31–32, 303, 308
  emulsification using, 31*f*
  flavor encapsulation by emulsification with, 308–309
Higuchi square root model, 280–281
Higuchi's equation, 275
Hixson–Crowell model, 280
hMSCs. *See* Human mesenchymal stem cells (hMSCs)
HNTs. *See* Halloysite nanotubes (HNTs)
Homogeneous erosion, 314
Homogenization, 28–29, 28*f*
Homogenizers
  emulsification using high-pressure homogenizer, 31*f*
  ultrasonication comparison with high energy, 31–33, 32*f*
HOPDMS, 162
Hot water bath, aroma extraction using, 298
HPH. *See* High-pressure homogenizers/homogenization (HPH)
HPV. *See* Human papillomavirus (HPV)
8-HQ. *See* 8-Hydroxyquinoline (8-HQ)
Human mesenchymal stem cells (hMSCs), 122–124
Human papillomavirus (HPV), 198
Hyaluronic acid, 118–120, 124–126, 189–190
Hydrogels, 115–116
  design aspect for encapsulation, 118–120
Hydrogen bonds, 87
Hydrolysable tannins, 107
Hydrolysis, 64, 64*f*
Hydrotalcite, 156–157
  pigments, 154–155
Hydrothermal method, 73
6-Hydroxydopamine (6-OHDA), 107
Hydroxypropylcellulose nanoparticles, 137
8-Hydroxyquinoline (8-HQ), 154–155
Hypnozoites, 330

**I**

Imaging of encapsulated materials, 12–15
Imidazole, 280
In situ emulsion polymerization, 163–164
In situ hydrogels, 120–121
In vitro digestion, 49
In vitro models, 97
In vitro multimedia drug release study, 216, 227–228
In vitro simulation systems, 88
Inclination effect, 72
Inclusion complexation, 326

Induced pluripotent stem cell–derived mesenchymal stem cells (iPSMSCs), 124–126
Injectable alginate–CaP load-bearing hydrogel, 124–126
Injectable drugs, 112–113
Injectable hydrogels, 118
Inorganic molecules, 65
Inorganic nanocapsules, 144
Insecticides, 134
Insects attack on crops, 264
Interfacial polyaddition/interfacial polycondensation, 162
Interfacial solvent, 160
Intermolecular interactions, 113–114
Intralipid emulsion, 10–12
Ionic gelation, 325
iPSMSCs. *See* Induced pluripotent stem cell–derived mesenchymal stem cells (iPSMSCs)
Iron oxide nanocontainers, 169
  material and methods, 165
  preparation by LBL, 165–170
  preparation by ultrasound, 165–167
Iron oxide nanoparticles (Iron oxide NPs), 56, 70–71, 144
  LBL synthesis of $FeO_3$ nanocontainers, 166*f*
  morphological studies, 167
  preparation by ultrasound, 165–167
Iron oxide NPs. *See* Iron oxide nanoparticles (Iron oxide NPs)
Isoquercitrin, 107

**K**

Kaempferol, 107
Ketoconazole (KET), 212
  apparent partition coefficient studies, 213–214
  drug content determination, 216, 228
  effect on phase diagram, 214
  equilibrium solubility studies, 212–213
  freeze-thawing, 216, 228
  liquid SMEDDS characterization, 215, 224–225
  liquid SMEDDS preparation, 214, 222–223
  physicochemical compatibility with excipients, 213, 218–228
  pseudo-ternary phase diagram construction, 214
  self-emulsification study, 213
  SMEDDS
    characterization, 215–216, 225–227
    formulations, 222*t*
    preparation, 225
  solid ketoconazole SMEDDS preparation, 215
  solubility study, 216–218, 217*f*
  in vitro multimedia drug release study, 216, 227–228
Korsmeyer model, 276
Korsmeyer–Peppas model, 280

**L**

Labrasol, 218
Laplace pressure, 26
Laponite nanoparticles, 124–126
Large bowel cancer. *See* Colon cancer
*Larix sibirica* Ledeb. *See* Siberian larch-tree (*Larix sibirica* Ledeb.)
Layer double hydroxides (LDHs), 155–157
  base micro-and nanocontainer, 156–157
Layer-by-layer approach (LbL approach), 35, 329–330
  assemblies, 168–169, 180–182
    for nanocontainer synthesis, 181*f*
    procedure for benzotriazole loading on ZnO nanocontainers, 181*f*
  container, 160–161
  deposition of emulsifiers on droplet surfaces, 34*f*
  LBL-assembled nanocontainers, 154
  preparation of iron oxide nanocontainer, 165–170
  synthesis of $FeO_3$ nanocontainers, 166*f*
LbL approach. *See* Layer-by-layer approach (LbL approach)
LDHs. *See* Layer double hydroxides (LDHs)
Leica DMI6000B, 98
*Leishmania*, 328–329
Leishmania parasite, 327–328
Leishmaniasis, 328–330
Leucopenia, 334
Ligand exchange method, 73

Light-scattering techniques, 146
Lignin, 264
Lignosulfonate, 154–155
Lipid nanoparticle encapsulation systems, 331–332
Lipolysis, 49–50
Lipophilic active ingredients, 37
Lipophilic compounds, 24, 33, 37
Liposomes, 76, 332–333
  drug encapsulation, 330
  entrapment, 326
Lipoxygenase, 106
Lumefantrine, 331–332
Lyophilization, 326
Lysine-based modifications, 202

M

Maceration process for aroma extraction, 299
Macro-encapsulation, 132–133
Macro-nutrients, 134
Macro-pathogens, 131
Maghemite ($\gamma$-$Fe_2O_3$), 70
Magnetic
  functionalities, 154
  iron oxides, 70–71
  nanoparticle, 283
  resonance imaging, 144
  stimuli-responsive encapsulation system, 122
Magnetite ($Fe_3O_4$), 70
  Magnetite fluid hyperthermia, 149
Malaria, 330–332
Malignant tissues, FR expression in, 236–237
Maltodextrin (MD), 37–38
Malvern Zetasizer (Nano ZS90), 215
Matrix, defined, 9–10
Matrix (monolithic) systems, models for, 273–275
  dispersion model, 275
  dissolution model, 273–275
MBI. *See* Mercaptobenzimidazole (MBI)
MCM-41. *See* Mobil Composition of Matter-41 (MCM-41)
MCM-50. *See* Mobil Composition of Matter-50 (MCM-50)
MD. *See* Maltodextrin (MD)
Mechanical encapsulation. *See* Physical encapsulation
Medicine, nanotechnology in, 53–55
Melarsoprol nanosuspensions, 333–334
Melting, flavor release mechanism by, 315
Membrane, 9–10
Membrane technology for flavor encapsulation, 312
Mercaptobenzimidazole (MBI), 183–184
Mesenchymal stem cells, 124–126
Mesoporous silica, 157–158, 182
Mesoporous silica nanoparticles (MSNs), 68, 122–124. *See also* Silica nanoparticles (Silica NPs)
  synthesis, 68–69
  types, 69*f*
Metabolic waste removal, 124–126
Metallic/metals, 134
  alkoxides, 63–64
  nanoparticles, 144
  salts, 63–64
  structures, 154
Mg–Al LDHs, 157
Micro-encapsulation, 132–133
Micro-or nanocontainer approach, 154–156
  self-healing materials, 156
    of human skin, 155*f*
Micro-pathogens, 131
Microbeads, 137
Microbicidal activity, 138
Microcapsules, 185
Microemulsion, 2–3, 73
Microencapsulation, 6–7, 116, 135, 137, 303, 303*f*, 324. *See also* Encapsulation
Microfluidization for nanoemulsions synthesis, 309–310
Microfluidizer, 309–310
  emulsification using, 31*f*, 32
Micronutrients, 134, 266
Microorganisms, 137
Micropatterns on scaffolds and hydrogels, 114–115
Microscale hydrogels, 122
Microspheres, 185, 329–330
Microwave energy, 74

Microwave irradiation, bioactive aroma extraction using, 299
Microwave-assisted extraction, 299
Mitogens, 122–124
Mixed-layer approach, 35–36
Mobil Composition of Matter-41 (MCM-41), 68
Mobil Composition of Matter-50 (MCM-50), 68
Modified plasma arc system, 145–146
Molecular dynamics simulation, 284
Molecular inclusion, flavor encapsulation by, 302
Molecular recognition of colon cancer cells, 234–235, 245
Molecular-dynamic modeling, 267–268
Molybdates compounds, 154–155
Mono-core capsule, 295
Monte Carlo simulation, 281
MOPS algorithm, 91
MSNs. *See* Mesoporous silica nanoparticles (MSNs)
Multifunctional cell/bioactive molecule encapsulation system, 120–122
Multifunctional nanoparticles, 55, 56*t*
Multilayer emulsions, 33–34
  applications, 36–38
  approaches for preparation
    layer-by-layer approach, 35
    mixed approach, 35–36
  for encapsulation of bioactive compounds, 33–38
  encapsulation of curcumin in, 45–50
  layer-by-layer deposition of emulsifiers on droplet surfaces, 34*f*
  ultrasonic-assisted synthesis, 38–45
Multiple emulsions, 23–24
Mycorrhizal fungi, 137
Myramistin, 276

**N**

*n*-hydroxysuccinimide (NHS), 202
Naked SPIONs, 74
Nano-biomaterial MCM-41, 331–332
Nano-precipitation method, 144
Nanocapsules, 142–144, 142*f*
  inorganic nanocapsules, 144
  polymeric nanocapsules, 143
  viral/virus-like nanocapsules, 144
Nanocarriers, 137–138, 259, 266–267
Nanocomposite scaffolds, 124–126
Nanocontainers, 178, 258
  active agents
    modeling of release, 268–281
    simulation of release, 281–284
  active molecules
    corrosion inhibitor, 187–188
    drug, 188
    perfume, 189–190
  applications of, 259*f*
  control parameters for nanocontainer applications
    encapsulation efficiency, 186
    pH-based response for release, 186
    release of active molecules, 186–187
    size of container, 185
    surface charge on container, 185–186
  halloysite, 183–185
  LbL assemblies, 180–182, 181*f*
  mechanism of release in, 258–268, 260*f*
    anticorrosive self-healing coatings, 261–262
    catalysis, 264–266
    drug delivery, 259–261
    encapsulants in food industry, 266–267
    fertilizers, 264
    gas storage, 267–268
    wastewater treatment, 262–264
  polymeric, 179–180
  preparations routes for designing delivery system, 184*f*
  shell, 154
  silica-based delivery system, 182–183
Nanoemulsion, 2–3, 307–312
  microfluidization for synthesis, 309–310
Nanoencapsulation, 16, 132–133, 178, 324
Nanofluids-based delivery system, 143, 145–148
  applications, 148–149
  encapsulation of nanomaterials, 143–145
  nanofluid stability evaluation methods, 146–148
  preparation, 145–146
  targeted drug delivery, 148
Nanomaterials, encapsulation of, 143–145

Nanometeric size, 132–133
Nanoparticles (NPs), 53–54, 142, 197–198
encapsulation techniques, 144–145
coacervation method, 144–145
emulsion diffusion method, 145
nano-precipitation method, 144
solvent evaporation method, 145
in pharmaceutical and medicine fields, 54*f*
Nanoprecipitation, 179–180
Nanorattles, 264–265
Nanoreactors, 170–171
Nanoscale core–shell emulsion encapsulation, 5–6
Nanoscience, 53–54
Nanospheres, 142, 142*f*
Nanostructure material, 156–157
Nanostructured lipid carriers (NLC), 331–332
Nanotechnology, 142, 197–198, 258–259
in pharmacy and medicine, 53–55
tools, 113
Nasal spray, 112–113
Natural antioxidants, 85
Natural biopolymers, 266–267
Natural polymers, 115–116, 118–120
NCE. *See* New chemical entities (NCE)
Neuroblastoma cells, 97, 99*t*
Neutropenia, 334
New chemical entities (NCE), 212
NHS. *See* *n*-hydroxysuccinimide (NHS)
Nifurtimox, 332
Ninosomes, 334–335
NLC. *See* Nanostructured lipid carriers (NLC)
Non-Fickian Case-II transport, 276–278
Nonenveloped VLPs (nVLPs), 199, 199*f*
Nonhydrolyzable organic groups, 65–66
Nonspecific binding of biomolecule and nanoparticle, 144
NP-based delivery vehicles, 200–201
NPs. *See* Nanoparticles (NPs)
Nutriceuticals
effective delivery strategies for, 89–92
possibility of encapsulating taxifolin into cyclodextrine, 89–90
nVLPs. *See* Nonenveloped VLPs (nVLPs)

## O

O/W emulsions. *See* Oil-in-water emulsions (O/W emulsions)
Octapod-shaped SPIONs, 71–72
6-OHDA. *See* 6-Hydroxydopamine (6-OHDA)
Oil-in-water emulsions (O/W emulsions), 23–24, 33
orientation of emulsifier molecules, 27*f*
One-dimensional two-pack epoxy system, 156
Optical microscopy of flavor encapsulated emulsion, 316
Optimization of sonication time, 40–41
Oral delivery systems, 245–249
development, 212
folic acid functionalization on, 238–240
Oral strips, 112–113
Organic molecules, 65
Organic polymers, 137
ORganically MOdified SILica matrix (ORMOSIL), 65–66
Organometallic precursors, 63–64
Ornithine decarboxylase, 104
Osteogenesis, 122–124
Osteogenic gene expression, 122–124
Ostwald ripening, 25
Oxalate, 157

## P

PAA. *See* Polyacrylic acid (PAA)
Parasites
encapsulation of drugs in, 328–335
need for encapsulated drugs against, 327–328
Parkinson's disease (PD), 104
Particle size
and distribution, 10–12
measurement, 39
Particles sedimentation velocity, 147–148
Passive targeting, 148, 245
Passively protecting coatings, 261
Pathogen recognition receptors (PRRs), 203–204
Payload phase, 9–10
PC. *See* Phosphatidyl choline (PC)
PCL. *See* Polycaprolactone (PCL)
PCR. *See* Polymerase chain reactions (PCR)
PD. *See* Parkinson's disease (PD)

Pd(II) porphyrins, 15
PDES, 162
pDNAs, 122–124
PE. *See* Primary emulsion (PE)
PEG. *See* Polyethylene glycol (PEG)
PEI. *See* Polyethylene imine (PEI)
Pentavalent antimonials ($Sb^V$), 328–329
Pentosam, 328–329
Peppermint oil, 189
Perfume, 189–190
Permeability of membrane, 115–116
Peroxide value measurement (PV measurement), 39–40
Pesticides, 134, 264
pH, 48–49, 147
pH-based response for release, 186
Pharmacy, nanotechnology in, 53–55
Phase inversion emulsification, flavor encapsulation by, 312
Phase-inversion composition (PIC), 311–312
Phase-inversion temperature (PIT), 311–312
Phase-separation technique. *See* Coacervation
Phenolic compounds, 104
Phenols, 37
3-Phenylpropyl silane functionalized MCM-41, 331–332
Phorbol-12-myristate-13-acetate (PMA), 97–98
Phosphate compounds, 154–155
Phosphatidyl choline (PC), 332–333
Photo initiators, 121
Photoactivated encapsulation systems, 121
Photocatalytic degradation, 262–263
Photopolymerization, 134
Physical cross-linking-based hydrogel fabrication, 118
Physical encapsulation, 2–3, 300
Physical entrapment of bioactive molecule in biomaterial scaffold, 113
Physical immobilization of bioactive molecules, 120
Physical methods of encapsulation. *See also* Chemical encapsulation techniques
  agglomeration or granulation process, 306
  extrusion method of flavor encapsulation, 304
  fluidized bed coating, 305–306
  freeze-drying and vacuum-drying method, 304–305
  spray chilling or spray cooling techniques, 305
  spray-drying method of flavor encapsulation, 303–304
Physical mixture, 219*f*, 220*f*, 223*f*, 225–227
Physicochemical characterization techniques
  bright-field STEM images, 15*f*
  crystallinity of encapsulation systems, 15–17
  imaging of encapsulated materials, 12–15
  particle size and distribution, 10–12
  rheology of encapsulated materials, 17–20
  surface charges, 12
PIC. *See* Phase-inversion composition (PIC)
Piceatannol, 104
*Pichia pastoris*, 198
Pickering emulsion container, 161
PIT. *See* Phase-inversion temperature (PIT)
PLA. *See* Poly(L-lactide) (PLA)
Plant virus–based VLPs, 201–202
Plants, 103
*Plasmodium falciparum*, 330
PLGA. *See* Poly(lactic-glycolic acid) (PLGA)
PMA. *See* Phorbol-12-myristate-13-acetate (PMA)
Pollutants, 262–263
Poly glycolic acid, 324
Poly lactide-*co*-glycolide, 324
Poly-ε-caprolactone, 143
Poly(acrylic acid), 188
Poly(allylamine), 183–184
Poly(allylamine hydrochloride), 180–181
Poly(diallyldimethyl-ammonium), 183–184
Poly(diallyldimethyl-ammonium chloride), 181–182
Poly(ethylene oxide)-*b*-poly(propylene oxide)-*b*-poly(ethylene oxide), 121
Poly(ethyleneimine), 183–184, 188
Poly(L-lactide) (PLA), 122–124, 324
Poly(lactic-*co*-glycolic acid) microparticles, 122–124
Poly(lactic-glycolic acid) (PLGA), 114
Poly(lactide), 143
Poly(lactide-*co*-glycolide), 143
Poly(*N*-vinylpyrrolidone), 143
Poly(sodium 4-styrenesulfonate), 181–182

Polyacrylic acid (PAA), 167
Polyamines, 105
Polycaprolactone (PCL), 114
Polycondensation, 135
Polydispersity, 10–12
Polyelectrolyte, 180
Polyethylene glycol (PEG), 73, 76, 148, 218
hydrogels, 12–15, 124–126
movement index of *C. elegans* in, 14*f*
Polyethylene imine (PEI), 157–158
*Polygonaceae*, 107
Polyhydroxybutyrate, 135
Polymer, 115–116
dissolution, 270–271
erosion, 278
polymer-based nanocarriers, 264
polymer-based NPs, 200–201
shell and polyelectrolyte with ceramic core container, 157–158
halloysite nanotube TEM image, 158*f*
stabilizers, 155–157
swelling mechanism, 276
Polymerase chain reactions (PCR), 149
Polymeric microstructures, 9–10
Polymeric nanocapsules, 143
Polymeric nanocarriers, 143–144
Polymeric nanocontainers, 179–180, 187–188
self-healing mechanism of self-healing coating, 180*f*
Polymeric nanoparticles, 143
Polymethacrylic acid, 157–158
Polyphenols, 87–88, 104
bioavailability, 88
biological interaction, 88, 88*f*
Polystyrene (PS), 179
Polystyrene sulfonate (PSS), 157–158, 180–181
Pore size, 68, 120
Porous materials, 68
Porous scaffolds, 111–112
Power law–based equation, 276
Praziquantel, 334–335
Pretreatment methods to bioactive flavor compounds, 297–298
Primaquine, 330
Primary emulsion (PE), 35, 40–41
encapsulation efficiency of, 47*f*
Propylene glycol, 218
PRRs. *See* Pathogen recognition receptors (PRRs)
PS. *See* Polystyrene (PS)
Pseudo-ternary phase diagram construction, 214, 221–222
*Pseudomonas*, 137
*Pseudonomous* exotoxin, 240–241
PSS. *See* Polystyrene sulfonate (PSS)
Pure ketoconazole, 219*f*, 220*f*, 223*f*, 225–227
API, 218–220
Putrescine, 105
PV measurement. *See* Peroxide value measurement (PV measurement)

**Q**

QSAR-analysis, 91
Quantum dots, 15, 15*f*
Quercetin, 106
glycosides, 107

**R**

Racemization, 86–87
Raman spectroscopy, 15–16
RCNMV. *See* Red clover necrotic mosaic virus (RCNMV)
Reactive oxygen species (ROS), 97, 105
blocking, 98
Red clover necrotic mosaic virus (RCNMV), 144
Release activity, 45–50
Reservoir system, models for, 271–273
Resveratrol, 37, 104
Rheology of encapsulated materials, 17–20
*Ricinus communis*, 136
RIG-I, 204
ROS. *See* Reactive oxygen species (ROS)
Rutin, 107

**S**

*S*-adenosylmethione, 236–237
S/CoS. *See* Surfactants, and cosurfactants (S/CoS)
SA. *See* Sodium alginate (SA)
Santa Barbara Amorphous material (SBA material), 68
SBA-15, 68

SBA material. *See* Santa Barbara Amorphous material (SBA material)
Scaffold-based biomolecule delivery, 112
Scanning electron microscope (SEM), 316
Scanning tunneling EM (STEM), 15
SCF-extraction technique. *See* Supercritical fluid extraction technique (SCF-extraction technique)
Schistosomiasis, 334–335
Schizophrenia, 104
Screening
  of cosurfactants, 218
  of oils, 217–218
  of surfactants, 218
SDS. *See* Sodium dodecyl sulfate (SDS)
Secondary emulsion (SE), 35, 47
  encapsulation efficiency of, 47*f*
Secondary metabolites, 104
Sedimentation, 147–148
Selective targeting of colon cancer cells, 234–235, 245
Self-assembling peptide hydrogels, 122–124
Self-emulsification study, 213, 220–221
Self-healing
  chemistries, 156
  materials, 156
  surface, 261
Self-microemulsifying drug delivery systems (SMEDDSs), 212
  characterization, 215–216, 225–227
  formulations, 222*t*
  grading system for visual assessment, 214*t*
  liquid SMEDDS characterization, 215, 224–225
  liquid SMEDDS preparation, 214, 222–223
  material, 212
  methods, 212–216
  preparation, 225
  solid ketoconazole SMEDDS preparation, 215
SEM. *See* Scanning electron microscope (SEM)
Separation index (SI), 39
Serotonin, 104
SF. *See* Silk fibroin (SF)
Shell material, 1–2, 9–10, 294
Sherwood number (*Sh*), 273–274
Shikimate pathway, 104
SHS. *See* Spherical hallow silica (SHS)
SI. *See* Separation index (SI)
Siam weed, 136
Siberian larch-tree (*Larix sibirica* Ledeb.), 86
SIF. *See* Simulated intestinal fluid (SIF)
Silanes for surface modification of silica NPs, 65–66, 67*t*
Silica ($SiO_2$), 54–55
  nanocapsules, 155–156
  silica-based delivery system, 182–183
  silica-based nanocontainer, 189–190
Silica nanoparticles (Silica NPs), 55–67, 57*t*, 59*t*, 60*t*, 164–165. *See also* Ceramic NPs
  sol–gel process versatility, 63*f*
  surface modification, 65–67
  synthesis, 62–64
Silica NPs. *See* Silica nanoparticles (Silica NPs)
Silk fibroin (SF), 122–124
Simple coacervation process, 302
Simple extrusion, 325
Simulated intestinal fluid (SIF), 46–47
Single-core capsule, 295
Single-walled carbon nanotubes (SWCNTs), 149, 268
SLN encapsulation. *See* Solid lipid nanoparticle encapsulation (SLN encapsulation)
Small-molecule surfactants, 27
Smart coatings, 170–171
Smart materials, 122
SMEDDSs. *See* Self-microemulsifying drug delivery systems (SMEDDSs)
SOD. *See* Superoxide dismutase (SOD)
Sodium alginate (SA), 35
Sodium dodecyl sulfate (SDS), 165
Sol–gel silica NPs synthesis, 63
Solid lipid nanoparticle encapsulation (SLN encapsulation), 331–332, 334–335
Solubility study, 216–218, 217*f*
Solution route, 63
Solvent displacement method, 312
Solvent evaporation method, 135, 145
Sonication
  horn, 310
  time
    effect on droplet size, 40*f*

optimization, 40–41
Sonochemical method, 73–74
taxifolin–β-cyclodextrine conjugates, 92–97
Soxhlet extraction method, 299
Soxhlet extractor, 298–299
Span, 10–12
Spectrophotometric method, 317
Spermidine, 105
Spherical hallow silica (SHS), 157–158
SPIONs. *See* Superparamagnetic iron oxide nanoparticles (SPIONs)
Spontaneous emulsification for flavor encapsulation, 311–312
Sporozoites, 327–328
Spray chilling. *See* Spray cooling
Spray cooling, 305, 325
Spray drying, 325
method of flavor encapsulation, 303–304
Spraymate spray dryer, 215
Stability and uniformity, of VLPs, 201
Stability measurements of food emulsions, 315
Stabilization of emulsions, 25–29
homogenization, 28–29, 28*f*
role of emulsifiers, 26–28
Stearylamine, 332–333
Steglich esterification. *See* DCC/DMAP conjugation chemistries
STEM. *See* Scanning tunneling EM (STEM)
Stem cells, 111–112
stem cell–based therapy, 115–116
Stimuli response with ceramic core, 158–159
Stimuli techniques, 113
Stimuli-responsive materials, 120–121
STING, 204
Stöber method, 64–65, 69
Sulfonic acid, 154–155
Supercritical fluid extraction technique (SCF-extraction technique), 300
Superexchange interactions, 72
Superoxide dismutase (SOD), 107
Superparamagnetic iron oxide nanoparticles (SPIONs), 54–55, 61*t*, 69–79
advantages and disadvantages, 75*t*
biocompatibility and toxicity of ceramic nanoparticles, 77–79
coating, 74–77
core–shell structures and properties, 77
protection, stabilization, and functionalization, 74–77
structure and magnetic properties, 70–72, 70*f*
synthesis, 69–70, 72–74
Superparamagnetism, 70–71
Surface
charges, 12
on container, 185–186
erosion, 278
functionalization, 202–203, 239–240
morphological study, 216, 227
Surfactants, and cosurfactants (S/CoS), 212–214, 221–222
Suspension, 134
SWCNTs. *See* Single-walled carbon nanotubes (SWCNTs)
Swelling
controlled release systems, models for, 276–278
flavor release mechanism by, 314–315
Synthetic polymers, 115–116, 118–120, 134

## T

Tachyzoites, 327–328
Tannins, 107
Targeted drug delivery, 148, 197–198
active targeting, 148
passive targeting, 148
Targeting functionalities, 154
Taxifolin, 86
chemical structure, 87*f*
effective delivery strategies for nutriceuticals, 89–92
properties and chemical interaction with components of food matrix, 86–88
spatial structure, 87*f*
taxifolin–β-cyclodextrine conjugates, 92–100
bioavailability and bioactivity of, 97–100
solubility, 95*t*
sonochemical methodology, 92–97
TE. *See* Tissue engineering (TE)
Tea, 104

TEM. *See* Transmission electron microscope (TEM)
Temperature, 48–49
temperature-responsive hydrogels, 121
Template-directed method, 69
Templated synthesis, 63
Tetraethyl orthosilicate (TEOS), 63–64
TGF-β1, 113–114, 122–124
Theranostics, 53–54
Therapeutics, virus-like particles as, 201–203
Thermal decomposition, 73
Thermal gelation, 135, 325
Thermophoresis on nanoparticles fraction, 143
Thiolated poly(methacrylic acid), 143
Three-dimensional (3D)
cell culture, 114–115
environment, 114–115
epoxy system, 156
printing, 114–115
Thrombocytopenia, 334
Tissue engineering (TE), 111–112
Tissue regeneration, 112–113
Titania ($TiO_2$), 54–55, 157–158, 262–263
Tobacco, 136
Top-down synthesis methods, 62, 62*f*, 62*t*
Total phenolic content estimation, 316
*Toxoplasma*, 334
Toxoplasmosis, 334
TP580 (polypeptide), 122–124
Transcutol P, 218
Transmission electron microscope (TEM), 316
Transmission test, 224
Transmittance test, 215
Tri block polymer, 121
*Trichoderma*, 137
Triggered mechanism. *See* Uncontrolled release mechanism
Trypanosome parasite, 327–328
Trypanosomiasis, 332–334
Trypomastigotes, 327–328
Tumor necrosis factor α, 106
2D surface patterns, 114–115

**U**

Ultrasonic
emulsification for flavor encapsulation, 310–311
treatment in food industry, 85
ultrasonic-assisted synthesis of multilayer emulsions, 38–45
experimental, 39–40
results and discussion, 40–44
waves, 29–30
Ultrasonication, 298
application for emulsification, 29–33
comparison with high energy homogenizers, 31–33, 32*f*
emulsification mechanism, 30–31
assisted emulsification process, 30*f*
influence on physical and oxidative stability of MEs, 43–44
Ultrasound, 96, 260–261, 308
cavitation, 30
technology, 310
ultrasound-assisted extraction of bioactive aroma compounds, 300, 301*f*
ultrasound-assisted miniemulsion, 3–4
encapsulation process, 3*f*
Ultraviolet (UV)
light cross-linking approach, 121
rays, 262
Uncontrolled flavor release modes, 315. *See also* Controlled flavor release
Uncontrolled release mechanism, 4

**V**

Vacuum-drying method for flavor encapsulation, 304–305
Vascular endothelia growth factor (VEGF), 113–114, 122–124
Vascularity, 122–124
Velocity values, 143
Viral/virus-like nanocapsules, 144
Virus-like particles (VLPs), 197–198
classification, 198
conjugation chemistries, 203*f*
current applications as targeted therapeutics, 204–206
expression systems for, 198*t*
immune responses induced by, 203–204

overcoming limitations of other therapeutic approaches, 200–201
prerequisite factors in designing, 201–203
role as good drug delivery vectors, 199–200
Visual inspection method, 213, 218
Vitamin B9, 234–235
Vitamins, 24, 37, 234–235
VLPs. *See* Virus-like particles (VLPs)

**W**

Wastewater treatment, 262–264
Water-in-oil emulsions (W/O emulsions), 23–24
Water-soluble vitamins, 234–235
Weight-loss method, 160–161
Wet agglomeration, 306
Whey protein isolate (WPI), 36–37
WPI. *See* Whey protein isolate (WPI)

**X**

X-ray diffraction (XRD), 15–16, 215–216, 227

**Y**

Yolk–shell nanoparticles, 264–265

**Z**

Zeolite potassium chabazite, 268
Zero-order kinetics, 260–261
release kinetics, 272–273
Zeta potential, 12, 315
analysis, 215, 225
of BSA, pepsin, hemoglobin, and myoglobin, 13*f*
measurement, 39, 146–147
Zinc oxide (ZnO), 54–55, 160–161
nanoparticles, 148
Zirconia ($ZrO_2$), 54–55
ZnS–CdSe quantum dots encapsulation, 15

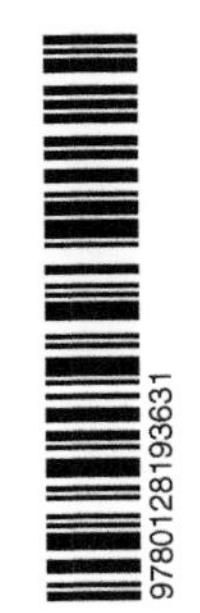
9780128193631